**Books are to be returned on or before
the last date below.**

ONE WEEK LOAN

LIBREX —

Advances in Mining Science and Technology, 2

Reclamation, Treatment and Utilization of Coal Mining Wastes

Advances in Mining Science and Technology

Advisory Editor: B.N. Whittaker
Mining Engineering Department, University of Nottingham,
Nottingham, Great Britain

Volume 1   Underground Mining Methods and Technology
(Szwilski and Richards, Editors)

Volume 2   Reclamation, Treatment and Utilization of Coal Mining Wastes
(Rainbow, Editor)

Advances in Mining Science and Technology, 2

Proceedings of the Second
International Conference on the

Reclamation, Treatment and Utilization of Coal Mining Wastes

Nottingham, England, September 7-11, 1987

Edited by

A.K.M. Rainbow
British Coal Corporation, Minestone Services, Philadelphia, Houghton-le-Spring, Tyne and Wear, England

ELSEVIER
Amsterdam — Oxford — New York — Tokyo 1987

ELSEVIER SCIENCE PUBLISHERS B.V.
Sara Burgerhartstraat 25
P.O. Box 211, 1000 AE Amsterdam, The Netherlands

Distributors for the United States and Canada:

ELSEVIER SCIENCE PUBLISHING COMPANY INC.
52, Vanderbilt Avenue
New York, NY 10017, U.S.A.

ISBN 0-444-42876-3 (Vol. 2)
ISBN 0-444-42846-1 (Series)

Printed in The Netherlands

FOREWORD

A.K.M. RAINBOW B.Sc., Ph.D., C.Eng., F.I.C.E., F.I.M.M., F.I.H.T.,
F.I.W.E.W., FGS

Head, British Coal Corporation, Minestone Services

Organiser and Editor:

2nd International Symposium on the Reclamation, Treatment and Utilization of
Coal Mining Wastes.

It is only in recent years that attention has been drawn to the proper
engineering use of the waste that results from the mining of coal.

Information on the properties and general uses of colliery spoils continues
to become widely disseminated, and use of the material by authorities and
contractors is still extending. Nevertheless there is a continuing need to
devote considerable effort to satisfying potential users that the available
materials are suitable for their applications, and to advise on how to set
about using them. Doubts are occasionally raised about the possibility of
spontaneous heating arising in unburnt spoil if used in particular
circumstances. Other doubts which have been raised relate to sulphate content,
frost susceptibility, compaction characteristics, moisture content, etc. In
most cases it has been possible to resolve them by providing further
information.

The use of colliery spoils in major highway works is of course now well
established, however, doubts about the suitability of particular spoils for
particular uses in other sectors of the construction industry are at present
less easy to resolve authoritatively. One reason for this is that the guidance

at present provided by Government in the Building Regulations and in Building Research Digests is ambiguous, and in certain respects prejudiced against colliery spoils.

Studies of the intrinsic properties of coal mining wastes has been frustrated because the nature of the material does not readily fit in with the sciences of soil or rock mechanics.

Most national standards governing the engineering utilization of minestone utilize existing specification clauses developed for engineering soils or rocks which are often inappropriate for minestone - this unsatisfactory state of affairs has been recognised by the United Nations who have currently on the working programme of the Group of Experts two items which will go some way to remedy the solution: these items are namely:-

(i) Elaboration of a Draft List of Parameters Characterising Properties of Waste from Coal Mining and Preparation as well as Methods of Determining those Parameters.

(ii) Elaboration of a Methodology of Economic Analysis Concerning the Replacement of Conventional Raw Materials by Waste from Coal Mining and Preparation.

It is interesting that a suggestion made by the Corporation some years ago that the appropriate British Standards should include details of the circumstances in which tests described could be meaningfully applied to colliery spoils now appears to be close to a reality in that there is a serious proposal to have a British Standard devoted to Minestone. This is necessary since engineering design cannot be accomplished without specific quantitative data on prior performance or a pre-determined evaluative regime to establish likely performance.

Empirical solutions based on prior performance go some way in the earlier phases of development of a science but ultimately these will be replaced by specific evaluative techniques related to specific, often unique, promotions.

That the science of 'minestone mechanics' is in its infancy is recognised in this symposium where there is a mixture of relating case histories - with the purpose of demonstrating satisfactory (or otherwise) performance - to postulating evaluative critiques for determining field and laboratory measurements.

Opening Address delivered by

SIR ROBERT HASLAM
Chairman
British Coal Corporation

CONTENTS

Foreword.. v
Keynote Address
        The Lord Graham of Edmonton.. xiii
The Economics of Minestone Utilization
        W. Sleeman.. 1
An Evaluative Framework for Assessment of Disposal Options for Colliery Spoil
        M.F. Noyce... 21
The Development and Utilization of "Stone Coal" ("Bituminous Shale") in China
        Li Yingxian.. 35
Utilization of Mining Operations and Coal Preparation Processes Wastes in the
USSR and the Principles of Their Classification
        V.A. Ruban and M.Ya. Shpirt.. 45
Research on Suitability of Coal Preparation Refuse in Civil Engineering in
the Federal Republic of Germany
        D. Leininger, J. Leonhard, W. Erdmann and T. Schieder................. 55
A Forecast of Composition of Coal Wastes and of their Directions of
Utilization during Explorations in Coal Deposits in the USSR
        V.R. Kler and M.Ya. Shpirt... 69
Coal Waste Industrialization in Brazil: the State of the Art
        E. Da Motta Singer and A.J. Marchi.................................. 79
Variability of Coal Refuse and its Significance in Geotechnical Engineering
Design
        S.C. Cheng and M.A. Usmen.. 95
The Characteristics and Use of Coal Wastes
        J.G. Cañibano and D. Leininger.......................................111
Technical Approaches and Related Policies for Land Reclamation and Treatment
in Chinese Coal Mines
        Yan Zhicai..123
Compaction Control and Testing of a Colliery Spoil for Landfill Sites
        W.M. Kirkpatrick and I.P. Webber.....................................133
House Building on Bog Land
        I.McD. Hart, A.S. Couper and D.C. Stephenson.........................145
Colliery Spoil in Urban Development
        W. Sleeman..163
Laboratory and Site Investigations on Weathering of Coal Mining Wastes as a
Fill Material in Earth Structures
        K.M. Skarzynska, H. Burda, E. Kozielska-Sroka and P. Michalski........179

X

Recovering Combustible Matter from Coal Mining Waste and Measures to
Extinguish Waste Pile Fire
  Shengchu Huang...197
Coal Waste in Civil Engineering Works. Two Case Histories from South Africa
  F.W. Solesbury..207
Minestone Impoundment Dams for Fluid Fly Ash Storage
  A.K.M. Rainbow and K.M. Skarzynska.......................................219
Fly Ash Ponds and the Dam Legislation in Finland
  J. Saarela..239
The Utilization of Coal Ash in Earth Works
  J. Havukainen...245
Biotechnical Methods in the Treatment and Restoration Use of Coal Mining
Waste
  P.J. Norton...253
Coal Mine Spoil Tips as a Large Area Source of Water Contamination
  J. Szczepanska and I. Twardowska...267
Minestone and Pollution Control
  M. Nutting..281
The Study of Saturated Coal Mining Wastes under the Influence of Long-Term
Loading
  K.M. Skarzynska and E. Zawisza...295
Suction Pressures in Colliery Embankment Surfaces
  R.K. Taylor, S.J. Billing, F.T. Bick and R.J. Simonds...................303
Deep Ripping: a More Effective and Flexible Method for Achieving Loose Soil
Profiles
  A.R. Bacon and R.N. Humphries..321
Combined Tipping and Opencast Coal Scheme at Bentinck Colliery
  D.M. Brown..331
The Design of Colliery Spoil Tips: Objectives and Techniques
  J.R. Talbot...343
The Utilization of Dirt from Coal Mines and Land Reclamation
  Gao Youlei, Han Guangxu and Sun Shaoxian.................................357
An Investigation into Cheaper Monolithic Packing Materials Utilizing
Colliery Tailings
  A.H. Zadeh, A. Barkhordarian, P.S. Mills, A.S. Atkins and R.N. Singh..369
Geotechnical Aspects of Fine Coal Waste Disposal in Lower Silesia, Poland
  B. Bros...381
The Application of the Multi-Roll Belt Filter to De-Watering of Fine Coal
Waste Slurries
  H.G. King...393

Pumpability of Coal Mine Tailings for Underground Disposal and for Regional
Support
    A.S. Atkins, R.N. Singh, A. Barkhordarian and A.H. Zadeh.............401
Improved Rock Paste. A Slow Hardening Bulk Fill Based on Colliery Spoil,
Pulverized Fuel Ash and Lime
    K.W. Cole and J. Figg..415
Present Situation, Problems and Development of Utilization of Coal Mining
Wastes in China
    Sun Mao Yuan...431
Direct Tree Seeding on Coal Mine Wastes in Britain. A Technique for the
Future
    A.G.R. Luke..441
Management of Opencast Restored Land for Cereal Production
    E.J. Evans...459
Pre-Mine Baseline Data: an Essential Tool in Reclamation of Colliery Wastes
    S.O. Adepoju and G. Fleming..471
Reclamation of Manners and Pewit Collieries, Ilkeston, Derbyshire
    P.E. Wright and K. Shipman...489
An Investigation into the Reclamation of Opencast Backfilled Sites Destined
for Road Construction in the United Kingdom
    R.N. Singh, F.I. Condon and S.M. Reed..............................501
The Cost-Effectiveness of Rehabilitating Colliery Sites through Coal
Recovery
    T. Macpherson..513
Short-Term Durability of Cement-Stabilized Minestone
    M.D.A. Thomas, R.J. Kettle and J.A. Morton.........................533
The Wetting Expansion of Cement-Stabilized Minestone – An Investigation of
the Causes and Ways of Reducing the Problem
    C.E. Carr and N.J. Withers...545
Assessing the Durability of Reinforcement Materials in Minestone
    D.C. Read and C.E. Carr..561
Polymeric Mesh Element Reinforcement of Reinforced Minestone
    R.K. Taylor, T.W. Finlay and D.A. Fernando.........................573
Reinforced Minestone Using Special Designed Reinforcement
    R.B. Singh, T.W. Finlay and A.K.M. Rainbow.........................583
The Sliding Resistance between Grid Reinforcement and Weathered Colliery
Waste
    R.W. Sarsby..587
Infilling Old Mine Workings and Shafts
    J.P. Hollingberry..597

XII

Sources Study. A Strategy for the Identification and Selection of Colliery
Spoil for Use in the Infilling of Abandoned Limestone Mines in the Black
Country
    D.W. Stevens and K.L. Seago..601
The Infilling of Limestone Mines with Rock Paste
    P.A. Braithwaite and T. Sklucki......................................615
Methods of Development above Ancient Shallow Pillar-and-Stall Coal Workings
    I.E. Higginbottom...639
Foundations for Sites over Natural Voids and Old Mine Workings
    P.F. Winfield..653
Building Materials from Industrial Waste of Coal Based Power Plant
    M.D. Desai and D.B. Raijiwala..661

KEYNOTE ADDRESS

A NON-GOVERNMENTAL PARLIAMENTARY VIEW OF THE RECLAMATION, TREATMENT AND
UTILIZATION OF COAL MINING WASTES - SET IN THE CONTEXT OF BRITAIN'S ENVIRONMENTAL
RENAISSANCE

THE LORD GRAHAM OF EDMONTON, BA., CSD., FBIM., FRSA
House of Lords, Palace of Westminster

There are appropriate and less appropriate times to seek to stimulate
public debate on such esoteric topics as that which form the core of your
interests here in Nottingham this week. - How to make better use of minestone
than we do, both here in Britain in 1987, but perhaps elsewhere throughout the
world. I can assure you that this is a particularly good time to debate the
issue and to exchange experiences, as far as Britain is concerned, for the
following reasons.

Firstly, we have just had a General Election which has resulted in the
return of a Conservative Government for a third term. This is appropriate and
timely as an ingredient in our discussions especially if, as I do, you value
and recognise the part to be played in these matters by Government. With an
eight year old track record behind it, few cannot be unaware of the broad
thrusts of the Government's attitudes in this and related matters. The
background is important, as is the knowledge that with a majority in our
Parliament of 100 seats there is little doubt that this newly elected
Government will serve for between 4 and 5 years.

Secondly, the recent past has seen a surge in public interest in matters
environmental and I cannot see that burgeoning interest doing other than
continuing to grow. Not only have we had the emergence of single-issue politics
in this country but the creation and strengthening of action by new political
parties; and the close attention by the established parties of environmental
and conservationist policies within their main Manifesto.

Thirdly, we have, and will continue to have a deepening public interest and
concern in the energy and nuclear aspects of national policy. The tragedy of
Chernobyl, the fraught and fragile world of oil, the miners strike of 1984/5,
the conservation and use of alternative energy sources here in Britain and
heightened public awareness of all things related to energy - and that means
coal.

My modest claim to engage your attention at this time stems in part from
the fact that I speak on the environment for Her Majesty's Official Opposition,

and too, by virtue of being born in Newcastle-on-Tyne and coming from mining
stock I know the mining scene as well as anyone not professionally engaged in it
as most of you are. I intend to deploy my experiences as a Parliamentarian,
which encompass ten years in the House of Commons and four in the House of Lords
as I seek to search, with you, for a better understanding by the public and by
government of the waste of a national asset - minestone. I have to say, however,
that the portends are not favourable, not least from your experiences over the
recent years, but also in my view from "The Shape of Things to Come".

Using Waste

Whether we call it spoil, refuse, dirt, slag, shale or waste, however we
define it - and I have read more than one definition - we are talking about the
utilization of an asset. Long since, we have lost the attitude that minestone
is useless, that it was just that part of a process for which there is no use.
We find ourselves here in Britain in 1987 forcing ourselves to examine practices
which formed part of our industrial heritage virtually undisturbed for many
years. Not only in the realm of utilizing minestone. I will be referring
to the heightened awareness and perception to avoid waste and to capitalise on
reclamation and recycling across a broad spectrum of waste products which has
been steady, not spectacular, here in Britain these past few years. As a non-
technical person with a natural desire to see that we maximise, for the public
advantage such waste products as we have - like minestone - I want to help in
the process of reducing rates and taxes, and of wringing the last ounce of value
out of coal. I have to confess that I did not appreciate how hard that task
would be until my attention was engaged and I raised the issue with ministers
in Parliament. After some thought, I have come to the conclusion that the issue
is influenced by political considerations. Not Party Political considerations,
but by those which could be called "lobbying". Sadly, I do not detect a
comparable pro-lobby for the use of minestone, and that needs to be remedied as
soon as possible. I have some encouraging signs to report.

Dual Tendering

Looming over the debate these last three years since the symposium at
Durham in 1984 has been the arguments surrounding "dual tendering" and we are on
the verge of a new phase in what I will call Government initiatives. You will
know of the deep division of opinion which emerged when the report of the
interdepartmental committee on the use of waste material for road fill
recommended that the system of dual tendering be abandoned. It always made
sense to me that tendering for major road and similar public works can provide a
unique opportunity to use large quantities of minestone, which, if left where it
is would simply call for public spending to make slag heaps aesthetically
acceptable. It also made sense to me that the relative cost of using minestone

compared with other materials is to find out the comparative cost of the two types of fill, the source of both being specified, and to expect that the cost most advantageous to the public purse would be used.

You know better than I of the frustration experienced of making "dual tendering" work - indeed there will be those in this audience who would argue that dual tendering was never given a fair crack of the whip. Both the Verney committee and the Flowers committee had much to say which encouraged that concept - wherever it made sense to do so. No-one to my knowledge has argued for positive discrimination in favour of minestone. The nearest argument I ever saw pleaded that if transport costs, as part of the global costs could be offset by the savings made by not paying reclamation grants if the minestone remained in place, it would make minestone more attractive. The main stumbling block remained that the dual tendering procedures were advisory and not mandatory, and my information is that - in the event - the suppliers of minestone - insufficiently, had opportunities to provide decision-makers with the choice of opting for minestone.

The interdepartmental committee seemed to me to show that it did not have the ability to be used as a mechanism to implement the fundamental concept of encouraging the use of wastes - indeed, by discontinuing "dual tendering" there is a danger of it being held that its use is being discouraged.

I have referred to powerful interests playing a significant part in shaping policy - a proper and honourable part in our pluralistic democracy. Prominent amongst those who opposed the abandonment of dual-tendering were the County Planning Officers, The Association of County Councils, District Councils, Metropolitan Councils - and we would expect - British Coal.

In favour of the proposal the Civil Engineering Contractors - and, as you might expect - the British Aggregate construction materials industries. Other important public bodies expressed strong and powerful reservations.

Meeting the Minister

The Minister for highways is Mr Peter Bottomley, a parliamentarian and a person for whom I have the greatest respect. We could be heartened when he announced his acceptance of the Interdepartmental Report by telling us that he remained committed to encouraging the use of waste material for road fill, but that the dual tendering procedures would be superceded by consultation and exchange of information. Experiences since that announcement in November last year show that there is yet still the need to maintain a pressure, even to make this new approach work. Let me say however, that Mr Bottomley has proved to be very willing to see that the new procedures work. Frustration by the Minestone Executive at a void in the time scale led to my taking the Head of Minestone Services, Dr Keith Rainbow to see Mr Bottomley with his officials, and I am pleased to say that the outcome was most satisfactory to Dr Rainbow.

I conclude my direct remarks on the prospects for the better use of minestone - acknowledging that even with an improvement we are looking at scratching the surface - 1000 million tons on the ground with perhaps 2 to 3 million tons in demand per year - by quoting the views of the Rural Town Planning Institute when it was consulted on the draft circular of the Department of the Environment earlier this year. It sums up well the realities which we would be foolish to ignore.

"With no compulsion on the use of alternative fill as might have been achieved even with dual-tendering, the option to use such alternative fill must depend to a great extent on financial issues, which are inevitably paramount in what will be a commercial situation. Unless some contribution towards costs of transporting or providing such material is made it seems unrealistic to expect contractors to use a more expensive 'secondary' or alternative source where a cheaper 'primary' source exists. This free market enterprise, together with the inherent attitude of many engineers to use only those traditional aggregates on which their usual experience is based, will ensure that little inroad is made into the large volume of theoretically usable waste".

So, what is on offer to us, now that the dual tendering procedures are being swept away even though it may not have worked, if at all well?

Firstly, that while the Government remains convinced that the use of waste material in a constructive and economic way is in the nations advantage it has concluded that this objective can best be achieved by the improved and early information exchanges recommended in the interdepartmental committee's report.

Secondly, at the earliest opportunity the highway authority in consultation with the minerals and Local Planning Authority and waste producers, will identify whether suitable waste material is likely to be available. The whole question of an economic radius for transporting such waste remains.

Thirdly, as soon as information is available on the likely requirement for fill, the Highway Authority will inform the minerals and Local Planning Authority and waste producers.

Finally, the Highway Authority in their instructions for tendering, will draw these points to the attention of tenderers. What we are left with is the hope that by a series of changes there is both the aspiration and the prospect that more minestone will be used. I have already reported that Dr Rainbow has expressed to me his satisfaction that he accepts that the Transport Minister, Mr Peter Bottomley will give the new procedures his positive support.

International Experience

I read with interest many, not all, of the papers which were before the Symposium in 1984, especially those from other countries. They have

arrangements which make better use of their minestone. It seemend to me that
in the Federal Republic of Germany much more thought, of research, of money,
was spent by Government, both regionally and locally in both using it and
treating it in an environmentally sympathetic manner. In Germany, Belgium and
Holland I found that minestone for civil engineering projects concerned got a
special boost after the floods of 1953, not only for its special comparative
qualities, but the special circumstances of the cheap cost of using the River
Rhine, the power of the Port of Rotterdam and the policy of most Local
Authorities to subsidise the removal of minestone tips. In France these past
25 years use of minestone as fill and base layers for roads and motorways has
seen the scope widen to include the decoration of green areas, used for sports
grounds, the manufacture of concrete and bricks. From America I learned that
minestone has a bright future in construction as the granular backfill that
is the major component of a constructional material - reinforced earth. South
Africa had a contribution to make. All is relative. I think we are in the
midst of a pioneer movement, almost universal, one pincer represented by
pressures from environmental considerations, the other from improved
technological methods which permit the second use of once discarded waste. It
is an exciting prospect, the discussion of which can do nothing but good.

What we need, on this as on many other energy or waste related issues is
to devote much more time and thought to how to articulate effectively, and by
that I mean influencing Government at all stages of policy formation. This
means getting inside Parliament as well as Whitehall. I can assure you that
parliamentarians are very receptive to briefings from interest groups,
especially one of such importance and with such a message as yours. Of course
you will be competing with hundreds of other interest groups and I have already
referred to the growth of many special interest groups and of the broad
consensus amongst the political parties that anti-pollution, reclamation, recycling and conservation measures are firmly on the agenda. We should be
reminded of what they are.

Other Wastes

The industry committee for packaging and the environment (INCPEN) seeks to
represent fillers, distributors and retailers of packaging. Its members are the
suppliers of glass, metal plastics and paper packaging. Pans, tins, foils,
glass bottles and jars, plastic bottles, containers, wraps and films, paper
and board, cartons, boxed cases. They are concerned to improve the public
perception of their industry, to encourage the public in avoiding and abating
litter by a programme of education via the Local Authority, augmented by such
national initiatives as Richard Branson's on litter control. It will come as
no surprise to be told that the costs of collecting and treating primarily

household and industrial waste must be realistic and that there is growing
evidence that waste disposal costs have been dramatically under estimated
not least because no allowance is made for replacement costs when existing land
fill sites are filled. The recycling movement has the merit of not only using
some of our materials more than once, it makes the consumer feel good. We have
some way to go to equal the position in parts of Germany, where householders
are provided with two bins - one for recyclables and one for other waste, but,
I believe the British public is slowly moving towards recognising that it is
sense to help in the process of recycling by paying for measures which either
aid, assist, or initiate processes which are designed to reduce pollution of
one form or another. Household refuse in the UK amounts to 15 million tonnes
a year - one third paper, one third kitchen waste, one third glass, packaging.
Gone are the days when half the weight came from dust and cinders. With our
changed life-style, our dustbins reflect central heating, self-service
supermarkets, higher disposable incomes, working wives, DIY, drinking at home,
car ownership and maintainance. This is a dramatic change not lost on the
consuming public.

I do sense that as far as household, domestic waste, the battle has been
joined with some imagination and flair, persuading us all of the values of
recycling, that it can:-
* Reduce the amount of waste produced
* Extend the life of landfill sites
* Provide secondary raw materials
* Conserve primary materials
* Save energy
* Remove hazardous materials from the waste stream
* Reduce environmental impact arising from the extraction of resources
* Make money, and thereby create business and jobs
* Increase environmental awareness in the general public
* Reduce imports.

Packaging helps manufacture to sell their products and provide the consumer
with brand and product information - including that required by law. For the
retailer, packaging facilitates the storage, stacking, and display of wares and
permits self-service reducing prices by limiting waste and spoilage. In modern
day retailing of consumer goods, packaging offers the shopper not only added
convenience but also a wider variety of goods of a higher consistent quality and
a lower price than could ever have been envisaged at the end of the last century
when pre-packaged food first appeared on the market.

Is there a case for considering employing some of the techniques used by
other pressure groups - like the packaging industry? There is a similarity.
Both have a secondary responsibility to either dispose of or find a secondary

use for an essential first use material which is not popular a heightened awareness of what can be done with it, and its place in the total environment, mixed with a commercial spin-off, not to be sniffed at.

What has developed in the glass industry is indicative of the opportunities for reclamation which exist. The total of bottle banks increased last year from 2500 to 2837, with an estimated 5.6 million people now returning glass through the bottle bank scheme every week. Britain recycled about 14% of its waste glass in 1986 - yet is still 10th in a European league table of 12 countries.

Pollution Control

Pollution control in its widest forms does exercise the mind and the attention of the British people to an increasing degree. This is nowhere more illustrated than by the identification within each Government and opposition Parties of a Minister who would have special responsibilities for 'Green Issues'. William Waldegrave made a significant impact in the time he occupied the "Green" Chair at the Department of Environment. Of course he was responsible for promoting and advocating Government policies - and did that with conviction, flair and panache. Yet he is only human - and from a remarkable candid report in a national newspaper - frustrated at the situation in which he had responsibility to clean up our environment.

Speaking to senior industrialists and environmentalists, he painted a depressing picture in which Britain's cautious approach to tightening regulations fails to stimulate innovation, leaving a domestic pollution control industry too weak and outdated to provide the improved technology which Government can insist on polluters using.

After referring to an estimated 1500 firms who constituted the pollution control industry, he said that it was loosing home markets because other countries had higher environmental standards or forced the pace of technological development by setting standards unchangable with existing technology. He illustrated this point by reporting that the British firm Davy McKee had been able to sell 30 flue gas desulphurisation plants for power stations abroad without one domestic sale. Yet, when, with other British Firms in the field it had developed an effective lobby which had made FGD the policy for future power stations, and had let to plans for retrofitting three existing stations at a cost now estimated at £780m. The export of water effluent treatment plant had fallen from a peak of £140m in 1983 to £50m in 1986, while the share of EEC exports of air pollution control equipment fell from 63 to 40 per cent between 1975 and 1985. He suggested that the Department of Environment as sponsor of the pollution industry should fund research and development, stimulate home demand by higher regulatory standards and other incentives, and point firms towards potentially profitable new investments.

European Year of the Environment

A very real reason why it is appropriate to discuss the developments
within the coal industry and minestone in particular is that we meet in what
has been designated as the European Year of the Environment. We could be
heartened that this is a sign throughout Europe, but here in Britain, in
particular, we want to take environmental problems more seriously in the future
than we have in the past. Just how serious do we want to take it?

As a Parliamentarian I have seen over the past 14 years the emergence of a
positive environmental lobby. There is an all-party conservation group with
more than 150 members from both Houses and from all parts. The tasks facing
that group are daunting. High on its agenda is adherance to the Bruntland
Report of 1983, the product of a United Nations initiative of 1983, when it
formed the World Commission on Environment and Development, charged with
reporting on the problems of protecting and enhancing the environment. That
Report comes up with two central themes, dangerous problems, both of which are
of mankind's own making and both of comparative recent origin. First there is
the growing shortage of living space. Already the world population of $2\frac{1}{2}$
billion people in 1950 has nearly doubled, and will double again in the early
years of the next century. Many life forms as we know them will not survive.

Secondly, for little over 100 years man has been feeding back into the
atmosphere the fuels that have accumulated over the past 500 million years, and
in the past 50 years we have added the waste products of the chemical era. The
records show that carbon dioxide levels in today's atmosphere have increased by
27 per cent since the mid 19th Century. Carbon dioxide holds heat close to the
earth and has a potential for melting the polar ice caps and raising the levels
of the oceans. Now that methane and chloro fluoro carbons have been added, it
means that we are tampering with our basic weather-making components. We do
not know the outcome. The risks are a warmer, flooded earth or a world seared
by the sun. In the meantime these extractions and additions are causing the
pollution and destruction of many life forms with which we have been involved.

I served on the Committee of the Wildlife and Countryside Bill, both in
the Commons and later in the Lords, when amendments were made. What I saw,
was told, learned, was a great revelation - as it was and is to millions of
uninformed members of the public who are often blissfully unaware of what is
happening in their world. Here in Britain it is a world where hedgerows are
in severe danger. In the first thirty years after the war, one quarter of
Britain's hedgerows were removed at the rate of 4,500 miles a year, a total of
120,000 miles. The rate since then has increased. The Dutch Elm disease
plague of the 1970's destroyed 11,000,000 trees.

Our ground water is polluted by nitrates and pesticides. Our rivers have become polluted; Current public spending policies allied to plans to privatise our water supplies pose special problems relating to the discharge of sewage effluent into water courses. Successive Governments have wrestled with making a reality of the principle of making the polluter pay. A difficulty we have is that the environmental responsibilities to prosecute a vigorous co-ordinated policy is thwarted when the responsibilities for doing so is fragmented between the Department of the Environment, The Ministry of Agriculture, Food and Fishing, Trade and Industry, Education and Science, Transport. The fact is that there is an imbalance between environmental and other considerations in reaching policy decisions because individual departments of Government are not required to consider the environmental consequences of their own policies and also because in some ways the Department of the Environment lacks clout in promoting environmental issues through Government. Thus, agricultural policy has tended to be decided rather more in the interests of the energy producers. Nor do we take account of environmental considerations when we formulate Government policy on foreign aid. One can readily see why both Labour and the Alliance proposed a Department of Environmental Protection to be responsible for planning, conservation, pollution control, leisure and recreation, and animal welfare. A seat in the Cabinet with a brief to co-ordinate these matters across departmental boundaries would surely ensure a more effective impact on improving our environment? Yet what is the Government response to this, Environment Year?

The response of the Government announced in March this year was to provide additional £$\frac{3}{4}$ million to support its own and other projects. This is how Lord Skelmersdale put it in the House of Lords:- "WE ARE ALSO CONTRIBUTING PROJECTS OF OUR OWN TO THE COMMITTEE'S PROGRAMME AND ARE USING ALL OPPORTUNITIES TO PROMOTE THE YEAR AND ITS GOALS. OTHER DEPARTMENTS ARE CONTRIBUTING FURTHER RESOURCES, PROJECTS AND ASSISTANCE IN PROMOTING THE YEAR. THIS IS EVIDENCE OF THE GOVERNMENT'S GENUINE COMMITMENT TO THE YEAR"

When you consider that the main aim of the year is to raise awareness of the importance of environmental protection, so that better progress can be made in conserving and improving the environment, the additional sum to do so strikes me as ludicrous.

Industry and local authorities have been urged to play a full part, and we all know there will be some good support from some quarters, but we do not yet, I fear fully appreciate in this country that time is slipping away. Tree planting, childrens essay competitions, anti-litter campaigns, special weeks for birds, woodland, water, beaches, most on a voluntary basis by national groups does not strike me as adding up to a national crusade. Here in Britain we have some excellent bodies giving the interested individual a home for this or her special interest in the environment. The National Trust is almost 100

years old, as is the Royal Society for the Protection of Birds. The World
Wildlife Fund was established in 1961. Friends of the Earth was formed in 1971,
campaigning on saving the whale, bottle banks, litter and recently on
conservation of land and nuclear related issues. It has 400 local groups. It
is an effective and respected campaigning lobbying group. Greenpeace is a body
with almost 100,000 members in the UK, adopting an aggressive stance, sailing
ships into nuclear testing areas, into whaling and seal situations, but it is
turning to the world-wide problems of pollution and of the influence of the
World Bank on developing countries.

If the European Year of the Environment limps along - as it has all the
signs of doing, it will be a wasted opportunity - but not the end of the world.
How we can avoid that is the sixty-four-thousand dollar question!

Politicians Promises

Not only this industry, but much of the industrial life of this country is
affected by the legislative framework laid down by the Government of the day:
and not only by acts of Parliament but in the many other ways that Government
influences the business and national environment in which your industry has to
operate. So, as you would expect with a General Election so fresh in our minds
we can see how the main political parties drew up their manifestos on the
central environmental matters of the day. Do not forget too, that the Party
shopwindow was set out in the way that it was based upon pressure from groups of
one kind or another - as well as the doctrinal beliefs of the party concerned.
Let me give you a sketch of the programme promised by the Parties before and
during the election.

Labour First

"Labour will establish a ministry of environmental protection to take
positive action to safeguard the quality and safety of life."

We will set up an environmental protection service and a wild life and
country service.

We will extend the planning system to cover agricultural forestry and water
development requiring them, and industry, to take account of environmental
considerations.

We will invest more in land reclamation and cleaning up, in recycling and
conservation, in development of new products, processes and pollution control
equipment. This will not only make the country cleaner, but will create jobs as
well.

We will take action to deal with acid rain.

We will stop radio-active discharges into our seas and oppose the dumping
of nuclear waste at sea.

We will provide for better monitoring, inspection and enforcement of
pollution control, to cover areas ranging from air pollution to beaches, from

hazardous chemicals to food additives, and from water quality to vehicle emmissions.

The Alliance had this to say:-

The Alliance will set up a new Department of the Environmental Protection headed by a Cabinet Minister, who will be responsible for environmental management, planning, conservation, and pollution control.

There will be powerful disincentives to polluters based on tougher penalties and implementation of a 'Polluter Pays' principle for cleaning up the damage backed by support for good practice. There must be the safest possible containment and disposal for industrial waste with recycling wherever feasible.

Clean air legislation setting new standards with tough measures to deal with acid rain and an acceleration of the phasing out of lead in petrol.

Introduction of a statutory duty for both private and public sector companies to publish annual statements on the impact of their activities on to the environment and of the measures they have taken to prevent, to reduce and eliminate their impact.

Continued modernisation and development of the coal industry, including new coal-fired power stations with measures to prevent acid rain and more help to areas affected by pit closures. The power to licence coal mines would be transferred from British Coal to the Department of Energy to prevent abuse of monopoly. These, however, are in the category of what might have been, but they are important in the context of surveying the broad background of political action, potential political action, which indicates that the environmental agenda is not the exclusive preserve of any one political party. When we look at the prospects put forward by the Conservative Party we can look back at what they claim to have done during office these past eight years, what they offered the electorate on June 11 and what appeared in the Queen's speech, to form the actual environmental agenda for this Parliament.

Government Record

The Conservatives say they are by instinct conservationists - committed to preserve all that is best of our Country's past. Since coming to office in 1979 they have:-

* More than doubled the area of specially protected green belt, and will continue to defend it against undesirable development;
* Established new arrangements, backed with public funds, to make farming more sensitive to wildlife and to conservation;
* Established a new powerful pollution inspectorate;
* Enacted new laws on the control of pesticides and implemented new controls on the pollution of water;
* Put in hand plans for cleaning up Britain's beaches costing over £300m over the next four years;

* More than doubled spending after allowing for inflation on countryside and
  nature conservation since 1979;
* Set in hand the establishment of the new Broads authority - a major
  environmental initiative;
* Established a huge programme costing over £4000m to clean up the environment
  of the Mersey Basin by the early years of the next century.

"Wherever possible we want to encourage large-scale developments to take
place on unused and neglected land in our towns and cities rather than in the
countryside. We want to improve on our performance in 1986 when nearly half of
all new development took place on reused land".

This is what the Conservatives promised the electorate if it was returned
for a third term:

* Continue our programme of £600m for modifying power stations, to combat acid
  rain;
* Adapt improved standards in concert with Europe for reducing pollution from
  cars;
* Will introduce new laws on air pollution and dangerous wastes;
* Double the funding for environmentally sensitive areas;
* Introduce new laws giving extra protection to the landscape of our national
  parks;
* Support scientifically justified international action to protect the
  atmosphere and the sea from pollutants;
* Establish a national River Authority to take over responsibilities for
  ensuring strict safeguards against the pollution of river and water courses
  and to pursue sound conservation policies. The water supply and sewage
  function of the water authorities would be transferred to the private sector;
* Set up safe facilities for disposing of radioactive waste from power stations,
  hospitals and other sources* and NIREX has been asked to come forward with
  proposals for deep disposal.

The broad outline of how these promises were to be given legislative form
appeared in the Queen's speech on June 25 it is stark and simple and was as
follows:-

"In all these policies, my Government will have special regard to the needs
of inner cities. Action will be taken to encourage investment and to increase
enterprise and employment in these areas......"

"Measures will be introduced to promote further competition in the
provision of local authorities services......."

"Legislation will be introduced to enable the water and sewage functions
of the water authorities in England and Wales to be privatised....."

If you seek for any more precise form of Election promises translated into
legislative detail you will have to be satisfied with - "other measures will be

laid before you."

Time will tell.

The Disposal of Nuclear Waste

The disposal of nuclear waste is undoubtedly a major matter of debate in Britain today. John Baker is the Chairman of UK NIREX Ltd, and in his annual report last year, he sought to put the minds of the British people at rest:

"To people who work regularly with radiation the extent of public worry about radioactive wastes often seem misplaced, they know that radioactive wastes must, of course, be respected and handled properly, but they believe that the safe management of the wastes is readily accomplished with existing technologies and that any safety problems are relatively trivial."

I have to say that that is the simplistic view, for earlier this year public interest in the sites and the manner of disposal of nuclear wastes was a major parliamentary occasion, accompanied with implications for many communities as well as illustrating the way public policy can be significantly affected by public agitation - well managed.

Some time ago, it was announced that NIREX was involved in a search for a shallow dump on land for nuclear waste, substantially to replace using a trench at Drigg, near Sellafield. A House of Commons Select Committee had commented adversely on this method. Indeed it had said that Britain's nuclear industry is far behind those in other countries in waste disposal. Unless it shared the professionalism shown in Sweden and Western Germany it was unlikely to find disposal sites anywhere in the UK without the most extraordinarily difficult and costly political battle and public objections. Events proved that to be the understatement of the year.

In February 1986 the Government announced that it was considering four possible sites for the disposal of low and intermediate level radioactive wastes: Bradwell, Elstow, Fulbreck and South Killingholme. Immediately there was extreme agitation in the areas named. Protest groups were formed, television pictures of local people stopping contractors getting on to sites for exploration filled our screens, it seemed every night, and the issues became much more than scientific or disposal orientated, they were elevated to new environmental and political planes. For, it turned out, all four sites were within the constituencies of Conservative MPs, three of them Ministers, one of them, the Government Chief Whip. Uninvolved citizens who had distained being involved in any public involvement - in anything - all their lives were galvanised into organised opposition. They were aided by the experience of campaigning groups like Friends of the Earth and Greenpeace, but also by the fact that each protesting group was sustained and supported by the activities of the other three. The drip, drip, drip of public rejection had an effect,

not least because it all took place in a pre-Election period. Within two weeks
of the actual announcement of the General Election, the Secretary of State for
the Environment made his capitulation statement to Parliament. He said:-

"I have received a letter from the Chairman of NIREX reporting on the results
hitherto of the investigations at Bradwell, Elstow, Fulbreck and South
Killingholme, and giving me his correct assessment of the economics of shallow
disposal of low-life waste; and of the alternatives of deep disposal together
with intermediate waste. He concludes that although a safe near surface
disposal facility could certainly be developed at any of the four sites
currently being investigated, the economic advantages of separate near surface
low level waste disposal are nothing like as great as NIREX originally thought.
Consequently, NIREX concludes that it would be preferable to develop a multi-
purpose deep site for low-waste and intermediate level waste rather than
proceed with the investigations for a near surface facility at any of the four
sites currently under investigation."

It was in the Daily Telegraph of that day - 1 May, before the announcement
was made, that the nature of one of the imperatives for that reversal of policy
was spelled out in this front page headline treatment:

"Threat to Four Tory Seats Lifted"

"Nuclear Dump Site Search is Abandoned"

"The surprise announcement follows intense public opposition to the drilling
programme to find a suitable place to deposit low level radioactive waste.
Ministers have been warned that the public outcry could cost the Conservatives
four seats at the next Election as the four sites under review are all in Tory-
held constituencies."

Thus, however one wishes to view the manner in which the current situation
has been handled, I think it is a clear case of the politicians being forced to
reflect and then revise their plans in the light of pressures. I do not for a
moment belittle the part that political considerations played. Part of what I
want to say deals directly with the proper interplay of politics in the
resolving of major policy matters in which the public interest is involved. I
strongly urge that every element in a situation in which public policy is made
has to take on board that these matters are not determined wholly on technical
grounds, and that the well-equipped public servant must have in his armoury a
knowledge of how pressure group politics work - I will have more to say on this
later.

Alternative Disposal

Of course, dismissing one method of disposing of low and intermediate level
radioactive waste, does not dispose of the problem. The Secretary of State
told Parliament that NIREX was already evaluating the relative merits of
techniques of deep burial of intermediate level waste in a repository on land,

tunnelling under the sea from the shore, and disposal into the seabed from a
sea based rig and that it will now extend this study to embrace low-level waste
as well. There is a growing feeling that in bowing to pressures, the ultimate
costs of finally resolving the matter could well prove extremely expensive.
More expensive engineering solutions will prove more costly than the use of
almost natural features. The main option for low and intermediate level wastes
are to drive straight into hard rock and excavate caverns, drill into the seabed
using adopted oil rig technology, or drill slantwise from a coastal site under
the seabed to chambers situated a mile off the coast. Max Wilkinson writing
in The Financial Times after the announcement said that disused mines would
clearly be a candidate, perhaps in Cornwall where granite would form a solid
protection. He then said:-

"The choice between these options is sure to depend as much on political
consideration as on the analysis of geologists and nuclear physicians over the
next few years, since there is little doubt that deep burial is safe and
feasible."

A further suggestion from the UK Centre for Economic and Environmental
Development argued for the long-term storage of nuclear waste and spent fuel
in underground repositories at each nuclear power station. It further argued
that the effectiveness of the local protest group was indisputable, and the
accummulated experience could be picked up by anyone who felt threatened by
future moves on nuclear waste disposal. Yet, when the possibility of using a
remote part of Caithness for dumping waste was mooted, it causes agitation, not
least from Scottish environmentalists, even though it had been suggested
because it was 15 miles from Dounreay with its fast breeder reactor.

The clearest indication that this issue was made on other than technological ground comes from the reaction to it by the Government's own
independent scientific advisers. The Radioactive Waste Management Advisory
Committee. The Acting Chairman, Professor John Greenman was sufficiently
angered to say that he is considering his position, having told The New
Scientist of his anger, frustration and surprise at the way Ministers had dealt
with the issue, asserting that the key decisions are being taken on political
and economic grounds, it no longer seeming to him that decisions are in need of
scientific advice.

Acid Rain

Acid rain is now a fact of life - or a fact of death. The acidification
of Scandanavian soils is now so bad that no matter what Britain does, the
situation will not improve for decades. What is that situation? Sulphur
emissions from Britain's power stations do not in the main fall on British
soil, but 70% of it falls elsewhere. For instance, a country like Norway which
suffers acutely from acidification is responsible for somewhat less than 10% of

the acid deposition defiling its own environment. There is a 30% club, of
countries which have pledged to reduce their sulphur emission by 30% from the
1983 level within a decade. This will not solve the problem, but it will
demonstrate a determined effort to act as good international neighbours, for
this is undoubtedly an international problem. Britain has so far declined to
join; much has been made of this, for even with the commission of the £600m
sulphur-emissions reduction programme to be completed by 1997, it will only
reduce UK emissions by 14%. We have a situation where the damage is beyond
dispute but where Britain refuses to comply with her obligations as a good
neighbour. Kevin Bishop is investigating surface water acidification at the
Department of Geography at Cambridge University. In an article in ECOS he
tells us that the reticence of our Government to join the 30% club is to do
with its fears that the continental concensus on the need for acid-rain control
could develop a momentum of its own which would result in international demand
for more drastic cuts in acid emissions. The fact is that until the Spring of
1986 the Government refused to concede that there was a proven link between
sulphur emissions and environmental damage in Britain - let alone abroad. Its
domestic support comes primarily from the Central Electricity Generating
Board's scientists, whose speciality has been finding reasons for discounting
the evidence of study after study which has found damage caused by acid rain.
That acknowledgement that UK emissions had caused some damage was made in a
meeting with the Norwegian Foreign Minister, which took place less than 24
hours after President Reagan made a similar concession to the Canadian Prime
Minister, Kevin Bishop says:-
"Coincidence like that suggests that it is primarily political expendiency and
not scientific evidence that is the controlling factor in Government Policy."
Using Westminster
        More than once I have sought to lead you to consider making Parliament
work for you as well as many of you working for Government, both locally and
nationally. We should also encompass the European Parliament if we are to
consider the part played in your affairs by Parliamentarians. I want you to
understand and accept that there are many ways in which you, your industry,
your company and your special interest can engage their attention, legitimately
and to the mutual advantage of all concerned.
        In my experience, your general scope and your special interest in Minestone
is a matter in which many Parliamentarians would be open to receiving your
gentle pressure, in particular those who have either a constituency or an
industrial interest regardless of politics. It is in the interest of MPs and
Peers - to be as well informed as possible on as wide a spectrum of matters as
he or she can be. When one considers that the mundane narrow interests of
minestone is concerned within the wider brief of the field of energyresources

Perhaps the most productive way in which those outside Parliament can advance
their interests is by the creation of a group within Parliament specialising
in that special interest subject. They can be geographical, and all-party.
For instance there are groups in Parliament where the officers are elected on
a bi-partisan basis, the purpose of the group being to foster good relations
between Britain and that country. Embassies look upon these groups as
providing valuable points of contact, the members being willing to be briefed
on matters which affect relationships - at present there are 110 different
Parliaments and Legislatives within the Commonwealth alone.

These have an effective umbrella organisation in the Commonwealth
Parliamentary Association, but there are groups concerned with Albania, Bahrain,
Belize, Botswana, China, Cuba, Falklands, Gibraltar, Hong Kong, Iceland, the
Ivory Coast, Manx, Mongolia, Nepal, Peru, Soviet Union, Uganda, Zambia, and
Zimbabwe. It is felt by very busy people that it is worth their time and a
tiny bit of their money to cultivate Parliamentarians, I suggest others can
profit from similar cultivation.

Special Interest Groups

Special groups abound at Westminster. When it comes to the special
interest groups we see more interesting issues which have emerged, sometimes
as the result of pressures from groups and individuals such as yourselves.
They make interesting reading!

Alcoholism, Animal Welfare, Book Publishing, Chemical Industry, Cotton
and Allied, Deposit Reform, Energy Strategies, Firearms, Fluoridation, Footwear,
Minerals Paper Industry, Pharmaceutical, Retail Trade, Rudolf Hess Campaign,
Space Committee, Lighting Industry.

These groups are what the members make of them. They meet as often as
they wish, often only to discuss specific matters, sometimes arising directly
out of projected piece of legislation or business before the House. Those in
industry, or who represents that industry will see that Parliamentarians are
supplied with backfround information, suggest questions to be asked, supply a
brief for a debate, invite the members to visit a location or some aspect of
the interest involved. As an active member and an officer of the All-party
Retail Grade Group I have benefitted from visiting the boardrooms and the shops
of John Sainsbury, Tesco, ASDA, visitedthe London Docklands Developments, met
senior executives, attended seminars - in general equipped myself to present
the face of retailing to Parliament in as kindly a light as possible.

On top of these, each political party has its own backbench committees,
comprising solely of its own political persuasion. These involve themselves
often in creating policy position papers for internal digestion, but they will
often invite an outside interest to meet them so that their discussion can be
better informed. Often, these groups prove most effective, for no punches are

pulled, frank speaking takes place, and the outside interests can speak frankly and openly to them. Never forget that today's backbenchers can be tomorrow's Ministers, nor that the Opposition of today can be Government of tomorrow.

What requires to be done by outside interest groups is to recognise that the ears o. he Parliamentarian are there to be bent – and that if the words that pour intu them are not yours, they will be those of someone else.

In Our Own Hands

One thing is for sure, your world will be in a state of flux for the forseeable future. You may well reply that it has been in that state for some time. I have sought to put to you my thoughts, primarily as a Parliamentarian, one for whom the periodic challenge of the ballot box holds no fears, yet one who is still very much aware of the power of the ballot box. It will not have escaped your attention that an initiative of some significance – to take the ownership of the country's mining industry out of public hands and place it in private hands – is now out in the open. It is not new, but it must now be seen against the background of the doctrine and philosophy of the recently elected Government, returned to power with a substantial majority. "We have no plans to do so at present" may well be a stock reply, but we know that whether it is on any hidden agenda or not, nothing can be ruled out. Thus, when we examine, as you will do over the next few days, the very wide range of matters which emanate from a discussion on the best uses of Minestone – even the very future of its uses at all – at least as far as Britain is concerned, we must be conscious that you could be in a whole new ball game over the next ten years. If it is feasible to comtemplate a change so fundamental as that envisaged if British Coal in some way was privatised with that impact on the future use of minestone, also keep in mind that in four years time we could have a Government with diametrically opposed views. All the more reason for there being the maximum consensus, but a consensus which was positive in its use of Minestone there is no lack of vibrancy in the attitudes of the Minestone Executive. I applaud initiatives such as that which was pursued in the attempt to sell minestone to Sweden to be used to fill the disused dock at Malmo, perhaps a contract worth £40m. I am well aware that the professionals engaged in the Minestone Executive and by British Coal itself require to be sustained in the knowledge that there is a national imperative to be served. I believe that a better parliamentary understanding of these imperatives will assist you all to discharge your tasks in as good an odour as possible. I tell you that there are a host of good friends of Minestone at Westminster, and I count myself fortunate to be one of them.

Reclamation, Treatment and Utilization of Coal Mining Wastes, edited by A.K.M. Rainbow
Elsevier Science Publishers B.V., Amsterdam, 1987 — Printed in The Netherlands

THE ECONOMICS OF MINESTONE UTILISATION

W. SLEEMAN B.A., MIHT., F.G.S.

Minestone Utilisation Manager, Minestone Services,
British Coal, Philadelphia, Houghton-le-Spring, Tyne and Wear.

SUMMARY

Since the formation of the Minestone Executive of the National Coal Board –
now British Coal Minestone Services - many millions of tonnes of Minestone and
Burn Minestone have been successfully utilised. Whilst the marketing of Minestone and Burnt Minestone has been carried out commercially in competition
against other materials, British Coal operations have often gained benefit from
the disposals. In addition, the country as a whole may gain from the reduction
in spoil on redundant tips thereby generally improving the environment. In
this paper the potential for increasing colliery spoil utilisation is
discussed and the financial implications assessed.

INTRODUCTION

With full mechanised mining it is inevitable that a proportion of colliery
spoil is produced with the coal. In 1979/80 67 million tonnes of colliery
spoil resulted from the production of 109 million tonnes of saleable coal from
National Coal Board mines. Table 1 (ref. 1.) details the sources and quantity
of colliery spoil produced.

TABLE 1

Sources and Quantity of Dirt Produced 1979/80.

Source of dirt	Quantity million tonnes
1. Coal face (a) Roof, seam, floor and geological disturbances	35
(b) Face rippings	10
2. Roadway repairs, back ripping and dinting	6
3. Ad hoc sources, special excavations, etc	6
4. Development drivages including surface drifts	10
TOTAL	67

These wastes derive from Coal Measure rocks consisting primarily of siltstones, mudstones, seatearths, sandstones and limestones. (Fig. 1.) (ref. 2.)
Of the 67 million tonnes spoil produced 64 million tonnes were brought to the
surface for disposal following separation from the coal in the coal preparation
plant. The remaining 3 million tonnes was retained underground in gateside
packs etc.

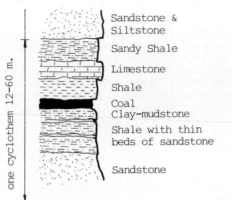

Coal Seams Occur in the Sedimentary Rock
Sequence Known as Cyclothem (after Davies).

Fig.1 Coal Measures Rocks.

In addition to the 50-60 million tonnes per annum colliery spoil produced
from current mining activities there is approximately 3000 million tonnes on the
surface resulting from previous operations.

These historic stockpiles impose an environmental as well as a financial
cost on the community.

Where there is no alternative to disposing of coal mine wastes at the
surface British Coal take due notice of increasing public environmental awareness in their planning applications for further surface disposals as well as
recognising the considerable public sensitivity to and concern over the environmental impact at existing operations. Accordingly millions of pounds sterling
are being spent annually on regrading and landscaping existing colliery spoil
heaps.

The cost of disposal of colliery waste is therefore high and any means of
reducing these costs have to be seriously considered.

The utilisation of the wastes derived from coal mining is obviously in the

interests of British Coal, since although there will still be considerable
quantities tipped, any utilisation will effectively mean that additional
tipping space is being made available on the local tips, resulting in considerable 'cost savings' to the industry if the alternative results in remote
tipping.

In addition, economic considerations make it likely that the cost of
natural aggregates will increase in those areas where they are already in short
supply, whereas waste materials are by their very nature available at low cost,
at least at source.

The utilisation of colliery spoil within the mining industry itself is
therefore most rewarding for as well as achieving considerable cost savings and
a reduction in environmental impact they afford an opportunity to reduce
expenditure on alternative aggregates.

There is a further consideration, no less important, if one considers the
wider issues; that is that it is obviously in the national interest to make use
of waste material as alternatives to naturally occurring aggregates, where
technically feasible, as this conserves the supply of good quality aggregates
for more appropriate uses. The environmental implication being a reduction in
the number of new applications to quarry which is an equally sensitive
consideration in the United Kingdom. Also, the number of borrow-pits,
essentially used for winning bulk-fill, along the routes of new roads could
certainly be reduced.

Plate 1. Typical Perceived View of Colliery Spoil Heap.

In this paper the economics of maximising the utilisation of colliery spoil is discussed.

There are two direct ways of gaining financial benefit from utilisation. The material can be marketed to provide a profit or to reduce costs of disposal by other methods. In addition the potential environmental and financial gains to the community as a whole are assessed.

APPLICATIONS OF COLLIERY SPOIL

Colliery spoil suitable for use can be classified into three basic categories:

1. Unburnt colliery spoil or MINESTONE
2. Burnt colliery spoil or BURNT MINESTONE
3. Modified/Processed colliery spoil.

In addition to materials from within these three categories, the fines discard produced from coal preparation plants and normally in the form of slurry or tailings are a major disposal problem to British Coal.

The majority of these fine discards are disposed of at high cost and inconvenience to the collieries. Small quantities in the form of pressed filtercake are used, primarily as a substitute for puddle clay.

Although research is being pursued on ways to utilise these fines, in general they are classified as materials unsuitable for any use and are dis-regarded when assessing sources of supply for markets. Some possible applications for the fines are briefly discussed in the section on modified/processed colliery spoil.

Applications and potential uses of materials from each of the above categories are discussed below.

Minestone

By far the greatest availability and potential is for Minestone.

Since early 1968, more than 60 million tonnes of Minestone have been used in the United Kingdom. The major use, has been as imported fill for which Minestone has largely taken over from Burnt Minestone.

While the characteristics of available Minestone vary from source to source, and sometimes between parts of a spoil heap, laboratory study and field experience have shown that the majority can be readily compacted into stable fills of high dry density. The variability between sources derives from differences in rock-types associated with the coals; differences in mining and coal preparation practice, which may particularly affect the size grading of the material freshly deposited on the spoil heap; the physical and, probably to a much smaller extent, chemical weathering processes that take place.

Plate 2. Highway Embankments utilising Minestone. M62 Motorway, Yorkshire.

For projects requiring large quantities, Minestone excavated from the spoil heap is the most desirable. Recent changes in Town and Country Planning legislation have restricted the availability of Minestone from spoil heaps and therefore greater consideration is now being given to wider use of material taken direct from discard bunkers. The principal advantage to the user is that vehicles can be loaded direct from the bunker avoiding the need to have loading plant on hand - particularly useful when only occasional loads are wanted. However, there are several potential drawbacks. Supplies may be interrupted if there is a stoppage anywhere in the mining and coal preparation sequence, so that vehicles may be kept waiting. The material will probably carry an excessive amount of surface-water picked up during passage through the preparation plant and this may cause nuisance during transport, though individual particles will usually still be dry inside so any degraded material produced locally during compaction may be too dry for satisfactory compaction. Material cannot, of course, be taken faster than it is produced and this may be serious on large projects. At times there could be large differences in the characteristics of successive loads - some could be mainly fine material with high moisture content, others large dry material. So in some situations it will be important to make a thorough assessment before using fresh-wrought material. For many decades Minestone has been used as fill, though probably with minimal control, at and around active collieries; but little is known of this in engineering circles. More recently the sites of many closed collieries have been converted for developments of many kinds by local authorities and others. Minestone from the disused spoil heaps, properly handled - usually by spreading and compacting in layers as for highway earthworks - provide good stable ground, strong enough to support many types of structure on suitable foundations, easily trenched for services etc.

During the same period there has also been increasing use of Minestone imported to other sites for similar purposes. These include: the elimination of surface irregularities on building sites; construction of temporary haul roads across roads and roadworks over low bearing ground; replacement of silts, peats, soft clays, water-logged and other unsuitable materials, to allow site development to proceed; back-filling of disused quarries, and gravel and clay pits to provide building or recreational land; raising of ground levels on low lying sites; blinding and covering of municipal tips; filling of disused canals and docks. (ref. 3.)

There are indications that the range of applications in which Minestone may be effectively used as fills is likely to be extended by the increasing use of ground treatment techniques.

Reinforced earth is a composite building material formed by the interaction of two essential ingredients; soil and reinforcing members. Minestone Services is currently researching the use of Minestone as a substitute for conventional soil in reinforced earth.

In an effort to establish the suitability of Minestone in the technique continuing research and development has resulted in the construction of a number of Reinforced Minestone Structures at Bedwas, Oxcroft, Newmarket-Silkstone, Prince of Wales, Donisthorpe and Blaenant Collieries.

Many of these structures have been subjected to exhaustive tests and instrumentation and are being continually monitored to check their integrity.

Although the structures have been used for research purposes they were all integrated into operational locations and were required to function satisfactorily under heavy loading. (ref. 4.)

Minestone is also the constructional material for many inland PFA lagoons, and is extensively used within the coal industry, to a strict code of practice, for the lagoons required to impound tailings and slurries (the suspensions of fines which are generated during coal preparation). Similarly Minestone has been used in the construction of dams and reservoirs and flood protection work. In this connection it has been found that selected Minestone can be compacted to extremely low permeabilities. This has suggested that flood protection works and perhaps reservoir walls could be satisfactorily constructed from compacted selected Minestone without clay cut-offs, and that similar compacted Minestone could be used for sealing disused quarries and other excavations to ensure that there could be no seepage.

Plate 3. Minestone used to form PFA Lagoons, Gale Common, Yorkshire.

Another related application is for the daily blinding of refuse tips -
particularly municipal refuse tips - and ultimately covering them to enable the
land to be brought back into use.

Vegetation can readily be established on Minestone when it or other
materials are used for highway embankments it is usual to top-soil exposed
surfaces and to grass or plant them. Similarly, if Minestone is used in
reclamation for raising site levels to up-grade land usage through development,
it would be usual to top-soil exposed areas before grassing or planting and
there is now a considerable range of experience on how to get the best results
when this is done.

However, experiments and experience have also demonstrated that it is not
always necessary to top-soil before seeding or planting. Indeed, it has been
suggested that in some circumstances it may be better to apply appropriate
treatments to the Minestone, and then sow or plant directly on it. Well
developed trees and shrubs on spoil heaps show how successful this can be.

Other experiments have shown that Minestone is a suitable material for
covering up and sealing off heavy metal wastes. When they are suitably fer-
tilized or improved with organic wastes, plant growth on Minestone is as good
as on alternative, much more expensive, materials and they also more effectively
prevent the heavy metals reaching the surface, where they could get into the
food cycle via grass and other vegetation.

Burnt Minestone

Burnt Minestone is an acceptable material for use as common fill. However,
it is now seen to be somewhat extravagant to use large quantities for this
purpose in most areas because they are becoming comparatively scarce. The
reasons for this are initially they were less plentiful than unburnt Minestone

and of course are not being added to; they occur in the older spoil heaps and hence in the older parts of the coalfields, so were conveniently on-the-spot when redevelopment of those areas began; quantities were buried under unburnt Minestone or incorporated into local authority and other reclamation schemes.

Since unburnt Minestone is at least equally acceptable for use as fill in most applications and is plentiful, widespread and more economic, it is usually advantageous to forego using burnt Minestone as common fill in large constructions. Thus remaining stocks are conserved for other applications in which they are technically more suitable than Minestone or are less costly than alternative materials.

Plate 4. Burnt Minestone used as Road Sub-Base.

These uses include: 'special fill material' - for example for use immediately below formation level in road works; free draining material in fills and in drainage layers; aggregate for temporary haul roads, car parks and hardstandings; an alternative to hard core: type 1 and type 2 sub-bases in roads and other works; for which a particularly strong aggregate is not required. For some applications specification limits have to be met. Some burnt Minestone is frost susceptible and would only be permitted as sub-base materials in special cases. The sulphate contents of some burnt Minestone may be too high for use as sub-base materials or for some other applications.

For many of the above applications the burnt Minestone may be used as lifted. For others it may be necessary to do some preparation - such as screening to meet a specified size grading (as with type 1 sub-base material), or yield a particular size fraction for a special application. For many years the smaller size fractions have been prepared in this way for use on hard tennis courts, running tracks and footpaths. A notable recent addition is the construction of speedway tracks, including those used for world championships, from specially prepared burnt Minestone.

Plate 5.   Burnt Minestone utilised as a 'free-draining' material in a
           Highway Embankment in Gateshead.

A few decades ago some use was made of suitably prepared burnt Minestone
as aggregates in, for example, concrete foundations, and for making pre-cast
concrete blocks.   Investigations by British Coal (then National Coal Board)
and other organisations have confirmed that selected burnt Minestone could
still be used for these applications even though some specifications are now
tighter.   Some prepared burnt Minestone meets the British Standard Specification for lightweight aggregates and attractive pre-cast blocks can be made
from them.

Burnt Minestone meeting the specification requirements of BE3/78 (ref. 5.)
is acceptable for use as fill in reinforced earth structures.   A number of
structures on public work schemes have used burnt Minestone.

Modified and Processed Colliery Spoil

While most Minestones are weaker than commercial quarried rocks, the range
of strengths and durabilities is considerable.   A few, because of their origins
have large sandstone contents which, when separated, are the equal of quarried
sandstones.

Use has already been made of some of the stronger Minestones as sub-base
and road-base in minor roads.   There may well be justsification for wider uses
as the need to conserve higher grade aggregates for commensurate applications –
already generally accepted as desirable - becomes more acutely unavoidable in
particular parts of the country.

It has been known and practiced for many years that granular materials

and soils may have their properties beneficially modified by the addition of a
stabilizing agent. Stabilizing agents generally fall into two broad groups -
Active stabilizers which produce a chemical reaction with the aggregate or soil
particles or Inert stabilizers which bind together and/or waterproof the
particles; although many stabilizers display various combinations of active
and inert characteristics.

Coarse colliery discard (Minestone; which falls between a rock and a soil)
from various sources has been investigated in conjunction with a number of the
more commonly available stabilizers; lime, cement and tar/bituminous based
substances and have shown that they respond to stabilization i.e. their
strength and durability are improved. In the United Kingdom, the addition of
ordinary Portland cement has been the most successful, whereas the use of a
bituminous binder appears to be preferred in the Federal Republic of Germany.

Cement bound Minestone has been used both by British Coal for in-house
construction and by commercial organisations, for the construction of car-
parks, hardstandings, sub-bases and road bases. (ref. 6-10)

Plate 6. Cement Bound Minestone used as Car Park pavement material at
          Gatwick Airport.

Coal Measures shales have long been recognised as sources of raw materials
for the production of bricks, etc. The use of MINESTONE, largely composed of
shales, can avoid unnecessary quarrying of like materials. A certain proportion of clay is required during the manufacture of cement, and selected Minestones are suitable for this use. (ref. 11.)

In addition to the above relatively simple processing, research has been carried out to prove the feasibility of modifying colliery spoil for use in a number of more technically demanding applications.

The manufacture of lightweight aggregate by sintering mixtures of clay and pulverised solid fuel is well known. Using a similar technique, colliery spoil has been converted into lightweight aggregate by first crushing the spoil then agglomerating and finally sintering by utilising the inherent coal content to provide all the necessary heat. The clinker thus produced is crushed into angular aggregate particles. Tests have shown that a wide variety of colliery spoils are suitable for use as feedstock for the production of lightweight aggregate. Sintering colliery spoil tends to produce tar fog emissions which are unacceptable by today's standards for environmental control. To eliminate the need for waste gas cleaning, colliery waste which had previously been heated to about 800°C has been used experimentally as the starting material for preparing manufactured aggregate. The feedstock is derived from the combustion of coal washery tailings in a fluidised bed which has been proposed as a means to increase the handleability of tailings for disposal purposes. This process produces carbon-free coarse ash from the bed and a finer ash containing unburnt carbon which is removed from the exhaust gases by cyclone separation. A feed-stock containing about 5% carbon was prepared by blending these two products and pelletised with water. The pellets were dried and sintered without generating tar fog and the resulting sinter cake could be readily separated into individual, spherical lightweight aggregate particles.

The profitability of such schemes depends upon many factors. e.g. proximity to the market, competition from other lightweight products, the buoyancy of the building industry, and it is very susceptible to subsidiary process costs, particularly combating air pollution from the heat treatment of spoils containing tar-producing carbonaceous material.

Pellets made from the crushed bed ash derived from the fluidised bed combustion of washery tailings have been kiln fired to produce a strong, dense aggregate. This product would be suitable to replace natural stone as the coarse aggregate in concrete. (ref. 12.)

With a few exceptions, all natural stone chippings used in road surfaces tend to be polished by traffic, thus increasing the risk of accident by skidding. The stone most resistant to polishing is calcined bauxite but this is too expensive for general use in the size grading required for roads. It has been found that when about 20% of fine (and hence cheap) bauxite was mixed with crushed burnt spoil, the mixture could be extruded, broken into chippings and fired in a kiln to give a product that could form a durable roadstone. The resistance to polishing of this stone (the Polished Stone Value) was 68 to 72, compared with a value of 80 or higher for calcined bauxite and 55 to 65 for most natural stones. (ref. 13.)

It has been estimated that a number of colliery spoils contain sufficient coal to provide all the heat to melt the spoil and form a slag. The properties of slag prepared from colliery spoil in a small gas-fired furnace were found to meet the specification for air-cooled blast furnace slag coarse aggregate for concrete. (ref. 14.)

It is not envisaged that all the current make of colliery waste could be converted into slag, even if this was technically possible. However, it is possible that the waste from a number of selected collieries could advanta-geously be converted to slag for use as aggregate. Small quantities of slag could also be converted directly into mineral fibres.

Thermal insulants are manufactured by melting crushed rock, mixed with coke as a fuel, and pouring the molten material through high speed jets of air or dry steam when long fine fibres are formed.

Unburnt spoil, burnt spoil, washery discard and tailings and fluidised combustor ash are possible raw materials suitable for this process.

A process has been devised for making plastics from tailings. The coal content of the tailings is converted into a plastic binder by digestion with a heavy coal tar oil, and in the course of this the water is evaporated. The mineral matter acts as a filler in the resulting mixture, known as tailings digest. It can be moulded, pressed or extruded into clear useful materials.

The tailings digest product is a bituminous material which has low grade properties compared with synthetic resins it is not, however, as brittle as pitch and generally is a mechanically more desirable material. (ref. 15.)

ALTERNATIVES TO UTILISATION

Ideally it would be desirable to utilise all the colliery spoil produced thereby making one hundred per cent use of a natural resource. Discounting the obvious technical reasons the huge quantity involved make the achievement of this objective difficult to attain.

Although full utilisation of current spoil production would make a major impact, the existing spoil heaps would still remain.

Colliery Spoil from Current Coal Production

In 1982 the Flowers Commission reported on the environmental effects of coal mining. (ref. 16.) They concluded that local tipping, that is depositing spoil directly adjacent to the colliery, would continue, as the most important method of disposal. Excluding commercial utilisation other alternatives to local tipping were:

a)   Back stowing;

b)   Leaving the dirt in the coal saleable product;

c)   Local land reclamation;

d) Remote land reclamation;

e) Marine disposal;

I propose to briefly comment on these alternatives as well as on local tipping.

Local Tipping. It is accepted that continued local disposal is inevitable at many collieries, however recent improvements in tip design and tipping practices including progressive restoration help to reduce the local impact of the tip.

Where land is to be restored to agricultural use the loss of agricultural output should be minimised by careful planning and supervision of restoration and after-care.

With the application of progressive restoration using the best restoration techniques, the use of land for tipping is temporary and the creation of the dereliction associated with the old tips is a thing of the past. Already, large areas of former tipping land have been restored to productive use and are now barely distinguishable from the surrounding countryside.

Backstowing. The feasibility of stowing colliery spoil underground has been investigated in a number of studies.

Storage of waste in the void created behind the coal face was common in the United Kingdom when less mechanised coal production was used. With modern mining methods and machinery the technical and economical feasibility of back-stowing is problematical.

Leaving the dirt in the coal saleable product. The majority of coal produced is supplied to Power Stations for electricity generation. An ash content of 13-18% is generally accepted, therefore a proportion of colliery spoil can be included in the product supplied to the Power Station.

In some cases, such as coal from the Selby mines, the run-of-mine product meets the specified requirements, however the majority of collieries require the 'dirt' level to be reduced in the coal product by washing and screening in coal preparation plants.

Research work on fluidised bed boilers is being carried out to enable coal with a much higher 'dirt' level to be used thereby reducing the production of colliery spoil.

Local and Remote Land Reclamation. Although these disposal methods were given as alternatives to commercial utilisation they are both uses of colliery spoil. In addition to providing a tipping site for the material they also benefit the community in reducing dereliction or reclaiming otherwise low quality land.

Marine Disposal. Marine disposal involves a relatively small quantity of material and is largely restricted to collieries on the coast of Durham and Northumberland. The disposal is effectively by local tipping into the sea

14

adjacent to the colliery.

Existing Spoil Heaps

The disposal of colliery spoil from working mines is essentially an operational problem for British Coal whilst the existing spoil heaps are not necessarily associated with active mining operations and indeed may be outside the control of British Coal. Several options are available for these non-operational sites other than utilisation:

   i)   Reclaim the site to agriculture

  ii)   Reclaim the site for Leisure use

 iii)   Reclaim the site for development

  iv)   Coal Recovery

   v)   Reclamation of voids created by mineral extraction

  vi)   Leave material on ground

    Reclamation of derelict sites. Alternatives (i), (ii) and (iii) are generally carried out by Local Authorities with the schemes being financed by Derelict Land Reclamation grants from government. Although considerable areas of dereliction have been treated in this manner the majority of sites have been restored to either low standard agricultural or leisure use and the creation of such sites by reclamation is restricted by the grant funds made available. Where development such as housing or industrial use can be provided the scheme could prove to be self financing.

Plate 7. Minestone used to Reclaim Derelict Steelworks site for Retail Park, Rotherham.

Coal Recovery. Many of the older spoil heaps contain significant quantities of coal. This results from the mining methods used and the market demands of time when these tips were constructed. Coal contents of up to 40% are not unknown, although a level of 8% to 15% is more likely.

Recovery of this coal at relatively low cost may enable the tip to be restored without the need for grant aid with the reclamation costs being absorbed into the operational costs of the coal recovery.

Reclamation of Mineral Workings. Where higher quality aggregates reserves such as limestone or sand and gravel are located adjacent to or below spoil heaps it is possible to recover them and restore the land form by filling the void with the colliery spoil. This serves two purposes:-

a) Aggregates are recovered

b) Dereliction is removed

It is also possible for quarry-prior-to-tipping operations to be carried out at working mines when suitable aggregates are located on or near the tipping site.

In addition opencast coal operations can provide voids for tipping adjacent to collieries or dispose of existing tips when they overly coal reserves to be extracted by opencast methods.

Leave material on ground. This is the least desirable long term option, however in cases where planned projects are expected to utilise the material then it can be justified to reserve the material for future use.

FINANCIAL ASPECTS OF UTILISATION

Colliery spoil utilisation can be a means of providing profit or used to reduce operational costs. When the wider cost aspects are taken into account, utilisation can contribute to the improvement of the environment by reducing or preventing further despoilation.

When market forces allow, the by-product of the mining operations that is colliery spoil is sold competitively to satisfy market demand. The profit resulting from these sales contributes toward the costs incurred by British Coal in promoting Minestone utilisation. These costs include the charges for the substantial research and development on colliery spoil use, sponsored by British Coal.

Where cost savings to colliery operations are possible or when alternative non-waste materials are marginally more financially attractive to users then colliery spoil may be offered at preferential rates to ensure its use.

The major demand for large quantities of fill materials has been for highway projects, and this demand is likely to continue.

Material for road fill has to be imported to the site when for any reason a road is to be on an embankment and when sufficient material from cuttings in other sections of the work is not available. Imported bulk fill comes from one main source - workings which are located near to the line of the road but outside its limits. These are known as borrow pits. Since they are outside the line of the road, planning permission is needed to use land to provide extra fill to build embankments and land onto which to tip surplus excavated material. Typically for such pits permission is given specifically for extraction of material in connection with a single contract for a limited period. On occasions waste material obtainable from colliery tips and other sources is used.

The general environmental argument against using borrow pit material when colliery spoil or other waste material is availble is given below: (ref. 17.)

   i)   It removes unsightly tips and makes it unnecessary to pay derelict land grant or for the operators to pay for restoration under the 1981 Minerals Act.

  ii)   It reduces the need for new tipping sites.

iii)   It avoids land take and loss of productive agricultural land.

 iv)   There is a gain of amenity where tips are removed.

  v)   It avoids borrow pits leaving unsightly scars on the countryside.

 vi)   It conserves finite resources such as sand and gravel which are found in borrow pits.

This case is further strengthened when colliery spoil is available close to a road scheme. Transportation costs are the main element effecting the price at which colliery spoil can be supplied. However, the costs of supplying colliery spoil is comparable to borrow pit material when hauled from sources up to 10 miles (16 kilometres) from the road site. When suitable wastes are available close to road schemes there is little justification for borrow pits.

In some cases the lack of availability of materials either from existing mineral workings or when a borrow pit is not feasible make the supply of colliery spoil viable over longer distances than the 10 miles quoted above. Market forces will then dictate the costs involved. When colliery spoil could be used but the transportation costs make it less attractive than alternatives then in the national interest the overall cost benefits of using colliery spoil should be examined.

If Minestone can be utilised from a spoil heap resulting in restoration or partial restoration of the tip then it is likely that a substantial cost saving of government grant aid funds could result.

This money made available to subsidise the hauling of material to the road scheme would enable its transport for a greater distance from the source. The benefits are obvious; the road scheme would be carried out without disrupting the environment with a borrow-pit and an area of dereliction would be improved at the tip site.

It is often argued that British Coal should provide subsidies to promote use of its materials. British Coal as a nationalised industry is charged with producing coal at the cheapest cost possible to the country. As a public body, costs incurred must either be covered by the sale of its coal products or if costs exceed income, subsidised from the public purse.

The community as a whole therefore pays the costs either directly or indirectly. As government controls the costs of reclaiming dereliction, it is preferable that it allocates any funds for that purpose.

CONCLUSION

It is obviously in the national interest to make use of waste materials as alternatives to naturally occurring aggregates, where technically feasible, both for environmental and economic reasons.

The Verney Committee (ref. 18.) recommended that:-

i)   To discourage unnecessarily high standard the Department of the Environment should publicise the results of studies on use of fine materials and poor quality aggregates and draw attention to the benefits to be gained from using such material.

ii) While there should be no relaxation in standard of safety and performance required by British Standards, Codes of Practice and Departmental Specifications should actively encourage use of new and tested waste materials.

iii) Nationalised Industries should promote the use of wastes, particularly in their own construction programmes. Where colliery spoil contains more carbon than is necessary to sustain the sintering process, aggregate manufacturing plant should be located so that colliery spoil can be mixed with other wastes so as to optimise the use of fuel content.

iv) The Government should put major and urgent emphasis on (a) the studies already in hand and (b) those we have now recommended with a view to maximising the contributions of waste materials to the supply of aggregates.

These recommendations have been supported by studies notably the 'Flowers' Commission on Energy and the Environment.

Whilst the sentiments of the Verney recommendations are generally applauded little positive action has been done to promote them. Indeed the recent events listed below would appear to be directly opposed to the spirit of Verney.

1. The new Department of Transport specification for Highways works (ref. 19.) raises standards and precludes the use of Minestone for some uses where it could be expected to perform adequately.

2. Suspension of the Dual Tendering Procedure by the Department of Transport, this procedure was originally introduced to promote the use of waste materials and does not appear to have been applied with any enthusiasm by the Department.

3. Introduction of revised general development order provisions within the Town and Country Planning Act. (ref. 20.) These revisions effectively require planning permission to be obtained for colliery spoil heaps to be worked for Minestone recovery. This obviously increases costs and additionally causes delays in the material being made available.

Notwithstanding the above difficulties Minestone must be recognised as a useful and viable material and its full utilisation can only benefit the nation.

REFERENCES

1 J.D. Blelloch, Waste Disposal and the Environment, Colliery Guardian, August, 1983, pp. 392-401.
2 Dr. A.K.M. Rainbow, Composition and Characteristics of Waste from Coal Mining and Preparation in the United Kingdom, United Nations E.S.C.E. Symposium on the Utilisation of Waste from Coal Mining and Preparation, Tatabanya, Hungary, 1983.
3 Minestone Services, Information Sheets, British Coal, London, 1986.
4 Dr. A.K.M. Rainbow, An Investigation of Some Factors Influencing the Suitability of Minestone as the Fill in Reinforced Earth Structures, National Coal Board, London, 1983.
5 Department of Transport, Reinforced Earth Retaining Walls and Bridge Abutments for Embankments, Technical Memorandum (Bridges) BE3/78, Department of Transport, London, 1978.
6 Dr. A.K.M. Rainbow et al, Cement Bound Minestone Users Guide for Pavement Construction, National Coal Board, London, 1983.
7 Minestone Services, Specification for Pavement Construction, "In Preparation", British Coal, London.
8 Minestone Services, Thickness Design Guide for Pavement Construction, "In Preparation", British Coal, London.
9 W. Sleeman, Practical Application of Cement Bound Minestone within the British Coal Mining Industry, Proceedings of the Symposium on the Reclamation, Treatment and Utilisation of Coal Mining Wastes, Durham, September, 1984.
10 Dr. A.K.M. Rainbow, Colliery Spoil - its production, properties and use in a cement stabilised form, Seminar on Waste Materials in Concrete, Cement and Concrete Association, Fulmar Grange, 1982.
11 D.A. Tanfield, Construction Uses for Colliery Spoil, Contract Journal, January, 1971.
12 C.E. Carr and M.J. Cooke, The Preparation of Mineral Products by the Heat Treatment of Coal Mining Wastes, Proceedings of the Symposium on the Reclamation, Treatment and Utilisation of Coal Mining Wastes, Durham, September, 1984.
13 Ditto.
14 Ditto.
15 J. Gibson, An Appraisal of the Utilisation Potential of Colliery Waste, Mining and Minerals Engineering, November, 1970, pp. 28-42.
16 Commission on Energy and the Environment, Coal and the Environment, HMSO, London, 1981.
17 Department of Transport, Report of the Interdepartmental Committee on the

use of Waste Material for Road Fill, Department of Transport, London, 1986.

18   Department of the Environment, Aggregates: The Way Ahead, Report of the Advisory Committee on Aggregates, HMSO, London, 1975.

19   Department of Transport, Specification for Highway Works, HMSO, London, 1986.

20   Department of the Environment, The Town and Country Planning General Development (Amendment) (No. 2) Order 1985, Department of the Environment, London, 1985.

ACKNOWLEDGEMENTS

The Author is grateful to British Coal for the support given in the preparation of this paper and for permission granted for its publication. The views expressed are those of the Author and not necessarily those of British Coal.

Reclamation, Treatment and Utilization of Coal Mining Wastes, edited by A.K.M. Rainbow
Elsevier Science Publishers B.V., Amsterdam, 1987 — Printed in The Netherlands

AN EVALUATIVE FRAMEWORK FOR ASSESSMENT OF DISPOSAL OPTIONS FOR

COLLIERY SPOIL

MALCOLM NOYCE

Project Director

Ove Arup and Partners

SUMMARY

Ove Arup and Partners was commissioned by the Department of the Environment to
design and develop an Evaluative Framework which would enable an overall
judgement to be made between the realistic disposal options for Colliery Spoil
taking due account of the economic costs and revenues, and balancing these
against the various environmental effects. The Evaluative Framework would
enable British Coal and the Mineral Planning Authority (MPA) to choose the best
disposal option. The MPA has the task of determining a planning application for
the scheme submitted by British Coal.

In addition Ove Arup and Partners was required to report on the current
procedures and methods of spoil disposal. This overview was presented in the
Final Report dated May 1986. However, this paper concentrates on an explanation
of the Evaluative Framework.

A.   DEPARTMENT OF THE ENVIRONMENT RESEARCH PROJECT

Between 1979 and 1981 a major investigation on Coal and the Environment was

conducted by the Commission on Energy and the Environment (CENE). The

Commission concluded that spoil disposal was one of the two most important

environmental impacts of deep mining, the other being subsidence.

At about the same time as the CENE investigation was taking place, spoil

disposal issues were under examination at two Public Inquiries. The first in

1979-80 centred on the proposed development of three new mines in the North

East Leicestershire Prospect (Vale of Belvoir). One of the main concerns raised

at the Inquiry and the reason for subsequent refusal of planning permission was

the expected impact of local spoil tipping, particularly at two of the

suggested locations. Similar circumstances surrounded the South Kirkby Public

Inquiry in 1981 which was the first Inquiry called specifically to deal with

spoil disposal objections.

The Government's response to these mounting concerns was given in May 1983 in the White Paper "Coal and the Environment". This stated that it had "initiated a major exercise in the Yorkshire, Nottinghamshire and Derbyshire coalfield on spoil disposal, designed to evaluate the main options and to establish a new framework within which spoil decisions can be taken". Ove Arup and Partners was commissioned by the Department of the Environment (DoE) in December 1983 to undertake a research project which formed part of that exercise.

The main aim of the project was to design and construct an Evaluative Framework to assist decision making on the selection of options for colliery spoil disposal. A Framework was to be drafted in the early stages of the project and then tested and successively refined by the examination of a series of case studies. These case studies had been selected to reflect a variety of disposal arrangements and to take account of different financial and environmental considerations.

A second strand of the project was to review current spoil disposal procedures. These vary for different types of spoil, by handling method and by location. A four part review has been completed in the Final Report covering technical, institutional, financial and environmental aspects.

To provide advice to the Consultant and to monitor project progress the DoE set up a steering group consisting of nearly 30 representatives oF British Coal, Mineral Planning Authorities (MPAs) in the Central Coalfield, Department of Energy, Ministry of Agriculture, Fisheries and Food (MAFF) and District Councils.

B.   WHAT IS THE EVALUATIVE FRAMEWORK?

For each particular spoil disposal requirement the Framework encourages the systematic investigation of a range of alternative schemes in order to compare the advantages and disadvantages of each option against various economic and environmental factors. The Evaluative Framework has been designed for use during the stage leading to the submission of a planning application by British Coal.

The Evaluative Framework is in three parts:-

Phase I   Selection of Options
Phase II   Costs and Impacts
              -     Economic Assessment
              -     Environmental Assessment
Phase III   Combined Evaluation

The main flow-chart illustrating the overall process is on the next page, and a fuller description is given in Section F.

The Framework is currently in its first release and it will need to be monitored and refined as experience is gained of its use in practice.

C.   WHY IS IT NEEDED?

British Coal must apply for planning permission to the relevant MPA for all new colliery spoil disposal schemes. During the 1970s there was a growing realisation amongst MPAs that the easy options for tipping, particularly on farmland adjoining collieries, were beginning to run out. This increasing pressure on land occurred together with a greater environmental awareness by the general public. There was a danger that a series of incremental extensions could cause problems resulting in delay and sometimes a conflict in views between British Coal and the MPA.

The development of the Evaluative Framework is in part a way of building upon and formalising the current ad-hoc approach and is a means of placing information before the MPA enabling it to be satisfied than the best disposal scheme has been selected.

Due to extensive advanced negotiations the majority of formal tipping applications determined by MPAs are approved. Some tipping still takes place as permitted development under the General Development Order, in which case no conditions have been attached. However, British Coal has been prepared to surrender most of these rights when putting forward proposals for new tipping on adjacent land.

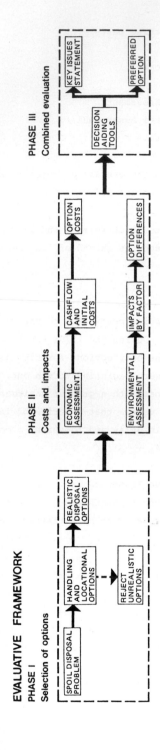

EVALUATIVE FRAMEWORK

PHASE I
Selection of options

PHASE II
Costs and impacts

PHASE III
Combined evaluation

SPOIL DISPOSAL PROBLEM

HANDLING AND LOCATIONAL OPTIONS

REJECT UNREALISTIC OPTIONS

REALISTIC DISPOSAL OPTIONS

ECONOMIC ASSESSMENT

CASHFLOW AND INITIAL COSTS

OPTION COSTS

ENVIRONMENTAL ASSESSMENT

IMPACTS BY FACTOR

OPTION DIFFERENCES

DECISION AIDING TOOLS

KEY ISSUES STATEMENT

PREFERRED OPTION

Land-based tipping accounts for over 90% of total annual spoil disposed nationally which amounts to nearly 50 million tonnes. Most of this is tipped on spoil heaps although a small amount is used in local land reclamation projects - for example, filling mineral working voids or raising the level of badly drained poor quality land. Marine disposal is restricted to North East England and accounts for about 9% of the spoil disposed.

The construction of coarse discard tips is a relatively straightforward mechanical handling operation. Fines discard is a more difficult substance to handle; it can either be pumped to settling lagoons or it can (at extra cost) be mechanically dewatered and then disposed of on the same tip as coarse discard.

There is increasing pressure to place the spoil in unrestored voids which invariably are remote from the colliery. Two fundamental questions arise. Are the environmental benefits gained by tipping into remote voids worth the additional economic costs incurred? If so, who should contribute the necessary finance?

Currently British Coal directly funds all spoil disposal operations to completion of tipping. However, in the historical context the DoE has funded reclamation of old inherited tips via its Derelict Land Grant. British Coal is required by central government to meet defined financial targets and local authorities have little available cash, leaving little flexibility for redistribution of disposal costs.

At the end of the day if British Coal is being persuaded to adopt a more expensive remote disposal option in preference to a cheaper local option then it will have to decide on a commercial basis whether or not to proceed with the more expensive scheme. This has clear implications for the viability of any affected colliery and particularly for employment prospects.

D.   WHO WILL USE IT?

The Framework requires that British Coal play the major role in completing it when there is a need to find a solution for spoil disposal for a particular colliery.

The MPA will be involved at certain stages during its completion, in particular for the identification of options, the joint selection of which ones to test, the joint decision on how much detail to include in the testing, the provision of some information for the environmental assessment and the joint use of the decision aiding tools.

Other Statutory Parties such as MAFF, the Water Authorities and the Nature Conservancy Council, need to be involved, particularly in the selection of options and in the environmental assessment.

At the wider level where the Framework might be used to test the viability of transporting spoil to a major reclamation site, it may be that the MPA might take the lead in completing the Framework, involving British Coal particularly in the cost estimate.

E.   WHEN WILL THEY USE IT?

The Evaluative Framework intentionally encourages a more co-ordinated approach to the spoil disposal problem. Joint exercises between British Coal and the MPA to establish expected spoil arisings and the various time horizons when new tipping arrangements will be required, will often be a forerunner to the use of the Framework. Its completion would comprise the major part of the background work necessary before the submission of a planning application.

Upon completion of the Framework British Coal will separately decide whether or not it proceeds to the submission of a formal planning application for a particular scheme. It will then be the responsibility of the MPA to respond to that application, taking account of the recommendation of its Officers. If planning permission is refused the Department of the Environment holds the ultimate power of determination of the application through appeals to the Secretary of State.

F.   DESCRIPTION OF THE EVALUATIVE FRAMEWORK

The Evaluative Framework is in 3 phases.

Phase I - Selection of Options

The purpose of Phase I is to identify all realistic options for disposal of
spoil from a particular colliery or development. It is therefore site specific
and requires British Coal to provide information where it exists on current
activity and to make available forecasts of future spoil make. The MPA or MPAs
may be able to help identify disposal locations and give guidance on planning
constraints.

Phase I is in the form of a series of questions which seek background
information to the exercise. It focuses initially on details of the colliery
(or groups of collieries) in question and the required tipping capacity for
spoil disposal. It then identifies areas of land which either singly or in
combination could satisfy the tipping volume requirement.

In selecting such options, attention is given not just to land in the vicinity
already owned by British Coal but also to sites with opportunities for
environmental gain. Options for disposal are then subjected to preliminary
testing in order to disregard unrealistic ones. Preliminary design and costings
work is then required for each of the feasible options as a basis for the more
detailed later analyses in Phase II. The number of options to be taken forward
into Phase II would normally be between 2 and 6, depending on circumstances.

Within an option there may still be alternative ways of working the site or
sites and alternative methods of ameliorating the level of impact which the two
parties should evaluate. Sections of the Phase II Framework assist in this
process.

Phase II - Economic Assessment

Phase II will take forward the limited number of realistic options and for each
will separately examine defined economic and environmental issues. Where
possible, all costs and factors will be quantified using separate indicators.

The objective of this phase is to identify the difference in costs between
options - not to define the absolute costs of tipping for any particular
option. The criteria for inclusion of costs should be whether the particular
item affects the decision between options or, in the case of one option,

whether to go ahead or not.

Costs for each option are summarised as follows :-

i)   Cashflow Summary. The costs of each activity are designated to specific
        years on an overall Cashflow Summary sheet. The flows of each year are
        then totalled.

ii)   Net Present Cost. The Net Present Cost is then calculated on the
        Cashflow Summary Sheet by discounting back to year 0 each annual
        cashflow by the appropriate factor, based on an annual rate of 5%.

iii)   Equivalent Annual Cost. The Equivalent Annual Cost is calculated on the
        Cashflow Summary Sheet and converted to a cost per cubic metre or per
        tonne of spoil by dividing by the average production of spoil to be
        disposed each year. Separate calculations can be made for each activity
        heading.

iv)   Initial Cost. The advance capital costs up to and including the first
        year of tipping are calculated by abstracting the relevant information
        from the questionnaire and transferring it to the Initial Cost Summary
        Sheet.

The end result of Phase II Economic is three sets of key indicators for each
option, namely:-

i)   The Equivalent Annual Cost per cubic metre or tonne of spoil disposed
        together with the EAC per activity heading.

ii)   The Cashflow Total as an indication of the overall scale of resources
        required.

iii)   The Initial Cost up to the first year of tipping.

The overall EAC and the Initial Cost are economic criteria carried forward into
Phase III; the EAC by activity and the cashflow total are purely indicative
measures.

Phase II - Environmental Assessment

There are no easy ways of measuring the environmental effects of spoil disposal schemes because the assessment is largely subjective and because environmental consciousness changes over time. Environmental effects will vary with the:-

-       form of disposal (surface tipping, lagooning, backfilling of voids, marine disposal), and

-       type of surroundings (existing colliery/mineral site or farmland, landscape quality, proximity to settlement, presence of any special features).

The impacts will vary over the life of a scheme, particularly between operational and after-use stages. Impacts are also likely to be perceived differently by different interest groups. Nevertheless the environmental effects can be grouped into five main categories:-

-       visual
-       land use
-       ecological/heritage/recreational
-       pollution risk
-       transport route.

The initial task for the environmental assessment involves the definition of the study area within which to assess the existing conditions and likely impact of the options under consideration. On maps of either 1:25,000 scale or 1:10,000 scale it is necessary to draw the approximate boundaries of the area from which the proposed site and scheme will be either visible or be affected in some way such as by extra noise disturbance, diversion or closure of routes or loss of amenity.

The environmental appraisal of each option entails both an assessment of the existing environmental quality of the proposed site with its surroundings and the impact which could be expected from each option. Different methods of working and scale of ameliorative measures may influence the expected impact of a given site. The exercise involves mainly desk-top work although site investigations will supplement the information derived from maps, plans and reports. Photographic records of the site area are of considerable benefit for information and display of impacts.

The assessment of environmental impacts consists of a series of questions which are grouped together under the different headings. The questionnaire has been designed for answers to be given alongside each question on the form. Many of the questions seek information in two parts, one relating to the site itself and the other to the surroundings. Likely data sources are described in the guidance notes.

The next stage consists of a set of tables in which the environmental impacts of an option can be summarised. Each table includes three columns to indicate the timing of the anticipated impact during the tipping scheme, the scale of features affected and the different parties affected. Environmental factors, grouped into four main headings, are listed down the left hand axis of the tables. The transport route section only needs completion for off-site options.

The final stage consists of a single table in which the differences, in terms of environmental impacts between the options being assessed, can be stated. To assist the evaluations in Phase III, the final columns of this table note whether differences between options are significant and if so, which option is least affected and which worst affected by each environmental factor. The conclusions from this table can then be taken forward for evaluation in Phase III.

Phase III - Combined Evaluation

Phase III brings together the economic and environmental aspects in a judgemental process. It is recognised that this area might give rise to differences of opinion between the parties. This will largely arise from the different emphasis each party might wish to attach to the economic costs and the various environmental factors.

The aim of Phase III is to present information on options in a manageable form to assist decision making. The method compares the outputs from Phase II Economic (which are definitive) with the outputs from Phase II Environmental (which are more subjective). The output from the economic assessment will be monetary units in terms of Equivalent Annual Cost per cubic metre or per tonne of spoil and initial cost for each option. Thus not only will there be an implied ranking for the options but the relative differences between the options will also have been established. The output from the environmental assessment has up to this Phase been measured in different units but in most cases may only be in descriptive form. Thus, although it is possible to rank the options on each factor by giving preference to those with less impact, there is often no easy and objective way to select an option on overall environmental grounds.

Before proceeding to the combination of the economic costs and environmental factors. Phase III invites the user to complete an opening statement identifying the key environmental factors in terms of comparing the particular set of options under consideration. This simplifies the issues to be tested using two decision aiding tools. These are intended to assist decision making by identifying the critical issues upon which subsequent negotiations can focus.

The first of the decision aiding tools (Option Rankings) provides a sieving technique. The second tool (Pairwise Comparison) provides a more detailed approach to comparing options on a factor by factor basis taking two options at a time.

Option Rankings, requires the user to indicate his preferences graphically by allocating stars between options on each economic and environmental factor. One star is given to the worst option, five to the best with the remainder within this range according to relative performance. Columns are then compared and if there is one option with a low incidence of stars it might be possible to eliminate it. On the other hand it might be that a dominant option might emerge to the satisfaction of all parties (indicated by a high incidence of stars).

The second tool, Pairwise Comparison, compares two options at random. For each factor the better option is chosen and the reasons noted. Then by comparison of all the factors the better option is compared with that remaining from the first test and the process of successive elimination continues until the preferred option remains. The preferred option should then be tested against the first option that was dismissed to ensure conformity of choice. If in comparing options, the preferred choice is not clear cut and a difficult trade-off is raised, that pairing can be left unresolved and each can be compared against the next option. At the end of the exercise there may need to be some reiteration to check that all options not eliminated have been paired against each other.

A variant on the use of Pairwise Comparison is to start with the least cost option and consider in turn and comment upon each environmental factor. Again the worse overall option is discarded. The next least expensive scheme could then be taken and its environmental effects compared with those of the lesser cost scheme. In this way a direct relationship could be established on an iterative basis for each option on each aspect. A further variant would to be take the best environmental option first and to compare it successively with other options.

Finally, the users of the Framework are required to complete a Key Issues Statement. In cases where a preferred option emerges this may mean briefly stating the reasons for this choice with reference to justify it against the economic costs. In other cases the statement should summarise the key trade-offs that need to be resolved by subsequent decision makers in selecting the preferred option.

For guidance worked examples of the summary sheets for the Phase II Economic and Environmental Assessments, together with a full example of the Phase III techniques are included in an Annex to the Framework. The tables reflect one of the case studies used in the Ove Arup and Partners' research study and are based on published data sources.

G.   IMPORTANT ASPECTS OF THE EVALUATIVE FRAMEWORK

The Framework encourages the identification and evaluation of alternative disposal options for a colliery or group of collieries. It is intended that British Coal should apply the procedure when it is formulating ideas for particular collieries. The Framework should not be used after a decision on a disposal option has been made, nor should it be used solely as a way of justifying a pre-determined scheme.

In general, the parties will initially find it useful to select jointly the options to be investigated and to agree the level of detail for the subsequent analyses appropriate to the problem in hand. The costings information might then generally be assembled by British Coal while some of the environmental information could more easily be provided by the MPA. The techniques included in the Combined Evaluation stage are designed to assist the choice of a preferred option.

Used in this way it is hoped that the Framework will result in earlier agreement between British Coal and the MPA on the choice of the preferred disposal option and thus shorten the timescale for granting planning consent. In the event that agreement cannot be reached and a Public Inquiry results, the Framework should at least identify the areas of disagreement from those of agreed fact and provide a structured basis on which investigations can proceed.

At the wide level the Framework is intended to assist the formulation of broad options for disposal; for example by testing the viability of transporting spoil from one or more collieries to a remote reception site. The initiative for such schemes is perhaps more likely to come from MPAs at a sub-regional or regional level or from joint working parties of different interests. British Coal may also find it useful to use the Framework at Area level to test the effects of concentration schemes and to assess options for new coalfield developments.

H.   THE EVALUATIVE FRAMEWORK IN USE

The Framework was developed and then tested against a series of case studies based upon historical data and proposed projects. In addition during the course of the project the evolving framework was used in real life to assist investigations of two disposal schemes. The first was to examine marine disposal from the North East Area and in the second case the Framework was used by British Coal and the local authorities to help in the assessment of possible spoil disposal options for the proposed South Warwickshire Prospect.

Lessons from the first use in practice were incorporated in the Evaluative Framework when presented in June 1986 as witnessed by the Appendix covering marine disposal. Lessons learned from the second use have been presented to the

DoE subsequent to the presentation. These lessons require fuller guidance notes in some instances and a requirement to assess alternative final forms and after uses, particularly in the case of tipping into voids. There is also a need to assist users in reducing the number of options which are taken into Phases II and III.

At the time of writing (September 1986) the DoE is organising a series of introductory seminars at regional level to assist the understanding, and hence implementation, of the Evaluative Framework. Thereafter the Framework will be monitored and further refined as it is used in practice.

The Framework itself is presented in a self contained Procedural Manual published by HMSO. A summary guide in leaflet form describes the essence of the Framework, why it is needed and who should use it. A full description of its purpose and development is contained in the Consultants' Final Report.

Reclamation, Treatment and Utilization of Coal Mining Wastes, edited by A.K.M. Rainbow
Elsevier Science Publishers B.V., Amsterdam, 1987 — Printed in The Netherlands

THE DEVELOPMENT AND UTILISATION OF "STONE COAL" ("BITUMINOUS SHALE") IN CHINA

Li Yingxian, Mining Engineer and Editor of "World Coal Technology", China

SUMMARY
This paper is on the origins, properties and geographical distribution of
"stone coal" ("bituminous shale"). It discusses and evaluates several of its
uses in China today and concludes that there are good prospects for its
utilisation there.

1. Origins and Distribution of "Stone Coal" ("Bituminous Shale")

"Stone coal" ("bituminous shale") is a kind of combustible, organic rock,
with low calorific value and high ash content, which is found in ancient
(geological) strata. In the early Palaeozoic Era, algae, fungi and plankton
lived in the warm, calm, shallow coastal waters of the ocean, where there was
plenty of sunshine, and they mixed with the remains of low-level organisms and
with minerals and sank to the sea-bed. Where the water became stagnant and
there was a lack of oxygen, and where they were covered over with other sediments, they underwent long and complex biochemical and physical processes at
high temperatures and pressures, and were transformed into a kind of very hard
sedimentary rock stratum, which has the external appearance of stone but which
can be burnt - i.e. "stone coal"("bituminous shale"). The position of the
stratum is relatively stable, there is little variation in thickness, it is
quite near the surface and it is easy to extract.

Stone coal (bituminous shale) is fairly widely distributed through 13
provinces in southern China. Total reserves amount to 6188 hundred million
tons. From north to south, four bands of it can be distinguished - Qinling,
Yangzi, Jiangnan ,and Dongnan.

The period of formation of Qinling stone coal (bituminous shale) was
comparatively long and the band stretches over southern Shaanxi, north-east
Sichuan, north-west Hubei, south-west Henan, northern Anhui etc. Its
geological structure is complex but the quality of the coal is good and it has
a high calorific value, generally 4000-5000 k/cals/kg. There are reserves of
27.4 hundred million tons.

The Yangzi stone coal(bituminous shale) band includes Sichuan, south-west
Hubei, north-west Hunan, central Hubei and the area close to Nanjing in Jiangsu.
It covers a wide area, but conditions for the formation of the coal were not
good, and the calorific value is low. Reserves are 8.1 hundred million tons,
i.e. 1.3% of the national total. Some 95% of these are concentrated in the

TABLE OF STONE COAL RESERVES IN SOUTHERN PROVINCES AND AUTONOMOUS REGIONS   UNITS: 10,000 tons

Type of deposit / Province	Total reserves	Known deposits			Comprehensive survey or deposits					
		Known deposits Total	Already examined and approved	Not yet examined and approved	Comp. Survey of deposits Total	Distinguished according to era				
						Sinian period	Early Cambrien	Silurian	Other	
Jiangsu					33700					
Zhejiang	1063530	123707			939823		939823			
Amhui	745910				745910		745910			
Fujian										
Jiangxi	083403				688403		633403			
Henan	43812				43812	33770	10112			
Hubei	259415	46421	17765	28656	109994	3542	170496	23979	12277	
Hunan	1871637	159272			1712365	56390	1653975			
Guangdong	509				509				509	
Guangxi	1287962	58389	4430	53959	1229573		1229573			
Guizhou	82928				82928	9423	73505			
Shaanxi	151564	1941	1810	131	149623		21731	103961	23931	
Total	6187670	389730	24005	82746	579740	103055	5530228	129740	36717	

References: 1.   Coal development & processing uses, 1984, 12.

2.   Report on a general examination of stone coal resources in Anhui province

3.   "Uses of stone coal", ed. by Zhejiang coal industry bureau.

4.   "Uses of gangue and stone coal" 1981, May.

5.   "Mining industry references   " 1986, 3

area between Hefeng and Yichang.

The Jiangnan[1] band of stone coal (bituminous shale) includes eastern Guizhou, northern Guangxi, western Hunan, central Hunan, eastern Hubei, northern Jiangxi, southern Anhui, western Zhejiang etc. The coal is found over an extensive area, and the position of the stratum is stable, but the calorific value is low (mainly 800-1000 k/cals/kg). There are extensive reserves reckoned at 578.4 hundred million tons, i.e. 96.2% of the national total.

The Dongnan[2] band of stone coal (bituminous shale) covers central Guangxi, eastern Guangxi, northern Guangdong, south-east Hunan and southern Jiangxi etc. It covers a large area but conditions for the formation of the coal were poor and it has a low calorific value. Reserves amount to 1.8 hundred million tons, i.e. 0.3% of the national total. (see table).

---

(1)   this means "south of the Yangtze river"

(2)   this means "south-east"

Extent		By Calorific Value			By Slant Depth	
Reliable	Possible	800-1200 cal/gr	1200-3000 cal/gr	3000 cal/gr	0-100m	100-300m
838778	101045	521658	418165		313257	626566
62411	683499	608154	137756		249587	496323
80141	603262	465066	218337		227801	455602
13812		10175	33637		30489	13323
130259	79735	146817	35798	27379	71171	438817
679704	1032661	1355434	356778	153	571461	1140904
509			509		170	339
912203	317370	1229573			409858	819715
82928		82928			27643	55285
17386	132237	11929	59300	78394	47968	101655
2848131	2949809	4431734	1260280	105926	1949406	3848529

2. Properties of Stone Coal (Bituminous Shale)

Stone coal (bituminous shale) is a high-ash, high-sulphur, low heat/value low-grade fuel, composed of both organic and inorganic matter. Organic matter makes up 10-15% of it, the major part coming from the transformation of low-level organisms, and this is the part which produces heat when burnt. Inorganic matter makes up the 85-90%, and consists mainly of minerals. This is what produces the matter left after the coal has been burnt.

Stone coal (bituminous shale) is normally produced in three forms - as lumps, as flakes or as powder. Its level of metamorphosis is greater than that of ordinary smokeless fuels, but the type of coal rock originates from highly-metamorphosed sapropelic smokeless coal. It has a faint lustre, few crevices, and it is hard and brittle; it has a high burning-point, of about 700-800°C, and when it burns it gives out little or no flames. It has a specific gravity of 2-2.4; and an ash-content as high as 70-80%. Its calorific value is generally 900-2000 k/cals.kg., and it can be split into three grades, according to variations in this value.

CHART 1

Grades of Stone Coal (Bituminous Shale)

Heat Value (k/cals/kg.)	800–1200	1200–3000	>3000
Classification	Low heat value stone coal, or inferior stone coal	Medium heat value stone coal or medium stone coal	High heat value stone coal or superior stone coal

CHART 2

Composition of Stone Coal (Bituminous Shale)

Component	Content (%)	Remarks
Carbon	12–90	could be higher
Hydrogen	0.3–0.4	Max. 1.7
Nitrogen	0.2–0.3	Max. 1.7
Oxygen	2–3	
SiO_2	70–80	
Al_2O_3	4–10	
Fe_2O_3	4–6	
CaO	1–7	
MgCl	0.8–1.2	
Sulphur	0.01–20.19	highly variable

There are quite a number of minerals and elements found associated with stone coal (bituminous shale) which are very useful. Initial findings have shown that there are more than 60 different useful associated elements, including vanadium, molybdenum, copper, gallium, cadmium, uranium, silver, rare earths, yttrium and samarium – all of which are of value to industry. Besides these there are also phosphorous, potassium, pyrites, barite, asbestos, carbon asphalt etc.

3. Use of Stone Coal (Bituminous Shale)

The extraction and utilisation of stone coal (bituminous shale) has a long history in the south of China. It started several hundred years ago but the coal has only been extracted and used on a large scale since 1970. According to figures for 1981 from the six important areas where stone coal (bituminous shale) is mined, there are altogether about 40 stone coal mines at county level and above. In the last two years, China's production of stone coal (bituminous shale) has reached 4–5 million tons. It is calculated that in 1986 the total will reach 5,370,000 tons. Most is mined in Zhejiang, where between 1972–82 a total of 26,300,000 tons was extracted and used. (This is

equivalent to 5,200,000 tons of raw coal). Hunan mines about 15,000 tons each
year. The other provinces produce only small amounts.

There are many ways in which stone coal (bituminous shale) can be used,
and these have been developed relatively swiftly over the last few years. They
have developed from non-industrial uses to industrial uses, from use only as a
fuel to use as raw material for construction materials, in the chemical
industry, in chemical fertilisers, and so on. People are now paying a good
deal of attention to it.

(a) Power Fuel. Because stone coal (bituminous shale) has a high ash
content, is very hard and has a low calorific value, when it is used as a
power fuel special boilers must be used. At present there are two main kinds
of industrial boiler in use in China - the fluidised-bed boiler and moving bed
boiler. There are more than 2000 fluidised-bed boilers in use, with a total
evaporative capacity of more than 1000 tons/hour. The maximum evaporative
capacity of an individual boiler is 130 t/h, and the minimum is 0.2 t/h.
Fluidised-bed boilers are very easy to adapt to various fuels and can have a
heat efficiency of 65%. The moving bed type of boiler is suitable for use in
small-scale industrial boilers which burn stone coal (bituminous shale) in
lumps. These are boilers of 0.1, 0.15, 0.2, 0.4 and 0.8 t/h etc. Zheijiang
province alone has 630 small-scale moving bed boilers with an evaporative
capacity of 225 t/h, and heat efficiency of 49-50%. In this type of special
boiler we can effectively control the temperature of the combustion zone and
eliminate occurrences such as partial slagging or combustion failure, due to
excessively low temperatures, caused in ordinary combustion when the "pile
ratio" is high, the fusion point of ash is low and there is a lot of draught
resistance. Only 1000 k/cals/kg. stone coal (bituminous shale) in lumps of
50mm can be burnt. After making improvements, Zheijiang has come up with a
method of burning 800 k/cals/kg. stone coal (bituminous shale), and has
extended the uses of low calorific-value coal.

But its evaporative capacity is low, and so is its heat efficiency, and
disposal of waste is problematic.

(b) Construction Materials. Reserves of stone coal (bituminous shale)
are plentiful, and the ash content is high. In the ash are chemical elements
similar to silicates, which are ideal raw materials for the construction
industry. At present, the most important materials being produced are cement,
lime, red bricks, and cement products etc. There are also some new materials,
such as ceramics, light aggregate, white cement etc.

When burning stone coal (bituminous shale) to make cement, the coal is
both fuel and raw material. The vanadium, phosphorous etc., which it contains
function as catalysts and help to prevent the cement clinker from pulverising

and increase the early strength of the cement. But when making cement, the calorific value of the stone coal (bituminous shale) must be more than 1750 k/cals/kg., and since most stone coal (bituminous shale) has an insufficiently high value, it is often necessary to add some bituminous coal. One of the most crucial things in the burning of stone coal (bituminous shale) to produce cement is the homogeneity and control of the quality of the coal. Zheijiang has more than 70 factories making cement from stone coal (bituminous shale), with an annual output of in excess of 800,000 tons (i.e. about one-fifth of the total output (of cement) of the province). In Ningxiang county in Hunan, the Stone Coal Utilisation Company, the Research Institute and other units concerned together use 13m³ small blast furnaces, and utilise local stone coal (bituminous shale), waste rock, limestone, dolomite, high-sulphur local coke and other raw materials to smelt white cement and manufacture slag sulphate white cement, with a strength of over 400, and with a whiteness of 74.1 - 79.1 degrees.

Because there is a lot of ash in stone coal (bituminous shale), after combustion there is 80% slag, and uses of stone coal slag are thus directly related to the future prospects of the utilisation of stone coal (bituminous shale). Using slag to make carbonated bricks and other silicate-like building materials is a simple technological process and needs little investment. Costs are low and it uses a lot of slag and can do away with fears about overflow disasters. The strength of carbonated bricks is proportional to the alumina content of stone coal (bituminous shale) slag, and the resistance to pressure and to breakage both exceed 100, while the resistance to cold is also high. Performance at -30° was very good. Zheijiang has 147 carbonated brick kilns, with an annual output of four hundred million bricks, i.e. 13% of the total brick output of the province. Besides these, there are also 490-plus earth standing kilns which burn stone coal (bituminous shale) to make lime, with an annual output of 1,200,000 tons, which is about 80% of the lime produced in the province (each year). And there are 98 kilns which burn stone coal (bituminous shale) to make bricks, with an annual output of five hundred million red bricks i.e. one-sixth of the total brick output of the province - and an annual consumption of 3,000,000 tons of stone coal (bituminous shale).

(c) Using Stone Coal (Bituminous Shale) to Generate Electricity. From the 1970's on, China has been using stone coal (bituminous shale) to generate electricity. This is an important way of utilising the calorific value of stone coal (bituminous shale) efficiently. The power station at Yiwu, in Zheijiang, has been experimenting with burning nothing but stone coal to generate electricity. They use fluidised-bed boilers of 6 t/h and burn local 100-1500 k/cals/kg. stone coal (bituminous shale), generating a steady 750-850 kw.

And in Hunan, the Yiyang stone coal (bituminous shale) power station uses fluidised-bed boilers with an evaporative capacity of 35 t/h, burning stone coal (bituminous shale) of 950-1200 k/cals/kg. with an installed generating capacity of 6000 kw., at normal operation.

Once a power station changes over to stone coal (bituminous shale), because the ash content is high, heat output low, and the heat efficiency of fluidised-bed boilers is also relatively low, consumption of stone coal (bituminous shale) and the amount of ash are far greater than boilers which burn bituminous coal.

(d) Extraction of Vanadium Pentoxide (Va$_2$0$_5$). The vanadium found in association with stone coal (bituminous shale) is China's most abundant potential source of the mineral. One of the important uses of stone coal (bituminous shale) is for the extraction of vanadium pentoxide and other rare materials.

Vanadium pentoxide is used in the manufacture of vanadium-iron, and vanadium-aluminium alloys and other, non-ferrous alloys. It can also be used as a catalyst, in sulphate factories, synthetic ammonia plants and in other synthesising technology. In the past few years the development of techniques for extracting vanadium pentoxide from stone coal (bituminous shale) has been quite rapid in China. Research has been successfully carried out into open-hearth kilns, fluidised-bed boilers, sodium-roasting, calcium-roasting, direct immersion in acid and other techniques. With present day production techniques best results have been gained from self-burning sodium roasting and this is becoming quite widespread. The purity of the vanadium extracted can be as high as 98%, with an overall recovery rate of 45-50%. The Jiande stone coal (bituminous shale) vanadium plant in Zheijiang produces 100 tons of refined vanadium pentoxide each year, all reaching national standards.

Stone coal (bituminous shale) slag can also be used in fertilisers. In the slag are many kinds of nutrient necessary for agricultural crops - e.g. silicon, calcium, gallium, phosphorous, sulphur, magnesium, iron, molybdenum, boron, zinc, copper, manganese. After absorption by the crops they can promote the growth and development of the plants, and their resistance to disease and insect pests. Thus the use of stone coal (bituminous shale) slag to make fertiliser is one way of extending the use of fertiliser, of supporting collective agricultural production, of using stone coal (bituminous shale) resources rationally and of promoting their various uses.

(e) Gasification of Stone Coal (Bituminous Shale). The gasification of stone coal (bituminous shale) is an important way to make rational use of its thermal energy, increase its heat efficiency and prevent pollution. The principle by which stone coal and stone waste gas boilers make gas is basically the same as when coal gas is made in ordinary boilers. Gas is made from stone coal (bituminous shale) by making the fixed carbon in the coal react with the oxygen in the air, at high temperatures, thus producing carbon monoxide, carbon dioxide and nitrogen. At the same time it has a dry distillation effect and produces small amounts of combustible gases like methane, hydrogen and hydrogen sulphide. The nitrogen fertiliser factory in Changan county has carried out experiments to produce gas using stone coal (bituminous shale) from Duanjiagou, in Ziyang (Shaanxi) and produced synthesis gas as the raw materials for the synthesis of ammonia, and these experiments have been basically successful. The volume of gas produced was 300 (standard) cubic metres per hour, and the gases were as in Chart 3.

CHART 3

Results of Constituent Analysis of Stone Coal (Bituminous Shale) Gas

	Upper Line	Lower Line
CO_2	17.60	21.78
CO	23.90	18.84
O_2	1.00	0.4
H_2	45.10	5
N_2	10.70	3.39
CH_4	1.70	1.61

In Hunan the Changde power station and Oaijiagang commune use stone coal (bituminous shale) with a calorific value of 1800 k/cals/kg., and in simply constructed coal gas boilers, they produce gas of 700-900 k/cals/cu.m. But the making of gas from stone coal is still at an early experimental stage and there are several problems which need further experimental research - e.g. low gasification rate, desulphurisation, etc.

(f) Ore-dressing (By Washing). In recent years, there have also been advances in ore-dressing technology. In the experimental stone coal ore-dressing plant at Kaihua in Zheijiang, they are using such technological processes as crushing, screening, float-and-sink, plane tables and rotation troughs in carrying out industrial-type experiments which show that stone coal (bituminous shale) in powder or flake form has a certain selectability. Through dressing the ore, the waste-rock content of stone coal (bituminous shale) is reduced and there is abundant vanadium content. The fine dressed coal can be used to burn for cement, medium-grade is used in fluidised-bed boilers as fuel and lowest-grade coal is used in brick-making burning.

4. CONCLUSION

Many years of practice have shown that stone coal (bituminous shale) not only gives comparatively good results as a fuel but also in metallurgy, for construction materials, in the chemical industry and so on. The utilisation of stone coal (bituminous shale) needs little capital investment, and produces quick results. It has advantages for saving large amounts of energy; for preserving cultivated land, and reducing environmental pollution. It is also of benefit to the rural economy and promotes the development of sideline stock-raising in rural areas.

The uses and development of stone coal (bituminous shale) are part of the seventh Five-Year Plan. From now on we must take futher steps to formulate general and specific policies and concrete plans for developing and using our resources of stone coal (bituminous shale), and to reinforce work on prospecting for resources and extending the range of uses of stone coal.

Following new technological advances in research and manufacture of light-boned building material, purification of rare metals and other elements and the use of the coal slag left over from fluidised-bed boilers in the generation of electricity for the extraction of vanadium, the utilisation of stone coal (bituminous shale) will play an ever greater part in the achievement of China's four modernisations.

Reclamation, Treatment and Utilization of Coal Mining Wastes, edited by A.K.M. Rainbow
Elsevier Science Publishers B.V., Amsterdam, 1987 — Printed in The Netherlands

UTILIZATION OF MINING OPERATIONS AND COAL PREPARATION PROCESSES
WASTES IN THE USSR AND THE PRINCIPLES OF THEIR CLASSIFICATION

V.A. RUBAN[1], M.Ya. SHPIRT[1]

[1]Fossil Fuels Institute, Ministry for Coal Industry of the USSR

SUMMARY

The conditions diversity of fossil fuels deposits formation
in the Soviet Union calls forth the variety of physico-mechancal
and physico-chemical properties both for salable coals (shales)
and for wastes from coal mining and preparation processes. Unique
methods for sampling of overburden rocks, wastes of coal preparation processes and rejects of refuse dumps have been developed
and used in practice. The analysis of the results confirms that
the coal preparation wastes are characterized by the highest
stability of the properties.

The main quantity of coal preparation wastes is used to substitute the earth for construction of artificial ground structures and for levelling of the relief. Approximately 1 million
tonnes per year of coal preparation wastes are used as a fuel-
mineral component which is added (10-15%) to the blend for
bricks production. The first brick production shop from the
wastes of coal preparation plants - the only component of the
blend - have been commissioned.

Technical and economic calculations, carried out on the basis
of pilot and industrial tests results confirm the high economic
expediency of coal preparation wastes utilization for production
of different building materials, as well as a feedstock in ferrous and non-ferrous metallurgy, chemical industry and in agriculture.

The generalization of experimental data gave the opportunity
to introduce a classification for solid wastes from mining operations and coal (shale) processing as a feedstock for industrial utilization.

INTRODUCTION

Judging from prognosis evaluations of the energy balance, for
the industrial and developed countries of the world the nearest
future will be characterized by the increase of the solid fuel
consumption, that will lead to the increase of coal mining output and the volumes of coal preparation. A considerable increase
of solid mineral $(C_o^d \leq 4\%)$ and organic mineral $(C_o^d > 4\%)$
wastes (overburden and the waste of mines, the reject of gravity
methods of preparation and floatation tailings) will be the consequence of this tendency.

In the Soviet Union (according to the average data) per 1 T
of coal mined by opencast, underground or subjected to preparation 4 T, 0.2-0.3 T of stripping, mining rocks or 0.25 T of solid
preparation wastes respectively are formed (ref. 1,2).

PROPERTIES AND UTILIZATION OF COAL WASTES
Due to some peculiarities of underground mining the lithologic-
mineralogic and chemical composition of coal mining rocks vary
considerably for each separate mine. Therefore it is very difficult to sample these products.

The special methods for sampling of overburden rocks, wastes
of coal preparation processes and rejects of refuse dumps have
been developed in the USSR and used for evaluation of composition
and properties of coal wastes in the most coalfields.

As a rule, the so-called mineral wastes, which combine the
great quantity of stripping and a part of wastes from underground
mines, practically do not differ by their properties from sedimentary rocks of analogous content (clays, sands, limestones and
others). But in an overwhelming majority the overburdens and the
wastes of underground mines present by themselves the mixture of
different lithologic contents. The selective mining is acceptable
only in those cases when the separate components of the wastes
are considered to be of great value.

A considerable quantity of mineral wastes (overburden stripping and mining wastes, nearly 48-49% from an each year yield)
are used for filling in opencast and underground mines, for production of filling materials, for covering of roads, construction
of dams and quarry railways.

The conducted investigations give the opportunity to draw
a conclusion, that the matter content of coal wastes depends on
many factors (conditions of formation and exploitation of coal
seams, coal preparation technology and others). In coal-containing wastes it was diagnosticated more than 50 minerals but the
content of many of them does not exceed 1-2% (for example, gar-
nets' anatases and others). It is evident, that many physical and
mechanical and also physical and chemical properties of organic-
mineral wastes will be determined by the content of their main
mineral components. Taking into account these facts, it is reasonable to classify the coal wastes on sandy (summary sandstones
content $\geq$ 30-40%), carbonate (in the residue after calcination
$CaO + MgO$ > 15-20%) and clayish (summary content of clay minerals

$\geqslant$60-70%). As it will be shown below, such a grouping will give
the opportunity to choose the most rational ways of coal wastes
utilization. It is of interest to note, that the main mass of
coal preparation wastes belongs to the clayish group. In contrast
to common clays they are like stones, with high strength[*] and
with lower moisture content. The coal preparation wastes plasticity depends on both lithologic-mineralogic content and the size
content. So, the clayish preparation wastes which are considered
to be non-plastic, as a rule, at the size of particles $>$3 mm
become medium plastic after crushing to -1 or -0.5 mm of the
cementing matter of the mineral clay particles.

The analysis of the sampling results confirms, that the coal
preparation wastes are characterized by the highest stability of
the properties. However the chemical composition of coal preparation wastes is different in various coalfields and can be
changed on a rather large scale even within the limit of coalfield (Table 1). But for each separate concrete plant the variations of chemical content between samples of one day are comparatively not high. That's why the coal preparation wastes of one
plant can be considered to be the material of stable content.
As it is clear from the stated data, the coal preparation wastes
of various coalfields differ greatly by sulfur content. The maximum sulfur content characterizes the coal wastes of Kizelovsk,
Podmoskovny and Donetsk coalfields.

Coal-containing wastes of mining operations even for one opencast, as a rule, differ greatly by their organic matter contents.
Therefore, for their utilization it is necessary either to produce blending or separation into two fractions: of high ash content and of low ash content.

The coal preparation wastes of the majority of plants are
more stable by carbon content, that is secured by the technology
of the production. So, as a rule, daily reject samples of gravity
methods of preparation of one and the same plant differ by ash
content within the limits of $\pm$5% (abs).

---

[*]Mechanical strength depends not on mineralogic-lithologic content of clayish wastes, but on properties of the cementing
matter. The coal wastes of a number of enterprizes are characterized by lower strength during storage because of cracking.

TABLE 1

Coal preparation wastes composition for main coalfields
of the Soviet Union

Content %	Coalfields (method of preparation)							
	Donbass		Kuzbass		Karaganda		Pechora	Ekibas-tuz[**]
	grav.	float.	grav.	float.	grav.	float.	grav.	grav.
A^d	67–85	53–75	59–85	42–76	67–86	64–73	59–86	60–75
C_t^d	7–21	10–28	6–22	10–47	8–23	18–25	9–28	12–22
S_t^d	<1–4.5	<1–3.8	<1–1.4	<1–1.1	<1–2	<1–2.3	<1–3.3	0.1–0.3
SiO_2[*]	50–62	51–58	57–78	50–67	53–63	54–56	61–65	53–64
Al_2O_3[*]	17–31	18–32	14–26	14–31	23–35	23–28	20–24	27–39
Fe_2O_3[*]	3–16	4–12	2–10	2–9	4–7	6–13	6–8	0.2–5
CaO[*]	0.3–5	2–4	1–7	1–10	1–6	1–5	1–2	0.4–2
MgO[*]	0.8–2	1–2	0.3–3	0.7–3	0.3–1	1–2	2–4	0.4–1

[*] In ash

[**] Coal-containing wastes of mining (opencast) operation and
coal preparation pilot plant.

According to our opinion, taking into consideration the set
of technical and economic parameters (the fuel part content,
stability of the composition and others), the coal preparation
wastes and coaly wastes of mining operations are very promising
for top priority utilization. These products (coaly wastes) are
in essence a new type of complex material, which is called an
organic-mineral raw material or coaly rocks. As it will be
shown below, these coaly wastes can be used according to their
physical, mechanical and physico-chemical properties in the
following directions (ref. 1,2):

- for building materials production, for highway constructions and artificial ground structures;

- for sulfur compounds and organic-mineral fertilizers production;

- as a feedstock in ferrous and non-ferrous metallurgy for silicon-aluminium alloys, oxygen compounds of aluminium, silicon-carbide materials and other valuable products;

- as a fuel for energy production (with preliminary preparation or without any).

The fuel and mineral components value of coal preparation wastes and coaly rocks is more completely realized when using them as a feed in the processes connected with the thermal treatment in the oxidation or reduction mediums.

To evaluate the principal expediency of coaly rocks application as a feedstock for reduction thermal treatment a thermal dynamic analysis of the corresponding processes has been carried out. To solve this task a special programme of chemical thermodynamics has been developed. It gives the possibility to evaluate the quantitative composition (up to 200 components) of silica, alumina, iron, calcium, magnesium, sulfur compounds and other elements, contained in coaly rocks. Depending on carbon content, temperature of treating and gas medium, as the data demonstrated, alumina concentrates (up to 70 – 90% on oxide) silicon carbide materials, silicon, alumina and iron melts, nitride-containing materials and other valuable products can be obtained.

Thermodynamic investigations of the system Si – 0 – Al – Fe – Ca – Mg – S in the temperature range of 1000 - 3000°K allowed to theoretically determine the process parameters of obtaining alumina concentrate from kaolinite-containing rocks.

However in many respects the real indices of the process depend on kinetic parameters of the proceeding reactions and on the deviation from ideal system state. In connection with this some experimental investigations were carried out to ascertain physico-chemical peculiarities of the process of obtaining alumina concentrates, silicon carbide materials and "sialones". These investigations confirmed as a rule correctness of thermodynamic evaluations (ref. 2,3).

The large-scale effective preparation wastes utilization is the production of porous aggregates for light-weight concrete, ceramic wall materials and cement. All these materials are produced by burning, that makes possible to realize the advantages of coal preparing wastes as a new type of fuel-containing feedstock. Different methods of porous aggregates production from blends, containing 75-100% preparation wastes (agglomeration, rotary kilning or burning in fluidized or fountain bed kilns) were studied. The agglomeration method of porous aggregates production is considered to be the most applicable due to some peculiarities of coal wastes properties (carbon content, particle size, magnitude of swelling, temperature of swelling etc.) and technical-economic reasons.

The technology of coal preparation wastes agglomeration was successfully tested in experimental-industrial conditions. Projection documentation concerning the erection of the corresponding industrial plants with different yearly output $(0.1-1.2 mil.m^3)$ of agglomerate) was developed.

As it was already mentioned above, in the Soviet Union coal preparation wastes are used on a large scale nearly one million tonnes yearly as a fuel-mineral additives (12-15%) to the blend in clay brick production. It allows to improve the quality of the salable bricks and to decrease total fuel consumption 25-30%.

In future it is planned to increase considerably the usage of coal wastes as a fuel-mineral additives.

The technologies of brick and drain tubes production from blends, consisting of 70-100% coal preparation wastes were also developed.

Depending on coal wastes properties (organic matter content, plasticity etc.) it is recommended to use either plastic moulding or semi-dry pressing. It should be noted that coal preparation wastes is a feedstock with practically low sensitivity to drying. It makes possible to reduce the time of drying and to improve the quality of the salable articles. Judging from the results of

experimental-industrial testing technological fuel consumption
is reduced by 70-75%, high strength ($\geqslant$150-250 kg/cm^2) frost resistance of the ready-made articles are secured.

The first brick production shop from coal preparation wastes
as the only component of the blend have been commissioned in Kuz-
netsk coalfield. Bricks are being produced by semidry pressing
and have good quality, frost resistance and strength.

Technical and economic calculations carried out on the basis
of pilot or industrial results confirm the high economic expediency of coal preparation wastes and coaly rocks utilization as
a feedstock in ferrous and non-ferrous metallurgy, chemical industry and in agriculture.

PRINCIPLES OF CLASSIFICATION

Summarizing all the data on chemical, lithological composition
and physical-chemical coal wastes properties, as well as the results of their application for producing various products makes
it possible to suggest the principles of their classification as
a feedstock for usage. The ten assumed (ref. 2,4) characteristic
parameters are summarized in Table 2 (source of production (A),
lithologo-mineralogic characteristic (B), organic carbon content
(C), degree of organic matter coalification (D)* and others).
Every characteristic parameter is subdivided into groups and (if
necessary) subgroups, marked by Arabic figures and Latin letters,
respectively. As an alternative indexing we suggest also the index marking system according to which each characteristic mark
is shown by an Arabic figure and its position in numerical index
corresponds to the position in numerical index of Table 2.

Zero indicates absence of data or their insignificance for
discussed trend of coal wastes utilization. For example, coal
wastes after preparation of bituminous coal +13 mm, coalinite,
C_o^d = 7%; S_t^d = 2.5%, $\Sigma CaO + MgO$ (in ash) = 2%; with unknown iron,
aluminium contents and plasticity can be indicated as
A2L3aB1aC2D2S3M1 or 3313203010.

The value of classification indices are determined after the
laboratory study of the considered coal waste of the certain enterprise. The most promising trend of certain types of coal

---

*It is measured either by carbon (organic) content on combustible
matter (C^c) or by vitrinite reflectance (R^o):
$$C^c = \frac{C_o^d}{100-M} \cdot 100,$$ where M,% - content of mineral matter.

wastes usage is selected depending on classification indices
guiding by special tables compiled according to the technical-
economic indices of coal wastes treatment.

TABLE 2

Index marking system of coal wastes classification
groups (ref. 2,4)

Characte-ristic parameter mark	Characteristic parameter, group, subgroup	Position in nume-rical index	Number of groups (subgroups) in numeri-cal index
1	2	3	4
A	Coal wastes source		
	1. Coal mining wastes		
	a. stripping rocks	1	1
	b. mining rocks	1	2
	2. Coal preparation wastes	1	3
	3. Waste heap rock	1	4
	4. Thermal treatment wastes		
	a. combustion ash-slag wastes	1	5
	b. gasification ash-slag wastes	1	6
	c. hydrogenation wastes (slurry)	1	7
L	Initial characteristic		
	1. Burnt rock	2	1
	2. Unburnt rock	2	2
	3. Coal preparation wastes		
	a. coarse rock (+13 or +25 mm)	2	3
	b. middle sized rock (0.5(1) - - 13(25) mm)	2	4
	c. fine rock or floatation wastes (-1(0.5) mm)	2	5
	4. Ash-slag wastes		
	a. fine fly ash (>1000 cm^2/g)	2	6
	b. coarse fly ash ($\leqslant 100$ cm^2/g)	2	7
	c. solid slag	2	8
	d. liquid slag	2	9
B	Lithologo-mineralogic characteristic		
	1. Clayish:		
	a. kaolinite	3	1
	b. hydro-micaceous	3	2
	c. montmorillonite	3	3
	2. Sandy	3	4
	3. Carbonate		
	a. calcite	3	5
	b. syderite	3	6

(To be continued)

1	2	3	4
C	Carbon (organic) content		
	1. Low carbonic (0-4% C_o^d)		
	a. below 2% C_o^d	4	1
	b. 2-4% C_o^d	4	2
	2. Little carbonic (4-8 % C_o^d)	4	3
	3. Middle carbonic (8-12% C_o^d)	4	4
	4. Carbonic (12-20% C_o^d)	4	5
	5. High carbonic ($>$20% C_o^d)	4	6
D	Organic matter coalification degree		
	1. Low rank ($C^c \leqslant 75\%$) $R° \leqslant 0.49\%$	5	1
	2. Middle rank (C^c-75-90%) $R° = 0.50\text{-}2.49\%$	5	2
	3. High rank ($C^c > 90\%$) $R° \geqslant 2.5\%$	5	3
Fe	Iron compounds content (in ash)		
	1. Low ferrous ($<$1.5% Fe_2O_3)	6	1
	2. Little ferrous (1.5-5% Fe_2O_3)	6	2
	3. Middle ferrous (5-12% Fe_2O_3)	6	3
	4. Ferrous (12-18% Fe_2O_3)	6	4
	5. High ferrous ($>$18% Fe_2O_3)	6	5
S	Sulfur content		
	1. Low sulfurous ($<$0.5% S_t^d)	7	1
	2. Little sulfurous (0.5-2.0% S_t^d)	7	2
	3. Sulfurous (2.0-3.0% S_t^d)	7	3
	4. Sulfurous (3-4% S_t^d)	7	4
	5. High sulfurous ($>$4% S_t^d)	7	5
Al	Aluminium compounds content (in ash)		
	1. Little aluminous ($<$15% Al_2O_3)	8	1
	2. Middle aluminous (15-28% Al_2O_3)	8	2
	3. High aluminous ($>$28% Al_2O_3)	8	3
M	Sum of calcium and magnesium compounds content (in ash)		
	1. Low calcium ($<$3% CaO+MgO)	9	1
	2. Middle calcium (3-6% CaO+MgO)	9	2
	3. Calcium (6-18% CaO+MgO)	9	3
	4. High calcium ($>$12% CaO+MgO)	9	4

(To be continued)

1	2	3	4
P	Plasticity		
	1. Non-plastic P=0	10	1
	2. Little plastic P$<$7	10	2
	3. Middle plastic P=7-14	10	3

REFERENCES

1. V.V. Lebedev, V.A.Ruban, M.Ya. Shpirt, Kompleksnoe ispolzova-
   nie uglei, Nedra, Moskva, 1980.
2. M.Ya. Shpirt, Bezotkhodnaya tekhnologiya. Utilizatsiya otkhodov dobychi i pererabotki tvyordogo topliva, Nedra, Moskva,
   1986.
3. L.A. Kost, N.N. Novikova, L.A. Sinkova, O poluchenii kontsen-
   tratov okisi alyuminiya, Khimiya tvyordogo topliva (Chemistry
   of Solid Fuel), 3 (1979) 182-189.
4. M.Ya. Shpirt, Yu.V. Itkin, Osnovnye printsipy klassifikatsii
   otkhodov dobychi i pererabotki uglei, Khimiya tvyordogo topliva (Chemistry of Solid Fuel), 2 (1980) 78-83.

Reclamation, Treatment and Utilization of Coal Mining Wastes, edited by A.K.M. Rainbow 55
Elsevier Science Publishers B.V., Amsterdam, 1987 — Printed in The Netherlands

RESEARCH ON SUITABILITY OF COAL PREPARATION REFUSE IN CIVIL
ENGINEERING IN THE FEDERAL REPUBLIC OF GERMANY*)

Dr.-Ing. Dieter Leininger[1], Dr.-Ing. Joachim Leonhard[2],
Dr.-Ing. Wilfried Erdmann[2], Dr.-Ing. Theo Schieder[2]

[1] Head of the Coal Preparation Service, Bergbau-Forschung GmbH,
Essen, Federal Republic of Germany
[2] Scientists with Bergbau-Forschung GmbH, Essen,
Federal Republic of Germany

SUMMARY

Notwithstanding the massive efforts of the West-German hardcoal
industry just about 15 M t/y refuse went during the few recent
years into road construction - predominantly as bulk material -,
into other soil amelioration projects as well as into the
construction of dams and embankments. This is roughly 20 % of
the total refuse yield. For this reason extensive research was
done in view of further possible applications of colliery shales.
Due to the more stringent regulations on thermal insulation
released in the Federal Republic of Germany, energy-saving
constructional designs were introduced in building construction
during the past few years. At this juncture there is a scarcity
of light-weight, small-size products for roughcast-and brickwork
binders so that the building industry has to recur to quartz
sand. It was tried therefore to produce light-weight sands from
crushed small sized washery refuse and pelletized flotation
tailings by means of thermal treatment in a fluidized bed furnace.
The products obtained proved their usefulness for thermal
insulation in different applications of the construction industry.
Further to this, injection of synthetic foam into the chambers
of light-weight concrete blocks substantially increased their
resistance to thermal transfer. In case of hollow blocks made
from lime-bound green wastes certain difficulties arose so that
further tests are necessary.
For checking the suitability of green refuse as a concrete
aggregate a wall was erected in which were used different types
of cement and reinforcing steel bars. The objective of the present
investigations was firstly to find out the compressive strength
of refuse-based concrete and, secondly, the carbonation and
corrosive behaviour of such wastes in presence for steel reinforcing bars. Said wall which was constructed two and a half years
ago has not shown so far any spallings nor other disintegrations.

---

*) This paper contains the results of R+D projects which were
Subsidized by the Commission of the European Communities and
the Ministry of Economy, Trade and Technology of North Rhine
Westfalia.

Afther the colouration of lime-bonded blocks from refuse by
means of ferrous oxide pigments - for facade decoration - had turned up to be quite expensive, tests on colouration were run with
red slurry form the aluminium industry. The results were satisfying
and production cost were considerably lower.

The studies on the suitability of washery refuse in underground
construction were extended to include also road construction.

In connection with investigations into the production of underground construction materials from cement, fluidized bed ash from
flotation tailings and swept material from green refuse shot-
creting test were run on a technical scale in a quarry. The outcome
was, however, only a mortar of late bearing. Other tests on blends
from cement, fluidized bed ash and limestone with addition of a
accelerator resulted in the make-up of an early-bearing construction material for underground use. It will be tried now to
increase the proportion of fluidized bed ash and replace part of
the limestone by green refuse.

1. UTILIZATION OF WASHERY REFUSE IN BUILDING CONSTRUCTION

1.1 Production of light-weight sand from washery refuse

The specifications for thermal insulation in building
construction became sensibly more stringent. The ordinary
rough-cast and brick-laying mortar ist of a substantially higher
thermal conductivity than heat-insulating wall bricks. Despite the
low proportion of joints of just 5 to 10 % the impact of mortar
on thermal insulation of brickwork walls appears to be high.

It has been suggested therefore to improve thermal insulation
also in the joint area by utilizing some mortar of low apparent
density. It has been for many years therefore that to avoid
so-called "cold bridges" in brickwork walls the appertaining
mortars were manufactured from light materials als expanding clay,
expanding sand and pumice instead of fine gravel and natural sand.
As expanding clay and sand as well as pumice are scarce materials
one has been using materials crushed below 6 mm, brick chippings
or calcareous sand although for several reasons (uncontollable
water absorption, somethimes high apparent density along with
reduced thermal insulation capacity) they do not work out too
well as light-weight mortar additives.

Another possibility of manufacturing fine light-weight construction materials is the ceramization of crushed small-size
refuse or granulated flotation tailings (of sand grain size)
in the fluidized bed. In this case, and unlike with clay, the
residual heat contained in the material is utilized so that

normally any external energy supply is not necessary. Appertaining
work therefore included processing and fractionation of flotation
tailings and small washery refuse to form a granulated material
which is amenable to be transformed into a burnt product for the
construction industry.

During the combustion tests in fluidized bed reactors of 300 mm
viz. 600 mm diameter there were used firstly crushed small
washery dirt and pelletized flotation tailings of 4 - 1 mm grain
size in a moistened condition. It turned up that due to the
"thermal shock" the feed underwent on entering the fluidized bed
there was a high amount of abrasion. The bed ash/fly ash ratio
thus was as high as 1 : 1. For this reason the experiments were
continued exclusively on air-dried small refuse or pelletized
tailings containing < 1 % moisture. Thanks to this and other
measures the bed ash to fly ash ratio was increased to 8 - 9 : 1.

Depending on grain size the apparent density of the feed was
1.40 - 1.55 kg/dm³ for crushed small refuse and 0.96 - 1.10 kg/dm³
for pelletized flotation tailings. During the tests the carbon
contents oscillated between 6.6 (small washery refuse) and 28.0 %
(pelletized flotation tailings). Combustion became self-maintaining
at a carbon content in the feed of roughly > 10 %.

In the course of combustion the apparent density of the
crushed refuse was reduced to about 0.96 - 1.08 kg/dm³ as a
function of fluidized bed temperature (850 - 950 °C on average)
and of a residence between 60 and 90 min; after combustion the
pelletized feed exhibited raw densities between 0,58 and
0.94 kg/dm³ with maximum concentration between 0.80 and
0.85 kg/dm³. The pellets were poorest in strength, exhibiting
apparent densities between 0.58 - 0.69 kg/dm³. The carbon content
of the final product was sensibly reduced during all of the combustion tests and oscillated between 0.1 and 0.3 % so that
combustion efficiency amounted to 98 - 99 %. The light-weight sands
were in any case superior to natural sand as far as their thermal
insulation at an apparent density between 1.60 - 1.80 kg/dm³ is
concerned. No expanding agents as red slurry,sulfide waste lye,
ferrous compounds etc. were used for the tests since with the
temperatures applied no expanding takes place anyway.

During the studies on practical application in building construction it was started by using the light-weight sands for preparing rough-cast- and brickwork binders of varying constitutions. The specified compressive strengths were attained and sometimes by far exceeded. Expanding and shrinking behaviour as well as the dry apparent density remained within the usual tolerances. Furthermore the thermal transfer index as well as the resistance to thermal transfer for different walls out of light concrete blocks joined by means of natural and light-weight mortar as well as rough-cast using mortar from natural sand or light sand were determined. With the other initial conditions maintained it turned out e.g. for one brickwork wall joined by natural sand mortar and coated on both sides by a rough-cast of the same material that it exhibited a resistance to thermal transfer of 0.87 $m^2 \cdot K/W$; whereas another wall joined by light-weight mortar and coated on both sides by a light sand rough-cast the resitance to thermal transfer was 1.12 $m^2 \cdot K/W$. The latter thus showed a considerably lower thermal conductivity and, as such, better thermal insulation properties.

A rather similar dependency of thermal conductivity was represented by composition floorings, blocks from limebound refuse as well as light concrete blocks and light-weight concrete whenever during the manufacturing of these materials natural sand was completely or in part replaced by light sand.

1.2 Chamber insulation of hollow blocks by injecting synthetic foam

Thermal insulation of external brickwork walls requires, in general, a considerable labour expenditure, thus implying high extra costs.

More recently it was possible to develop a method of injecting synthetic foam into the hollow wall bricks after completion of the main construction work. A critical point with this method are the interconnections between adjacent block faces in combination with holes at the upper faces which in many hollow building blocks are provided in the form of so-called thumb holes. Once the bricks are laid the foam, being injected through one hole, spreads into the neighbouring bricks by penetrating through the lateral channel openings and through the holes in the upper faces of underlying bricks.

Practical experiments with the method were run using light-weight concrete hollow blocks manufactured under addition of 60 % ceramized coal mine wastes.

The blocks were laid on a foundation plate to form a brick wall consisting of 6 layers of a horizontal step formation on both sides. The faces of the steps were coated both horizontally and vertically by glass slabs serving as inspection windows. Starting from 2 holes all of the brick work wall was then injected with insulation foam used for machinery. As was observed already during injection the foam filled the block cavities uniformly.

The upper 3 layers of the brickwork there then removed and the blocks thus obtained cut apart in parallel and normal to their longitudinal axis at 10 cm spacings. So all the block cavities were open to visual inspection of the foam. As the block material was of the required porosity there was no problem with air displacement from the hollow spaces. Any and all cavities in the blocks and at their joints were filled with insulation material so that the foaming was complete (fig. 1). The calculated thermal transfer coefficient k of the insulated wall, for a wall structure of 24 cm block thickness, internal mortar of 1.5 cm and external coating of ordinary mortar of 2.0 cm thickness, amounted to 0.3 W/m$^2 \cdot$K. Slabs of an appropriate profile are by now available for manufacturing this type of hollow blocks so that the necessary interconnections of cavities are brought about during compaction and shaping of the bricks.

Development in the field of cement-bonded light-weight hollow blocks from burnt refuse has thus reached operational maturity.

By contrast the development of lime-bonded blocks from crushed refuse sand for foam injection was not without problems.

First it was tried to prepare appropriate mixtures from crushed refuse sand and other substances which, besides the specified minimum compressive strength of hollow blocks from lime-bonded refuse of $\geq$ 12 N/mm^2, were at the same time of sufficient porosity to comply with foam injection requirements. To this end crushed refuse sand, natural sand, boiler ash, and ceramized sand were selected.

Fig. 1. Removal of the upper rows for visual inspection
         of cavity filling

During the experiments it turned out that as far as
compressive strength of the blocks is concerned an aggregate mix
of crushed refuse sand and boiler ash in a 55 : 45 % ratio was
optimal. Lime addition amounted to 12 %. The shaping moisture
was adjusted to 3 %. In these conditions 13.1 N/mm² of average
strength of the hollow blocks was attained. After this the blocks
were laid to form a wall so as to ensure interconnection of the
block cavities when joining them with a lime mortar. The wall was
then foamed as described above. When removing the upper rows it
became visible that in some areas individual cavities were filled
up in part only and in such a way that bigger hollow spaces were
left over in parallel to the longitudinal cavity axis. In this

specific case foam had just formed a coating on the cavity walls.
When examining these poorly filled blocks it turned out that their
walls and bridge sections were highly porous.

Examination results are indicative of the necessity of having
to modify the bridge sections of the moulds in order to obtain —
at adequate strength of the freshly shaped blocks - a more
consistent material distribution in the mould.

1.3 Erection of a steel concrete wall using "green" wastes as aggregate

To widen the application range of wastes in hydraulically
setting materials a silencing wall out of reinforced concrete
from refuse was developed. The silencing wall panels from
prefabricated concrete elements were manufactured using different
types of cement and reinforcing steel bars.

The investigations were aimed at testing concrete out of
"green" washery refuse for compressive strength, apparent
density, carbonate formation, water absorption, corrosion
behaviour and, thus, the influence of sulphate and chloride
contents of the refuse on both the concrete and its reinforcement.

When manufacturing the sample plates it was used coarse
refuse of > 60 mm grain size which was reduced in an impact
crusher and the fractionized into 16 - 8, 8 - 3.15 and
3.15 - 0 mm. The 16 - 3.15 mm fractions were used as aggregate.
The 3.15 - 0 mm fraction was substituted for by natural sand of
4 - 0 mm grain size. The proportion of cement amounted to
340 kg/m^3. This water/cement ratio matched the K 3 consistency
range, i.e. the moisture content of the compacted concrete was
adjusted to 190 dm^3/m^3 which corresponds to a water/cement
factor of 0.51.

As any experience on the behaviour of reinforcing steel in
concrete from refuse was lacking we inserted a test program for
determining the influence of possible chemical reactions of said
concrete on the steel reinforcement.

The following types of reinforcing steel and cement were
used:

> PZ 35 Portland cement and HOZ35 blast-furnace cement
> as well as constructional steel ST.U.III (galvanized)
> and ST.IIIK (galvanized).

One column, 4 plane plates (12 cm thick) and a tub for receiving plants whose exterior wall was made of concrete with aggregate (from burnt waste) exposed by washing, were erected. Construction of the elements from refuse concrete containing the aggregates described went smoothly. The elements were kept on an open-air storage yard for several months and then installed early in 1984 on the premises of Bergbau-Forschung GmbH in Essen (fig. 2).

Fig. 2. View on the wall

The untreated concrete surfaces do not show any corrosions or other alterations so far.

The concrete elements complied with the following character-istics (table 1):

PZ35 Portland cement	HOZ35 blast-furnace cement
Apparent density (kg/dm³) 2.15 - 2.30	Apparent density (kg/dm³) 2.14 - 2.29
Compressive strength (N/mm²) 53.9 - 48.0	Compressive strength (N/mm²) 34.5 - 30.0
Water absorption (M%) 7.4 - 7.6	Water absorption (M%) 8.4 - 9.0
Extent of carbonate formation (mm) 1 - 2	Extent of carbonate formation (mm) 3

TABLE 1   Characteristics of the concrete slabs from refuse, including different types of reinforcement

When reexposed after a certain while none of the steel types used for reinforcement showed the slightest signs of corrosion. Long-term testing is scheduled to go on for another couple fo years.

1.4 Colouration of lime-bonded finishing bricks from refuse using red slurry

In connection with previous building structures from refuse-based bricks it had already been demonstrated that thin plates out of lime-bonded refuse bricks were useful as decorative material on facades. The contribution to the "1st International Symposium on the Reclamation, Treatment and Utilization of Coal Mining Wastes", Durham, from September 10 - 14, 1984, entitled "Recent Developments in the Federal Republic of Germany as to Utilization of Washery Tailings" - deals with this subject. Whilst these bricks called "rock face" have a rough surface, their colour does, however, not differ too much from current sandlime bricks. It was tried therefore to colour them to make the exterior of houses look more decorative.

When using natural sand the pigments are added, in general,
only colours the binder of the material. Crushed sand from refuse
is an exeption from this in so far as the individual grains are
susceptible of being impregnated by the pigment which consists of
different shades of ferrous oxide. The pigmentation tests on lime-
bonded bricks from refuse using Bayer pigments have been completed
successfully. We were successful in manufacturing on a commercial
shale thoroughly coloured bricks with about 1,0 % proportion of
red pigment. It has to be admitted though that even minor
variations in operational conditions would affect the uniformity
of pigmentation. To arrive at a uniform brick colouration one has
to use therefore a pigment quantitiy of > 1 % referred to the dry
mixture. Inspite of the modest proportion of pigments and technically impeccable manufacturing results the cost of such bricks
was boosted considerably by the addition of ferrous oxide
pigments so that we were prompted to look for complete or at
least partial replacement of this natural pigment by some cheaper
one. Red slurry yielded as a waste product of aluminium production
may be considered a cost-effective substitute.

The objective of the above tests was to find out about the
required red slurry volume to be added to arrive both at a
visually satisfying colouration and resistance to weathering
(namely by live steam hardening) as well as at some assessment
of the cutting faces and the incidence of pigment admixture on
the compressive strength and density of the brick material. We
started by testing the colouration intensity and stability at
temperature levels up to 200 °C. Starting with 3 % red slurry
addition this proportion was increased by 1 % increments up to
7 %. Each of this series was tested by exposing it to temperatures
between 160 and 220 °C (at 10 °C increments).

Pigment saturation was arrived at between 4 - 5 % of red slurry
content. The following positive aspects speak in favour of the
use of dried red slurry:
- The pigment is stable and supports hardening in the autoclave.
- As the pigment is very cheap nothing speaks against its
  addition up to the saturation limit.
- Dry admixture of the red slurry does not cause any inconvenience
  and therefore does not require any expensive dose-feeders as in
  the case of liquid substances.

Some disadvantages were observed as well:
- With the given size of the red slurry particles only external parts of the minestone sand gets coloured.
- By the sole use of red slurry one cannot produce any variations in shades. Intensive colour impregnation and effect was attained by using a combination of 4 or 5 % red slurry with addition of 0.4 % genuine pigment which made a wide variation of shades possible.

By now some sections of the exterior wall of a test building on the Bergbau-Forschung GmbH premises were lined with split lime-bonded bricks from refuse coloured by means of red slurry and genuine pigment, for testing their weathering and colour resistance (fig. 3).

Fig. 3.   Facade of lime-bonded bricks from refuse coloured by red slurry

2. USE OF GREEN MINE WASTE IN UNDERGROUND CONSTRUCTION

At the 1st International Symposium on the Reclamation, Treatment and Utilization of Coal Mining Wastes held in Durham from December 10 - 14, 1984, we had already an opportunity of presenting you the paper entitled "Recent developments in the Federal Republic of Germany as to Utilization of washery tailings" which dealt with the utilization of "green" waste of varying bonding types as road bases. The test roads which initially had been arranged only on the premises of mining complexes were so successful that now the first sections of public roads are constructed using washery refuse of the bituminous or hydraulic binder type for the upper road layers.

3. PRODUCTION OF CONSTRUCTION MATERIAL FOR UNDERGROUND MINES, USING THERMALLY TREATED MINE WASTE

Further studies dealt with the production of construction material for underground mines were calcinated waste was used. First trials were run to test mixtures from cement and ceramized waste as well as light-weight sand. The Department of Mine Support And Rock Mechanics with Bergbau-Forschung GmbH found out, however, that the material caused relatively high wear in the pipelines due to impact and sliding effects. Subsequent trials on mixtures from 46 % PZ45 Portland cement, 46 % fluidized bed burn-up from flotation tailings and 8 % swept material from "green" waste, including or without accelerator, provided a construction material of either delayed or early bearing strength and of high resistance to impact and slide effects so that a guniting test on an operational scale appeared to be justified.

During said test which was run in a quarry and where at the beginning there was some dust development the following was observed upon using a pre-wetting apparatus:
- It was possible to pneumatically convey and inject, without dust molestation, the material wetted to a moisture content of about 14 %;
- It was possible to apply to a vertical wall a 55 cm thick and unsupported mass of the material in one guniting operation
- No segregation nor dry nests were observed in the to be gunited material.

Pursuant to the compressive strength measurements carried out after varying setting times the material could not be classified but as being of delayed bearing strength. Maybe this is due to the cold weather prevailing during the guniting test since cement hardens later in these conditions. A different mixture was used for another test at the underground of a colliery in the eastern Ruhr coalfield. It consisted of 30 % cement, 30 % fluidized bed ash, and 40 % limestone under addition of an accelerator. This guniting mortar met any and all requirements to be put on an early-bearing construction material.

Studies in view of producing underground construction materials are ongoing. Their main objective is to increase the proportion of fluidized bed ash and replace the limestone at least partly by waste.

4. CONCLUSIONS

In order to reduce waste disposal reqirements the German hard coal industry makes strong efforts to find alternative applications of utilization of colliery wastes. As part of this study, investigations have been carried out on

- thermal insulation of external walls by use of light-weight sands from burned small washery refuse and pelletised flotation tailings for preparing rough-cast-and brickwork binders and injection of synthetic foam into the cavities of light-weight lime-bonded and concrete blocks

- checking the suitability of green wastes as a concrete aggregate

- colouration of lime-bonded blocks from refuse by means of ferrous oxide pigments and red slurry from the aluminium industry

- suitability of washery refuse in road construction and

- production of mortars for underground use which consist of cement, fluidized-bed ash as well as fines from green waste.

The results obtained were partly satisfied or further tests are necessary.

Reclamation, Treatment and Utilization of Coal Mining Wastes, edited by A.K.M. Rainbow   69
Elsevier Science Publishers B.V., Amsterdam, 1987 — Printed in The Netherlands

A FORECAST OF COMPOSITION OF COAL WASTES AND OF THEIR DIRECTIONS
OF UTILIZATION DURING EXPLORATIONS IN COAL DEPOSITS IN THE USSR

V.R. Kler[1] and M.Ya. Shpirt[2]

[1]Institute of Litosphere, Academy of Sciences of the USSR

[2]Fossil Fuels Institute, Ministry for Coal Industry of the USSR

SUMMARY
   The production methods used in the USSR as well as the investigations of the composition, technological properties and wastes reserves from mining operations, coal preparation and coal
utilization at the process of deposits surveying have been considered in the report. The principles and specifications, used
for the evaluation of associated with coal materials, mineral
components and rare elements, contained in solid fuels including
the methods for investigation of the elements, hazardous for the
environment, have been presented. The directions and specifications (norms) for the utilization of overburden sands and gravel,
clays, carbonates and other rocks, coaly rocks of internal ripping material, coal preparation wastes, ashes and slags have
been considered.

INTRODUCTION
   For solving the problem of the complete utilization of the
coal deposits the forecast of compositions and yields of coal
mining preparation and burning wastes are carried out in USSR.
These data are nessasary for designing the processes with low
yields of wastes. The investigation and evaluation of the coal
associated minerals and valuable components, the coal mining,
preparation and processing wastes, the simultaneously produced
mine methan and water are obliged and envisaged by the legislative acts, instructions of state Comissions on Reserves of va-
luablefossils of the Council of Ministers of the USSR (GCR USSR)
[I] , braneh instructions confirmed by the Ministry of Geology
and Ministry of Coal industry of the USSR [2-5, IO, II]. According to these acts the surveying and prospecting are carried out
wifh evaluating of all valuable minerals and wastes, which can
be utilized profitably and the forecast of their yield, reserves
in the earth has to be done. The methods of many-sided investigations of the associated feed stock and wastes are summarised
in some publications [7-9] .

<u>Forecast of directions of coal wastes utilization.</u>

The works are carried out in the six following directions:

I. The investigations and evaluation of coal mining and preparation wastes mainly as a feed stock for the production of building materials (fig.I, 2). The solid coal mining wastes are evaluated as a material for reclamation, building of roads, dams and ofher artificial ground structures; the burnt rocks in the heap for road building. The coal preparation wastes are investigated and evaluated mainly as fuel-mineral additives to clay or the main component of blend for the production of brics,drain tubes [6, I2] and aggloporite (light porous aggregates). The laboratory and pilot tests are carried out for the wastes of the exploited coal deposits. The modele wastes are tested for prospected coal mines and designing preparation plants.

2. It is known many possibilities of slag and fly ash utilization (fig.3). The special sampling works are not carried out. The yield, composition tehnological properties of slag and fly ash, their applications are forecasted by the data of technological investigations of power-generating coals according to ash content, its composition and thermal properties.

3. The investigation of content of trace (minor) elements and their evaluation as valiable or dangerous for environment components of solid fuels are carried out always during the prospecting of coal deposits. The main method semiquantitive spectral analyse of the duplicates of the samples. The special methods are applied for the elements (Hg, F, Sb, Re, Au and ofher) which cannot be detected with high sensitivity by spectral methods. The quantitative determinations are carried out on the sampls from the zones of coal bearing strata where the high concentrations of elements are discovered. Then the average contents of the elements are evaluated and the possibility of their recovery or polution environmental action are evaluated. The high local concentrations in coals (IO-IOO times exceeding the average contents) are found for Ge, Be, W, U, Se, Mo, Re, Zn, Pb, V, Hg, As, Bi, Au, Ag. These elements are accumulating mainly on sorption, reduction, hydrosulfide barriers. Ge, W, V, U, Se, Mo, Ag are considered now as valuable or potentially valuable components. Be, F, As, Hg, Se, U — as a potentially polutants of the environment.

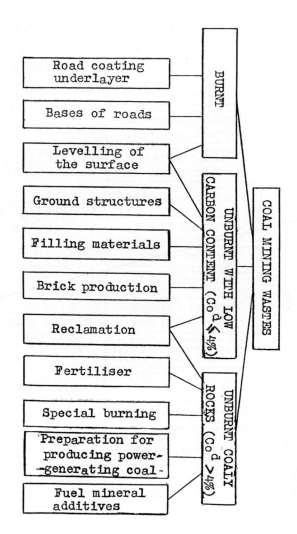

Fig. 1. The ways of utilization of coal mining wastes.

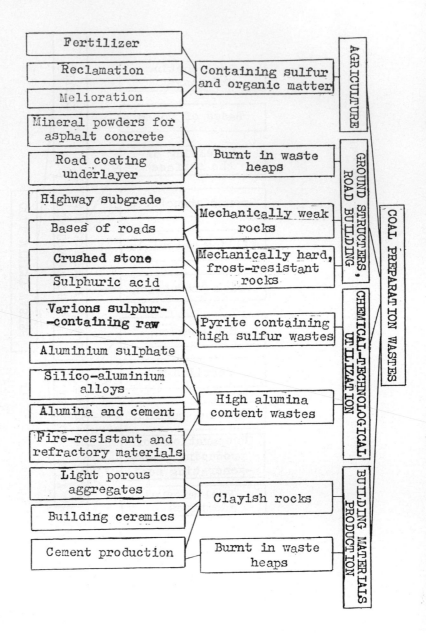

Fig. 2. The ways of utilization of coal preparation wastes.

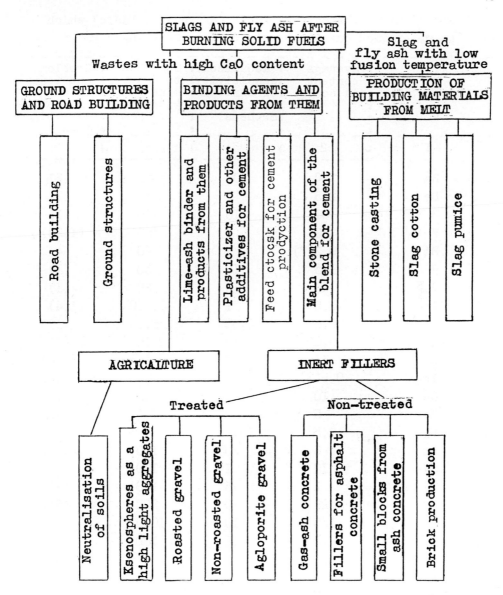

Fig. 3. The main ways of power plants wastes utilization.

Contents of minor elements of fuels (g/t on dry basic) which have to determined quantitatively during surveying

Element	Average content in coals	Required sencitivity of surveying ahalyticol methods	Contents to determine quantitatively	Problems and feature of quantitative investigation
I	2	3	4	5
Ba	I50	I00	I000	(3)
Be	2,5	I	50	(2)
B	80	I0	200	(I)
V	30	5	I00	(I,28)
Bi	0,2(?)	I	20	-
W	I,5	I0	50	(I)
Ga	I0	I0	20	(I)
Ge	I,5	I	3, I0	(I,4)
Au	0,0	0,05*	0,I	(I)
In	0,02	I0*	I0	(7)
Y	I0	I0	по ΣTR	(I,3)
Yb	0,9(?)	3	-"-	(I,3)
Cd	I0	I0*	I0	(I,3)
Co	5	I0	I00	(7)
La	I,5	I0(7)	по ΣTR	(I,3)
Li	6	30*	I00	(3, I0)
Cu	I0	I	I00	(7)
Mo	2	5	I00	(I)
As	25	I0	300	(2)
Mn	I50	I0	I000	(2)
Ni	I0	I0	I00	(2)
Nb	I,2	I	(7)	-
Sn	I	7	50	(I)
Rb	I7	I0	I00	(3)
Re	0,06	0,I*	I	(I)
Hg	0,05	0,05*	I	(2)
Pb	I5	I0	50	(I,2)
Ag	0,I	0,I	2	(I)
Sc	I,8	30	по ΣTR	(I,3)

I	2	3	4	5
Sr	80	IO	IOOO	(3)
Sb	2(?)	IOO	300	(2)
Se	0,5(?)	5*	50	(I,2)
Tl	0,5		IO?	-
Ta	0,2	30	(7)	-
Ti	600	IO(7)	(5)	(5)
F	IOO	IOO*	500	(2)
Cr	I8	30	IOO	(2)
Ce	-	IOO	по ΣTR	(I,3)
Zn	35	IOO	IOO	(I)
Zr	50	IO(7)	500	(3)
Cs	I,5		IOO	(3)

- no data; ? - data have to be precised; * - content cannot be determined with semi-quantitative spectral methods;

Problems and peculiarity of quantitative investigation:

I - valued or potentially valued components;

2 - harmful or potentially harmful components;

3 - fir problemy of geology or geochemistry;

4 - Ge ⩾ 3 g/t for coking coals; Ge ⩾ IO g/t for pover-
-generating coals;

5 - contents determined quantitatively during technological investigation of ash composition;

6 - quantitative investigation is carried out for the sim-
mary of La, Y, Yb, Ce, Sc it ΣTR+Sc ⩾ 500 g/t;

7 - high concentrations of elements are not detected;

8 - the vanadium content is ⩾ 500 g/t simultaeons quantitative determination of Ni and Cr has to be done;

9 - simultaneous quantitative determination of Li, Rb, Cs
are carried out.

76

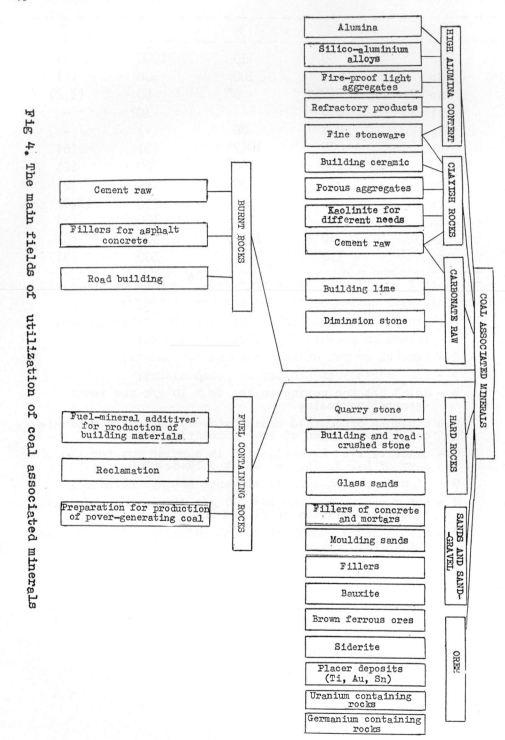

Fig 4. The main fields of utilization of coal associated minerals

4. The investigation and evaluation of associated minerals in
enclosing coal rocks (fig.4) are obligatory during the prospecting of the coal deposits, which will be exploited by opencast
method. Only the scarce and extramely valuable associated minerals are prospected for coal mines. As associated minerals are
evaluated now refractory feed stock, kaolinite clays for ceramics and porous aggregates, burnt rocks in heaps and opencast
collieries, carbonate feed stock, ferrous ores, sands (building,
glass, moulding), stone building materials. The prospecting of
reserves of associated minerals are done if there are industrial
demand for them and a possibility of their simultaneous excavation by the equipment used in coal opencast collieries. Only forecasted reserves are evaluated if there is no industrial demands for associated minerals.

5. The associated waters produced during the explotation of
the coal deposits (drainaged and other) are considered as a source of waters for technical needs and for recovery from them
valuabled components. Their acidity and mineralisation are evaluated to prevent the possible pollution of the environment,
which can occure if the content of the harmful components in
them are higher than the permissable contents.

6. The investigation of the associated hydrocarbon gases is
carried out by special methods during the surveying og the gas-
bearing bituminous coal deposits [2].

REFERENCES
I. Временное руководство по определению объема и номенклатуры
исходных данных для составления мероприятий по утилизации вскрыш-
ных и вмещающих пород. ВНИИОСУголь МУП СССР, г.Пермь,1983, 30 с.
2. Инструкция по определению и прогнозу газоносности угольных
пластов и вмещающих пород при геологоразведочных работах. М.Нед-
ра, 1977, 86 с.
3. Инструкция по изучению и оценке попутных твердых полезных
ископаемых и компонентов при разведке месторождений угля и горю-
чих сланцев. М. Наука, 1986.
4. Инструкция по изучению токсичных компонентов при разведке
угольных и сланцевых месторождений. АН СССР. Мингео СССР, М.
1982, 84 с.
5. Использование отходов угольной промышленности в качестве
сырья для производства керамических стеновых изделий. ВНИИЭСМ
М., 1976, 42 с.

6. Клер В.Р. Изучение сопутствующих полезных ископаемых при разведке угольных месторождений. М., Недра, 1979, 272 с.

7. Лебедев В.В., Рубан В.А., Шпирт М.Я. Комплексное использование углей. М., Недра, 1980, 239 с.

8. Металлогения и геохимия угленосных и сланцесодержащих толщ СССР. Геохимия элементов. В.Р.Клер и др. М., Наука, 1987.

9. Методические рекомендации по изучению и оценке попутных вод месторождений полезных ископаемых в целях их использования в качестве гидроминерального сырья. М. ВСЕГИНГЕО, 1985, 97 с.

10.Методика опробования текущих отходов обогащения углей и породных отвалов угольных шахт и углеобогатительных фабрик.Пермь, ВНИИОСУголь МУП СССР, 1982, 42 с.

11.Рекомендации по использованию топливосодержащих промышленных отходов в качестве добавки при производстве стеновых керамических изделий. ВНИИСТРОМ им.П.П.Будникова, М.; 1977, 15 с.

12.Сборник руководящих материалов по геолого-экономической оценке месторождений полезных ископаемых. ГКЗ СССР, том I, М., 1985, 576 с.

Reclamation, Treatment and Utilization of Coal Mining Wastes, edited by A.K.M. Rainbow
Elsevier Science Publishers B.V., Amsterdam, 1987 — Printed in The Netherlands

COAL WASTE INDUSTRIALIZATION IN BRAZIL: THE STATE OF THE ART

EUGENIO DA MOTTA SINGER and ALDÉRICO JOSÉ MARCHI
Mineral Resources Division, Companhia Energética de São Paulo, CESP.
Alameda Ministro Rocha Azevedo 25, São Paulo, SP, Brazil, 01410

SUMMARY
   High ash and sulfur coal has being produced in Brazil for energy generation
and steel production. Recently, there was a significant increase in the
Brazilian coal reserves as a result of financial resources application by the
Federal Government.
   The feasibility of a national coal utilization plan shall consider not only
the energy content of this resource, but also the environmental costs associated
with coal utilization.
   This study focuses on the industrialization of coal wastes generated during
the coal mining and coal utilization. Special emphasis is given to the
industrialization of pyritic wastes in the largest area of coal production in
Brazil. The industrialization of coal ash is also considered.

INTRODUCTION

   Coal utilization in Brazil until the early 70's was geographically restricted
to the production coal areas in the South, namely, Rio Grande do Sul and Santa
Catarina. While the former region was driven towards energy generation only,
Santa Catarina coal area was producing both steam coal to supply Jorge Lacerda
Power Plant (482MW) and steel coal to supply the Southeast coke plant at
Companhia Siderúrgica Nacional (CSN). Figure 1 shows the Brazilian coal areas
and their respective coal reserves.

   The Brazilian coal is classified as sub-betuminous or betuminous coal, with
high volatiles content, with ash content within the range of 35 to 65% and
sulfur content within the range of 0.5 to 10%.

   Since 1970, large amounts of financial resources have been supplied by the
Federal Government in order to change the utilization scenario described above.
Most of these financial resources were applied to coal prospection and resulted
in substantial increases in the national coal reserves, as indicated in Figure 2.
From 2.7 billion tonnes in 1970 to 31.0 billion tonnes in 1985 the Brazilian
coal reserves increased elevenfold. However, coal internal supply had not
evidenced such a proportional increase for the same period. From 2.4 million
tonnes in 1970, the coal internal supply jumped to 7.7 million tonnes in 1985.

   The main contribution for the increase in the Brazilian coal demand was the
substitution policy moving from oil to coal consumption, which has been adopted
by the cement industrial sector as illustrated in Figure 3. From a zero coal

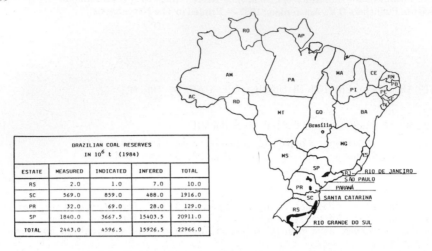

Figure 1. Brazilian coal reserves.

consumption in 1976, the cement industrial sector jumped to 2.2 million tonnes consumption in 1985.

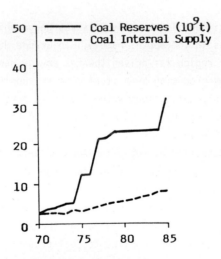

Figure 2. Brazilian coal reserves and coal internal supply.

Considering the Brazilian coal quality and its restrictive uses, a large amount of wastes is generated during its preparation and utilization. Coal preparation wastes average 69% of the run of mine (ROM) for the whole Brazilian coal production. This high average of waste generated is not only due to ash

10^6tEP
6
4
2
0
70
73
76
79
82
85
Chemical
and others
Cement
Transformation

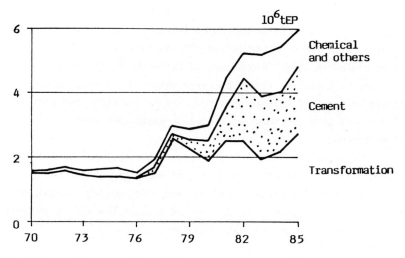

Figure 3. Evolution of steam coal total consumption in Brazil.

and sulfur contents in the coal seams, but also due to the intercalated siltstones and shale beds which are usually mined altogether with the coal, as illustrated in Figure 4, which represents Barro Branco coal seam in Santa Catarina.

TH
NAME
LITHOLOGY
35
FORRO
Pyritic coal with bright strips with shale intercalations.
60
QUADRAÇÃO
Dark shales and siltstones with thin coal beds.
06
CORINGA
Slightly pyritic coal
28
BARRO BRANCO
Clear grey to dark siltstones slightly stripped.
67
BANCO
Pyritic coal with bright strips intercalated with siltstones beds and dark grey shale.
SHALE
SILTSTONE
COAL

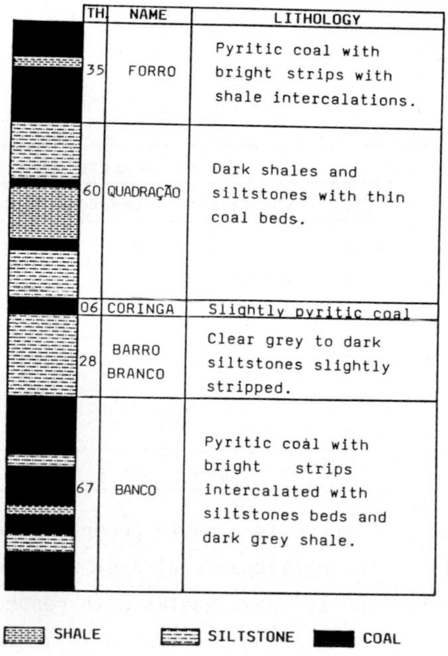

Figure 4. Barro Branco coal seam in Santa Catarina.

The wastes produced from the three coal areas have markedly different characteristics. Santa Catarina coal area is by large the main waste producer contributing with 76% of waste in the ROM, and almost 15 million tonnes of wastes generated in 1985. Rio Grande do Sul with 46% of wastes in the ROM produces around 2 million tonnes per year while Paraná waste production is negligible in terms of weight, when compared to the other two areas production. Paraná produces only 150 thousand tonnes of wastes per year, but is very high in sulfur content.

Due to different geological conditions, either the coal or its associated waste production is substantially different among the three coal areas. There is an increasing sulfur content gradient and a decreasing ash content gradient Northword. The characteristics and the end uses of the wastes generated are discussed in detail in the following sectors.

THE PRODUCTION OF COAL WASTES IN BRAZIL

As stated in the introduction of this paper, the Brazilian coal production is concentrated in the South of the country. Coal proporties are significantly different among the three coal areas, and the variability of ash and sulfur contents in the Brazilian coal characterizes the processes involved in the coal preparation and consequently its products and by-products.

The saleable coal products are steam coal and steel coal and their specifications are shown in Table 1.

TABLE 1
Saleable coal specifications.

CHARACTERISTICS	COAL PRODUCTS								
	CE 6000	CE 5200	CE 4700	CE 4500	CE 4200	CE 3700	CE 3300	CE 3100	CM
Heat Content (kcal/kg)	5.700	5.200	4.700	4.500	4.200	3.700	3.150	2.930	7.800
Granulometry	35 x 0	25 x 0	50 x 0	25 x 0	75 x 0	25 x 0	500 x 0	75 x 0	25 x 0
Moisture Content (%)	18	10	19	12	17	19	14	17	10
Ash Content Maximum (%)	25	35	35	42	40	47	54	57	16
Sulfur Content Maximum (%)	5,0	2,5	2,5	4,0	2,0	2,0	1,2	2,0	1,8
Free Swelling Index FSI	-	2	-	-	-	-	-	-	5 - 6
Area of Coal Production	PR	SC	RC	SC-PR	RS	RS	RS	RS	SC

Associated with coal production there is a waste generation which averages 69% of the run of mine (ROM). Each individual coal area produces different wastes. The percentiles composition of these wastes with respect to the ROM according to 1985 coal production data are all illustrated in Table 2.

TABLE 2

Coal and waste production in Brazil (1985)

COAL AREAS	RUN OF MINE (ROM) 10^3t	STEEL COAL 10^3t	STEAM COAL 10^3t	WASTES 10^3t	WASTE % IN ROM
Rio Grande do Sul	4594	nil	2496	2098	46
Santa Catarina	19602	1417	3324	14821	76
Paraná	423	nil	267	156	33
Total	24619	1417	6087	17075	69

* 1% of the ROM is lost in handling and transportation.

Coal utilization by several industrial sectors in Brazil generates considerable amount of coal ash due to the high ash content of the Brazilian coal (37.5% average). From the overall coal production of 7.7 million tonnes, 7.5 million tonnes of coal were consumed in 1985. The national sectorial consumption for the steam and steel coal is described in Figure 5 and Table 3.

* A - STEEL
* B - THERMOELECTRICITY
* C - CEMENT
* D - PULP AND PAPER
* E - FOOD AND BEVERAGES
* F - CHEMICAL
* G - OTHERS

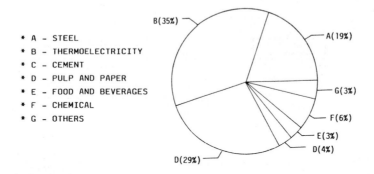

Figure 5. National Sectorial Consumption of coal in 1985.

The coal ash generated in the above industrial sectors sums up to 28 million tonnes. The major contribution comes from thermoelectricity generation, and 67% of the total ash generated is pulverized fly ash (PFA) which is entirely consumed either by the cement industry or by raw material(pozolane)to dams construction. All the ash generated in the cement industry is incorporated into the clinker, during the cement production process, and 67% of the ash

TABLE 3

Brazilian Sectorial Coal Consumption and coal ash production (1985)

SECTOR	COAL USED 10^6 t	% OF TOTAL COAL USED	ASH GENERATED $(10^6 t)$	% OF TOTAL ASH GENERATED
Thermoelectricity	2.66	35.5	1.24	44.3
Cement	2.20	29.3	0.78	27.9
Coke and Steel	1.40	18.7	0,23	8.2
Chemical	0.48	6.4		
Pulp and Paper	0.27	3.6	0.55	19.6
Food and Beverages	0.23	3.0		
Others	0.26	3.5		
total	7.50	100.0	2.80	100%

generated at Companhia Petroquímica do Sul (COPESUL) is also PFA which is
consumed by the cementeries too. The slag which is the waste from the blast
furnaces of the steel industry accounts for 8% of the total and actually it is
being used by the cement sector.

The flow diagram and a mass balance of coal waste generation in Brazil is
presented in Figure 6.

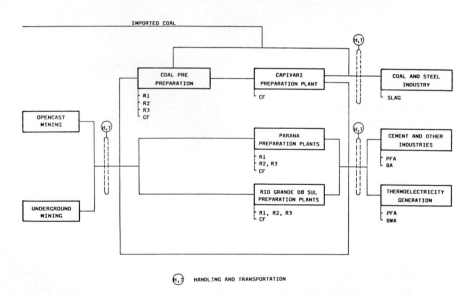

Figure 6. Flow diagram of coal waste generation in Brazil.

The coal mining wastes are classified basically into four categories known
as: pyritic wastes (R1), shale wastes (R2), clay wastes (R3) and coal fines(CF).

The coal utilization wastes which are considered in this study are pulverized fly ash, bottom and wet ash and slag. Further descriptions on the quantity and characteristics of both coal mining wastes and coal utilization wastes are presented in the next sections.

Coal mining wastes

The three coal areas in the South of Brazil produce a variety of coal products as shown in Table 1. They are altogether 9 products each with their own specifications. Likewise, different coal mining wastes are produced within these coal areas. As illustrated in Figure 6, the coal preparation is distinct among the three coal areas. Santa Catarina coal area has a peculiar difference from the other two coal areas. All the coal produced in Santa Catarina is pre-prepared at the collieries sites. At these pre-preparation plants all four types of wastes are generated, and the coal products are sent to a central coal-preparation plant called Capivari preparation plant. Furthermore, Santa Catarina is by far the largest coal producer in Brazil with 63% participation in the coal internal supply. Considering that the Barro Branco coal seam has several siltstones and shale beds intercalated, the dirt volume in the ROM is even greater. Santa Catarina averages 76% of waste material in the ROM. Therefore, it is in Santa Catarina coal area that 87% of the total coal wastes in Brazil is generated.

From the other two areas, Rio Grande do Sul contributes with 12% of waste generation in Brazil and Paraná with only 1%. On the one hand Rio Grande do Sul coal has a very low sulfur content (1%) and high ash content (45 - 60%), on the other hand, Paraná coal has the highest sulfur content (10%).

Santa Catarina coal mining wastes. Emphasis is given to the Santa Catarina coal area as most of the wastes are generated in this area, as well as all four types of wastes are produced there. Figure 7 is a flow diagram for this area under analysis.

Four kinds of wastes are generated at the pre-preparation plants, and the characteristics of these wastes are presented in Table 4.

The wastes of pre-preparation plants, although being classified in four categories are not all segregated at the plants. The first cut is for the R1 (pyritic wastes), and the second cut is for the R2 (shale wastes), R3 (clay wastes) and coal fines are disposed in sedimentation ponds and then recovered for industrial use. At Capivari preparation plant 2% of coal fines are generated resulting in 90000t/year of coal fines disposed at lagoons near the site.

High sulfur content in the R1 wastes, transportation facilities and environmental impacts of this waste type have contributed to industrialization

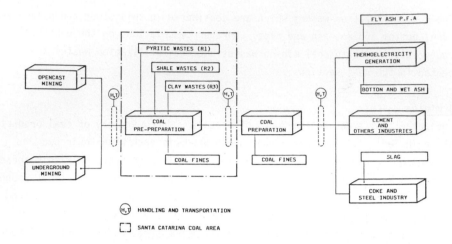

(H,T) HANDLING AND TRANSPORTATION

☐ SANTA CATARINA COAL AREA

Figure 7. Flow diagram for Santa Catarina coal mining wastes.

of the pyritic wastes.

Coal fines from the pre-preparation plants and Capivari preparation plant are now being used as fuel in fluidized bed boilers.

TABLE 4

Characteristics of coal mine wastes in Brazil.

CHARACTERISTICS / WASTES	SiO_2 %	Al_2O_3 %	Fe_2O_3 %	TOTAL MOISTURE CONTENT %	POROSITY %	VOID RATIO	SULFUR CONTENT %	RUN OF MINE %	WASTE PRODUCED $10^6 t$
. PYRITIC WASTES (R1)									
Rio Grande do Sul (1)	nd	nd	nd	4.2	na	nd	0.8	46.0	2.1
Santa Catarina	43.3	33.3	12.7	9.7	37.3	0.6	10.0	25.0	4.9
Paraná (1)	nd	nd	nd	na	nd	nd	10.0	33.0	0.2
. SHALE WASTES (R2)									
Santa Catarina	59.4	33.5	3.2	7.2	32.8	0.5	nd	35.0	6.9
. CLAY WASTES (R3)									
Santa Catarina	58.2	34.4	3.0	7.2	33.2	0.5	nd	15.0	2.9
. COAL FINES (CF)									
Rio Grande do Sul	nd	nd	nd	na	nd	nd	3.0		0.2
Santa Catarina	nd	nd	nd	na	nd	nd	2.0	5	0.3
Paraná	nd	nd	nd	na	nd	nd	3.0	3	0.1

(1) Rio Grande do Sul and Paraná coal mining wastes comprehends R1, R2 and R3.
na not available
nd not detemined

Rio Grande do Sul and Paraná coal mining wastes. Few attention
has been paid to the wastes generated in these two coal areas. In
Rio Grande do Sul, coal mining wastes are mostly shale and clay
wastes, and these wastes have being disposed as landfill materials
close to the colliery sites. In Paraná the wastes are basically
pyritic wastes and there is not a significant amount of R2 and R3
as the coal mining is very selective. The flow diagram for these
two areas is presented in Figure 8. The characteristics of the
wastes from these areas are also presented in Table 4.

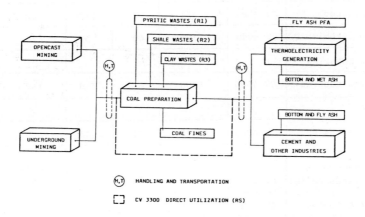

Figure 8. Flow diagram for Rio Grande do Sul and Paraná coal mining
wastes.

Coal utilization wastes

There are several different coal products and consequently a
variety of uses for these coal products but in general they are
classified as steam coal and steel coal. The coal utilization pie
as shown in Figure 5, covers 81% of steam coal and 19% of steel
coal.

The steam coal utilization covers thermoelectricity generation
(35%), cement fabrication (29%), chemical (6%) and other industries
(30%) including food and beverages, pulp and paper, ceramic, transportation, textile, etc.

Coal utilization in thermal power plants generates two products:
pulverized fly ash (PFA) and bottom ash (BA). In most thermal
power plant the bottom ash is transported hydraulically to
settlement ponds and the fly ash is stocked in bunkers for
commercialization. An average of 69% of fly ash is generated with

TABLE 5

Production of coal utilization wastes.

ASH PRODUCTION / INDUSTRIAL SECTOR	COAL PRODUCT	COAL CONSUMPTION 10^6 t/YEAR	ASH CONTENT %	P.F.A.[1] 10^3 t/YEAR	BOTTOM AND WET ASH[2] 10^3 t/YEAR
. THERMOELECTRICITY Rio Grande do Sul					
Candiota PP	CV 3300	711	54	257	119
Charqueadas PP	CV 3100	451	57	172	80
São Jerônimo PP	CV 4200	122	40	37	15
Santa Catarina					
Jorge Lacerda PP	CV 4500	1280	42	360	167
Paraná					
Figueiras PP	CV 6000	75	25	13	6
. CEMENT					
Rio Grande do Sul	CV 4700	535	35	–	187 (3)
Santa Catarina	CV 5200	1698	35	–	594 (3)
. PETROCHEMICAL					
Rio Grande do Sul	CV 3700	483	47	152	70
. OTHERS					
Rio Grande do Sul					
RS1	CV 3100	16	57	–	9
RS2	CV 3300	385	54	–	208
RS3	CV 4200	69	40	–	27
RS4	CV 5900	1	22	–	1
Santa Catarna					
SC1	CV 4500	69	45	–	31
Paraná					
PR1	CV 6000	192	25	–	48

(1) 67% is PFA, assuming 69% FLY ASH with 98% removal efficiency by ESP

(2) 31% is BWA, assuming 25% ash from the boilers, 5% from the economizer and 1% from the heaters.

98% removal efficiency by the eletrostatic precipitation. 31% of ash is wet ash removed at the bottom of the boilers, economizers, and heaters.

The coal used in the cement industrial sector yields a considerable amount of ash which is incorporated into the clinker as raw material.

Others industrial sectors using coal represent only 10% of total consumption and the main by-product is bottom ash from the boilers system where they are used.

Table 5 presents a detailed scenario of coal utilization in Brazil and its respective coal utilization wastes. Also the characteristics of PFA and bottom ash are presented in Table 6.

The next section discusses the industrialization of wastes generated during coal mining and coal utilization.

TABLE 6
Characteristics of coal utilization wastes in Brazil.

CHARACTERISTICS / WASTES	SiO_2	Al_2O_3	Fe_2O_3	CaO	MgO	Na_2O	K_2O	TiO_2	P_2O_5	MnO_2	S	C
. PULVERIZED FLY ASH												
Rio Grande do Sul	64.7	24.6	4.8	1.5	0.5	0.5	1.5	1.2	0.1	–	0.2	0.5
Santa Catarina	56.2	28.3	7.4	1.4	0.9	0.2	2.3	nd	nd	nd	0.3	1.4
Paraná	33.1	11.7	32.7	1.6	1.1	0.9	1.5	nd	nd	nd	1.24	16.9
. BOTTON AND WET ASH												
Rio Grande do Sul	60.2	21.8	7.4	2.1	0.6	0.3	1.5	0.8	0.1	0.1	0.3	0.5
Santa Catarina	12.1	8.6	60.8	0.2	nd	nd	nd	nd	0.1	0.1	0.2	1.5
Paraná	na	na	na	na	na	na	na	na	na	na	na	na

na   not available
nd   not determined

THE INDUSTRIALIZATION OF COAL WASTES IN BRAZIL

Coal waste industrialization in Brazil is concentrated in pyritic wastes (Rl) and pulverized fly ash (PFA). Pyritic waste industrialization is taking place in Santa Catarina coal area where 4.9 million tonnes of Rl are generated with an average sulfur content of 10%. Pulverized fly ash industrialization is well developed in Rio Grande do Sul coal area where half million tonnes, is at present being produced annually with forecast of a double production equivalent to 1 million tonnes of PFA by 1990. Paraná coal area does not contributes yet to industrialization of coal wastes in Brazil.

Industrialization of coal mining wastes

The environmental problems which have being faced by the Brazilian coal industrial sector, as well as the lack of raw material such as sulfur and its derivatives, in addition to the potential uses for coal products have markedly contributed to the industrialization process of coal mining wastes in Brazil.

Figure 9 illustrates the potential uses of coal mining wastes.

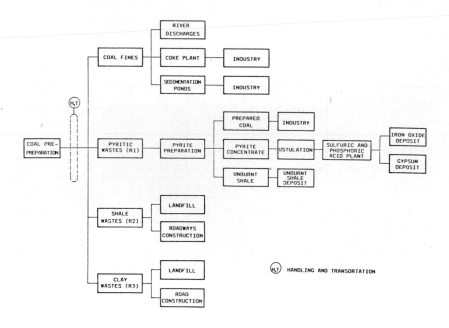

Figure 9. Potential uses for coal mining wastes.

Pyritic wastes. Two very important steps were taken by the Brazilian Government to face the environment problems of high sulfur coal mining and lack of sulfur products. First, due to accumulated reserves of pyritic reserves in Santa Catarina wetlands and availability of dock facilities at Imbituba port (40 miles from Criciuma) a sulfuric acid plant (300000t/year),and a phosphoric

acid plant (115000t/year) were installed within the limits of Santa
Catarina coal area. For the sulfuric acid plant the raw material
was gathered in the region disposal sites until 1983. The raw
material for the phosphoric acid fabrication was the sulfuric acid
produced and the phosphatic rocks from Central Brazil. Second, in
1983 a pyrite concentration plant was installed close to Criciúma
colliery sites to produce a pyrite concentrate from the R1 wastes.
These wastes with average 10% sulfur content were prepared and
resulted in a concentrate with 44% sulfur content, 40% Iron and 8%
Carbon. The nominal supply capacity for the pyrite concentration
plant (PCP) is at present 440t/h of R1. This is half the
production of R1 wastes in Santa Catarina.

In addition to a 55t/h production rate of pyrite concentrate,
the PCP produces two other additional by-products: 13t/h of
prepared coal and 372t/h of shale wastes. The shale wastes are, in
turn, classified into coarse wastes (317t/h) and fine wastes (55t/h).
Characteristics of these wastes are presented in Table 7. All
these wastes by-products are to be disposed at lagoons near the
PCP at annual rates of one million cubic meters of coarse material
and 200000 cubic meters of fines at respective costs of US$ $0.6/m^3$
(Busch and Abrão).

TABLE 7
Pyrite Concentrate Plant products and by-products kcal/kg.

CHARACTERISTICS WASTES	PRODUCTION RATE t/h	SULFUR CONTENT %	ASH CONTENT %	CARBON CONTENT %	HEAT CONTENT kcal/kg	GRANULO- METRY mm
Pyrite Concentrate	55	44	-	8	1800	
Pre prepared coal	13	1.6	35		3800	25 - 0
Coarse Shale waste	317	5	-	7	na	7.5
Fine shale waste	55	9	-	16	na	0.5

The pyrite concentrate from the PCP is then transported after preparation to
the sulfuric acid plant in Imbituba. The concentrate is then ustulated in fluid
bed kiln, following the reaction below, and producing a substantial amount of
purple ore,

$$4\,FeS_2 + 11\,O_2 \quad 2\,Fe_2O_3 + 8\,SO_2 \tag{1}$$

(720t)   (450t)

and then 900t of H_2SO_4 are produced as follows:

$$2\ SO_2\ +\ O_2 \qquad SO_3 \tag{2}$$

$$SO_3\ +\ H_2O \qquad H_2SO_4 \tag{3}$$

$$C\ +\ O_2 \qquad CO_2 \tag{4}$$

For H_2SO_4 production, pulverized iron oxide is produced at a daily rate of 450t, posing additional environmental problems for its disposal. Presently the iron oxide is being disposed in appropriate sites and commercial utilization is being investigated for pig iron production. Iron Oxide to be used in this project is being predicted at 350t/day or almost 80% of the daily production.

During the ustulation process, thermal generation is achieved with a 10.7 MW thermal electricity power plant, thus supplying 2/3 of the power required for the entire plant system.

All the production of sulfuric acid is consumed by the phosphoric acid fabrication, wich is described as follows:

$$Ca_3\ (PO_4)_2\ +\ H_2SO_4\ +\ H_2O \qquad H_3PO_4\ +\ CaSO_4\ .\ 2\ H_2O \tag{5}$$

(1200t/day)   (900t/day)   (360t/day)   (1800t/day)

For H_3PO_4 production, 1800t/day of gypsum is produced and disposed near the plant. Half of this production is being commercialized and at the moment studies for further utilization are being developed by the Nuclear industry and the Navy Research Institute.

An additional doubling capacity PCP and sulfuric acid plant is nowadays being investigated by a joint venture among fertilizers producers including PETROFERTIL, which is the present owner of the PCP, sulfuric acid plant and phosphoric acid plant.

Coal fines. Coal fines which are either produced at the pre-preparation plants and preparation plants and which were either disposed in sedimentation ponds or discharged into the rivers in the past, are now being consumed as raw fuel in fluidized bed boilers, mainly by the ceramic industry. Actually there is an installed capacity for 14t/h coal fines consumption with the same characteristics as described in Table 4. The coal fines production in Santa Catarina is seven times greater than the present demand. There is a potential demand for use of the entire coal fines production capacity at Santa Catarina.

Shale and clay wastes. Very few studies have been developed for utilization
of these wastes. Basically these wastes are being used for landfill, spoil dams
and access road constructions. These uses have not been investigated on a
scientific basis. Rio Grande do Sul scientific and technological foundation
CIENTEC has just started a program for utilization of R2 and R3 wastes as civil
construction materials for several uses such as parking lots and road
construction.

Large amounts of burnt shales have been disposed all around the colliery
sites. These spoil heaps are kept at these sites without any reclamation.

Industrialization of coal utilization wastes

Industrialization of coal ash generated after coal utilization in thermal
power plants or industrial boilers has achieved significant advancement.
Pulverized Fly Ash is by large the product with the most extensive application.
The pozolanic characteristics of PFA has justified the use of PFA in concrete
structures such as hydroelectric dams for long time. The use of by the cement
industrial sector as raw material being incorporated into the clinker is a
reality and now most of PFA has this fate.

The more recent applications of PFA are in materials used for civil
construction,in lightweight aggregates, roof tiles industry or in road
construction.

Figure 10 indicates the most common uses for PFA and bottom ash from
thermoelectricity generation.

It is undoubtfully in Rio Grande do Sul coal area that industrialization of
coal ash has obtained the best results. Several studies have been developed
at Technology ans Science Foundation (CIENTEC).

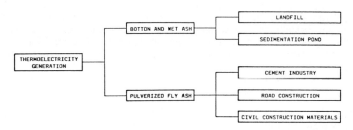

Figure 10. Coal Ash uses from thermoelectricity generation.

In 1978, a 1 km section of an important Brazilian motorway(BR 101) was built,
with the soil stabilization by adding lime with a proportion of 6% for 20% of
PFA. This material has substituted the road base of graded rock economically.
According to the study of Gonçalves et alli, the cost of pozolanic base is 32%
cheaper than conventional graded rock road base. Now a diversion in the same
motorway will be tested with heavy traffic load for a more realistic evaluation.

However, just a few and preliminary studies were performed to use the bottom ash in civil construction materials. CIENTEC is investigating the production of cement blocks with this material.

CONCLUSIONS

The Brazilian coal program may be considered an advancement for the national energy plan. During the oil crisis in 1973 and 1978 consumption of the national reserves in substitution of oil has contributed strategically for the nation's development. However, Brazilian coal is a low grade coal and its use must be redirected and reorganized. Environmental considerations must not be misregarded and the rationale exploration of the national coal reserves must be emphasized.

Coal utilization must be consistent with the harmonic utilization of nonrenewable resources. Thus, the industrialization of coal wastes contributes benefically to society, improving the overall efficiency of the coal mining industry.

Brazil has launched the programme for pyritic wastes utilization, and it is consuming a significant amount of coal ash generated in the coal utilization processes. Further studies to improve mining efficiency taking in account the societal costs of the environmental impacts associated with coal mining and utilization, are being developed. In this direction a more balanced utilization of coal resources shall be achieved.

REFERENCES

1 Busch G.B. and Abrão P.C., O tratamento dos rejeitos de carvão Revista Minérios PP 54 - 58, Outubro 1984.
2 Centrais Elétricas de Santa Catarina, Aspectos sobre Carvão Mineral, 1985.
3 D'Avila M.L.A., Chies F. and Zwonok O., Caracterização Geotécnica de cinzas de Carvão Mineral do Estado do Rio Grande do Sul, in: VIII Congresso Brasileiro de Mecânica dos Solos e Engenharia de Fundações, Porto Alegre, 1986.
4 Fundação de Ciência e Tecnologia, Carvões Minerais do Brasil, Características de carvões brutos do Rio Grande do Sul, Porto Alegre, 1980.
5 Ministério de Minas e Energia, Balanço Energético Nacional, Brasília, 1986.
6 Ministério de Minas e Energia, Informativo Anual Indústria Carbonífera DNPM, Brasília, 1984.
7 Montenegro Danilo A.F., Industrialização de Rejeitos de Carvão Brasil Mineral nº 28, PP 48 SC, Março de 1986.

VARIABILITY OF COAL REFUSE AND ITS SIGNIFICANCE IN GEOTECHNICAL ENGINEERING DESIGN

S.C. CHENG[1] and M.A. USMEN[2]

[1]Department of Civil Engineering, Lafayette College, Easton PA 18042, U.S.A.

[2]Department of Civil Engineering, West Virginia University, Morgantown WV 26506, U.S.A.

ABSTRACT

The variability encountered in coal refuse geotechnical properties such as water content, density, friction angle, and cohesion are documented in this paper in terms of the mean and coefficient of variation values. Point Estimate Method (PEM) was employed in the determination of the reliability and risk assessment of the geotechnical structure constructed by coal refuse. It is shown that the uncertainty around the material directly reflect upon the estimate of the factor of safety, and the probabilistic method can assist the geotechnical engineer in accounting for variability and uncertainty through quantification of the risk involved in design. Furthermore, a Monte-Carlo simulation technique for risk analysis with correlated random variables is also performed on one of the examples. Excellent agreement is achieved for the first two statistical moments between the two probabilistic methods.

INTRODUCTION

Variability encountered in the properties of coal refuse is a major source of uncertainty encountered by the geotechnical engineer in material and site characterization, analysis and design. In general, the true values of the geotechnical parameters can not be known or determined with full accuracy; they can only be estimated from a discrete number of laboratory and/or field tests. The scatter that is almost always observed in the test results is attributable to the inherent heterogenity of the materials as well as the errors arising from the sampling, testing and human judgement (ref. 1).

Coal refuse is a waste material generated in the mining and preparation of coal (refs. 2 and 3). Substantial quantities of coal refuse are produced every year in the United States as well as in the rest of the world, and the effective disposal and utilization of this material presents an increasingly important problem to the coal industry in terms of safety, economics and environmental acceptability.

The traditional approach to the assessment of the safety of a coal refuse disposal facility as part of the design processes has been the use of a deterministic factor of safety, which can be obtained by the use of a single value for each of those material properties considered. The single value used is often a conservative point estimate that is intended to account for the uncertainty of the true value.

Recently, considerable interest has focused on the use of probability and statistics (refs. 1 and 4) in the analysis and design of geotechnical structures to account for the variabilities and uncertainties associated with input parameters. It has been shown that probabilistic reliability risk assessment techniques can be valuable supplements or complements to the existing deterministic procedures. In the probabilistic approach, the geotechnical parameters are treated as random variables having a mean value, standard deviation and a probability density function (PDF), as opposed to single-valued quantities. This enables the engineer to incorporate the variability or uncertainty of relevant parameters into computations, and with proper probabilistic risk assessment techniques, to have a better understanding of the reliability of the proposed geotechnical structure.

A study was conducted by the authors to collect and document the geotechnical engineering properties of coal refuse and investigate the extent of variability that may be expected in the relevant test properties. An effort was then undertaken to implement and computerize a probabilistic safety assessment of a geotechnical structure constructed by using coal refuse. A parametric study was performed to assess how different levels of variability or uncertainty that may be encountered in input parameters may impact the engineer's ability to estimate the factor of safety of the structure. A Monte-Carlo simulation analysis including the correlations between input variables was also performed on one of the example. Comparisons were made on the results of these two probabilistic risk analysis.

VARIABILITY OF COAL REFUSE

Variability of geotechnical engineering test properties can be investigated by treating each property as a random variable described by a mean value, a standard deviation value and a probability density function. The coefficient of variation, which is obtained by taking the ratio of the standard deviation to the mean has particular merit in describing variability. This is because it measures the spread in the data in terms of the mean value which will be different from property to property and frequently from group to group for a single property. The coefficient of variation is a dimensionless property which is often expressed as a percentage.

Tables 1 and 2 summarize the index and engineering properties of coarse and fine coal refuse materials in terms of the mean, $\bar{x}$, and coefficient of variation, CV (expressed as a percentage) for each property. The sample size, n, and the reference or source used to obtain the data are also shown in these tables. Each line in the tables represents a single site with the exception of the last line which was an unpublished study conducted by the authors. Therefore, the variability indicated mainly is for that site in question.

An examination of Table 1 reveals that among the index properties considered natural water content exhibits higher coefficients of variation than specific gravity and in-situ dry density for both coarse and fine coal refuse. Regarding the variability of engineering properties coarse and fine coal refuse documented in Table 2, one can readily observe that cohesion values indicate substantially higher coefficients of variation than friction angles, and permeability is by far the most variable test property, yielding excessively high coefficients of variation. A comparison of the data presented in Tables 1 and 2 clearly shows that the engineering properties are significantly more variable than the index properties for coarse as well as fine coal refuse.

TABLE 1. Variability of Index Properties for Coal Refuse

Type of Refuse	Natural Water Content, %			Specific Gravity			In-situ Dry Density, pcf			Reference or Source
	$\bar{x}$	CV	n	$\bar{x}$	CV	n	$\bar{x}$	CV	n	
	6.1	21.5	9	1.99	10.1	9	90.6	9.8	9	ref. 2
	5.5	26.0	6	2.19	2.7	6	92.2	17.5	6	ref. 2
Coarse	7.2	13.9	5	1.88	9.0	5	88.6	9.8	5	ref. 2
Coal Refuse	9.6	34.8	122	1.69	1.0	21	98.1	11.3	122	DNR[1]
	5.9	51.1	80	1.16	5.7	8	98.9	8.4	80	DNR[1]
	10.1	59.3	93	2.25	-	1	100.7	12.8	93	DNR[1]
				2.04	14.2	47	93.2	14.1	47	
Fine Coal Refuse	54.2	8.6	15	1.57	4.4	15	52.1	5.2	15	ref. 5
	48.8	32.1	13	1.69	10.4	13	52.5	7.9	13	ref. 5
	30.2	41.6	19	1.50	7.3	19	53.9	13.4	19	ref. 5
	30.5	22.6	11	1.83	9.7	11	62.4	10.7	11	ref. 6
	13.4	28.8	4	1.48	3.6	10	49.7	6.7	4	ref. 7
				1.62	14.2	49	56.2	16.7	49	

(1) Data obtained from the files of West Virginia Department of Natural Resources, Charleston, West Virginia.

TABLE 2. Variability of Engineering Properties for Coal Refuse

Type of Refuse	Cohesion c', tsf			Friction Angle, ϕ'			Permeability 10^{-4} cm/sec			Reference or Source
	x	CV	n	x	CV	n	x	CV	n	
Coarse Coal Refuse	0.20	57.9	10	32.8	4.9	10	1.38	105.6	10	ref. 2
	0.50	40.0	15	39.0	6.7	15				ref. 8
	0.30	68.0	12	28.0	7.9	12				ref. 8
	0.05	100.3	9	38.0	17.9	9				DNR [1]
	0.12	139.3	6	36.3	7.6	8				DNR [1]
	0.16	101.4	47	32.6	16.4	47				
Fine Coal Refuse	0.20	45.4	6	29.8	8.6	6	0.068	209.2	15	ref. 5
	0.13	86.7	6	30.5	21.0	6	4.22	244.3	13	ref. 5
	0.17	71.6	8	35.0	11.1	8	8.70	221.7	19	ref. 6
	0.18	79.3	5	31.2	7.2	5	1.31	143.7	10	ref. 6
	0 [2]	-	12	31.2	10.0	12				ref. 7
	0 [2]	-	16	36.9	3.2	16				ref. 7
	0.21	95.8	49	29.0	26.7	49				

(1) Data obtained from the files of West Virginia Department of Natural Resources, Charleston, West Virginia.

(2) Mohr-Coulomb envelopes forced through the origin.

RELIABILITY AND RISK ASSESSMENT

The safety of coal refuse disposal facilities is conventionally assessed by a deterministic factor of safety which is calculated by using single-valued geotechnical properties for the refuse materials (generally estimated from a few tests or from previous experience), a deterministic foundation profile (established from subsurface explorations), a conservative deterministic estimate of pore water pressure conditions (i.e. a high phreatic line), and a deterministic disposal facility geometry (generally an embankment containing zones of coarse and fine refuse and having a slope of 2 to 1 or flatter to assure a sufficiently high factor of safety). In essence, however, all of these parameters are likely to be random variables rather than deterministic values, and their variability can not be systematically incorporated in the calculation of the conventional factor of safety. Furthermore, the factor of safety does not properly measure the level of risk associated with designing the disposal facility under many uncertainties.

These shortcomings can be overcome by a probabilistic approach in which the safety of the disposal facility is assessed by alternative parameters, such as the reliability index or the probability of low factor of safety. The probabilistic approach is not necessarily suggested as a total replacement or substitute to the deterministic procedure but rather to serve as a supplement to it in sharpening the decision making process.

By considering the input parameters as random variables, the mean, standard deviation of the factor of safety can be determined by employing the point estimate method (PEM) (ref. 9). Furthermore, the probability density function (PDF) of the factor of safety can be determined by the use of Pearson's system. (ref. 10). Brief description of these techniques are presented in the following sections.

The Point Estimate Method

Consider $y = f(x)$ to be the probability distribution of the random variable x. Refer to Figure 1, where the continuous distribution is replaced by two discrete point estimates, P_+ and P_-, acting at x_+ and x_-. Then we have:

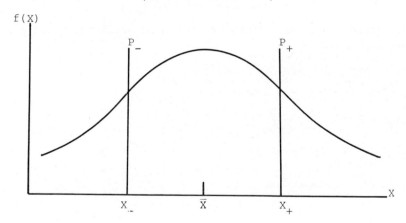

Figure 1. Point Estimate Approxiation

$$P_+ + P_- = 1 \tag{1}$$

$$P_+ x + P_- x_- = E[P(x)] = \bar{x} \tag{2}$$

$$P_+(x_+ - \bar{x})^2 + P_-(x_- - \bar{x})^2 = s_x^2 \tag{3}$$

$$P_+(x_+ - \bar{x})^3 + P_-(x_- - \bar{x})^3 = \beta_1 s_x^3 \tag{4}$$

where $\bar{x}$ is mean value, $s_x{}^2$ is variance, and β_1 is coefficient of skewness for variable x, and whose solution is

$$P_+ = (1/2) \left[1 \pm \sqrt{1 - 1 / (1 + (\beta_1/2))} \right] \tag{5a}$$

$$P_- = 1 - P_+ \tag{5b}$$

$$x_+ = \bar{x} + s_x \sqrt{P_-/P_+} \tag{5c}$$

$$x_- = \bar{x} - s_x \sqrt{P_+/P_-} \tag{5d}$$

If f(x) is symmetrical $\beta_1 = 0$ and equation (5) can be reduce to:

$$P_+ = P_- = 1/2 \tag{6a}$$

$$x_+ = \bar{x} + s_x \tag{6b}$$

$$x_- = \bar{x} - s_x \tag{6c}$$

With the distribution f(x) approximated by the point estimates P_+ and P_-, the moments of y = f(x) are

$$E[y] = \bar{y} = P_+ y_+ + P_- y_- \tag{7a}$$

$$E[y^2] = P_+ y_+{}^2 + P_- y_-{}^2 \tag{7b}$$

or in general

$$E[y^n] = P_+ y_+{}^n + P_- y_-{}^n \tag{7c}$$

For a function of four correlated random variables, say $y = f(x_1, x_2, x_3, x_4)$ or safety factor = f (unit weight, friction angle, cohesion, pore pressure), then it can be expressed as follow:

$$E[y^n] = P_{++++} \; y_{++++}{}^n + P_{+++-} \; y_{+++-}{}^n + P_{++--} \; y_{++--}{}^n \ldots$$

$$\ldots + P_{---+} \; y_{---+}{}^n + P_{----} \; y_{----}{}^n \tag{8}$$

where, $y(\pm\pm\pm\pm) = f[\bar{x}_1 \pm s_{x1}, \; \bar{x}_2 \pm s_{x2}, \; \bar{x}_3 \pm s_{x3}, \; \bar{x}_4 \pm s_{x4}]$ and $\tag{9}$

$$P(++++)=P(----)=\frac{1}{2^4}[1+\rho_{12}+\rho_{13}+\rho_{14}+\rho_{23}+\rho_{24}+\rho_{34}]$$

$$P(+++-)=P(---+)=\frac{1}{2^4}[1+\rho_{12}+\rho_{13}-\rho_{14}+\rho_{23}-\rho_{24}-\rho_{34}]$$

$$P(++--)=P(--++)=\frac{1}{2^4}[1+\rho_{12}-\rho_{13}-\rho_{14}-\rho_{23}-\rho_{24}+\rho_{34}]$$

$$P(+---)=P(-+++)=\frac{1}{2^4}[1-\rho_{12}-\rho_{13}-\rho_{14}+\rho_{23}+\rho_{24}+\rho_{34}]$$

$$P(+--+)=P(-++-)=\frac{1}{2^4}[1-\rho_{12}-\rho_{13}+\rho_{14}+\rho_{23}-\rho_{24}-\rho_{34}]$$

$$P(+-+-)=P(-+-+)=\frac{1}{2^4}[1-\rho_{12}+\rho_{13}-\rho_{14}-\rho_{23}+\rho_{24}-\rho_{34}]$$

$$P(+-++)=P(-+--)=\frac{1}{2^4}[1-\rho_{12}+\rho_{13}+\rho_{13}-\rho_{23}-\rho_{24}+\rho_{34}]$$

$$P(++-+)=P(--+-)=\frac{1}{2^4}[1+\rho_{12}-\rho_{13}+\rho_{14}-\rho_{23}+\rho_{24}-\rho_{34}] \tag{10}$$

Here s_{xi} is the estimated standard deviation of x_i, ρ_{ij} is the coefficient of correlation between x_i and x_j. The sign of ρ_{ij} is determined by the multiplication rule of ij; i.e. i=(-), j=(+) yield ij=(-)(+)=(-). Of course when all the parameters are uncorrelated, then all the P terms become equal to $1/16 = 1/2^n$.

Pearson's System

Based on the results from the PEM, Pearson's system (refs. 4 and 11) can be employed to determine the probability density function. Pearson's system is based on the finding that a majority of the continuous probability distributions f(x) can be generated from the differential equation

$$\frac{df(x)}{dx} = \frac{(a_0+x)f(x)}{b_0+b_1x+b_2x^2}$$

by the proper selection of the four constants a_0, b_0, b_1, and b_2 (ref. 10). the PDF is then decided upon the K-criterion, expressed as

$$K = \frac{\beta_1(\beta_2+3)^2}{4(2\beta_2-3\beta_1-6)(4\beta_2-3\beta_1)} \tag{11}$$

Where, β_1 and β_2 are the coefficient of skewness and of kurtosis. This criterion is illustrated in Figure 2

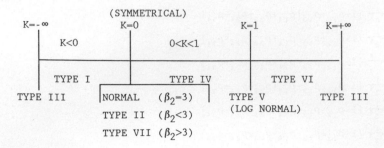

FIGURE 2. The Criterion K (after Eldertown and Johnson 1969)

Reliability Index for Factor of Safety

Armed with the results from the PEM and Pearson's system, one can easily establish a "Reliability Index" (ref. 12) for the factor of safety defined as:

$$\beta = \frac{\overline{FS} - 1.0}{S_{FS}} \tag{12}$$

where $\overline{FS}$ and S_{FS} respectively denote the estimated mean and standard deviation of the factor of safety of the disposal facility. The reliability index β indicates the normalized distance between $\overline{FS}$ (i.e. best estimate of FS) and the nominal failure value 1.0 in terms of the standard deviation. For example, $\beta=3$ means that the best estimate of FS is 3 standard deviation higher than the limiting value. The higher the β value, the higher is the reliability of the geotechnical structure in consideration.

Probability of Low Factor of Safety

Armed with the probability density function determined from Pearson's System, one can alternatively establish a "probability of low factor of safety" defined as

$$PLFS = P[FS \geq 1.1] = F_{FS}[FS=1.0]$$
$$= \int_{-\infty}^{1.0} f_{FS}(x) \, dx \tag{13}$$

where, F_{FS} = cumulative density function for FS; f_{FS} = PDF for FS; and x = a dummy variable for integration.

The probability of low factor of safety, PLFS, indicates the probability of the computed factor of safety being less than a limiting value (e.g. 1.0). This parameter has also been called "probability of failure" by some investigators (ref. 13) and it has been denoted as P_f. The complement of P_f

the β or PLFS parameters, one can replace the nominal failure value of unity by any other appropriate value adopted as the allowable factor of safety.

While the computation of β is straight-forward and can be done as a direct comuputation in the analysis, it will be necessary to incorporate statistical analysis software packages (i.e. SAS, SPSS, etc.) or manual computations using statistical tables to obtain PLFS.

Monte-Carlo Simulation of Correlated Variables

Monte-Carlo simulation is a powerful tool in terms of verifying or validating approximate analytical solution methods although it may be limited by constraints of economy and computer capability. A key task in the application of Monte-Carlo simulation is the generation of the appropriate values of the random variables (i.e., random numbers) in accordance with the respective prescribed probability density function. Based on the assumption of normal distribution for the input geotechnical parameters in this study, a computer program was developed to perform the random number generation of correlated variables. The computer simulation involves the use of an equation of the type (ref. 14)

$$Y = \sigma X + m \tag{14}$$

in which X is a standardized variate with mean equal to zero and standard deviation of σ corresponding to the variate Y with mean m and standard deviation σ. Therefore, the attention is focused in the generation of random vector $X = (x_1, x_2, x_3, \ldots\ldots, x_n)$ from $N(0,A)$, where A is the variance-co-variance matrix in the form of

$$A = \begin{bmatrix} \sigma_{11} & \cdots\cdots & \sigma_{1n} \\ \vdots & & \vdots \\ \sigma_{n1} & \cdots\cdots & \sigma_{nn} \end{bmatrix} \tag{15}$$

Let Z be distribute $N(0, I_n)$ (I_n is the unit matrix of size n) and let $X = CZ$. Then X is distributed $N(0, CC')$ in which $CC'=A$. The elements of C are determined recursively as follows:

$$c_{i1} = \sigma_{i1} / \sqrt{\sigma_{11}} \; ; \qquad\qquad \text{when } 1 \leq i \leq n$$

$$c_{ii} = \sqrt{\sigma_{ii} - \Sigma c_{ik}^2} \; ; \qquad\qquad \text{when } 1 < i \leq n, \text{ and } k = 1 \text{ to } i-1$$

$$c_{ij} = [\sigma_{ij} - \Sigma c_{ik} c_{jk}] / c_{jj} \; ; \qquad \text{when } 1 < j < i \leq n, \text{ and } k = 1 \text{ to } j-1$$

$$c_{ij} = 0 \qquad\qquad\qquad\qquad \text{when } i < j \leq n \tag{16}$$

once the c_{ij}'s have been determined the x_i's are obtained as

$$x_i = \Sigma c_{ij} z_i \; ; \qquad\qquad \text{where i = 1 to n, and j = 1 to i} \qquad (17)$$

and eventurally from equation (14), we have

$$y_i = \sigma_i x_i + m_i \qquad\qquad\qquad (18)$$

SAMPLE ANALYSIS

A case study is presented herein to demonstrate how the variability of the geotechnical engineering test properties can influence the assessment of the safety (i.e. risk analysis) of a coal refuse disposal facility. This can be best done by studying the relationship between the means and coefficient of variations of the relevant parameters and the probability of low factor of safety (PLFS) or the reliability index (β) of a coal refuse disposal facility. The facility in question is given in figure 3 (ref. 15). Consolidated-undrained triaxial compression tests with pore pressure measurements have indicated that the coarse coal refuse has a ϕ' angle of 37.3° and c' = 0 at a wet density of 115 pcf, while the fine coal refuse has ϕ' = 33.5° and c' = 0 at a wet density of 80 pcf. The minimum (static) factor of safety computed by the modified Bishop method using these deterministic parameters and the phreatic line shown in Figure 3 is 1.673.

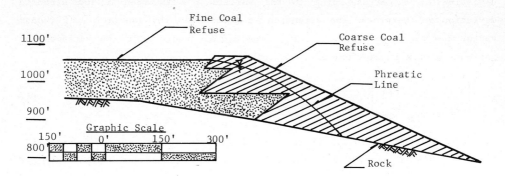

Figure 3. A Coal Refuse Disposal Facility Constructed by the Upstream Method (After Cowherd 1977)

A probabilistic analysis of the safety of the same disposal facility was conducted by the authors using the Point Estimate Method that was discussed in this paper. The results of the parametric studies are summarized in Table 3. The input geotechnical parameters were considered to be normal random variates.

Three levels of mean values for ϕ angle ($\bar{\phi}_{cr}$, $\bar{\phi}_{fr}$), and four levels of coefficients of variation (CV) for each of these parameters ($CV_{\phi cr}$, $CV_{\phi fr}$, $CV_{\gamma cr}$, $CV_{\gamma fr}$) were incorporated in the analysis as shown in the upper portion of Table 3. The disposal facility geometry and the position of the phreatic line given in Figure 3 were assumed to be deterministic.

TABLE 3. Input variables and output results for the reliability analysis of a coal refuse disposal facility.

INPUT

	CASE I				CASE II				CASE III			
c	0				0				0			
ϕ_{cr}	37.3				32.0				32.5			
ϕ_{fr}	33.5				32.5				30.0			
$\bar{\gamma}_{cr}{}^{1}$	115				115				115			
$\bar{\gamma}_{fr}{}^{1}$	80				80				80			
$CV_{\phi cr}$	5	10	15	25	5	10	15	25	5	10	15	25
$CV_{\phi fr}$	5	10	15	25	5	10	15	25	5	10	15	15
$CV_{\gamma cr}$	5	7.5	10	15	5	7.5	10	15	5	7.5	10	15
$CV_{\gamma fr}$	5	7.5	10	15	5	7.5	10	15	5	7.5	10	15
V_{GWT}	0				0				0			

OUTPUT

	CASE I				CASE II				CASE III			
FS_{min}	1.673				1.543				1.404			
$\overline{FS}$	1.675	1.683	1.696	1.739	1.545	1.551	1.561	1.593	1.406	1.410	1.416	1.441
S_{FS}	0.110	0.220	0.331	0.560	0.098	0.196	0.295	0.497	0.078	0.172	0.260	0.434
CV_{FS}	6.6	13.1	19.5	32.2	6.3	12.6	18.9	31.2	5.5	12.2	18.4	30.1
β	6.14	3.10	2.10	1.32	5.56	2.81	1.90	1.18	5.21	2.38	1.60	1.02
PLFS[2]	≈ 0	≈ 0.1	1.0	10.0	≈ 0	≈ 0.1	2.5	13.0	≈ 0	0.4	5.0	17.5
PLFS[3]	≈ 0	0.1	1.8	9.4	≈ 0	0.3	1.9	12.0	≈ 0	0.9	5.5	15.0
PDF	Beta (Type I-∩)				Beta (Type I-∩)				Beta (Type I-∩)			

1. In pcf
2. Based on Beta Distribution
3. Based on Normal distribution.

Shown in the lower portion of Table 3 and in Figure 4 are output results of this probabilistic analysis. FS_{min} values indicate the minimum factors of safety computed for the deterministic case, where the mean values of the input parameters have been taken as the design values. It is generally agreed that a minimum deterministic factor of safety of 1.5 must be attained in a coal refuse disposal facility . This facility would thus be regarded as "safe". However, a risk is involved in designing this facility with the indicated safety margin. The probabilistic analysis offers the geotechnical engineer a practical tool for quantifying this risk by first computing the mean factor of safety ($\overline{FS}$) and its standard deviation (S_{FS}) or coefficient of variation (CV_{FS}), and then establishing reliability index (β) and/or PLFS based on these parameters and the probability density function (PDF).

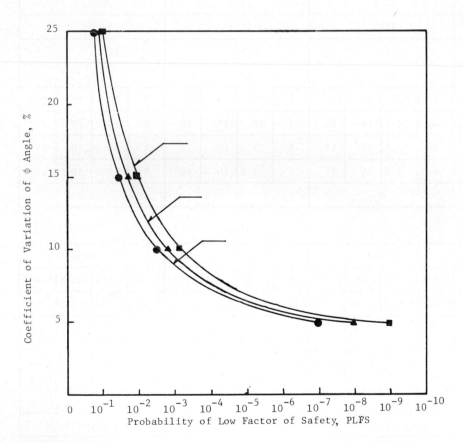

Figure 4. Relationship Between The Coefficient of Variation of ϕ Angle and The Probability of Low Factor of Safety

The quantitative results given in Table 3 for FS indicate that the diffferences between these values and the deterministic FS_{min} values are relatively insignificant. In light of this observation, the risk associated with the structure will clearly depend upon the uncertainty around the mean value as influenced by the uncertainties around the design parameters. The higher the uncertainties (or variabilities) are around the input design variables, the higher the uncertainty will be around the output factor of safety. This is clearly illustrated by the results presented in Table 3 and Figure 4. Considering that the ϕ angle is the key parameter controlling the safety of the structure, one can note that both its mean value and coefficient of variation has an effect on the PLFS and β values. The impact of the coefficient of variation on the PLFS is particularly drastic as shown in Figure 4.

The β and PLFS values listed in Table 3 provide indications of the reliability of the disposal facility or the level of risks implied in the design. High reliability index values signify high reliability in the design and low PLFS values designate low risk of failure. It can thus be observed from the table that as the variability of input design parameters become larger, a reduced reliability and increased risk are effected.

There is no commonly accepted design criteria relative to the minimum value of β or the maximum value of PLFS. According to Baecher, 1986, common geotechnical design is typically based on β values varying between 2 and 3. In a probabilistic slope stability analysis performed by McGuffey, et. al. 1982, a limiting value of 2 percent was considered realistic for P_f (termed PLFS in this study); however, lower values are generally desired. Based on these considerations, the β and PLFS derived with high variability in the input parameters from the parametric study indicates relatively low reliabilities and high risks of failure even though the mean factor of safety (FS) values are well within accepted range.

Monte-Carlo simulation of correlated variables were performed on Case II. 4000 sets of random numbers were generated based on the mean, standard deviation, and PDF given in Table 3. Slope stability analysis was then performed by using those generated random numbers as input parameters. The mean, variance, skewness, and kurtosis of factor of safety was then computed. Table 4 lists the results of the Monte-Carlo simulation on the same disposal facility. The results were also compared with PEM results. It is shown that in all levels of uncertainty both methods result in excellent aggreement in first two statistical moments. The skewness and kurtosis values shows considerable difference between the two methods as it was also observed in ref. 14.

TABLE 4. Statistical moments results between point estimate method and Monte-Carlo simulation (4000 run) for Reliability Analysis Case II

	Method	Mean	Variance	Skewness	Kurtosis
Variability 1	PEM	1.545	0.010	0.001	0.894
	Monte-Carlo	1.547	0.010	0.115	3.125
Variability 2	PEM	1.551	0.041	0.002	0.889
	Monte-Carlo	1.554	0.040	0.256	3.266
Variability 3	PEM	1.561	0.093	0.002	0.888
	Monte-Carlo	1.567	0.092	0.413	3.546
Variability 4	PEM	1.563	0.264	0.003	0.887
	Monte-Carlo	1.606	0.286	0.855	5.104

SUMMARY AND CONCLUSIONS

The study presented in this paper shows that Coal refuse materials exhibit appreciable variability in geotechnical engineering properties. For both coarse and fine coal refuse, water content is more variable than specific gravity and density, cohesion is more variable than the friction angle, and permeability is by far the most variable property. In general, higher levels of variability are expected in the engineering properties than in index properties. Meanwhile, these variabilities can exert a significant impact on the assessment of safety of a coal refuse disposal facility by slope stability analysis. These variability/uncertainty factors directly reflect upon the factor of safety; that is, if the input design parameters are random variables rather than deterministic quantities, the factor of safety also turns out as a random variable. The higher is the uncertainty around the input design parameters, the higher will be the uncertainty around the factor of safety.

Probabilistic methods can be very valuable tools for the geotechnical engineer in analysis and design as shown by the sample application of the probabilistic slope stability analysis herein. Such methods can serve as supplements or complements to the presently employed deterministic techniques in accounting for the variability and uncertainty through quantification of the risk involved in design. The parameters such as β and PLFS used in probabilistic methods are very useful measures of safety because they balance the safety implied by a best estimate of facility performance against the uncertainty in that prediction. Using these parameters, one can distinguish between the case of high estimated factor of safety with correspondingly low uncertainty. This will sharpen the engineer's decision making ability in the design process.

REFERENCES

1 Lumb P., 1974 "Application of Statistics in Soil Mechanics," Chapter 3 in Soil Mechanics - New Horizons, I. K. Lee ed., Newness-Butterworths, London.

2 Busch, R. A., Backer, R. R., and Atkins, L. A., 1985 "Physical Property Data on Fine Coal Refuse," U. S. Department of the Interior, Bureau of Mines, Report of Investigation 7964.

3 D'Appolonia Consulting Engineers, Inc. 1976, Engineering and Design Manual-Coal Refuse Disposal Facilities, U. S. Department of the Interior, Mining Enforcement and Safety Administration.

4 Harr, M. E., 1984. "Reliability-based Design in Civil Engineering." Twentieth Henry M. Shaw lecture in Civil Engineering. School of Engineering, North Carolina State University.

5 Busch, R. A., Backer, R. R., and Atkins, L. A., 1975 "Physical Property Data on Fine Coal Refuse," U. S. Department of the Interior, Bureau of Mines, Report of Investigation 80602.

6 Backer, R. R., Busch, R. A., and Atkins, L. A., 1977 "Physical Properties of Western Coal Waste Materials," U. S. Department of the Interior, Bureau of Mines. Report of Investigation 8216.

7 Siriwardant, H. J., and Liong, K. H., 1984 "Constitutive Modeling of a Granular Mine Tailings Material." Report CE/GEO-84-1, Department of Civil Engineering, West Virginia University, Morgantown, West Virginia.

8 Pierre, J. J., and Thompson, C. M., December 1979 "User's Manual, Coal-MIne Refuse in Highway Embankments," Prepared for Federal Highway Administration, Implementation Division.

9 Rosenblueth, E., 1975 "Point Estimates for Probability Moments," Proceedings, National Academy of Science, U. S. A., Vol. 72, No. 10, Math., pp. 3812-3814.

10 Elderton, W. P., and Johnson, N. L., 1969 "Systems of Frequency Curves," Cambridge University Press.

11 Harr, M. E. 1984. "Reliability-based Design in Civil Engineering". Twentieth Henry M. Shaw lecture in Civil Engineering, North Carolina State University.

12 Baecher, B. B., January 1986. "Geotechnical Error Analysis". Paper presented at the 65th annual meeting of the Transportation Research Board, Washington, D. D., January.

13 McGuffey, V., A-Grivas, D., Iori, J., and Kyfor, Z., October 1982 "Conventional and Probabilistic Embankment Design," Journal of the Geotechnical Engineering Division, ASCE, Vol. 108, GT 10, pp. 1246-1254

14 Nguyen, V. U., and Chowdhury, R. N., 1985 "Simulation For Risk Analysis With Correlated Variables," Geotechnique 35, No. 1 46-58.

15 Cowherd, D. C., June 1977 "Geotechnical Characteristics of Coal Mine Waste," Proceedings, Conference on Geotechnical Practice for Disposal of Solid Waste Materials, ASCE, University of Michigan, Ann Arbor, Michigan, pp. 384-406.

THE CHARACTERISTICS AND USE OF COAL WASTES

J. GONZALEZ CAÑIBANO[1] and D. LEININGER[2]

[1] Dirección Diversificación y Desarrollo, HUNOSA, Avda. de Galicia, 44, 33.005 Oviedo (Spain)

[2] Steinkohlenbergbauverein, Postfach 130140, 4300 Essen 13, West Germany

SUMMARY
In view of the grave problems posed by coal waste disposal, a majority of coal producing countries are actively investigating or undertaking programmes designed to minimize or resolve them.

As the overall characteristics of the coal waste determine the limiting factors of the investigation, this paper summarises the petrographic, mineralogical and chemical characteristics of the same, together with their physical and mechanical properties (size analysis, calorific value, thermal behaviour, moisture content, ash content, bulk density, specific gravity and plasticity) of the wastes of various coal-producing countries. These data are significant not only as regards the material under study, but also because the results obtained through this research relative to different countries may, should they coincide, furnish data applicable on an industrial scale to coal wastes in general.

Possible use of coal wastes are also examined in this paper.

INTRODUCTION

As is well known, the underground coal extractive industry generates large amounts of waste yearly which have been accumulated since long ago in tailings ponds or in tips.

Although the study of solving problems associated to coal wastes (1) may be undertaken from many points of view, the most adequate seems to be the approach directed towards using them as raw material for different industrial applications. This could give rise to some added value, to the creation of new companies and work positions and also to important fuel savings as a consequence of the carbon content of coal wastes.

From this perspective, it is necessary to study the characteristics and behaviour of these materials, which in turn provides a basis to determine their possible ways of application. Therefore, in this work the mineralogical and chemical characteristics of coal wastes from different coal producing countries - are summarized. The objective is mainly to compare the properties of different coal wastes since, if these properties coincide, the results of the researches carried out in a particular country could be extrapolated to the rest of the countries. Also, a scheme for the possibilities of use of coal wastes in different applications is suggested.

PETROLOGICAL AND MINERALOGICAL COMPOSITIONS

From the petrological point of view, coal wastes from the majority of coal-fields are sedimentary rocks with very variable quartz/clay ratios; the rocks composed of clay minerals, generally, being more frequent than quartz rocks (2) as can be seen en Fig. 1, where the types of rocks present in washery wastes from the Ruhr basin are given (3). However in some basins, as in the case of South Yakutia (USSR), sandstones are predominant reaching up to 60-70%. Coal wastes also present minor amounts of ferruginous, carbonous, etc., rocks.

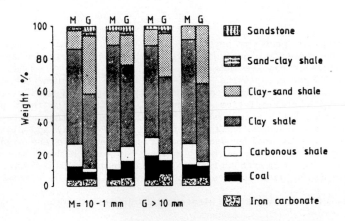

FIG 1 - ROCKS IN WASHERY COAL WASTES OF WEST GERMANY

Concerning the mineral composition, coal wastes are mixtures of different components which may vary largely from one basin to another although the predominant ones are those giving rise to clay rocks, such as illite, kaolinite, chlorite, etc.. Quartz appears in larger proportions in coarse(150-1 mm)fractions than in the fine ones (minus 1 mm). Other frequently found minerals are pyrites, carbonates, etc. (2). In Fig. 2 the mineralogical composition of coal washery wastes from the North of Spain is schematised (4).

CHEMICAL COMPOSITION

In Table 1 the intervals of variation or the mean values of concentration (on dry or on calcined basis) are given for the main elements present in coal wastes from different countries.

It is deduced from Table 1 that coal wastes from the various coalfields exhibit similar chemical compositions. Silica (SiO_2) and alumina (Al_2O_3) are the main components, followed by Fe_2O_3, K_2O, etc.. Also, coal wastes contain substantial amounts of carbon which increase when going from coarse particle sizes to the finest ones (2) (5).

TABLE I

Chemical composition of coal wastes

| | F.R. Germany | | Australia | | | Czechoslovakia | | | |
| | | | Northern | | Washery of | Ostrava-Karviná | | Kladno | |
	150–10 mm	10–1 mm	Washery	Red	Coalcliff	Coarse	Tailings	Coarse	Tailings
SiO_2	(57,7)	(50,0)	(64,0)	(65,5)	(67,4)	50,6	52,6	52,1	35,9
Al_2O_3	(24,0)	(22,9)	(24,8)	(25,1)	(24,6)	19,2	27,8	24,2	21,8
Fe_2O_3	(7,4)	(9,4)	(3,0)	(3,6)	(2,8)	5,7	5,1	3,5	2,0
TiO_2	(1,0)	(0,9)	(0,9)	(0,8)	–	0,8	0,9	0,6	0,6
CaO	⎱(3,4)	⎱(5,4)	(0,3)	(0,3)	(0,3)	1,1	1,7	0,4	0,3
MnO	⎰	⎰	–	–	–	0,1	0,1	0,1	–
Na_2O			(0,2)	(0,2)	(0,2)	0,8	0,2	0,5	0,1
K_2O	(4,5)	(4,4)	(2,6)	(2,8)	(2,3)	3,0	3,0	1,8	2,0
MgO	(1,7)	(1,4)	(0,7)	(1,0)	(0,6)	1,6	0,6	1,0	0,5
V_2O_5	–	–	–	–	–	–	–	–	–
S (total)	(0,7)	(1,4)	0,3	(0,1)	0,2	1,0	0,7	0,8	0,7
C	6,2	12,5	30,4	(0,2)	24,0	10,4	34,2	8,4	28,6.
Others	–	–	–	–	–	–	–	–	–
Ignition loss	–	–	–	–	–	16,6	36,0	15,7	39,7

( ) on burnt sample

TABLA I (continuation)

Chemical composition of coal wastes

	Spain			France			Poland		United Kingdom	USSR	
	150–10 mm	10–1 mm	– 1 mm	Washery	Red	Roof	Floor	Red		Donbas	Donetsk
SiO_2	49.2(56.5)	47.1(55.6)	43.1(56.0)	(45-55)	52	47-53	40-46	49-64	28-67	(50-60)	(50-62)
Al_2O_3	21.7(24,9)	23.7(27.9)	23.2(30.2)	(25-30)	30	29-32	19-24	18-22	15-27	(20-30)	(12-31)
Fe_2O_3	6.8(7.8)	5.3(6.2)	4.6(6.0)	(5-8)	7	1-3	3-8	5-10	3-10	(8-15)	(2.6-19)
TiO_2	1.1(1.3)	1.2(1.4)	1.1(1.6)	(1)	–	–	–	–	1	–	–
CaO	1.4(1.6)	1.1(1.3)	2.0(2.5)	(0.5-1.5)	1.5	1-2[a]	4-8[a]	2-6[a]	0.2-4	(0.3-8.6)	(1.5-12)
MnO	–	–	–	–	–	–	–	–	–	–	–
Na_2O	0.4(0.5)	0.4(0.5)	0.3(0.4)	(3-6)	4	1-3	–	–	1.0	(0.5-2.5)	–
K_2O	3.0(3.3)	3.2(3.8)	2.9(3.8)				2-4	–	1-5	(0.9-4.7)	3-15
MgO	1.5(1.7)	1.4(1.6)	1.2(1.5)	(1-2)	1.5	–	–	–	0.5-3.5	(0.8-5.3)	–
V_2O_5	0.2(0.3)	0.2(0.3)	0.2(0.3)	–	–	–	–	–	–	–	–
S (total)	0.6	1.4	1.0	0-1	1.0	–	–	–	0.5-1.9	(1-7)	(0.2-7)
C	3.5	5.5	11.4	–	–	–	–	–	–	8-22	3-15
Others	–	–	–	3	3	–	–	–	–	–	–
Ignition loss	–	–	–	0.2-12	–	–	–	1-6	–	–	–

( ) on burnt sample

[a] CaO+MgO

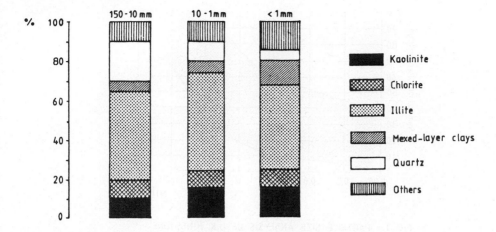

FIG 2 - MINERALOGICAL COMPOSITION OF WASHERY COAL WASTES
FROM NORTHERN SPAIN

SIZE DISTRIBUTION

The particle sizes generated could exclude coal wastes from some uses, or establish the need to effect a size operation or a grinding procedure, or reflect the waste content profitable for a given applications, etc. So it is convenient to know the size distribution of coal wastes which usually vary within a continuous band comprised between o and 200 mm. This can be seen in Table II and Fig. 3, where the particle size distribution for coal washery wastes from F.R. Germany (6) and for coal tipo wastes (7), respectively, are reported. It is hoped that the corresponding values for other coal basins should not be very different from these, since the procedures followed for size cuts and for coal washing are very similar in washeries of different coal producing countries.

An interesting fact to be noted is the existence of a trend towards increase of concentration of fine particules, mainly due to mechanization of mines (8).

CALORIFIC VALUE

One of the essential characteristics of coal wastes is that they contain variable amounts of carbon, coming from two sources:
. impossibility of a drastic separation of coal from waste in the washing process.
. presence of coal incrustations in the wastes during their formation.
This content of coal makes wastes a very important material to be used as feedstock in processes where a fuel consumption is necessary, since they may con-

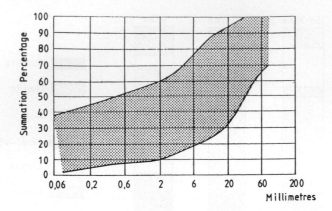

FIG. 3 - PARTICLE SIZE ANALYSIS OF U.K. MINESTONE

TABLE II

Mean Size Distribution of washery wastes of West Germany

Grain sizes (mm)	Proportion of size range % (by weight)	
+ 63	13,1	
63 - 20	29,3	71,0
20 - 10	28,6	
10 - 6,3	4,3	
6,3- 2	16,2	24,0
2 - 0,75	3,5	
0,75 -0,5	1,2	
-0,5	3,8	5,0

tribute to a part the fuel consumptions thus leading to an important energy saving as well to taking advantage of an energy source which was being squardered.

In Table III the high calorific values of different types of coal waste, corresponding to various coal producing countries are reported. Three facts may be observed:

a) The calorific values increase when going from coarse particles to fines, thus coinciding with carbon contents reported in Table I.

b) Generally, the calorific values for each type of coal waste are lower in Eastern countries than in Western countries or Australia.

c) For a given type of coal waste there exist appreciable differences among values for each country, coincidental with the dispersion observed for carbon content (Table I) and fuel material (Table IV) (8).

TABLE III

Calorific value of coal wastes (kJ/kg)

Country	150–10 mm	10–1 mm	–1 mm
F.R.Germany	2000	4000	6000
Australia	10350		18360
Czechoslovakia	3100	4300	6700
Spain	2200	3500	5500
U.K.	5000		8000
USSR	4400		6600
France	1700	–	5000

TABLE IV

Fuel material content

Country	150–10 mm	10–1 mm	–1 mm
F.R. Germany	4–8	11–12	15–22
Belgium	12	23	23
Czechoslovakia	15,6	22	35
Spain	5	11	15
France	3	5	12
Poland	10	13	25
U.K.	20		30
USSR	20,3	26	29

FUEL MATERIAL CONTENT

In Table IV the contents of fuel material for the different types of coal washery wastes from different coal producing countries are reported. Conclusions similar to those one for the calorific value can be reached.

THERMAL BEHAVIOUR

This is an aspect to be taken into account when considering the use of coal wastes as a feedstock for process in which a thermal treatment must be carried out.

Thermal analysis

The behaviour of coal waste ramples from Bryankowskaya washery in the USSR is shown in TGA and DTA curves of Fig. 4. A rapid loss in weight commenced at 400° C; the first exothermic effect appeared at the same temperature. The second exotermic effect was observed at 550° C. The first one was due to sulfur and the second one to carbon. Every sample displayed an exothermic effect at 880–890° C, associated with the breakdown of crystal hydrates; this can explain the swelling behaviour observed at temperatures above 900°C

118

(9). Similar curves, with logical variatious in neatness of effects depending on sulfur and/or carbon contents are obtained with coal wastes from other countries (10).

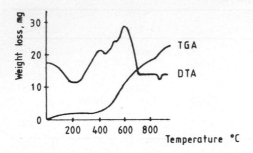

FIG.4 - DTA AND TGA CURVES FOR COAL TAILINGS OF BRYANKOVSKAYA

Heating microscopy

From studies by this technique it is deduced that succesive expansions and contractions suffered by samples made from coal wastes during their baking process, especially in the low and medium temperature zones, are normal for this type of materials. No sudden melting deformation processes have been detected (8); this fact being important for some applications of coal waste.

OTHER PROPERTIES

In this section some characteristics of coal wastes are described that, although in most cases do not per se determine a decision about whether they are or not useful for a particular application, may provide a deeper knowledge about these materials and serve as a complement to the rest of properties described.

Moisture content

In Tabla V the moisture contents of several types of coal washery wastes from different countries are reported. It may be observed that the moisture contents are very similar for the different samples, suggesting that washing techniques do not vary very much from one country to another.

Ash

Ash content values for coal wastes from different countries are given in Table VI. For a given type of coal waste, the percentages are higher for Western than for Eastern countries of Europe and Australia. This agrees with carbon contents (Table I) and calorific values (Table III) (8).

TABLE V

Moisture content (%)

Country	Washery wastes 150–10 mm	10–1 mm	– 1 mm [a]
F.R. Germany	6	10	25
Belgium	–	–	30
Czechoslovakia	3–6	8–11	20–25
France	–	–	25–35
Poland	4	10	20
Spain	5	9–10	25
UK	12		25
USSR	5–10		25

[a]After dehydratation with vacuum filters, etc..

TABLE VI

Ashes

Country	Washery wastes 150–10 mm	10–1 mm	– 1 mm
F.R. Germany	85	85	65
Australia	65		43
Czechoslovakia	82–85	60–63	
France	88–90	86–88	75–80
Spain	87	85	75
UK	80	65	
USSR	85	70	

Bulk density and specific gravity

Depending on their type, the mean value for bulk density of coal wastes varies from 1.3 to 1.7 g/cm^3.

Specific gravities of coarse particles (+ 1 mm) vary between 2.4 and 2.6, the lowest values corresponding to those coal wastes with a highest carbon content.

Plasticity

Fig. 5 fives values obtained in different samples of coal wastes from the United Kingdom. It may be deduced that this material is the medium plasticity. In other cases, such as washery and tip wastes from the North of Spain, it is deduced from the corresponding studies that they have low or no plasticity at all (11).

POSSIBILITIES OF USE

The mineralogical and chemical compositions of coal wastes, together with

120

mechanical properties, permit the establishment of the starting point of researches conducted towards finding new possibilities for coal wastes use.

Althoug, as we have previously seen, the composition of these materials
is variable (depending on their coalfield of origin), they contain in most cases a predominant clay fraction. They also contain a high proportion of silica
and alumina as well as coal spread all over their mass.

This point is a very important one, since it offers the advantage of supplying the whole amount or of at least a part of the energy needed in those
processes where coal wastes are used as feedstock and requiring a fuel consumption.

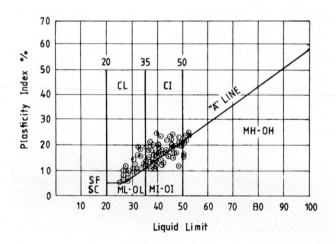

FIG. 5 – DISTRIBUTION OF LIQUID LIMIT VS - PLASTICITY INDEX FOR
U.K. MINESTONES AS EXCAVATED

Due to these facts and to the problems posed by coal wastes, the coal producing countries have been supporting researches for several years in order
to find industrial applications for these materials. From the experience of
the authors and from existing literature (6) (12) (19) different possible uses
for coal wastes are proposed and compiled en Fig. 6.

It is evident that some of these applications are considered merely from
a theoretical point of view it being necessary to carry out the corresponding
studies in order to ascertain their achievement in practice. Also, the domain
of application could be broadened from the results of these researches.

In the other hand, sin some processes coal wastes could be used alone, while in other cases it would be necessary to mix them with another feedstock in
different proportions according to the specific applications or to the required
characteristics.

121

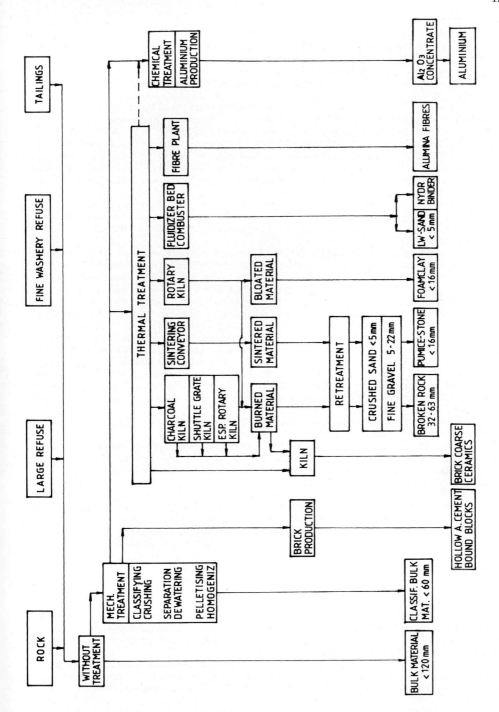

FIG 6 - DEVELOPMENT LINES OF MINESTONE PROCESSING AND UTILISATION

REFERENCES

1 J. González Cañibano, Los estériles del carbón. Definición y clasificación, Industria Minera, Octubre (1985) 45-53.
2 D. Leininger, J. González Cañibano, Características mineralógicas y químicas de los estériles del carbón de Europa, Energía, Enero-Febrero (1986) 85-89.
3 H. Knatz, H. Plogmann, L'utilisation des schistes houillers du point de vue matières premières, Journées d'information "Nouveaux procedés de valorisation du charbon", Luxembourg, September 26-28, 1979, Vol. 1, 307.
4 M. García, J. González Cañibano, Valorización de los estériles del carbón. La materia prima y sus utilizaciones, Energía, Nº 1, (1983) 65-72.
5 L. Anderson, The Pechiney H+ process and New South Wales coal washery refuse as a source of alumina, Department of Mineral Resources, New South Wales, June (1980).
6 D. Leininger, W. Erdmann, R. Köhling, R. Petry, Th. Schieder, Recent development in the utilization of Preparation Refuse, Symposium on the Utilization of Waste from Coal Mining and Preparation, Tatábánya, Hungary, October 23-28, 1983, Vol. V
7 A.K.M. Rainbow, D. Turnbull, Composition and characteristics of waste from coal mining and preparation, Symposium on the Utilization of Waste from Coal Mining and Preparation, Tatábánya, Hungary, October 23-28, 1983, Vol. I.
8 J. González Cañibano, D. Leininger, Propiedades físicas y mecánicas de los estériles del carbón de Europa, Energía, Mayo-Junio (1985) 115-125.
9 Y.M. Rubin, N.V. Gavrik, Y.I. Khvastukhin, A.P. Didenko, Combustion of cleaning-plant wastes, Coke and Chemistry, No. 12 (1977) 64.
10 J.A. Fernández Valcarce, J. González Cañibano, Coal refuse: raw material for the manufacture of ceramic products, 5th International Ceramic Congress on Research for Production, Rimini, Italy, September 24-27, 1983, 155-159.
11 J. González Cañibano, M. García, J.A. Fernández Valcarce, Empleo de los estériles del carbón en la construcción de terraplenes y rellenos, Symposium sobre Terraplenes, Pedraplenes y otros Rellenos, Madrid, Spain, Marzo 4-5 (1986).
12 J. González Cañibano, Aplicaciones de los estériles del carbón, Industria Minera, Nº 227 (1983) 5-12.
13 J. Gonzalez Cañibano, M. García, J.A. Fernández Valcarce, Utilización como materia prima y aprovechamiento de la energía de los residuos del carbón, VII Congreso Nacional de Química "Energías básicas y complementarias", Murcia, Spain, Octubre 17-19 (1985).
14 R. Chauvin, Les schistes houillers, sourçe de materiaux pour la construction et le genie civil, Journées d'information "Nouveaux procedés de valorisation du charbon", Luxembourg, September 26-28, 1979, Vol. 2.
15 J. Vejpustokova, J. González Cañibano, M. Líčka, Aplicaciones y características de los estériles del carbón en Checoslovaquia, Industria Minera "in press"
16 F. Knor, M. Kosina, P. Novotny, Utilization of waste from coal-mining and coal preparation by fluid bed firing, Symposium on the Utilization of Waste from Coal Mining and Preparation, Tatábánya, Hungary, October, 23-28, 1983, Vol. IV
17 V.R. Kler, N.J. Eremin, V.J. Panin, The coal wastes of coal-mining and coal-preparation as raw material base for producing aluminiun compounds, Symposium on the Utilization of Waste from Coal Mining and Preparation, Tatábánya, Hungary, October 23-28, 1983, Vol. V.
18 S. Rideraver, I. Stocker, Production of complex ferrosilico-aluminium alloy of aluminium-silicates containing combustibles, Symposium on the Utilization of Waste from Coal Mining and Preparation, Tatábánya, Hungary, October 23-28 1983, Vol. V.
19 M. Ya. Shpirt, Physicochemical principles and ecological problems of utilizing mining wastes and processing natural solid fuels, Coke and Chemistry, No. (1981) 103-107.

Reclamation, Treatment and Utilization of Coal Mining Wastes, edited by A.K.M. Rainbow 123
Elsevier Science Publishers B.V., Amsterdam, 1987 — Printed in The Netherlands

TECHNICAL APPROACHES AND RELATED POLICIES FOR LAND RECLAMATION AND TREATMENT IN CHINESE COAL MINES

YAN ZHICAI

Senior Engineer, Technical Committee & Consultancy, Ministry of Coal Industry, PRC

Summary
    This paper describes briefly the importance and the present status of land reclamation and renovation of subsided areas in Chinese coal mines. It concentrates on process and techniques of land reclamation; techniques for building houses on the reclaimed land and the results; process and techniques of backfilling of susided areas with fly ash from power station and experience with planting trees and crops in the reclaimed site. It also deals with investigation of approaches and policies for reclamation of subsided areas.

1. IMPORTANCE AND PRESENT STATUS OF RECLAMATION OF SUBSIDED AREAS IN COAL MINES

    Since the founding of new China, coal production has developed rapidly. It increased from 32 million tons to 850 million tons of coal in 1985, ranking the second in the world. Coal is a major energy source in China, which accounts for over 73% in primary energy consumption structure.

    With continuous expansion of mining area, extent of surface subsidence increases. And this problem is especially serious in the eastern plain area. Based on statistics of 40 mine areas in Huainan, Huaibei, Xuzhou, Kailuan, etc., by 1984, the cumulative subsided area was up to 1,356,000 mu* (or 9,030, 960,000 m^2); on the average, subside area per 10,000 tons of coal was 2.87 mu (or 1,869.5 m^2); the depth of subsided area was 0.67 - 18.5 m, or 5.69 m in average. In Xuzhou mine area, in average, 4.5 mu of land subsided for producing 10,000 tons of coal. Most of the subsided area collects water. Currently, there are 75 production brigades with an average cultivated land less than 0.5 mu (333 m^2) per capita in this mine area.

    As is well known, China possesses a large population and a little cultivated land. And the cultivated land now available is about 1.49 billion mu, or 1.5 mu per capita, which is much less 4.5 mu per capita in the world. As a result of coal extraction, vast land was disturbed, which intensified the contraditions between agriculture and industries. On one hand, the farmers lost their land which, in turn, affected production and welfare of farmers,

(* Note: Mu Chinese area unit 1 Mu = 666 m^2)

on the other hand, it made the coal mines very difficult to take over land for production and construction, or to move the villages to other places. The farmland and gardens were distrube and the environment of mine area was affected. People felt devastated, and the ecological balance was also distur- bed.

In recent years, the Chinese "Land Act", "Environmental Protection Act" and "Mineral Resources Act" have stipulated, "Lands disturbed by surface and underground coal mining shall be restored to suit local conditions by revege- tation, forestation and restoration of Landscape...". However, coal production cost is high and the selling price is low, and the coal producers can not afford to take liability to land reclamation and conservation. Therefore, at present, trails have been made in a few mine areas for comprehensive utili- zation of wastes and land reclamation, such as backfilling of mining wastes and fly ash from power plant, leveling and, development of breeding etc.

2. TECHNICAL APPROACHES FOR LAND RECLAMATION AND RENOVATION
2.1 Using wastes for land reclamation

The amount of cumulated wastes in Chinese coal mines have alredy reached over 1.2 billion tons, and each year over 100 mt of wastes are still produ- cing. In the future, in addition to extensive utilization of wastes, the min- ing wastes will mainly be used as materials for backfilling of subsided area, so that the disturbed lands can be partly restored, and less land be occupied by spoil heaps, and its effect on environmental pullution be eliminated.

2.1.1. Method for land reclamation by backfilling of wastes

There are three ways for land reclamation by backfilling of wastes, name- ly, land reclamation with newly disposed wastes; leveling of old waste heaps; and disposal of wastes over lands prior to their subsidence.

Backfilling of newly produced wastes means that spoils will no longer be piled up, and they are transported directly by trucks or railways to the subsided area and then leveled by bulldozer for revegetation. This is the most economic and rational method for reclamation, and it deserves recommen- dation. After leveling, the lands can be used for building houses or for reve- getation. If the disposal site is used for building, each layer of fill, not in excess of 0.4 m in thickness, should be compacted. Construction can be started before topsoil is covered. In general, a layer of topsoil, normally, 0.2 m or so thick, is sufficient for revegetation. If the disposal site is for planting crops, there is no strict requirements to compacting. However, the thickness of topsoil must be more than 0.5 m. If the disposal site is for planting tress, pits can be dug and backfilled with topsoil, and the site may or may not be covered with a layer of soil less than 0.2 m thick.

Leveling of existing spoil heaps is quite simple, and the requirements to grading depend upon the use of land. If the site is for building houses, dynamic compacting shall be used for treating foundations. A four storey building, a swimming pool and a playground have been built and used in Huaibei mine area for more than two years with satisfactory results. In general, leveling of existing spoil heaps is done to meet the requirements of environmental protection. When it reaches a certain contour, revegetation can be carried out. Of course, the land can be restored to such an extent that crops can be palnted. However, it is labour-consuming and sometimes, may not be feasible economically.

Disposal of wastes over land prior to subsidence is that wastes from development and initial stage of production are transported to the surface above the initial mining areas before occurrence of subsidence or in the process of subsidence. Based on prediction of surface subsidence by isopleth diagram, wastes are disposed to site where subsidence is expected. Topsoil is removed in advance and piled arround the site. When subsidence ceased topsoil is backfilled. The main purpose of this method is to use the restored land for establishing buildings for the enterprises, and for building new villages and towns for people who have moved here from other coal producing area, such that, coal producers might take over less lands and might extract the stagnant coal. The procedure for land reclamation by treating wastes, for example, is removing of topsoil--- loading and conveyance of wastes--- backfilling of subsided areas---leveling and compacting--- covering of topsoil.

Trucks, excavator and scrapers can be used for land reclamation. And hydraulic excavators or small sized dredgers are used for digging soils from water-logged subsided areas.

2.1.2. Land reclamation in some mine areas is shown in Table 1.

In Xuzhou mine area, 1,253 mu of land were restored, of which, 1.013 mu was for expansion of mine site and for other buildings. Xiaqiao Colliery of this area built a three-storey office building on the reclaimed land. No crack or damage was found in the past six years. In 1983, another house of 10,000 m^2 was built, which was a reinforced framed structure designed to meet antiseismic requirements.

In Huaibei mine area, except for building houses, the restored land was used to build a brick-making works with an annual capacity of 20 million bricks, which was put into operation for manyu years and no damage was found. Bulldozers were employed in Chazhuang Colliery in Feicheng to compact the layers in the waste backfilled area and a two-storey hospital was built. Currently, most of the backfilled lands are used to build single-storey buildings to accommodate people to be moved from other villages to the site. Each building has 3 to 6 rooms. Those buildings are arranged in rows. The ground in front or at the back of buildings are covered with a layer of 0.5m thick soil. Vegetables, trees are planted. They are growing well as if in cultivated land. And this is well received by the formers.

Practical experience has indicated that this method is simple in technology, and needs less equipment, which can be handled by the mining administrations in general. And experience with establishing buildings less than four storeies, establishment of various types of works and single-storey houses on the backfilled site is successful, which can be introduced in the future for providing new houses to accommodate people who moved from other village or town in order to extract the stagnant coal underneath the village.

The cost for land reclamation by backfilling wastes in coal mines in plain area is much less than that for taking over land. If coal, troilite, etc. can be recovered from wastes, the economic effectiveness would be even better.

TABLE 1

Name of mining Administrations	Land restored by backfilling of wastes		Use of reclaimed land
	Area Mu	Thickness of top soil layer, cm	
Xuzhou mine area	1,253	15 - 50	site for surface structures, multi- and signle- storey buildings, buildings for removal of villages
Datun mine area	418	15 -40	Buildings for removal of villages
Huaibei mine	643	15 - 40	Site for surface structures, multi- and single-storey buildings, brick-making works, buildings for removal of villages, swimming pool, and playground
Kailuan mine	250	15 - 50	Site for surface structures, single storey building, & planting trees

2.2 Land reclamation by useing fly ash from power plants

Coal ash from power plant is also one of major material sources for backfilling the subsided area. In Hauibei and Feicheng mine areas coal ash from pit-mouth power plants was used to reclaim 2,011 mu of land with good results by planting forrests, vegetables and crops or by building warehouses.

2.2.1. Method of using fly ash for land reclamation

This method is simple, economic and safe. The funds and necessary equipment for building fly ash stockyards can be listed in the capital construction programme, and coal mines can provide subsided area free of charge. The power plant is close to coal mines, in general, 10 to 15 km away from coal mines. By adding slurry transport pipeline to existing equipment of power plant, fly ash can be conveyed directly to subsided area for backfilling. Usually, there is an impermeable layer in subsided area. And water acumulated in the subsided area will not infiltrate into underground workings, so, it is safe. The procedure for land reclamation by backfilling fly ash is as the following: building ash stockyard--- hydraulic transportation of coal ash--- settlement and water drainage--- covering topsoil.

a, Building of ash stockyard. To increase the volume of ash stockyard built in the subsided area, usually, bulldozers, scrapers, trucks are used to transport the cultivated soil in the subsided area to the periphery of subsided area, the soil is then compacted into an embankment and an ash stockyard is thus formed. If the subsided area is full of water, dredgers are employed to dig the soil and put it around the subsided area for future use.

b, Hydraulic transportation of ash. The existing pumps of the power plant are applied to pump fly ash to stockyard via two slurry pipelines.

c, Settlement and water drainage. With constant deposition of ash in the stockyard, the water is drained from water outlet of stockyard into a drain ditch, and, finally, into river or lake. Since the quality of water is fairly good, (PH value is less than 9), it can be used for utility and for industrial purposes.

d, Covering topsoil. When the deposited layer of fly ash reaches the designed elevation, backfilling of fly ash will be stoped and water is drained. The site is then covered with a layer of topsoil, and its thickness, depending upon the need for vegetation, is generally 20 - 50 cm.

2.2.2. Lands reclaimed by backfilling of fly ash from power plant are shown in Table 2.

2.3. Land reclamation by backfilling mixture of lake mud, fluviatile mud and mining wastes

The subsided areas in coal mines near lake or river, can be restored by backfilling mixture of lake mud or fluviatile mud and wastes.

First of all, the spoils are dumped to the bottom of subsided area. Lake mud or fluviatile mud is dug by dredgers and dumped onto the wastes. When mud becomes dry, bulldozers are used to grade the land. And the soil is further ameliorated for revegetation.

To sum up, combination of disposal and utilization of mining wastes, fly ash, lake mud and fluviatile mud, etc. with reclamation of subsided area will eliminate hazards and bring in the following benefits: 1, the disturbed land can be restored onto useful land resources; 2, the subsided area can be used for building villages, industrial facilities, and less land to be taken over by mines; 3, economic results will be improved. Based on statistics, the cost for land reclamation is 50 to 70% of the cost for taking over the land; the cost for building ash stockyard in plain or in valley by the power plant is about 1.78 to 2.5 fold of that for building an ash stockyard in subsided areas.

TABLE 2

Type of ash stockyard	Land reclaimed		Uses and results
	Area Mu	Thickness of covering soil cm	
Ash stockyard in subsided area	2,011	20 - 40	The cotton was coming fine with many bolls. Each corn had 20 to 3 corn-cobs 20 - 30 cm long. The yield of soybean per Mu was 150 kg. Four single-storey house were built, and 120 Mu of forest was planted
stockyard in low-lying land	233	20 - 50	The yield of lint per mu was 50 kg, soybean 125 kg. Yield of wheat, corn, peanuts, sweet-potato, water melon, fruits was increased. Workshops and ware-houses were built.
Stockyard in valley flat	1,853	20 - 30	Yield of wheat per mu was 300 to 350 kg. Production of rape, peanuts, corn and vegetables per mu was similar to local production. Single-storey building and warehouse were built on the site.
Stockyard in plain	1,360	20 - 30	Yields of paddy rice and wheat per mu were 375 kg and 350 kg respectively.
Stockyard in valley	540	15 - 30	Yield of paddy rice per Mu was 250 to 350 kg, and yields of wheat, corn, beans, sweet potato cotton and vegetables were higher than those of ordinary cultivated land.

2.4 Comprehensive treatment of land and development of diversified economy

General speaking, whenever coal extraction takes place in a plain area, vast land will be distrubed, and only 15 - 20% of land can be restored by backfilling the disposed wastes, the rest of disturbed land will be backfilled by fly ash from pit-mouth power plant, mud from rivers and lakes and other wastes. However, the subsided areas still can not be fully stowed. Therefore, considerations should be given to other sources of stowing materials. An overall planning for land reclamation on the principle of using local materials to suit the local need is always necessary, so as to restore as much land as possible, and to develop diversified economy, and to arrange properly the life of farmers whose land are taken over by coal mines.

2.4.1 Grading of land

a, Changing paddy land into dry land or vice versa

If after coal extraction, a number of craks are found in subsided area with low settlement height and if heavy water leakage occurs even after grading, the paddy land should be changed into dry land, If the subsided area is flat with high ground water table, the dry land can be changed into paddy land according to the local situation, or to develop both paddy and dry lands alternately.

b,Establishing terraced fields

In shallow subsided areas, or in areas with low ground water-table, the site will be graded and built into terrace according to the local conditions. Economic allowance will be given to farmers so that they can make use of slack seasons in farming, or in the process of ploughing to build terraced fields step by step.

5,475 Mu of land in Luan, 204 Mu in Feicheng and 1,308 Mu in Pingdingshan mine areas were graded, the land in rest of mine areas was graded and changed into terraced fields, and cultivation was restored in some of the fields.

2.4.2 Digging deep land to backfil the shallow

Soil is taken from the deep area of subsided areas to backfill the shallow areas for growing crops, such as paddy rice. The deep area may be use as fish pond, and the shallow area is used to breed aquatic product. Trees and grasses are planted at the embankment. Irrigation system is also established. In a drought, water from fish pond can be pumped to irrigate the land, and water can be drained into river in a flood.

2.4.3 Developing aquatic products and breeding

If the surface land subsides over a large area, to a certain depth, and if the subsided area is water-logged, there is no need to restore it into farm lands, the water can be used for raising aquatic products, or plants, or for industrial or agricultural purposes. In some of the subsided areas in Huainan, Huaibei, Xuzhou, fish was raised at deeper part, and water plants, such as lotus and water chesnuts and reeds were raised at shallow part with good results. And the value of economic income in these areas was higher than that of agriculture production.

Many mine areas have already become industrial cities and towns. The old spoil heaps can be graded to a certain contour and trees can be plannted. In some subsided areas where it is not economic to restore the site into cultivated land, trees and grasses can be plannted to beautify the environment, so that the site will become a place for rest and entertainment and for increased income. For example, Chaili Colliery in Zaozhuang built a park in the subsided area, which ws welcome by the miners and inhabitants.

3. STUDY OF POLICIES FOR LAND RECLAMATION

Since purchase of subsided land, compensation of disturbed land, moving of villages, land reclamation and renovation involve the interests of state, enterprises and the farmers and are related to many departments, such as coal mining industry, municipal construction, electricity and power, water conservancy and irrigation, environmental protection, and state land bureau; since many problems have been remained unsolved in the history, and conditions are rather complicated, it is necessary to reach a common understanding, and to solve these problems in aspects of policy, act, organizations, funds, planning and design, management and research. A series of policies shall be made to provide criteria for establishing "Regulations for Coal Mine Land Reclamation and Renovation". The major contents will include the following points:

3.1 Land reclamation and renovation must be listed in mine planning and construction programme

Land reclamation in the mining territory is an inseparable part of mine planning programme. An overall planning and arrangement shall be made to specify the target and principles for land reclamation under the sponsorship and in close cooperation with local government and land administration departments. The planned items for land reclamation as specified in the plan or design shall be listed in the annual plan of the mine, and funds, equipment and labour organizations must be arranged accordingly.

3.2 Funds sources for land reclamation

The funds for land reclamation in the construction stage will be included in capital investment, and in the production period, in cost per ton of coal. Funds for disposal of spoil heaps and land reclamation in existing mines will be provided by the state, local government in the form of financial allowance.

3.3 Considerations in land reclamation

It is clearly specified that coal producers hold liability for land reclamation from beginning of coal production to the date when mines are abandoned, and it is implemented as a State's policy. The general target is that land reclamation shall combine with development of diversified economy so as to increase income, to improve and beautify the environment and coal mines, to enhance the working and living conditions, and the living standards of far-mers. In land reclamation, attention shall be given to the following four aspects: a, overall planning of disposal of spoils, fly ash and wastes and purification of mine water, shall combine with environmental protection, and it is not allowed to leave permanent spoil heaps; b, land reclamation shall be carried out in conjunction with land cultivation, development of diversi-fied economy, improvement of soil, land and harnessing of river and seaports; c, land reclamation is undertaken in conjunction with disposal of fly ash from power plant; d, land reclamation is conducted in conjunction with mine construction, moving of village for extracting the stagnant coal.

3.4 Intitiatives in land reclamation

Attention shall be attached, first of all, to the initiatives of local governments and the initiatives of enterprises. The State's "Regulation" speci-fy that the funds paid by coal producers for land reclamation shall be pro-perly used, so that the coal producers can get the land for building structure and for accommodating people who moved from other villages, and for improve-ment of mine area environment. Farmer's initiatives for land reclamation should also be promoted, so that the farmers can use the reclaimed land for estab-lishing revegetation and fish breeding, and to increase their income. It is stipulated that the reclaimed land will be exempted from taxes or the taxes can be reduced in certain period by the State.

3.5 Strengthening R & D for land reclamation

Land reclamation is a muti-disciplinary problem which involves many profes-sion, a number of research projects sponsored by the State of local government should be undertaken by related departments, depending upon their importance. The projects, which should be studied at the present time, are: building long-span structure on the land reclaimed by backfilling of wastes, breeding of aquatic product, thickness of topsoil to be recovered for vegetation and etc.. The projects, which should be studied step by step in the future are improve-ment of soil, analysis and treatment of toxic elements, effect of land recla-mation on quality of underground water, growing seedlings of tress, etc.

Reclamation, Treatment and Utilization of Coal Mining Wastes, edited by A.K.M. Rainbow
Elsevier Science Publishers B.V., Amsterdam, 1987 — Printed in The Netherlands

COMPACTION CONTROL AND TESTING OF A COLLIERY SPOIL

FOR LANDFILL SITES

Dr. W.M. KIRKPATRICK[1] and I.P. WEBBER[2]

(1) Kirkpatrick Geotech, 7-15 Dean Bank Lane, Edinburgh EH3 5BS

(2) Kirkpatrick Geotech. Now with Ove Arup & Partners

SUMMARY
     The paper describes the compaction, field testing and quality
control for a site at Bathgate, West Lothian in which a low lying
area was infilled with colliery spoil from neighbouring coal
bings. The area of the site ran to about 44 hectares and over
350,000 m³ of fill was placed.
     In view of the intended use of the site for housing develop-
ments stringent acceptance standards were necessarily imposed on
the structural quality and combustible potential of the fill.
     The paper is concerned with the site testing and quality
control, much of which was introduced for the special needs of the
materials and the site. Descriptions of these test methods are
given and discussed along with the relevance of current codes of
practice on testing of colliery spoil.

INTRODUCTION

     Increasing use has been made of colliery spoil as an in-
expensive source of bulk fill for civil engineering works and in
so doing has allowed large areas of land previously derelict due
to its occupancy by colliery spoil tips to be reclaimed. The
Little Boghead/Easton Bing project restored two previously
derelict sites in Bathgate by removing spoil from Easton Bings to
improve the site at Little Boghead.

     Before restoration the Little Boghead site consisted of 44
hectares of low lying land that was subjected to periodical
flooding. The ground conditions within the site generally consisted of peat up to 6 m thick overlying glacial till. The
restoration project consisted of removing the peat and replacing
it with colliery spoil from Easton Bing, and re-landscaping the
area to prevent future flooding. The details of the development
are described in the paper by Hart and Couper and Stephenson[1]
contained in these proceedings.

     Kirkpatrick Geotech is the Geotechnical Engineering arm of
J.A. Kirkpatrick & Partners, Consulting Civil Engineers who were
employed by the Scottish Development Agency in an advisory
capacity on the infilling at Little Boghead and to provide

independent engineering reports on the site for the National House Building Council.

This paper describes the special requirements of the contract to control the materials to allow the proposed use of the site for housing to be realised. These requirements relate to the nature of the fill materials and the nature of the site.

INITIAL TESTING

The properties of the colliery spoil at Easton Bing have been investigated over a long period of time. The Transport and Road Research Laboratory Publication LR125[2] reported on the properties of burnt colliery spoil taken from Easton.

In the early 1980's, two trial embankments were constructed at Little Boghead, using a method specification for compaction similar to the Specification for Roads and Bridge Works[3]. Monitoring of these embankments after additional dead load showed that the colliery spoil from Easton Bing was capable of providing a fill with good structural properties. Details of the trial embankment are presented by Hart, Couper and Stephenson[1].

In 1984 site investigations were carried out to determine the quantities of burnt and unburnt colliery spoil within the bing and further assess their suitability for use as fill material.

These investigations consisted of trail pits and boreholes taken within the two bings. A full range of standard laboratory tests were carried out including moisture content, specific gravity, particle size distribution, compaction and California bearing ratio tests. Chemical tests for pH and sulphate content were also carried out along with tests for coal content and combustible characteristics of the spoil.

Moisture contents of the spoil ranged from 6% to 28% with a mean of 14% compared to optimum moisture contents ranging between 7% and 15% with a mean of 11%. However all but one of the samples tested were wet of optimum (by an average 3%). Specific gravities of the unburnt spoil ranged from 1.94 to 2.35 averaging 2.21, whilst values for the burnt spoil varied between 2.45 and 2.57. Particle size distributions of the unburnt spoil showed it to be well graded and predominantly granular (Figure 1). Moisture condition value (MCV) tests on the material showed it to be an "all weather" material. Coal contents were measured at less than 10% with loss on ignition of less than 30% for the unburnt

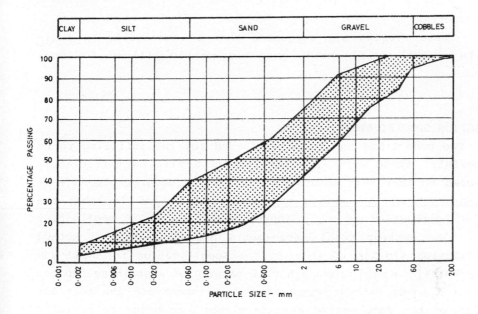

Figure 1. Particle Size Distribution Range: Unburnt Spoil Easton Bing Bathgate

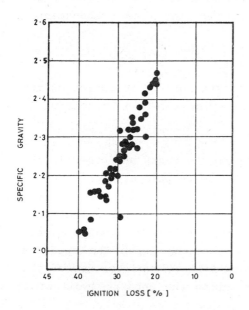

Figure 2. Relationship Between Specific Gravity and Carbonaceous Content
(after Rainbow 1985)

bing material.

These results all suggested that the materials in their in-situ states would provide good compactable fill of low coal content.

COMPACTION SPECIFICATION

Two conditions had to be satisfied for the infilling to be successful. These were (a) that completed fill be sufficiently compacted so that settlements in low rise structures placed near the surface would be within tolerable limits, and (b) that the risk of combustion of the unburnt spoil would be negligable.

The earlier trial embankments had shown that a method specification similar to that in the Specification for Road and Bridge Works[3] could produce a fill capable of sustaining typical foundation loads without significant settlement and the contract was let on the same method specification. In order to minimise the risks of combustion of the fill, a further performance specification limiting the air voids within the compacted unburnt spoil to 10%, was adopted. This was similar to other specifications used for compacting minestone in road embankments.

CONTROL TESTING

Control testing consisted of testing for the selection of the material at the borrow pit and the testing involved in control of the placement at the fill site.

(a) Material Selection. In addition to the tests performed in the preconstruction site investigation MCV tests were also employed at the borrow pit during the execution of the works to control material coming on to the fill site. Although a lower limit MCV of 8.5 was specified for suitability in practice much higher values were always obtained. This, together with the apparent continuing high quality of the placed fill allowed the use of the MCV test to be discontinued in selecting the unburnt spoil at a fairly early stage of the contract.

(b) Placement Control. The two conditions recognised in the previous section as being central to the successful development of the site referred firstly to the structural quality and secondly to the risk of combustion of the fill. Tests adopted to control the placing of the fill therefore should have relevance to these conditions and if they are to be useful in control they must be capable of providing reliable results quickly in order that steps can be taken to provide remedial action where it is seen that placed material is not up to standard.

In practice the structural quality was measured on the basis of the state of compaction and on the strength/compressibility of the material placed.

States of compaction were measured on site using sand replacement tests (BS 1377 (1975)[4] test 15B) to estimate in-situ dry density. This density was compared with the dry density in a one point laboratory compaction test on the same material at the same moisture content, using the 4.5 kg hammer (Test No. 13, Ref. 4).

These operations yielded a relative density for the site material as a percentage of that from the laboratory compaction test.

The calculation of dry density is also dependant on moisture content. This quantity measured using the BS 1377 oven methods would normally be expected to take more than a working day to be obtained. However to ensure good control in the field it was necessary to obtain a rapid measure of moisture content. Trials were carried out using both a "rapid" type moisture meter and using a microwave oven. The "Rapid" moisture meter gave variable and inaccurate results and was discarded. Initial trials with the microwave oven tended to cause the unburnt spoil to ignite. Even when used at full power for short periods, gases were released from the spoil. However on low power settings, using five minute cycles of power and cooling, repeatable and accurate results were obtainable within 40 minutes.

Since high relative compactions in themselves do not guarantee adequate structural quality further tests to measure this property more specifically were considered necessary. An in-situ loading test of some type was considered desirable for this purpose but since it was necessary to measure the properties of the material within the whole layer of placement and also so as results would not be unduly influenced by the presence of larger particles (> 20 mm) a plate bearing test was adopted in preference to a penetration test or a small scale bearing test such as the in-situ CBR.

The bearing test equipment, purchased commercially, was selected with a 300 mm diameter loading plate, this being of the same order as the thickness of the deposited layer. The reaction for the loading jack was provided by a loaded tipper truck which had a special bracket welded to its chassis for the purpose.

The general conditions of the test and procedures were as recommended in ANSI/ASTM Test, Ref. D1194/72[5]. The loading was taken up to give a maximum contact pressure of 200 kN/m² which was almost three times the maximum loading pressure likely to be applied by foundations in the development. The maximum was reached in two increments allowing the vertical deflections to reduce to zero in each stage. The granular nature of the fill was confirmed by the rapidity in which settlements were completed. With the two stage procedure the loading tests could be set up and completed within about one hour, so from the time aspect this was suitable for control.

Regarding the risk of combustion of the fill the standard required air content of the placed fill to be measured. Information from the sand replacement tests could be used for this purpose but in addition to density estimations of specific gravity and moisture content were required.

Guidance in testing colliery spoil is given in the National Coal Board Technical Memorandum - "Application of BS 1377 (1967) to the Testing Colliery Spoil" dated 1971[6] to be read in conjunction with BS 1377 (1967)[7]. As has been stated, to be able to use the results to control placement they must be repeatable and capable of being obtained quickly. With regard to the measurement of specific gravity neither of these requirements were met by the method recommended in the National Coal Board Memorandum[6]. Specific gravities of unburnt spoil are very dependant on coal contents, and Figure 2 shows the reduction in specific gravity with increased carbonaceous content for a colliery spoil in South Wales[8]. For the small quantity of material recommended in the NCB/BS 1377[6] test, it is very difficult to obtain a truly representative sample and small changes in coal content between two samples can give appreciably different specific gravities.

In an attempt to improve the matter of repeatability specific gravities were also measured by a method similar to that given by Lambe[9]. In this method a dried ground sample of spoil weighing approximately 500 gm (compared to 50 gm in the NCB/BS 1377 method) is boiled in de-aired water and then subjected to partial vacuum to remove entrapped air. This latter step is repeated until the volume of the spoil - water mixture is found to remain constant. After this the volumes are made up and the masses measured as in other methods.

A comparison of 200 tests using both methods showed that specific gravities of the unburnt spoil from the above modified method were generally repeatable and were consistently slightly higher than those obtained by the NCB/BS 1377 method. Results from the modified method could be obtained in about 3 hours whereas those from the NCB/BS 1377 method took about 72 hours.

After this trial period apart from occasional checks every 20 tests or so using the NCB/BS method, specific gravities were measured exclusively using the modified method. The slightly higher values given were accepted since they resulted in slightly more conservative estimated values of air voids.

Specific gravity values between 1.90 and 2.66 were obtained on the unburnt spoil and in view of this wide range and the sensitivity of the air voids to the value of specific gravity, the specific gravity of the spoil in every sand replacement test was measured.

During the progress of the contract it became apparent that the specified air voids of less than 10% could be obtained provided that the moisture content of the unburnt spoil was above 10% and that field compaction of 95% or over of the laboratory compaction was achieved. On the basis of this observation therefore, using sand replacement tests and microwave oven drying, it was possible to pass or condemn an area of fill within $1\frac{1}{2}$ hours of the material being placed. Further, using the modified method of measuring specific gravities it was possible to have a full set of results including air voids, on the same day as the material was compacted.

With regard to the plate bearing tests it was found in the early stages that the structural qualities of the compacted fill were very good and that other criteria such as relative compaction and air voids as mentioned above were more critical in site control. As a result plate loading tests were discontinued as site control for placement of unburnt spoil on a regular basis. Occasional further plate bearing tests were conducted however to provide continuing evidence of the quality of the material. The plate bearing tests on the whole showed settlements in compacted unburnt and burnt spoil to be less than 2 to 3 mm at pressures of 200 kN/m².

As has been stated above the basis of site placement control was reduced to two quantities, namely relative compaction and percentage air voids. For each placement point on the site examined a group of tests had to be performed to provide these quantities. The group of tests consisted of the in-situ sand replacement test, the one point laboratory compaction test, the moisture content test and the specific gravity test.

During the contract approximately 900 location points were examined with the equivalent number of test groups being performed, resulting in approximately 1 test group per 400 m³ of fill placed. The laboratory staffed with two technicians was capable of an output of 4 to 6 test groups per day and the above testing intensity could be met relatively comfortably.

OTHER RELEVANT FEATURES OF THE SPOIL

Prior to commencing site works several characteristics common to many colliery spoils were identified, investigated, and where necessary solutions were adopted within the design to overcome potential problems.

a) Swelling. Swelling is mainly caused by water uptake of the clay minerals within the spoil. This is particularly a problem with montomorillonitic spoils and to a lesser extent in illitic and illitic/vermiculitic spoils. BS 6543: 1985 considers swelling a possibility if unburnt colliery spoil is placed when excessively dry and then subsequently becomes wet. It recommends placing the material close to its optimum moisture content. The dominant clay mineral in the Scottish coal fields is known to be Kaolinite (Rodin[10]).

Scanning electron microscope and X-ray defraction tests on samples of the unburnt spoil showed this mineralogy to be the case and that swelling clay minerals were not identified. Swelling tests were also performed on the above samples. These showed no significant volume changes occurring. No precautions against swelling were therefore considered necessary.

The spoil however proved to be frost susceptible as was shown by the heaves that occurred during frost susceptibility tests.

b) <u>Aggressive Chemical Contents</u>. High sulphate contents can result from the oxidation of iron pyrites to form soluble sulphates. The susceptibility of concrete to sulphate attack is well documented. Sulphate contents ranging up to 2.1% by dry weight have been recorded on burnt spoil taken from Easton, and this puts the material as high as Class 5 according to BRE digest 250[11] although the burnt spoil tested would more characteristically be placed in Class 3.

c) <u>Combustion</u>. Unburnt colliery spoil found at Easton is combustible, and it was therefore necessary to isolate the fill from any source of combustion. The Inter-Departmental Committee on the Re-Development of Contaminated Land (ICRCL)[12] suggests that a possible solution is "to cover the combustible material from the effects of a severe surface fire " ... and ... " it is essential to lay all services to the site in inert material". The Fire Research Station suggests that an appropriate depth for such a layer would be 1 m.

The solution of providing a layer of inert, non-frost susceptible, non-combustible fill over the colliery spoil as adopted to solve all three problems a, b and c. The top 400 mm of colliery spoil was limited to only well burnt material and this in turn was capped by a further 600 mm of glacial till obtained from within the site. Topsoil 600 mm thick was spread on the glacial till.

This solution meant that no spoil was within 1 m of the surface and so should not be affected by frost in United Kingdom conditions. Furthermore foundations for the proposed housing would be within the glacial till cover and not directly in contact with the colliery spoil.

To check the protection given by the 1 m cover to the unburnt spoil against sources of heating a large bonfire was lit on the surface of the till. Temperature monitoring beneath the fire showed no significant rise in temperature at the surface of the unburnt spoil.

Spontaneous combustion is a question that is often considered relevant to the use of colliery spoil. However, current British Standards state that spontaneous combustion can be discounted if unburnt spoil is adequately compacted.

VARIABILITY OF SPOIL

The materials encountered on opening the bings proved to be much more variable than had been indicated in the preconstruction site investigation. It was found that the bings had been partially burnt out, but the burning had occurred within vertical "chimneys" and so within horizontal planes the material changed rapidly between burnt, unburnt and partially burnt spoil. This made material selection very difficult.

Moisture contents within the spoil were frequently well below optimum and it was necessary to "wet" the material throughout the Summer of 1985 in order to obtain satisfactory compaction and air voids.

Also significant deposits of spoil were found with coal contents of up to 20%. Whilst most of these materials were rejected because of the high coal contents, it was inevitable that some material was placed at Little Boghead with coal contents in excess of 10%.

The site investigation suggested that the material to be used from Easton Bing was typical of the spoils used in other reported civil engineering works. Testing of the spoils during site works did however show significant differences. When dealing with colliery spoils it should be noted that their composition has changed over the years of mining. As explained by Glover[13], colliery spoil was virtually pure coal of small particle size in the early days of mining. Later, mechanisation caused the particle size to be increased and the coal content to be decreased. The trend since then has been for colliery spoil to become generally finer, lower in coal content and with a higher moisture content. Since the 1950's all materials raised from coal mines have gone through coal preparation plant which has also significantly increased the average moisture content of the spoil.

Easton Bings were old tips which were in existence prior to 1913. It therefore follows that the spoil will be coarser, have high coal content and lower moisture content than more recent spoils. It is therefore questionable that the properties of colliery spoil reported elsewhere are applicable to the materials found at Easton and when planning uses of colliery spoil care must be taken in using information published for other bings.

CONCLUSIONS

Unburnt colliery spoil is increasingly used as a bulk fill for civil engineering works. As exampled by the proceedings of this symposium large amount of literature and experience now exists on the successful use of colliery spoil as a fill material. However colliery spoil exists in a large variety of forms dependant on age, mining method, processing and location. These factors effect the coal contents, particle sizes, moisture content and mineralogy of the resulting spoil. Whilst it is reasonable to assume that when compacted to the requirements of the Specification for Roads and Bridge Works (including recommendations on moisture contents) a colliery spoil will provide acceptable structural fill, it is less reasonable to quote generalisations about other properties such as swelling, frost susceptability, liability to combustion and chemical composition particularly as the majority of published information refers to spoils less than 25 years old.

Current British Standards (BS 6543[14], BS 6031[15]) make no distinctions between minestones regardless of their age or origin and care must be exercised in applying the general recommendations contained in these codes.

This paper has outlined some of the problems encountered when using spoil from an old bing for a landfill project, and has attempted to highlight some of the differences between what was found and what has been reported elsewhere. It describes special test methods developed to achieve a rapid appraisal of the fill after compaction and describes special measures required due to the use of colliery spoil.

With regard to site control and assessment the variability of material in the present works has shown that the frequency of testing has been generally necessary in view of the intended use of the site. For similar developments using bings of similar background and history a comparable level of testing is considered justified.

ACKNOWLEDGEMENTS

The Authors are grateful to the Scottish Development Agency for permission to publish this paper and to Lothian Regional Council for their co-operation during the site works.

REFERENCES

1. I.M. Hart, A.S. Couper and D. Stephenson, Building on Bogland, 2nd Symposium on the Reclamation and Utilisation of Coal Mining Wastes (1987).
2. C.K. Fraser and J.R. Lake, A Laboratory Investigation of the Physical and Chemical Properties of Burnt Shale. Ministry of Transport Road Research Laboratory Report LR125 (1967).
3. Department of Transport, Specification for Roads and Bridge Works HMSO (1976).
4. British Standards Institute BS 1377 (1975). Methods of Testing Soils for Civil Engineering Purposes.
5. ANSI/ASTM Ref. D1194/72. The Bearing Capacity of Soil for Static Load on Spread Foundations.
6. National Coal Board, Application of BS 1377 (1967) to the Testing of Colliery Spoil, Technical Memorandum (1971).
7. British Standard Institute BS 1377 (1967), Methods of Testing Soils for Civil Engineering Purposes.
8. A.K.M. Rainbow, Private Correspondence (1985).
9. T.W. Lambe, Soil Testing for Engineers, Wiley (1951).
10. S. Rodin, Review of Research on Properties of Spoil Tip Materials, NCB Headquarters Research Project, Wimpey Laboratories Report S7303 (1972).
11. Building Research Establishment Digest 250, Concrete in Sulphate Bearing Soils and Groundwaters (1981).
12. Inter-Departmental Committee on the Re-development of Contaminated Land, Notes on the Fire Hazard of Contaminated Land, BRE (1984).
13. H.G. Glover, Environmental Effects of Coal Mining Waste Utilisation, 1st Symposium in the Reclamation and Utilisation of Coal Mining Wastes, (1984).
14. British Standards Institute BS 6543 (1985), Use of Industrial By-Products and Waste Materials in Building and Civil Engineering.
15. British Standards Institute BS 6031 (1981) Earthworks.

Reclamation, Treatment and Utilization of Coal Mining Wastes, edited by A.K.M. Rainbow   145
Elsevier Science Publishers B.V., Amsterdam, 1987 — Printed in The Netherlands

HOUSE BUILDING ON BOG LAND

I McD Hart BSc CEng MICE, Land Engineering, Scottish Development Agency

A S Couper, Dip Arch, RIBA, Dip LD   Landscape Development Unit,
D C Stephenson BSc (Hons) CEng MICE Lothian Regional Council

SYNOPSIS

The paper describes a project financed by the Scottish Development Agency
(£1.65M) and designed by Lothian Regional Council which is now almost
complete. It utilised some 500,000 tonnes of minestone to create 20ha of
new land for private housing (750 units) plus a further 48ha for
agriculture, forestry, and amenity space.

A low lying area of bog land subject to flooding and underlain by a mixture
of poor soils and peat was reclaimed by removing the substandard soils and
subsequently replacing and upfilling them using minestone from an adjacent
colliery tip two kilometres away.

As a result of discussions with The National House Building Council, British
Coal and other interested institutions a comprehensive and rigorous
specification was adopted for selecting, placing and compacting the
minestone and finishing layers. Continuous monitoring has taken place
during the contract resulting in some refinement to the specification and
achieving a more consistent standard of compaction.

It is hoped that the site will be marketed in the next few months and that
development on the site will commence in 1988.

1. INTRODUCTION

The Bathgate area of Lothian Region (West Lothian District) set in
the centre of Scotland almost halfway between Edinburgh and Glasgow has,
in recent years seen a very serious decline in employment, culminating in
the closure of the British Leyland works.

Industrial growth in the area in the 19th century was mainly based on
the extractive industries of coal, oil shale and fire clay. The
population increased rapidly and though past its peak, over 30 collieries
were working in 1951. Shale mining and the associated refineries had all
closed by 1962 and with the closure of Polkemmet in 1986 all deep mine
coal production has now ceased.

In 1962 the government of the day designated the area around the
village of Livingston (8Km from Bathgate) to be Scotland's 3rd New Town
and at the same time encouraged the then British Motor Corporation to
establish new works in the Bathgate area.

For a town of its size and being the administrative centre for West
Lothian District, Bathgate is relatively unknown outside Scotland.
Within Scotland it tends to be associated with the decline of the shale
oil and coal industries and past difficulties at the British Leyland
works. The town centre, however, is relatively bright and pleasing, but
to the West and South of the town the coal industry has left its visible
legacy of major surface dereliction coupled with unstable and uncertain
underground conditions. Bathgate's main asset is its geographical
position, lying within 2km of the M8 motorway and 20km of Edinburgh
airport. It was recognised that one of the main problems with the area
was the pervading atmosphere of general dereliction. The value of land
renewal was therefore recognised by all parties and incorporated within
the Lothian Structure Plan. Ref (1).

From a land use planning aspect there was a need to initiate and
promote the development of these derelict areas for private residential
development. The largest single site in the area was called Little
Boghead, an unused low lying bog area underlain with poor soils including
peat. Despite its poor condition this site was the only major area zoned
for housing within the Bathgate Local Plan (Ref 2). By improving these
sites and by opening up the private housing market, it was the intention
to improve the position of Bathgate as a commuter town.

The concept of combining the removal and rehabilitation of the huge
Easton Colliery waste tip and using the material to improve the phyiscal
environment and stability of the nearby unused Little Boghead site for
future housing was considered to be the most effective way forward.

As well as poor ground conditions a major constraint to the
development at Little Boghead and in fact Bathgate generally was the Bog
Burn. A sluggish, almost flat and channel-less burn, it was originally
considered, in the 1960's to require major re-construction to overcome
the perennial problems of flash flooding of property in the town. The
scale of investment required always deterred progress and in consequence
new housing development steered away from the town centre and in
particular the Little Boghead site.

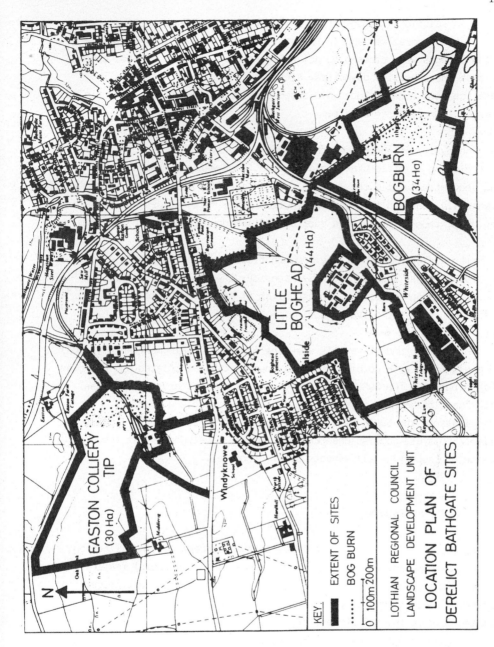

Plan 1 Showing the Three Sites

As a solution an imaginative low cost alternative was devised and developed on derelict land to the south of the town centre. A series of flood lagoons were created to act as reservoirs in times of peak storms so that the sensitive downstream areas would not be threatened. The derelict land area included a disused colliery, a domestic refuse tip and an iron foundry waste sand tip.

The hydraulic concept of the scheme was very simple, using constricting throttle pipes which caused flood water to back up the burn and sequentially fill the lagoons through overflow weirs. The water eventually subsided through small outlet pipes once the peak of the storm was past, thus maintaining a controlled flow downstream during a storm.

With the completion of the Bogburn flood lagoons in 1983 at a cost of £430,000 the way was open to fully consider the concept of the Easton Bing/Little Boghead project to establish if minestone could be used to infill bog land for private housing.

In order to undertake this work the Scottish Development Agency (SDA) in conjunction with Lothian Regional Council (LRC) initiated and promoted this proposal. The SDA is empowered amongst many other things to deal with the clearance of derelict land throughout Scotland. (Ref 3). In this instance the SDA's Land Engineering Division acting as Client/Employer, commissioned LRC's Landscape Development Unit (LDU) to design and supervise all aspects of the proposal.

2. ECONOMIC ASSESSMENT

To justify the economic viability of this concept of rehabilitation to the SDA's Board, it was necessary to undertake a financial feasibility study of the overall project including the housing development stage. To assist in this Coopers & Lybrand (Chartered Accountants) were commissioned to investigate and report on:-

1. The rehabilitation of the Easton Bing and Little Boghead as self-contained projects providing agricultural, forestry and amenity afteruses.

2. Rehabilitation through linking Easton Bing to Little Boghead, entailing the transfer of large quantities of minestone to create a substantial area suitable for private housing at Little Boghead with agriculture and forestry at Easton.

The objective was to provide an economic appraisal of the options and to recommend the most cost effective solution. To help in this assessment, LRC provided Coopers & Lybrand with all the financial and technical information connected with the physical aspects of the rehabilitation. Coopers & Lybrand also assessed the local private housing market with reference to planning, house builders reaction, supply and demand of housing in the area, and the potential land values. (Ref 4).

Planning permission was not a problem at Little Boghead as this was the only area designated in the local plan for private housing in the Greater Bathgate area. The future demand for housing was difficult to estimate but it was generally considered that the area at Little Boghead would satisfy all demand over a ten year period.

A number of house builders were approached who all expressed an interest in the development with most indicating a willingness to acquire parts of the site. Of those approached, (including some major nationals) all advised that the site was too large to be developed by one builder and should be marketed in packages.

Development land values were assessed and views sought from not only the builders but the Regional Council, the District Council and the S.D.A. The valuations ranged from £10,000 per acre to £30,000 per acre with the balance of opinion falling around £20,000 per acre. This latter figure was used for the projects financial assessment.

It should be noted that this figure (£20,000) assumed that at the completion of the rehabilitation works a platform suitable for traditional housing development would have been created, with no abnormal costs. No internal services i.e. road, sewers, water etc would be provided. These would be the responsibility of the individual developers.

None of the options considered produced a positive return with the net costs ranging from £215,000 to £825,000. (Fig 1)

FINANCIAL APPRAISAL - FIG NO. 1

| Option | EASTON BING | | | LITTLE BOGHEAD | | | | INCOME | | NET COST |
| | Afteruse (Area ha) | | Material Cut (Volume M³) | Afteruse Areas (Ha) | | Material Transported (Volume M³) | Cost £ | Additional Housing Area Created (ha) | Value of Land £ | £ |
	Agriculture	Town		Forestry Amenity	Housing					
1.	17.7	17.2	700,000	35.8	8.2*	-	825,000	-	-	825,000
2.	17.7	15.2	440,00	19.0	25.0	290,000	1,085,000	16.8	840,000	245,000
3.	17.7	15.2	440,000	15.4	28.6	385,000	1,235,000	20.4	1,020,000	215,000

Option 1   Self-contained schemes at both Easton Bing & Little Boghead.

Option 2   Linked scheme to prove housing land at Little Boghead

Option 3   As 2 but maximising development potential at Little Boghead

NOTES

1. All costs are at 1982 levels.

2. * The figure of 8.2ha. of housing land represents a part of Little Boghead site that could be released for development with no reclamation work required. The land was subsequently sold to Wimpey in 1984 for £20,000 per acre.

The study showed conclusively in this instance that rehabilitating derelict 'unbuildable' areas to maximise the development potential would achieve considerable savings - £610,000 over the more traditional reclamation afteruses.

In this economic exercise several factors were omitted e.g. maintenance costs for amenity space, long term return from forestry, and rating income from housing development. It was considered, however, to be sufficiently accurate to give a comparative guide and events have proven that the original costs were correct, i.e. adjacent land of a similar condition was sold to Wimpey for around £20,000/acre.

The SDA's Board in August 1984, approved option 3. (Fig 1).

3   NATIONAL HOUSE BUILDERS COUNCIL'S ROLE

When the ultimate landuse of the Little Boghead site had been determined and approved as housing, it was obvious that the backing of the National House Building Council (NHBC) was required. A meeting was therefore arranged at an early stage between the NHBC and the SDA to discuss their requirements.

The main points resulting from this meeting were:

1.   No minestone, either burnt or unburnt, could be used within 600mm of the underside of house foundations.

2.   The material above this level was to be generally inert and structurally adequate.

3.   As LRC owned the land and therefore had a vested interest in the project, an Independent Engineer had to be commissioned to oversee the infill operations and effectively guarantee that they had been carried out in a manner that would not be prejudicial to the future housing development.

The consequences of this meeting with the NHBC could have been very costly. It meant an additional source of infill material had to be found to satisfy these new requirements. This fortunately was resolved at little extra cost by discovering suitable clay material in sufficient quantities at a nearby site. However consultancy costs were greatly increased as a result of item 3 above.

J A Kirkpatrick and Partners were commissioned to carry out this role, i.e. to report on all selection, infilling and compaction operations connected with the future housing development, and thereafter to certify that the results would allow the completed site to be used for private housing.

Despite further discussions the NHBC were insistent that no material with a high sulphate content i.e. any minestone, could be placed within 600mm of the underside of the foundations. The view of both the SDA and LRC was that minestone could be used up to the underside of the foundations and that either sulphate resisting cement or an impermeable membrane could be used to protect the foundation concrete. Unfortunately, the NHBC were not prepared to alter their view and the proposals had to be amended accordingly.

4.   TRIAL EMBANKMENTS

In order to assess the suitability of upfilling the site with minestone two trial embankments were constructed.

Each embankment was some 30m x 30m in plan and located in an area known to represent some of the poorest ground conditions on site. Trial embankment No. 1 was constructed by excavating unsuitable material, consisting of topsoil, peat and sandy clay down to a firm strata, with infill material placed directly onto this formation. Trial embankment No. 2 was formed by placing a geogrid on the existing vegetation with the embankment constructed directly above. The sites were chosen so that the two embankments were situated on comparable ground. Nine settlement gauges were constructed on each embankment, four to monitor the base of the embankment and five to monitor the surface. (Fig 2)

The material in the trial embankments was a mixture of unburnt and burnt minestone extracted from the Easton tip compacted in layers of 200mm. Once the embankments were completed the settlement gauges were precisely levelled, initially on a daily basis, and thereafter on a monthly basis. After a period of six weeks a simulated house loading was placed on the embankments, consisting of concrete blocks giving a loading equivalent to that imposed by a two storey dwelling of some 50KN/m^2.

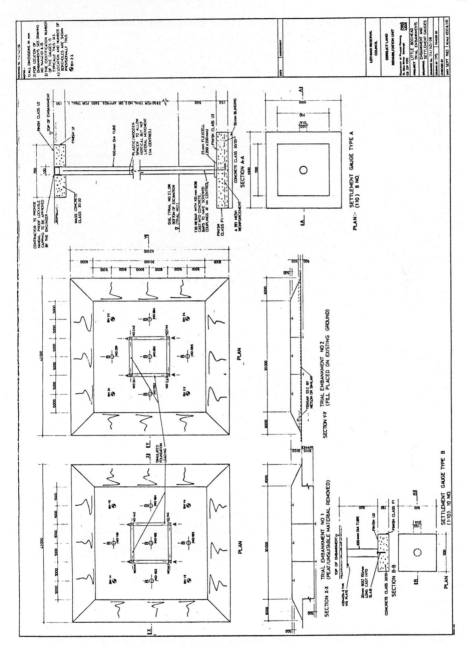

Fig No. 2 Construction of Trial Embankments

154

The settlements obtained from the trial embankments are illustrated
in Fig 3.

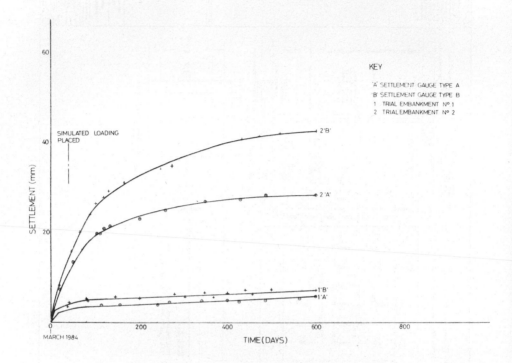

Fig No. 3 Trial Embankment Settlement Analysis

From Fig 3 it can be seen that the overall settlement on embankment
No. 1 up to Sept 1986 is 8mm with the majority of settlement taking place
within the first two months. The difference between the two readings
obviously records the settlement of the actual embankment rather than the
subgrade, and the consistent difference between the two settlement lines
in the later stages illustrates that the actual settlement of the
material in the embankment is minimal. The settlement curve for
embankment No. 2 shows a much greater overall settlement of 40mm.

From the results of the trial embankment contract it was concluded
that the only feasible course of action if the site was to be developed
for housing, was to remove the unsuitable material on the site and
replace this with properly compacted material. The trial embankment
contract also illustrated that the minestone was structurally capable of
being used for this operation.

5. MAIN CONTRACT

Make up of Infill

During the course of the trial contract and other site investigation work it was noted that the source of infill, the Easton bing, contained a great variety of materials ranging from well burnt through to unburnt. In order to restrict the variability of the material used and to limit the possible problems inherent in using different materials with variable properties, it was decided to use only well burnt or unburnt material. Also British Coal had suggested that if partially burnt minestone was used then swelling of the material might be a problem. Well burnt material could be readily identified as being bright red in colour, with the unburnt being black in its natural state unaffected by the heating found in parts of the bing.

Unburnt minestone when available was placed at the base of the excavation, on top of which a 400mm layer of burnt minestone was laid. This was capped with 600mm of structural fill. Finally, 600mm of the previously excavated unsuitable material was placed on top to protect the structural infill.

600mm Unsuitable
600mm Borrowpit
400mm Minimun Burnt Minestone
Infill of Burnt or Unburnt
    Minestone

INFILL CROSS-SECTION

Selection of Material at Easton

It was initially intended to carry out selection of the material at the bing using Moisture Condition Value apparatus. However, this proved to be unsatisfactory as the material was granular and selection was eventually carried out by visual inspection. This was rather time consuming and meant that a materials technician had to be on the bing throughout the contract ensuring that no partially burnt or other unsuitable material was loaded onto the vehicles.

Infill Criteria

It was agreed that to achieve a satisfactory fill on the site the following criteria should be adopted as a target for compaction.

1) Air voids ratio should be less than 10%.
2) Compaction for unburnt minestone 90% (4.5kg Rammer test to B.S. 1377).
3) Compaction for burnt minestone 95% (4.5kg Rammer test to B.S. 1377).
4) Coal Content should be less than 10%.
5) Moisture Condition Value should be greater than 8.5.
6) Settlement: by Plate Bearing test at
   100 KN/M^2 8mm max after 15 minutes
   200 KN/M^2 16mm max after 15 minutes

Excavation and Placing of Material at Little Boghead

Peat, sand, clay and topsoil was excavated from the site down to a reasonably firm formation level which was tested by a 'Mexe' cone penetrometer and by visual inspection. The formation level was then rolled where possible and infilled with minestone. The unsuitable material was either placed in temporary stockpiles or spread directly on top of the structural fill.

The minestone was placed in layers of 200mm thickness and rolled six times by a vibrating roller giving a force of 3170kg per metre width. Once the minestone was placed up to its required level, clay was placed on top in three layers of 200mm thickness giving an overall depth of 600mm. Each layer was given six passes of the vibrating roller as per the minestone. The excavated unsuitable material and topsoil was then placed in one unrolled layer of 600mm thickness to protect the structural fill.

Testing

A laboratory was set up on site to monitor target criteria. This involved carrying out the following tests:-

1) Moisture Content (Test 1 BS 1377) - (Ref 5)
2) Sand Replacement (Test 15B BS 1377)
3) Specific Gravity (Test 6(B)) B S 1377 and ASTM Test) (Ref 6)
4) 4.5Kg Rammer (Test 13 B S 1377)
5) Moisture Condition Value (TRRL Method)
6) Plate Bearing (ASTM Method)
7) Coal Content (NCB Procedure)
8) Organic Content (B S 1377 Test 8)

Site Control - Staffing

A full time site staff of six people were employed during the contract period. These staff were:-

1 Engineer's Representative
2 Assistant to the Engineer's Representative
3 Senior Laboratory Technician
4 Three Materials Testing Technicians

The Engineer's Representative on site was in overall charge of the management of the project generally supervising the progress of the works. His assistant was responsible for surveying all excavations, measurement and locating the position of all tests. One technician was employed at Easton bing and supervised the selection of infill material. The other three technicians carried out testing at Little Boghead and supervised the infilling operations. During the course of the works the site staff carried out approximately 900 sand replacement and associated tests, and an equivalent number of specific gravity tests. Some 200 Moisture Condition Value, and 100 Plate Bearing Tests were also undertaken.

The laboratory on site was equipped to carry out tests 1 to 6. Tests 7, 8 and other specialised tests were carried out under contract. The laboratory testing is described more fully in the following Paper "Compaction Control and Testing for Colliery Spoil for Landfill Sites". (Ref 7)

6. EXPERIENCES DURING THE MAIN CONTRACT

Selection of Material at Easton Bing

Selection as stated earlier was carried out on a visual basis at Easton bing. However, as it was very difficult at times to distinguish between coal and shale, unburnt minestone which had a high coal content could be inadvertently selected. Since the coal content test took time to be carried out, no satisfactory method was determined of ensuring material being excavated had a coal content less than the specified target. With experience of handling the material the problem was reduced but never entirely eliminated, and as the majority of the unburnt minestone was placed below the water table the coal content was considered to be less critical. Approximately 47% of the minestone used on the infilling works was unburnt.

A further problem with the selection was that the unsuitable material occurred in a random manner throughout the bing. The bing was still hot in places and burning had occurred in a series of vertical fires. Thus, it was possible for the quality of material to change rapidly throughout, to the extent that a face would appear to be satisfactory but a subsequent cut into it would then expose unsuitable material.

The amount of unsuitable material in the bing was greater than anticipated and to obtain the required infill of some 320,000m^3 an additional 340,000m^3 of unsuitable material had to be removed. The cost of moving this unsuitable minestone almost doubled the cost of obtaining the fill material.

Moisture Content

A site investigation carried out on the bing in 1984, indicated that the material on site was generally wetter than its optimum moisure content, showing similar results to previous studies. (Ref 8). However, during the initial stages of the contract it became apparent that the material once excavated and transported had a moisture content below its optimum value. Wetting was considered necessary and was carried out using two water bowsers which sprayed the material when initially deposited on site. This continued from May 1985 to November 1985.

Fire Test

The question of possible combustion in the infilled unburnt minestone at Little Boghead had been raised and after discussion with the Fire Research Establishment and the Unit of Fire Safety Engineering at Edinburgh University a combustion test was carried out in order to demonstrate the ability of the infill to withstand a typical house fire. An area representing the worst fire risk was selected where the compaction was lower than the target figure, and where unburnt minestone with a high coal content had been placed. A test fire, comprising 20 tonnes of timber, equivalent to the size of a typical house fire, was constructed on top of the structural fill layer. Thermocouples were installed in the infill layers to measure heat rise. The results of this test demonstrated that negligible heat was transmitted through the structural fill layer, and it was possible to conclude with some confidence that the black blaes could not be ignited by a surface fire of credible proportions.

Placing of Material at Little Boghead

Problems were experienced as the formation level was often below the water table. Although pumps were used to keep the excavation dry, water was at times drawn through the subgrade into the fill. This problem was exacerbated by the use of the vibrating roller and it was found that the only solution was often to leave the first layer of infill for a day to allow the excess water to drain out and be pumped away. A drainage layer of coarse burnt minestone 300mm thick was also placed on the subgrade to try to solve this problem.

The degree of compaction specified was found to be adequate under most circumstances, however occasionally it was necessary as a result of tests to carry out another two passes of the roller.

DISCUSSION

This contract illustrated that although old disused colliery waste tips can offer a source of cheap infill, the costs of extraction can be greatly increased as a result of the very variable nature of the material. Modern tips which have been properly compacted would not have the same degree of problems with partially burnt material.

With hindsight it would have been beneficial to have carried out a more extensive borehole investigation at Easton bing prior to commencing the contract. Although, it would still have been difficult to predict the amount of unsuitable material encountered it should have given a better indication of the likely problem.

The initial site investigation showed that the coal contents in the tip were generally 6% or less. A fuller investigation should have shown a truer picture indicating higher coal contents and indeed areas where the coal content was in excess of 20%.

Similarly, a better indication of the problems that were to be experienced due to the lack of moisture in the minestone might have been indicated. The initial site investigation consisted of thirteen boreholes and it is suggested that the size of the site investigation could have been doubled. This would have cost an extra £20,000 and whilst the results might have highlighted the problems to be encountered, it probably would not have changed the manner in which the contract was carried out. One advantage would obviously have been that the contract would have been drawn up with these points included in the tender documents and this would have limited the extra payments that had to be subsequently negotiated. Furthermore, if the amount of coal discovered during the contract had been forseen it might have been feasible to promote a coal recovery scheme on the site prior to or in conjunction with the main contract, and the laboratory on site might also have been equipped to carry out coal content tests.

As the majority of material condemned was partially burnt it would have been useful if more investigation had been undertaken to assess the properties to determine if some of this could have been used in the infilling operations.

The programming of the works was affected by the structural fill layer, the material being unworkable in adverse weather conditions. This meant that the unsuitable material excavated at Little Boghead during the works had to be temporarily stockpiled until areas had been capped with the structural fill. At times problems arose due to the amount of ground required for stockpiling.

The compaction specified in the contract documents was adequate apart from a small number of isolated cases where an extra two passes of the roller were required. The original compaction requirement was determined after experiences on the trial embankment contract and represents a 50% increase in the compactive effort over that required by the Department of Transport's Specification for Road and Bridge Works (1976).

Despite the problems outlined here it should be emphasised that the contract was carried out successfully, by Hewden (Contracts) Ltd, and without undue difficulty. The burnt minestone was found to be a very suitable fill material that could be worked in almost any weather conditions. The unburnt minestone was slightly more weather susceptible and became difficult to place in very wet conditions. It nevertheless, proved to be a suitable fill material.

7. CONCLUSIONS

The concept of rehabilitation adopted here combines environmental improvement with the creation of development land for housing.

The advantages of this method are as follows:-

1) Allows development of a site with poor ground conditions.

2) Reduces the pressure to develop greenfield sites.

3) Enhances the image of the area and improves the environment.

4)   Generates private investment within a site.

5)   Encourages outside investment within an area. In this case a
     local railway line (Edinburgh - Bathgate) was re-opened and part
     of the economic justification used was influx in excess of 700
     new private households to the area.

6)   Generates general financial benefits to Local Authorities.

The disadvantages are:

1)   An increase in construction costs including fees, royalties to
     British Coal and rates payable to the Local Authority.

2)   Risk in not satisfying physical specification i.e. compaction,
     combustion, chemical.

3)   Risk associated with the resale of land.

4)   An extended time period of construction nuisance.

8. SUMMARY

If the original engineering, planning and financial assessments are
encouraging then the increased risks of using minestone as bulk infill to
create housing land are worth taking because of the perceived advantages.

The need to satisfy NHBC requirements relating to inert structural
fill could have proven very costly in this project, and for the future it
would be very advantageous to have the restriction on placement of this
removed or reduced.

Ultimate success depends on the net cost of providing a suitable and
acceptable platform for housing development, and if the proposal is
economically viable then it is suggested that the route followed and
outlined in this paper is the way forward.

ACKNOWLEDGEMENTS

This paper represents the views of the Authors which are not
necessarily those of the Lothian Regional Council or Scottish Development
Agency. Gratitude is expressed to the Lothian Regional Council and
Scottish Development Agency for providing facilities and for allowing
publication of this paper. Thanks are also expressed to T P Simons the
Resident Engineer for his assistance and helpful comments.

162

REFERENCES:-

1) Lothian Regional "Structure Plan" (1985) As approved 1986

2) Bathgate Working Party "Economic Prospects and Employment in the Bathgate Area" (1984).

3) Scottish Development Agency Act (1975)

4) Coopers and Lybrand Associates - "Easton Bing/Little Boghead Bathgate Feasibility Study" (April 1983).

5) British Standards Institution - "BS 1377 (1975) Methods of Test for Soils for Civil Engineering Purposes".

6) American Society for Testing of Materials   D854 - 58 "Specific Gravity for Soils".

        D1194 (1972) "Bearing Capacity of Soil for Static Load on Spread Footings.

7) Dr. Wm Kirkpatrick and I P Webber - "Compaction Control and Testing of Colliery Spoil for Landfill Sites". (1986).

8) British Standards Institute - B.S. 6543 (1985) "Use of Industrial By Products and Waste Materials in Building and Civil Engineering".

Reclamation, Treatment and Utilization of Coal Mining Wastes, edited by A.K.M. Rainbow
Elsevier Science Publishers B.V., Amsterdam, 1987 — Printed in The Netherlands

COLLIERY SPOIL IN URBAN DEVELOPMENT

W. SLEEMAN B.A., MIHT., F.G.S.

Minestone Utilization Manager, Minestone Services,
British Coal, Philadelphia, Houghton-le-Spring, Tyne and Wear.

SUMMARY

For many years colliery spoil - the residual by-product of mining coal -
has been used to fill in holes in the ground, either to dispose of the waste
or to reclaim marginal or derelict land. Occasionally these sites are over-
built. In the early days houses and light industrial development was built
on sites containing colliery spoil with a total lack of technical knowledge as
to the consequence. The Paper will describe the problems that can arise as a
result of an incomplete understanding of the fill's performance and aggressiveness.

INTRODUCTION

In the United Kingdom colliery spoil is currently being produced at an
annual rate of 50-60 million tonnes. The increase in spoil production compared to saleable coal from 1920 to 1980 is shown in Fig. 1.(ref.1.)

The majority of the spoil goes directly to tips adjacent to the collieries,
being placed in a safe state to ensure the stability of the tip.

These spoil heaps will either be progressively restored by British Coal
for agricultural use or will often be acquired by Local Authorities for
reclamation once no longer required by British Coal.

In some instances, material is disposed, by the colliery, directly to
reclaim a derelict site or disused quarry working. These sites will usually
be filled as a tip to ensure a safe and stable structure.

In addition to current tipping, approximately 3000 million tonnes of
colliery spoil are on the surface. Some of the deposits on the surface are in
locations suitable for housing or industrial development. Construction has
taken place over colliery spoil which was either on the site prior to development or in some cases, where the spoil was imported as a general site fill
material. The stocks of colliery spoil (burnt and unburnt) are in the coalfields, not only at working collieries but also near closed collieries. The
locations and extent of the coalfields are shown on Fig. 2. (ref.2)

In many instances development has taken place without due regard or understanding of the nature of the colliery spoil. Necessary precautions in the
design and construction of workon the site may not have been observed with the
consequent risk of problems occurring therefore being greatly increased.

In this paper the properties of colliery spoil are discussed together with the potential problems that can arise when development is carried out without a full understanding of the likely performance of the material.

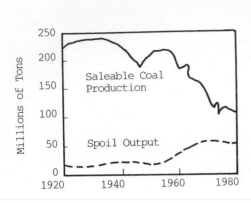

Fig 1. Saleable Coal Production
Compared to Spoil Output

Fig 2. British Coalfields

PROPERTIES OF COLLIERY SPOIL

Colliery Spoil is a by-product of mining coal and of its preparation for markets. They are derived from the rocks - mainly siltstones and mudstones, with seat-earths and sometimes sandstones, limestones and other rock types - lying above, below and sometimes within, the coal seams in the coal measures. (Fig. 3.) (ref.3). During mining operations quantities of these rocks unavoidably extracted with the coal, or in driving the tunnels (roads) which give access to the coal faces, are brought to the surface with the coal.

It is then generally necessary to remove some or all of this material to yield a coal product of the required quality. This separation is carried out with other operations such as crushing, sizing and de-watering, inthe coal preparation plants, generally by making use of the higher density of the colliery spoil - roughly between 1½ and 2 times that of the coal. There is usually no immediate use for the separated non-coal material, so it is tipped onto a spoil heap.

Until about twenty years ago, and more particularly before the post-World War II advance in mining technology, nearly all colliery spoil heaps were constructed by tipping the spoil loose from transporting containers - aerial flights, skips, wagons or trams. The resulting spoil heaps were nearly always loose and open textured so that air could readily penetrate. If for any reason the residual coal in parts of the spoil heaps became ignited, air could readily reach the burning section and sustain and extend the heating. Before World War II, nearly all collieries were steam powered and had to dispose of hot boiler ashes. There was then no well-established market for this clinker as lightweight aggregate, so it was often tipped on the spoil heaps, which were ignited and eventually burned through. The residues of calcined rocks left in these spoil heaps are burnt colliery spoils.

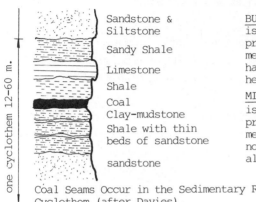

Sandstone & Siltstone

Sandy Shale

Limestone

Shale

Coal

Clay-mudstone

Shale with thin beds of sandstone

sandstone

BURNT COLLIERY SPOIL is made up of various proportions of coal measure rock type which have been altered by heat.

MINESTONE is made up of various proportions of coal measure rocks which have not been subjected to heat alteration.

Coal Seams Occur in the Sedimentary Rock Sequence known as Cyclothem (after Davies).

Fig. 3 Coal Measures Rocks.

Colliery discards are presently made up of about 50 M tonnes of coarse discard, largely disposed of in tips, and 5 M tonnes of fine discards (tailings and slurry), which are mainly discharged into lagoons.

Coarse and fine discards comprise mainly comminuted debris from the roof and floor rocks associated with the coal seams extracted by underground mining methods. The high coal content of lagoon sediments, represented by an average organic carbon content of 47.5% by weight, is due to the periodic discharge of preparation plant overflows, flushings and other effluents into the lagoon. Some very old tips contained large quantities of unsaleable small coal, although the national average coal content of tips is about 11%.

The mineral components of the predominantly argillaceous rock fragments of discards are dominated by clay minerals; illite, expandable mixed-layer clay, kaolinite and minor or trace amounts of chlorite. These are followed by smaller amounts of quartz and minor quantities of carbonate and pyrite, ferrous sulphide, which weathers rapidly.

Particle sizes of coarse discards range from silty sand/gravel to medium and coarse gravel with cobbles.

Although lagoon sediments are finer-grained than coarse discards they span a wider size range than might have been expected from their 0.5 mm top size coal content. Lagoon samples range from clay/silt sizes to sandy fine and medium gravel sizes. On the Casagrande Plasticity Chart the finer fractions of coarse and fine discards classify mainly as silts and inorganic clays of low to medium plasticity. However, it is estimated that up to 50% of all discards are non-plastic.

Fundamental properties of the two discard types are significantly different because of the high coal and moisture contents of slurry and tailings. The mean natural moisture content of tip samples is 11.0% and the bulk density is 1.902 Mg/m³. This compares with 34.8% and 1.457 Mg/m³ for lagoon sediments.

Bulk samples of burnt discard are generally higher in water soluble and acid soluble sulphates than unburnt types. Some 71% of unburnt and 49% of burnt types contain less than 0.2% water soluble and 1.0% acid soluble sulphate by weight.

Appreciable breakdown of discard takes place during transportation from the washery and during spreading on a heap. Discards associated with low rank coals appear to be the most susceptible, particularly if they are of a clayey nature (high seatearth content). Compaction can cause a small degree of further breakdown of the rock fragments.

Frost action is believed to be a very minor factor in the breakdown and degradation of coarse discards.

Subsequent to emplacement and burial, little if any further physical disintegration of unburnt discard occurs.

Chemical weathering processes in unurnt discards are a function of oxidation. Pyrite is the mineral most susceptible to oxidation and weathering. Its breakdown leads to acidity. Fresh discard is neutral or alkaline, becoming progressively more acidic with time.

In old, unburnt spoil heaps intense oxidation is limited to the outer zone of less than about 1m in depth.

Heating and ignition in old heaps was due to two different causes:
  (i) accidental: tipping of hot boiler ash, lighting of fires, and from braziers on a tip;
  (ii) exothermic oxidation of the (waste) coal content of discard; of other extraneous carbonaceous materials, and to a lesser extent, the exothermic breakdown of pyrite.

POTENTIAL PROBLEMS

Colliery spoil encountered in developments will either be in-situ in a previously placed deposit such as a spoil heap or it will be used as an imported construction material. In the latter case selected minestone will generally be assessed prior to use and unsuitable materials will not be made available. Suitable colliery spoils are referred to as MINESTONE and will be selected by Minestone Services from British Coal sources.

Previously deposited spoils can exist in a wide range of conditions. The potential problems on such developments are obviously much greater than those where a controlled fill has been utilised.

The conditions outlined in this section generally apply to engineered fills and in-situ deposits. However, greater caution is likely to be required on some aspects for development on in-situ deposits.

The largest single use of minestone in Britain (approximately 7M tonnes per annum) has been as common fill in the construction of road and railway embankments and the filling in of derelict and partially derelict sites for building or other purposes.

Plate 1.   Reclamation of Derelict Site, Newcastle-upon-Tyne by infilling with Minestone.

Because of the manner in which the material was used the level of technical understanding of the material's properties was minimal. Further attempts to promote wider utilisation for minestone ran into difficulties, perhaps the greatest being the necessity to comply with existing specifications which were designed as a diagnostic classification of the more conventional materials.

Engineers classify deposits as either a soil or as a rock on the basis of an arbitrary division involving an assessment of strength, related physical properties and use. The problem of classification is exacerbated because many weak rocks may behave as soils and conversely some very stiff or 'hard' soils may behave as rocks.

Existing classification systems for rocks or soils cannot, therefore, be adequately applied to colliery spoil as it often overlaps both systems.

168

Because the range of overlap is often difficult to define and colliery spoil, as a generic description, falls outside the narrowly prescribed parameter ranges generally regarded as sacrosanct by civil engineering specification drafters, a specific classification system has to be developed for minestone when used for civil engineering purposes. (ref.4)

However, it has been found that by careful preselection and refinement, some colliery spoils can meet existing specifications and have been used as alternatives for more conventional materials in civil engineering projects.

Fig. 4 shows the ranges of particle size distributions for colliery discards, together with the range for MINESTONE.

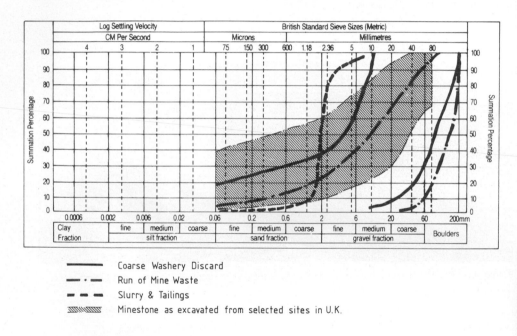

——————   Coarse Washery Discard
—·—   Run of Mine Waste
— — —   Slurry & Tailings
〰〰   Minestone as excavated from selected sites in U.K.

Fig. 4   Particle Size Distributions for Spoil and Minestone

Spontaneous Combustibility

There has been the fear that if unburnt colliery spoil (MINESTONE) is used for Civil Engineering applications then there would be a risk of burning occurring. It has been argued by British Coal that prior to 1960 all spoil heaps were formed of loosely tipped material and that, coincidentally, nearly all collieries were steam powered. There is much evidence to indicate that the hot boiler ashes were disposed of on the spoil heaps. It is probable that the residual coal became ignited as a result and because of the open-textured nature of the heaps air could readily penetrate and reach the burning sections to sustain and extend the burning. Since vesting date

British Coal have been charged with the responsibility to prohibit any burning occurring in new heaps and it is significant that no heating has occurred in these heaps.

Alternatively, there is evidence that spontaneous heating in loosely tipped spoils results from an atmospheric oxidation (exothermic) process.

However, spontaneous combustion is not considered to be a property of well-compacted spoil fills, since neither oxidation, spontaneous heating nor burning will occur in the absence of air, and it is unlikely that a minor event such as a grass fire would ignite a well-compacted spoil fill.

If the material has not ignited when in a loose state on the tip, it is highly unlikely that the same material would be susceptible to spontaneous heating when used in other well-compacted structures.

Although British Coal always perform a loss-on-ignition test in their evaluation of colliery spoil, there is some doubt whether the loss-on-ignition test is a fair measure of the risk of spontaneous combustion since the mineral matter in coal decomposes considerably below 500°C and consequently determination of the loss-on-iginition from a shale sample at this temperature would not yield a reliable estimate of combustible content. (ref. 5.)

Moisture. Wet fill will not develop heating. Likewise very dry fill does not appear to do so. The optimum conditions seem to be slight dampness and a supply of humid air.

Particle Size Voids Ratio and Specific Surface. The ease with which air passes through spoil containing carbonaceous material determines the rate at which heat generated by oxidation is carried away. With large size material and large air voids, the movement of air is usually sufficient to carry away any heat generated by oxidation and to cool the material. With well graded or fine material having small air voids, the air remains stagnant and the heat generated is retained in the mass; but when the available oxygen is consumed the heating stops.

Volume. In originally, loosely placed colliery spoil of appropriate material, heating might develop almost anywhere, whereas with respread material these same materials lose too much heat by conduction to be the cause of heating.

Compaction. Spontaneous heating relies on a steady supply of air to propagate the oxidation processes and to sustain heating. Therefore, well-compacted fills reduce the level of permeability and effectively exclude the supply of air.

It has been well established that colliery spoil could be utilised as a Civil Engineering fill material, but before it was possible to utilise the material in the motorway programme of the 1960's, the Ministry of Transport had to be satisfied that the material would not be prone to spontaneous combustion, indeed the D.O.T. Specification for Road and Bridgeworks prohibits the use of material susceptible to spontaneous combustion. To this end the National Coal Board commenced a programme to demonstrate and establish that the material could be safely used.

There is no suitable direct method of measuring or predicting whether a particular colliery spoil would be more or less liable to spontaneous heating than any other, however there is general agreement that if air can be excluded from the stockpile then any risk is minimal.

Compaction

There is considerable variation in the compaction of existing spoil in tips.

The compaction specified to form tips is primarily to achieve stability of the tip and the methods used can satisfy the compaction requirements without utilising specific compaction plant.

The level of compaction achieved is therefore generally below that required to satisfy the minimum air voids content as would be obtained when using the Department of Transport Specification as recommended by Minestone Services.

Water Table and Moisture Movement

Colliery spoil in keeping with all civil engineering soils and materials will express different degrees of integrity with different degrees of moisture content.

The material is unlikely to be any more susceptible to moisture movement than other natural or engineering material.

Load Bearing Capacity

Selected minestone, when placed and compacted in a controlled manner to meet a specification requirement, will result in a stable consistent fill area capable of carrying predictable design loads.

In the case of sites where the spoil is in-situ prior to development, the likely inconsistent nature of the deposit and the often unknown detailed knowledge of the site ground condition make the establishing of bearing capacity difficult.

In addition, with increasing land use and restoration pressures, temporary lagoons are now commonly overtipped with coarse discard so that they can be incorporated in a composite waste heap disposal structure. (ref. 6.)

Plate 2. Loosely placed fill in a Spoil Heap.

Plate 3. Timber debris buried within a Spoil Heap.

Plate 4. Detail of Timber debris with Spoil Heap.

Where a slip movement has occurred in a tip or foundations, the slip surface forms a permanent plane of weakness. The shear strength that can be mobilised along this plane corresponds to the residual shear strength of the material, which may be significantly lower than the peak strength. Comminution of the material along the shear zone may also occur as a result of a large shear movement. This may produce a layer consisting of material that has a lower permeability than the original material. Hence, where new construction is to take place over an existing tip, it is important to obtain an accurate picture of the history and condition of the existing tip and its foundations, and to look for any signs of a slip movement.

The action of wetting, drying or frost may change the physical properties of the spoil on the surface of an existing tip. Where new construction is to take place over an existing tip it may be necessary for the site investigation to include the shear strength and permeability characteristics of the weathered material. If the weathered material forms a layer of weakness or an impermeable zone that is critical to the stability of the new construction, it may be preferable to remove it. Possible problems can arise if slip movement has occurred in a deposit or through changes to the surface layer resulting from weathering.

Presence of Gas

The majority of methane associated with coal and coal measures strata is released and dispersed when the coal is disturbed during extraction underground. A relatively small percentage of the methane generated is absorbed and retained by the spoil material removed from the mine to the spoil heap. Any methane not released during the extraction or processing and washing procedures generally dissipates within a short period of being placed on the tip.

It is, therefore, highly unlikely that methane exists or will be produced in significant quantities if the depth of spoil is relatively shallow.

Methane is the major gas of concern, however the levels of any other potentially dangerous emissions are likely to be similarly insignificant.

Frost Susceptibility

In the U.K. severe frosts do occur and experience has shown that roads fail as a result of frost penetration and resultant heave effects with considerable loss of strength during the thaw.

Many colliery spoils are within the permissible limits of frost heave criteria and therefore generally non-frost susceptible, however it is essential to assess all materials which could be liable to frost action.

Sulphate

The presence of sulphate occurring naturally in rocks and soils is well known and particularly high concentrations have been reported in colliery shales. In considering sulphate attack on concrete structures it is usually sufficient to determine the sulphate content and pH of the soil and ground-water. If the sulphate content is found to be in excess of the permitted amounts, in terms of percentage SO_3, then steps have to be taken to eliminate the adverse effects on adjacent concrete structures. (ref. 7-9)

Occurence. The chief form of sulphur in freshly dug colliery shale occurs as iron pyrites (FeS_2). This mineral when exposed to the action of air and water becomes oxidised to ferrous sulphate and sulphuric acid (this reaction is strongly exothermic and is one of the causes of spontaneous combustion in shale heaps and goafs). The ferrous sulphate oxidises further to ferric sulphate but eventually in the presence of alkaline-earth compounds (clay minerals) most of the sulphate ions released by the oxidation of pyrites become the sulphates of CALCIUM, MAGNESIUM, SODIUM and POTASSIUM.

Attack on Concrete Structures. Sulphates in solution react with the four principle components of Portland cement (i.e. tricalcium silicate, dicalcium silicate, tricalcium aluminate and tricalcium aluminoferrate) to form insoluble calcium sulphate and calcium sulpho-aluminate.

Crystallisation of the new compounds are accompanied by an increase in molecular volume which causes expansion and disintegration of concrete structures that come into contact with aqueous solutions of sulphate. Dis-integration takes place initially at the surface, further exposing fresh areas to attack, which in turn can be attacked by a progressive flow of sulphates in the groundwaters. If the rate of flow is constant, complete disintegration can be very rapid.

If the groundwater is absolutely static, attack does not penetrate beyond the outer skin of concrete, nevertheless upward flow of water by capillarity due to climatic conditions or drying action of domestic heating or furnaces cannot be overlooked.

Easily soluble sulphates such as magnesium, sodium and potassium are more aggressive and therefore have greater significance than the less soluble calcium sulphate.

Insoluble sulphates do not contribute to attack.

Magnesium sulphate is especially harmful because it is readily soluble in water and is able to react with all four components of cement.

In all cases of sulphate attack, water is an essential part of the reaction and excess water has to be present for any attack to occur.

The solubilities of the three most abundant salts (i.e. calcium, magnesium, potassium), expressed as grammes of sulphur trioxide (SO_3) per litre of solution, differ considerably, calcium being only slightly soluble. The SO_3 content of groundwaters therefore gives an indication whether one (or both) of the more soluble sulphates are present.

The total sulphate content of the soil may be obtained by extraction with hot dilute hydrochloric acid, and is expressed as the percentage of SO_3 per dry weight of soil. Where the total sulphate level is high it is recommended that the water soluble sulphate level should be determined in a limited water extract to distinguish between calcium and the more soluble sulphates.

Acid-soluble sulphate contents represent all the sulphate that is present and that could, in theory, be leached out by water over a long time interval. In practice this amount is unlikely to be removed.

Factors influencing Sulphate Attack. The following factors influence the rate of attack of concrete structures that are subject to aqueous solution of sulphates:-

(a)   The amount and nature of the sulphate present;

(b)   The level of the water table and its seasonal variation;

(c)   The flow of groundwater and soil porosity;

(d)   The form of construction;

(e)   The quality of the concrete.

The tolerable concentration of the sulphate ion in the soil water depends on the nature of the concrete structures with which it will come into contact. If sulphates cannot be prevented from reaching the structure the only defence against attack lies in the control of factor (e). This is done by adopting either suphate resisting Portland cement, or admixtures of Ordinary Portland Cement with slag, p.f.a. or in extreme situations using SRPC with a protective coating against water.

Colliery shale used in embankment construction should not cause undue concern as long as good drainage is provided (e.g. on top of an embankment).

Where sulphate content is high within an embankment all concrete pipes, bridge abutements or cemented road materials should include sulphate resistant cements or adequate protective coatings of inert material such as asphalt or bituminous emulsions reinforced with fibreglass membranes.

It has been demonstrated that excessive amounts of sulphate may exist within colliery shale. These sulphates when taken into solution by flowing ground-water, and allowed to come into contact with concrete structures, will result in partial or total disintegration of the concrete.

Plate 5. Disintegration of Concrete Floor Slab due to Sulphate in underlying
          Burnt Minestone.

It is therefore necessary to test all colliery shale to be used as fill
for both total sulphate and water soluble sulphate concentrations. Should the
sulphate concentrations be in excess of the specified amounts then adequate
precautions should be taken to eliminate sulphate attack.

Chlorides

Sodium chloride and calcium chloride salts can produce reductions in concrete alkalinity. It has been suggested that the magnitude of the pH reduction
is sufficient to enhance the influence of chlorides, one of the primary agents
causing corrosion of concrete reinforcement. The determination of chloride
content is therefore of importance in assessing the corrosiveness of minestone
to embedded metallic reinforcement especially in poorly drained fill.

CONCLUSION

For many decades Minestone has been used as fill, though probably with
minimal control, at and around active collieries. MOre recently the sites of
many closed collieries have been converted for developments of many kinds by
local authorities and others. Minestone from the disused spoil heaps,
properly handled - usually by spreading and compacting in layers as for highway earthworks - provide good stable ground, strong enough to support many
types of structure on suitable foundations, easily trenched for services etc.

Plate 6.   Industrial Development on reclaimed Colliery Spoil Heap in
           Sheffield.

During the same period there has also been increasing use of Minestone
imported to other sites for similar purposes. These include; the elimination
of surface irregularities on building sites; construction of temporary haul
roads across roads and roadworks over low bearing ground; replacement of silts,
peats, soft clays, water-logged and other unsuitable materials, to allow site
development to proceed; back-filling of disused quarries, and gravel and clay
pits to provide building or recreational land; raising of ground levels on low
lying sites; blinding and covering of municipal tips; filling of disused
canals and docks.

Plate 7.   Burnt Minestone used as Road Sub-base within Chemical Works site,
           Teesside.

Undoubtedly, Minestone will continue to be increasingly used in these ways and particularly in bringing derelict or low grade land into better use.

In this paper, I have discussed the possible problems that should be considered when Minestone is encountered in a development. The satisfactory utilisation of the material over a number of years demonstrates that with a good understanding of the material, it can be used to the benefit of the project and the environment as a whole.

REFERENCES

1 R.K. Taylor, Composition and engineering properties of British colliery discards, National Coal Board, London, 1984.
2 Minestone Services, Information Sheets, British Coal, London 1986.
3 Dr. A.K.M. Rainbow, Composition and Characteristics of Waste from Coal Mining and Preparation in the United Kingdom, United Nations E.S.C.E. Symposium on the Utilization of Waste from Coal Mining and Preparation, Tatabanya, Hungary, 1983.
4 Dr. A.K.M. Rainbow, The use of a Problematic Rock-Soil Material, Civil Engineering, July 1986, pp. 43-46.
5 Dr. A.K.M. Rainbow, A Review of the State of the Art in Respect to the risk of Spontaneous Ccmbustion when using Minestone in Construction Works in the United Kingdom, United Nations E.S.C.E. Symposium on the Utilization of Waste from Coal Mining and Preparation, Tatabanya, Hungary, 1983.
6 National Coal Board, Technical Handbook, Spoil Heaps and Lagoons, National Coal Board, London, 1970.
7 R.A.Scott, The Occurrence and Effects of Sulphate in Colliery Spoil, Internal Report, Minestone Executive, National Coal Board, 1981.
8 P.T. Sherwood and M.D. Ryley, The effects of Sulphates in Colliery Shale on its use for Road Making, RRL Report LR324, Department of Transport, London, 1970.
9 Building Research Establishment, Concrete in Sulphate-bearing soils and Groundwaters, Digest 250, Building Research Station, Garston, Watford, 1981.

ACKNOWLEDGEMENTS

The Author is grateful to British Coal for the support given in the preparation of this paper and for permission granted for its publication. The views expressed are those of the Author and not necessarily those of British Coal.

Reclamation, Treatment and Utilization of Coal Mining Wastes, edited by A.K.M. Rainbow
Elsevier Science Publishers B.V., Amsterdam, 1987 — Printed in The Netherlands

LABORATORY AND SITE INVESTIGATIONS ON WEATHERING OF COAL MINING
WASTES AS A FILL MATERIAL IN EARTH STRUCTURES

K.M. SKARŻYŃSKA, H. BURDA, E. KOZIELSKA-SROKA, P. MICHALSKI
Department of Soil Mechanics and Earthworks, Agricultural
University /Poland/

ABSTRACT
    This paper presents the results of laboratory and site investi-
gations on weathering of coal mining wastes to determine the main
factors influencing minestone disintegration and the consequent effects on changes in basic geotechnical properties. Laboratory tests
included wet and dry freezing, slaking and model weathering test in
a box; site investigations were performed on flood embankments constructed from minestone. Laboratory tests were undertaken to define minestone susceptibility to these particular weathering factors
depending on the time and petrographic composition; site investigations were aimed to determine time-dependent changes on the surface and at various depths of earth structures occurring under natural atmospheric conditions.

INTRODUCTION
    The utilization of mining wastes for the purposes of earth stru-
cture construction requires knowledge of their geotechnical properties, crucial to the stability and safety of the construction, and
the behaviour of this material while the completed structure is in
use. A characteristic feature of mining wastes, which are a mixture
of argillaceous rocks, is a considerable susceptibility to external
factors due to which the rocks disintegrate, resultant fragmentation of the material, and changes in its geotechnical parameters.
The condition for correct material selection and designing of the
construction is that the susceptibility of this material to weathering processes and the direction of associated changes in geotechnical parameters should be determined. Owing to the shortage of literature on the behaviour of mining wastes in time and changes in
their geotechnical properties, comprehensive investigations were
planned and carried out in the Department of Soil Mechanics and E-
arthworks of the Agricultural University in Kraków; these were to
determine the effect of particular external factors on the disintegration of unburnt coal mining wastes of the Upper Silesian Industrial Region.

NATURE OF THE WEATHERING PROCESSES

The waste material from coal mines is a mixture of fragments of the rocks which accompany coal seams such as claystones, shales, siltstones, carbon shales and sandstones with a small admixture of coal. Depending on the type of seam worked and on the mine in question, the proportion of particular rocks in the mixture varies, this influencing the properties of the given type of waste.

After having been brought to the surface and then being dumped on a heap or built into an earth structure, this material is subjected to the action of external factors, which cause its disintegration and fragmentation. Broadly speaking, these factors can be divided into physical and chemical ones.

Processes such as the following can be considered to be physical weathering:
- decompression of material after extraction from larger depths to the surface
- thermal processes which cause internal tension between the individual components of the particular rock, due to changes in temperature
- the effect of sub-freezing temperatures, and cyclical processes of freezing and thawing of the water which fills the pores and cracks
- the cyclical wetting and drying of the material which causes "air breakage". This is caused by the considerable rise in air pressure in the pores, which is compressed by the water forced through by forces of capillary suction.
Susceptibility to physical weathering is connected with the type of rock and its structure, and is also strongly dependent on the hardness of rocks, this being evidenced by the relationship found between the impact strength index and the degree of fragmentation.

Chemical weathering consists mainly in processes of dissolving and washing out of readily water-soluble substances, and also oxidation and hydration. Most minerals are soluble in water, particularly when it contains dissolved oxygen and carbon dioxide /e.g. rainwater/. Groundwater containing carbonates, organic acids and sulphates can be equally active.

Physical weathering proceeds fairly quickly, and its effects are apparent in months, and sometimes even in weeks. Chemical processes take place much more slowly, and appreciable changes take many years. Both these processes take place concurrently, and their intensity is governed to a considerable degree by the ability of water and air to penetrate down into the waste material. The disintegration of rocks facilitates chemical weathering on the one hand, as it leads to increasing the surface area of the material, on the other hand it limits the action of air and water, since the tightness of the material increases in the process. Studies carried out to date /ref. 1 and 2/ have demonstrated that the effects of physical and chemical weathering disappear at depths of about 1-2 m.

PURPOSE AND METHODS OF INVESTIGATIONS

The aim of the present study was to determine the effect of physical and chemical weathering on the rate and character of the disintegration of rocks constituting coal mining wastes, and associated changes in basic geotechnical parameters. In order to determine the effect of particular factors on the disintegration of minestone, laboratory tests were carried out on model weathering in a box, on slaking, and on freezing under wet and dry conditions. In order to find out how the effects of weathering progressed with time and depth, tests were made on the flood embankments constructed from minestone, under natural conditions. The laboratory tests were carried out on material derived from four different sources,and the field ones on 3 flood embankments with differing petrographic composition, in order to determine the effect this composition has on susceptibility to weathering.

Independently of the above, freezing tests were carried out separately on each kind of rock constituting the coal mining wastes, in order to determine the resistance of particular rocks to frost weathering.

In the laboratory tests carried out, the initial step was to determine the grain size distribution of the material investigated using wet sieve analysis supplemented by the sedimentation method where necessary. These determinations were repeated later after a certain perid of investigation or a certain number of freezing cycles.

In laboratory tests, the basic geotechnical parameters were determined on the initial material and following the tests, such as the maximum dry density and moisture content, the coefficient of permeability, the angle of internal friction and cohesion.

In field investigation, determinations of grain size distribution and coefficient of permeability were made, being repeated later for samples taken at different times and from different depths.

Comparison of the differences in material grading caused by the various disintegration factors made it possible to determine the susceptibility of the material to the given factor, depending on how long it acted for and on the petrographic composition of the material.

TEST PROCEDURE

Model weathering in a box

Model investigation of weathering consisted in placing samples of material taken from the Makoszowy mine, whose properties were known, into two boxes and exposing them to the action of natural atmospheric conditions. Samples weighing 40 and 60 kg were compacted in boxes in layers of 0.10 and 0.16 m relatively, and placed on the roof of the building for a period of 22 months. After this time, the grain size distribution and the basic geotechnical parameters were again determined. Selected rock types, placed in separate containers, were observed simultaneously.

Slaking tests

Slaking tests were carried out on 10-20 kg samples of the material placed in containers of water. For these tests material from the Makoszowy, Zabrze, and Gliwice mines was selected, and also from the body of the ash lagoon dyke in Przezchlebie[1].

Prior to the study, the grain size distribution was determined for all the samples, this being subsequently repeated after 1, 12 and 36 months of saturation. Moreover, after the tests had been completed, determinations of the basic geotechnical parameters were made on the material of the Makoszowy mine.

Freezing tests

Investigations on frost susceptibility were made by the "wet" and "dry" methods.

The "dry" method consisted in alternately freezing and thawing the material under air-dry conditions, without the access of water.

"Wet" freezing was conducted on fully water-saturated samples. Each testing cycle involved freezing at a temperature of $-20^{\circ}C$ for a period of 4 hours, followed by thawing in water of ambient atmospheric temperature.

Two samples were selected for the study, each weighing 10 kg,

---

[1] The body of the dyke is formed of a mixture of coal mining wastes from 5 different sources.

of almost identical grain size distribution, taken from the Makoszowy mine. Following 10 and 25 cycles of freezing the grain size
distribution was determined, and the losses from particular fractions, by weight. Additionally, after 25 cycles of "wet" freezing,
the shearing resistance of the material tested was determined.

Natural weathering site investigations

Investigations of natural weathering were carried out on constructed river embankments. Three study structures were selected,
namely the embankment of the River Bierawka made from material of
the Szczygłowice mine, the embankment of the River Kłodnica, made
of wastes of the Sośnica mine, and the embankment of the River
Brynica, made of mixed wastes of the Andaluzja and Julian mines.
As part of all these investigations, test pits were made in the
bodies of these dykes, and samples were taken from various levels,
beginning from the land surface down to a depth of 1.1 m. Determinations of grain size distribution, and coefficient of permeability were made on the samples. On the embankment of the Bierawka,
these tests were carried out 2, 8 and 12 years after the structure
had been completed, 4 years from completion on the River Kłodnica,
and following one year of use on the River Brynica.

TEST RESULTS AND INTERPRETATION

Model weathering in the box

The model weathering test was carried out on material from the
Makoszowy mine. The petrographic compositiom of the coal mining
wastes subjected to laboratory tests is given in Table 1; it can
be seen that the main rocks constituting the material tested are
claystones and shales. Table 2 gives the results of sieve analyses

TABLE 1
Petrographic composition of material subjected to laboratory tests.
/percentage by volume/

Source of the material \ kind of rocks	claystones and shales	carbon shales	silt-stones	sand-stones	pure coal
Makoszowy coal mine	69	10	9	7	5
Zabrze coal mine	60	15	15	7	3
Gliwice coal mine	34	12	48	3	3
Przezchlebie dam	77	8	5	7	3

on two samples before and after the weathering test. It is easily noticeable that the result of degradation is a considerable reduction in coarse fraction content and an increase in fine fractions /sand, silt and clay/. The uniformity coefficient rose as a result of grading changes, this being of very great importance for the compactibility of this material. The different degradation of two samples of the same material can be explained by their differing thickness - material compacted in a thicker layer /sample II/ was subject to the penetration of atmospheric factors to a lesser degree, therefore its disintegration is less than that of sample I. The intensity of claystone and shale weathering with time is clearly apparent in the pictures /Fig. 1/ showing the disintegration of these rocks exposed to natural atmospheric conditions. In less than one year all the boulder fraction degrades to gravel.

TABLE 2
Results of model weathering tests on the material of Makoszowy mine.

Sample	Kind of material	Fraction content %%				D_{10} [mm]	D_{60} [mm]	U.C. [-]
		bould.	gravel	sand	silt and clay			
I Thickness of 0.11 m	- original	38	58	3	1	5	39	7
	- after 22 months	6	61	20	13	0.01	7	740
II Thickness of 0.16 m	- original	40	54	5	1	4.4	40	9
	- after 22 months	14	65	17	4	0.8	12	15

Slaking tests

Material subjected to slaking tests came from 4 sources. Its petrographic composition is given in Table 1. The dominating rocks were claystones and shales with the exception of wastes from Gliwice mine, where siltstones predominated. The results of slaking tests in the form of changes in grading ranges for all materials tested are shown in Figure 2. It can easily be noticed that most grading changes occur within the first year of slaking and further progress in disintegration is very small.

The influence of the petrographic composition on the effects of slaking is presented in Figure 3a where variations in D_{60} values

Fig. 1. Model weathering of claystones; a/ original material,
b/ after 1 month, c/ after 9.5 months.

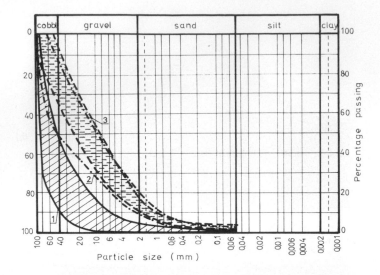

Fig. 2. Variations in grading ranges of minestone subjected to slaking tests /material from Makoszowy, Gliwice, Zabrze mines and Przezchlebie dam/, 1 - original material, 2 - after 1 year, 3 - after 3 years.

with time are shown for all the materials. Thus the greatest degradation expressed by the greatest reduction in D_{60} value takes place in wastes with the highest content of claystones and shales /Makoszowy mine and Przezchlebie dam/, the smallest in the material with a predominance of siltstones /Gliwice mine/. This Figure also confirms that degradation processes occur most intensively within the first year of slaking and have practically disappeared after 2 years.

Similarly, variations in uniformity coefficient shown in Figure 3b, after the rapid increase in its value in the first year of slaking, stabilize later but at much higher levels than in the original material.

Freezing tests

Tests on frost resistance of the material from Makoszowy mine were carried out using "dry" freezing and "wet" freezing methods. Test results given as changes of minestone grading are presented in Table 3. The original grain size distribution of samples subjected to both freezing tests was very similar but the results obta-

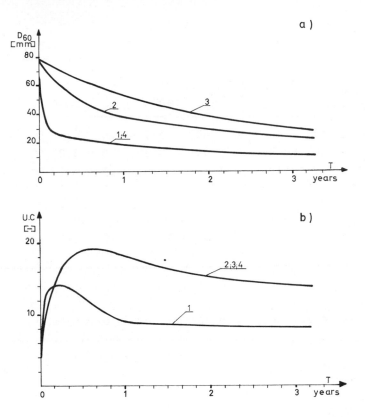

Fig. 3. Variations in the minestone grading with time in slaking tests, expressed as changes in a/ D_{60} and b/ Uniformity Coefficient, 1 - material from Makoszowy mine, 2 - Zabrze mine, 3 - Gliwice mine, 4 - Przezchlebie dam.

ined are totally different. The test results revealed that "dry" freezing caused almost no disintegration but "wet" freezing is a very strong degrading process resulting in considerable material disintegration. This process has the greatest effects on the coarse fractions, giving a 14-fold reduction in boulder fraction content and the same increase in sand fraction content after 25 test cycles. Changes in the fine fraction are smaller because silt and clay fraction content rose 5 times. The value of the uniformity coefficient changes relatively little.

The susceptibility of particular rocks constituting coal mining wastes to frost weathering is shown in Table 4. Analysis of results obtained reveals that the most susceptible to frost weathering are claystones and shales because the boulder fraction degrades completely to sand after 10 test cycles.

TABLE 3

Results of freezing tests on the material of Makoszowy mine.

Kind of test	Kind of material	Fraction content %%				D_{10} [mm]	D_{60} [mm]	U.C. [-]
		bould.	gravel	sand	silt and clay			
dry-freezing	- original	50	45	3.5	1.5	6	50	8.3
	- after 10 cycles	50	44	4.5	1.5	5.2	50	9.6
	- after 25 cycles	50	44	4.5	1.5	4	50	11.5
wet-freezing	- original	42	54	3.5	0.5	5.5	47	7.6
	- after 10 cycles	15	45	38	2	0.45	7	15.5
	- after 25 cycles	3	37	57.5	2.5	0.3	2	6.5

TABLE 4

Results of wet-freezing tests for various kinds of rocks.

Type of rocks	Kind of material tested	Fraction content %%			
		bould.	gravel	sand	silt and clay
Claystones	- original	100	-	-	-
	- after 5 cycles	-	45	54	1
	- after 10 cycles	-	16	82	2
Siltstones	- original	98	1	1	-
	- after 5 cycles	31	56	13	-
	- after 10 cycles	20	55	25	-
Shales	- original	100	-	-	-
	- after 5 cycles	-	30	69	1
	- after 10 cycles	-	4	94	2
Carbon shales	- original	100	-	-	-
	- after 5 cycles	67	29	4	-
	- after 10 cycles	21	71	8	-
Pure coal	- original	100	-	-	-
	- after 5 cycles	95	5	-	-
	- after 10 cycles	94	6	-	-

Siltstones and carbon shales are more resistant, after 10 cycles a 5-fold reduction in the boulder fraction was observed and the gravel fraction becomes predominant.

Pure coal is the most resistant to frost weathering, as after

10 cycles no tendency to disintegrate was observed.

Natural weathering site investigation

Site investigations on weathering under natural conditions have been carried out on 3 embankments constructed from minestone taken from 4 different sources. All these wastes had a similar petrographic composition with predominance of claystones and shales /Tab. 5/. A detailed list of material grading changes with time and depth below embankment surface on particular test structures is presented in Table 6, where the relative values of the coefficient of permeability are also given.

TABLE 5
Petrographic composition of minestone subjected to natural weathering in embankments /percentage by volume/.

Coal mine \ Kind of rock	Claystones and shales	Carbon shales	Siltstones	Sandstones	Pure coal
Szczygłowice	80	12	–	3	5
Sośnica	70	10	10	6	4
Julian	95	–	2	1	2
Andaluzja	80	19	–	1	–

Analysis of test results shows that as was expected, the quickest and most intensive degradation occurs on the embankment surface, in the form of total reduction of boulder fraction content and a very significant increase in the percentage of particles finer than 2 mm. At the same time it is very easy to notice that the intensity of these changes declines very quickly with depth. Change characteristics are shown in the graphs in Figures 4 and 5. In Figure 4 changes in D_{60} and uniformity coefficient with time at various depths are shown and Figure 5 gives changes in D_{10}, D_{60} and U.C. values with depth after 4 years' weathering. It is apparent that the most intensive decrease in D_{60} value occurs within the period of 2 years and the magnitude of these changes decrease with depth /Fig. 4a/. The intensity of the U.C. value changes depends also on the depth; initially, in the surface layer this parameter rises rapidly and then decreases slightly. In deeper layers a si-

TABLE 6

Changes in minestone grading and coefficient of permeability as a result of natural weathering on river embankments.

Source of material investigated	Kind of material	Fraction content %%				Coefficient of permeability k_{10} [m/s]
		bould.	gravel	sand	silt and clay	
Coal mine "Szczygłowice", embankment of the River Bierawka	- original	45	46	6	3	1.1×10^{-5}
	- after 2 years					
	0.0-0.1 m	3	49	27	21	-
	0.3-0.6 m	16	60	16	8	2.9×10^{-6}
	0.8-1.1 m	25	50	16	9	1.2×10^{-6}
	- after 8 years					
	0.0-0.1 m	5	52	23	20	-
	0.3-0.6 m	19	55	14	12	1.2×10^{-7}
	0.8-1.1 m	26	54	12	8	4.7×10^{-6}
	- after 12 years					
	0.3-0.6 m	6	33	32	29	-
	0.8-1.1 m	11	68	12	9	1.5×10^{-6}
Coal mine "Sośnica", embankment of the River Kłodnica	- original	33	52	13	2	1.0×10^{-3}
	- after 4 years					
	0.0-0.1 m	4	58	23	15	-
	0.1-0.3 m	11	54	17	18	2.5×10^{-6}
	0.3-0.6 m	19	58	13	10	3.2×10^{-5}
	0.8-1.1 m	30	49	12	9	4.5×10^{-5}
Coal mine "Julian" and "Andaluzja", embankment of the River Brynica	- original	42	43	9	6	1.2×10^{-4}
	- after 1 year					
	0.0-0.1 m	-	55	26	19	-
	0.1-0.3 m	15	51	20	14	1.6×10^{-7}
	0.3-0.6 m	21	49	18	12	6.7×10^{-7}
	0.7-1.1 m	12	55	19	14	2.4×10^{-6}

milar tendency is observed but is less pronounced /Fig. 4b/. Analysis of changes in D_{10}, D_{60} and U.C. values with depth after 4 years of weathering shows that these parameters are "reverting" to the values of the original material but with various intensity /Fig. 5a, b, c/.

A general view of the embankment slope after weathering is shown in the picture /Fig. 6/ where can be seen that the gravel fraction dominates and the boulder fraction has almost completely disappeared.

Changes in permeability are similar because values of the coefficient of permeability change with changes of minestone grading

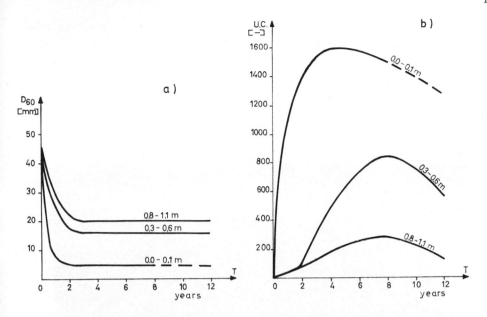

Fig. 4. Variations in the minestone grading with time at different depths below surface of embankment, expressed as changes in a/ D_{60}, b/ Uniformity Coefficient /material from Szczygłowice mine/.

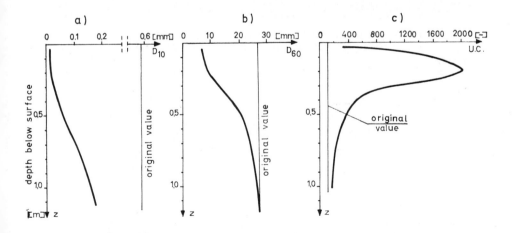

Fig. 5. Variations in the minestone grading with the depth below surface of embankment after 4-year natural weathering, expressed as changes in a/ D_{10}, b/ D_{60}, c/ Uniformity Coefficient /material from Sośnica mine/.

/Tab. 4/. In surface layers the k_{10} value falls sharply after relatively short time of weathering, even by 2-3 orders of magnitude, as compared with the original material, but in deeper layers the size of this reduction declines /ref. 3, 4, 5, 6/.

Fig. 6. Example of natural weathering of minestone /from Sośnica mine/ on the surface of embankment slope after about 4 years.

DISCUSSION

The investigations presented above revealed that the kind of weathering process, time and susceptibility of the material to weathering have the most influence on minestone degradation.

A comparison of the effects of the particular weathering processes, expressed by minestone grading, is shown in Fig. 7. It is clearly visible that of the laboratory tests the most intensive process is "wet" freezing.

In natural air conditions, the effects of each of the particular disintegrating processes are cumulated and moreover "air brekage" occurs, so disintegration of minestone is the greatest on the surface. As a result, the boulder fraction content decreases 12-fold,

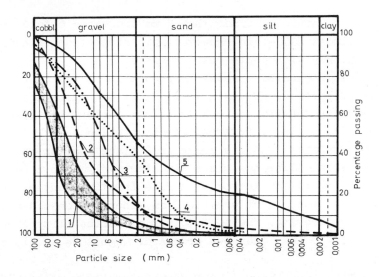

Fig. 7. Influence of various factors on minestone disintegration.
1 - grading range for original material /Makoszowy and Szczygłowice mine/, 2 - after 3-year slaking test /Makoszowy mine/, 3 - after
22-month model weathering test /Makoszowy mine/, 4 - after 10-cyc-
le wet-freezing test /Makoszowy mine/, 5 - after 2-year natural weathering in embankment /layer 0.0-0.1 m, Szczygłowice mine/.

TABLE 7

Changes in certain geotechnical parameters of minestone after disintegration processes in laboratory tests. Material from Makoszowy mine.

Kind of material	Max.dry density γ_{ds} [kN/m^3]	Opt.moistu-re content w_{opt} [%]	Coefficient of permeabi-lity k_{10} [m/s]	Angle of internal friction ϕ [o]	Cohe-sion c [kN/m^2]
Original	18.5	9.8	5.0×10^{-4}	46	31
After model weathering /22 months/	18.2	11.2	1.8×10^{-7}	31	34
After sla-king tests /22 months/	18.4	11.1	9.5×10^{-8}	33.5	37.5
After wet-freezing test /25 cycles/	-	-	-	24	51

the content of particles finer that 2 mm increases 6-fold and D_{10} is reduced almost 2000-fold. However, disintegration effects disappear very quickly with depth below surface.

Various disintegration processes result in change in basic geotechnical properties, given in Table 7. It can be noticed that values of max. dry density and opt. moisture content are almost unchanged but the angle of internal friction ϕ drops markedly and after a "wet" freezing test its value is 2 times smaller than in the original material. Cohesion rises simultaneously with a reduction in ϕ value owing to the increase in fine fraction contents so the stability of an embankment erected from minestone should not be affected by weathering. The coefficient of permeability drops very clearly and it is less than 3-4 orders of magnitude of its original value.

Disintegration processes occur very intensively within 1-2 years only and then the influence of the time factor disappears.

Slaking and "wet" freezing tests revealed that claystones and shales are very susceptible and siltstones more resistant to weathering.

CONCLUSIONS

From the investigations presented above the main conclusions to be drawn are as follows:
- Weathering causes a breakdown and disintegration of coal mining wastes which results in significant decrease in coarse fraction content and increase in fine fractions,
- Physical weathering is more intensive and quicker that the chemical kind,
- Minestone with a high content of claystones and shales is more susceptible to weathering than that with a prevalance of siltstones,
- Time-dependent changes occur within 1-2 years,
- Due to disintegration, the angle of internal friction decreases, cohesion rises and permeability of the material drops significantly,
- Disintegration on the surface reduces penetration of weathering factors down into the structure constructed from minestone and it can be predicted that weathering should not exceed a depth of 1-2 m,
- Generally, weathering of minestone is advantageous from the point of view of hydraulic structures.

REFERENCES

1 A.K.M. Rainbow, An investigation of some factors influencing
   the suitability of minestone as the fill in reinforced earth
   structures, Ph.D. Thesis, University of Nottingham, 1983.
2 R.K. Taylor, Colliery spoil heap materials-time dependent changes, Ground Engineering, Vol. 7, No 4, July 1974.
3 K. Skarżyńska, Wpływ wietrzenia i zagęszczenia na zmiany niek-
   tórych własności odpadów górnictwa węglowego /An influence of
   weathering and compaction on changes of certain properties of
   coal mining wastes/, Zeszyty Naukowe AR, Nr 160, Kraków 1979.
4 K. Skarżyńska, H. Burda, E. Kozielska-Sroka and J. Kurleto,
   Analiza stanu technicznego wybranych obiektów budownictwa ziemnego wykonanych z materiałów odpadowych pod kątem prognozy za-
   chowania się ich w czasie eksploatacji /Study on the technical
     state of certain earth structures constructed from coal mining
   wastes in the aspect of predicting their behaviour with time/,
   Zakład Mechaniki Gruntów i Budownictwa Ziemnego AR, Manuscript,
   Kraków 1983.
5 K. Skarżyńska, E. Kozielska-Sroka and J. Setmajer, Proces wietrzenia materiału odpadowego w świetle przeprowadzonych badań
   /Weathering process of waste material in the light of investigations carried out/, Zakład Mechaniki Gruntów i Budownictwa
   Ziemnego AR, Manuscript, Kraków 1983.
6 K. Skarżyńska, H. Burda, E. Kozielska-Sroka, S. Łacheta and M.
   Porębska, Analiza zmian materiału odpadowego w czasie na skutek
   działania czynników zewnętrznych oraz kryteria przydatności od-
     padów powęglowych do budownictwa wodnego /Study on changes with
   time of the waste material due to outside factors and criteria
   of applicability of the coal mining wastes for hydraulic struc-
   tures/, Zakład Mechaniki Gruntów i Budownictwa Ziemnego AR,
   Manuscript, Kraków 1983.

Reclamation, Treatment and Utilization of Coal Mining Wastes, edited by A.K.M. Rainbow
Elsevier Science Publishers B.V., Amsterdam, 1987 — Printed in The Netherlands

Recovering Combustible Matter from Coal Mining Waste and Measures to Extinguish Waste Pile Fire

by Shengchu Huang, Min.Eng.
Editorial Office, "World Coal Technology", China

ABSTRACT

The amount of material in the waste pile at China's coal mines is now more than 1000 Mt, of which only 20 per cent are utilized. Large amount of wastes not only occupies vast land and arises the environmental problems, but it is a waste of reources as well. Hence, the treatment and utilization of coal mining wastes have aroused people's attention in China.

The paper analyzes the causes   of the spontaneous combustion of the wastes, gives some examples of such combustion and its pollution of the environment. Finally, it has come to a conclusion that the pyrite and coal in the wastes must be recovered to prevent wastes combustion. The article laies emphasis on the methods of extinguishing the pilefire with lime sludge. Moreover, expériences of the prevention of fire and pollution of the waste pile in othercountries are also described.

INTRODUCTION

Coal, an important source of energy in China, has reached a total output of 870 Mt of which 406 Mt are produced by the Coal Ministry-owned mines in 1985. As a result of long-term extensive mining, however, the amount of materials in the waste piles at China's coal mines is more than 1,000 Mt and the area of tipping land is about 37.15 Mm^2. Under the present coal mining scale, the annual production of waste will be more than 90 Mt, of which only20 per cent are utilized. Large amount of discarded waste not only occupies vast land and arised the environmental problems, but it is a waste ofresources as well. Hence,great importance should be attached to the treatment and utilization of coal mine waste. That is why the treatment of waste residue has entered as an article in China's environment protection and mininglaw.

For many years, schools and colleges, institutions of coal research, local coal mines and other relevent establishments in China

have been making unremittingefforts in the comprehensive treatment
and utilization of coal waste and great achievements have been made.
The statistics show that the utilized coal waste has amounted to
14.51 Mt in 1985. Some wastes with a higher calorific value (more
than 1500 kCal/kg) have been used for fluidized-bed combustion,
some made into construction materials and some others are used for
recla-mating or constructing road. Experiences have also been made
in the research work to eliminate waste spontaneous combustion and
environment pollution. In the meantime, however, some 200 piles
throughout the country are still found to be in a state of combustion, causing great danger to the environment. The paper will try
to give a syntherical approach to the problem.

1. The waste Resources and the Problem of its Pollution in China
   Table 1 shows the amount of waste and instances of waste combu-
stion:
TABLE 1

	Amount of			Piles			·Area Occupied (Mt)
	Waste (Mt)	Utilized Waste (Mt)	Cumulated Waste (Mt)	Total Amount	Combusted	In Combustion %	
1984	90.88	13.42	984	857	193	22.52	28.16
1985	73.09	14.51	1092	1008	232	23.02	37.15

The ash content in the waste is generally 70-80 per cent, while
the calorific value is 800-1500 kCal/kg expect for a few cases in
which the calorific value is up to 2500 kCal/kg. According to an
investigation made in Beijing,Hebei, Heilongjinag, Liaonin, Shandong, Jiangsu, Anhui, there are 600 waste piles in these provinces
and municipality. Table 2 shows the distribution of the calorific
value of these wastes. The waste with the calorific value up to
1500-2000 kCal/kg can be used for fluidized-bed combustion.
TABLE 2

Calorific Value(kCal/kg)	Waste Volume(Mt)	Per cent
Total Waste Volume	593.62	100
Less than 500	201.32	33.9
500-1000	208.15	35.1
1000-2000	157.60	26.5
More than 2000	26.55	4.5

Another main content in the waste is sulphur, the amount of which is generally less than 1 per cent. But there are some exceptions. In Nantong Coal Mine Administration in Sichuan Province, for example, the sulphur contained in the waste is as mush as 15-20 per cent.

Substances such as Al_2O_3, Fe, V, Ge and Ti can also be found in the waste.

In1985, there are 1008 piles in China and 232 (23.02 percent) are in a state of spontaneous combustion, emitting large amount of SO_2 gas and acid drainage and causing great environmental problem. Cases of heavy waste fire damaging some constructions and industrial facilities and threatening people's health and life have occured in some places.

2.   Some Causes of Waste Fire

Analysis shows that nodular pyrite and other flamable materials like   coal and wood contained in the waste are the root causes of the fire,

The pyrite is an autoignition substance, with the ignition point of 280°C (lower than that of the coal) and the difference between the reducing and oxidizing forms of ignition points of more than 40°C.

The pyrite in the waste is the first to be oxidized and will give out heat enough for the ignition of the flamable materials such as coal and wood. Investigations show that spontaneous combustion is more likely to occur when more than 3 per cent of sulphur is contained in the waste.

Water and oxygen are two essential factors for the waste combustion. When the water content in the waste is lower than 15-20 per cent, the waste will absorb more oxygen as the water content increases, the temperature for ignition will decrease at the same time. That is why waste combustion is more likely to happen in wet season.

3.   Recover Pyrite and Coal in the Waste

As analyzed above that the waste combustion is caused by the flamable materials like pyrite and coal, thus to recover these materials is the key to the prevention of waste combustion.

3. 1.   Technique of Recovering Pyrite from Coal Waste

It is the main practice for the coal preparation plants in many

200

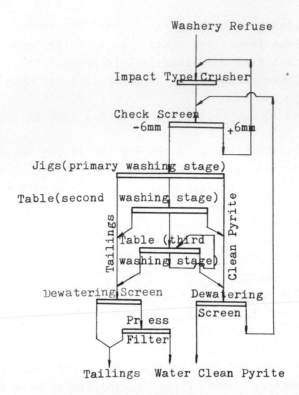

FIGURE 1: Pyrite Recovering Process in Pyrite Washing Section of the Nantong Preparation Plant

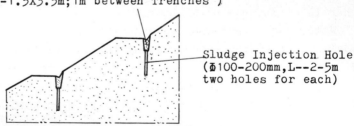

FIGURE 2: Fire Prevention with Sludge

countries to employ jigs, table concentors, water-only and heavy
medium cyclones in recovering pyrite by gravity method. The incorporation of water cyclone with tables has been highly evaluated
because of its effect.

The desinged capacity for the two preparation plants affiliated
to the Nantong Coal Mines Administration is 1.1 Mt. The sulphur
content (mostly pyrite) in the raw coal is 3.5-6 per cent. After
the coal has been processed, most of the pyrite are left in the refuse, making the content of sulphur as much as 15-23 per cent.
As s consequence, whenever a fire starts out in the waste piles,
large amount of SO_2 gas and acid drainage will be caused.

The two plants got additional pyrite washing sections in 1980
in order to recover pyrite. Figure 1 shows the preparation process:
jig--the primary washing equipment and flat tables are used in the
second and third stages.

The two pyrite washing sections can recover pyrite concentrate
14,000t of which the sulphur content is 31.37 per cent. The recovery rate is 73.8 per cent, making the cost as low as 19.2 RMB Yuan
per ton. It has been well proved by the fact that it is an efficient and economic way to seperate sulphur from the washery refuse
with higher sulphur content by gravity preparation method. The
disadvantage of this method is that 7-8 per cent of sulphur is still
left in the tailings. However, the amount of tailings is greatly
reduced.

Some other preparation plants throughout the country have also
established sulphur washing sections. Meanwhile, the coal mines
and preparation plants also seperate the pyrite from the waste by
hand picking or other means. In 1984, 148,990 t of pyrite was recovered by the coal industry of China. (Table 3)

TABLE 3

Unit:Ton

Provinces & Cities	Total	Amount of	
		Pyrite	Sulphur Concentrate
Total in China	148990	111811	30879
Shaanxi	238	238	
Hebei	5226	2434	2792
Liaoning	14744	9693	5051
Zhejiang	4498	4998	
Jiangxi	13269	13269	
Shandong	71694	71694	
Hunan	160	160	
Sicuan	2400	2400	
Chongqing	39261	13225	23036

3.2.   Recover Coal from the Waste

According to   imcomplete statistics, there are more than 20
plants separating coal from waste in China, with an annual separating capacity of 11 Mt and can recover 1.5 Mt of coal in a year.
There are five separating methods included(Table 4):

TABLE 4

Recovering Method	Annual Washing Rate(Mt)	Annual Recover Rate(Mt)	Plant or Coal Mine
Simply-constructed Waste Washing plant:with incl-ined trough medhod	1.0	0.2	Laohutai, Liujiang, Shangjing, Dujiang, Jingjing
Simply-constructed Waste Washing plant:with water cyclone	0.3	0.05	Majiaguo, Zhaogezhuang
Recovering Coal While Separating pyrite	1.4	0.25	Jianxing, Nantong, Ganbasi, Fengcheng, Caitun, Dangjiazhuang
Using Exsisting Jigs in Coal Washing Plant	2.5	0.30	Shengli, Taiji, Sanbao, Didao
Separating by Hand Picking, screen and Chute	5.8	0.70	
Total	11	1.5	

4.   Method of Extinguishing Waste Pile Fire and Of Preventing
Pollution

4.1.   Extinguishing Pile Fire with Lime Sludge

It is one of the methods of fire prevention to cover a layer of
soil or vegetation above the pile to seal the latter from the air.
However, since the air can pass through the layer and the layer can
be easily washed by the rain, the fire can
not be effectively prevented. Moreover, soil transporting and covering are very hard tasks. For this reason, the experiment of
fire prevention with lime sludge, a method which has achieved de-
sireable results in some coal mines in China, is introduced here.

SO_2 gas will be emitted when a waste pile with a high sulphur
content is combusted. The application of lime sludge can neutralize SO_2 gas, and $CaSO_3$ or $CaSO_4$ will be formed, making a strong
seal layer above the fire area and separating the pile from the air.

As a result, the waste fire can be extinguished.

The newly piled up Hexigou waste pile in Liangdu Coal Mine of Fenxi Coal Administration has a weight of 276000t, accumulated since 1970, and the face area is 19000m^2. The calorific value of the waste is 1340 kCal/kg with a sulphur content of 3.25 per cent. A waste pile fire was started in 1977, covering an area of 6000m^2 (31.58 per cent of the total area) with a temparature of up to 600° C. Sulphur crystal has been separated out during the fire and large amount of SO$_2$ gas is emitted into the air. The volume of the gas in some cases has been so great that it has surpassed 52 times the permissible standard volume of the regulation set up by the government, and is 9 times in average.

The Administration carried out experiments in 1981 to stop the fire with calcium lime and water in the proportion of 1:2-3. The mixture, dissolved and residuum removed, is pumped by a sludger to be sprayed to the pile fire area.

Besides, lime sludge are also filled into the pile through trenches and holes drilled or driven out along the side of the pile.

The fire is finally eliminated by spraying lime sludge for 10-20 times according the size of the fire.

An area of 5100m^2 has been covered with lime sludge and the fire is totally eliminated while the temperature within the pile drops from the original 600°C to 50-160°C. The substance covered with lime sludge has an average sulphur content of 8.5 per cent, and the maximum is 14.08 per cent. Sulphur crystal substances are found to be separated out under a hard layer formed by lime sludge and SO$_2$ gas is also greatly reduced and can meet the Standard of II Category of the State Regulation of SO$_2$ gas Release.

The cost for extinguishing fire with lime sludge is relatively low to an average of only RMB 2.31 Yuan/m^2.

To eliminate the pile combustion in the Lizijia Coal Mine of Sichuan Province, a new experimental method with lime sludge and so - dium hydroxide is employed for one waste pile and only lime sludge for another, for the purpose of comparasion. The former, as a result, can achieve a better effect than the latter because the mixture of lime sludge and sodium hydroxide has a better quality of chemical activity in neutralizing SO$_2$ to form a harder layer of seal which is superior to that of the latter as sealing is converned.

The sucessful experiment of the Fenxi Coal Mines Administration has aroused the great interest of many other Administrations like

Xinwen, Fengfeng, Xishan, Datong and Yangquan and they have decided to employ the same method to eliminate the pile combustion.

At present, 73 piles (31.1 per cent of all the piles in combustion in China) have been treated and utilized in varying degrees.

Coal mines such as "Red Star" inclined Shaft No 12 and the Shaft
No6 of Trudovskaya in the USSR, have also carried out experiments
to eliminate the waste fire with lime sludge. The experiments have
been proved sucessful and it is believed that the method has the
advantage of low cost and that the flying up of dust from the waste
piles can be prevented by the cemented hard layer of lime sludge.
At Shaft No 12 of "Red Star", pipes of 100mm diameter were used to
suck the SO_2 gas given out during combustion from a depth of 4-6 m
inside the pile with the help of vacuum pump to reduce the gas
releasing into the air.

Moreover, combustion can be effectively prevented if wastes are
treated with lime sludge before they are piled together. The Za-
peravalinaya Coal Mine in the USSR has been employing this method
for years and incidents of waste pile fire have ont yet occured
in this area.

The coal mines in Czechoslovakia have effectively prevented the
fire by covering the sludge formed by power fly ash. (See Figure 2)

4.1. Other Methods for the Prevention of Fire and Pollution of
the Waste Pile

A research carried out by Bobrov, a candidate of biological science of USSR, shows that the anaerobic sediments formed by biological purification can also be used to eliminate waste fire. After
the waste pile is covered by the sediments, a stable area is formed
when the organic substance that absorbs the anaerobic sediments is
fermented in anaerobic environment,and hence eliminate the fire by
preventing the free radical from reacting with the oxygen in the
air. It is determined that after it is covered by the mixture of
the sediments and water in the proportion of 1:1-10, the pile will
have its chemical activity reduced 20-25 per cent as such is a
proof that anaerobic sediments can prevent waste fire.

Besides, after the waste is treated with the sediments, it can
be turned into soil of high fertility through ferment and being fa-
vorable to reclamation and vegetation above the pile.

A research by the Bureau of Mines of US has found that surfactants can drastically reduce or prevent acid mine drainage from waste piles by inhibiting the growth of acid-producing bacteria, i.e.

by inhibiting Thiobacillus ferrooxidans, a type of bacteria that can promote pyrite oxidized to form an acid. The Thiobacillus ferrooxidans is protected by an outer membrane that enables it to survive in arid environment. Anionic surfactants containing negatively charged ions, can be used to kill the bacteria and thus slow down the oxidation of acid-forming pyrite by destroying the membrane.

Among the anionic surfantants tested to date, sodium lauryl sulphate appears to be the most effective as a bactericide. The other possible alternatives are alpha olefin sulphonate and alkyl benzene sulphonate.

In fact, these surfactants are contained in the common household necessities such as laundry detergent, shampoo and toothpaste. They are, therefore, readily available, inexpensive, biodegradable and present no environmental problems at low concentration. To treat a pile occupying an area of 1acre, 20-60 gallons of 50 P.P.m solution of 30 per cent sodium lauryl sulphate will be used.

The experiment carried out by the U.S. Bureau of Mines high oxidation pile in Belkley of West Virginia shows that the quality of the water has been greatly improved within one month of the sodium lauryl sulphate application. Acidity, sulphate and iron concentrations were reduced by up to 95 per cent four months after treatment.

CONCLUSION

The recovery of pyrite and coal from the waste can not only eliminate pile combustion as a whole, useful resources can be recovered as well. It is a feasible technology to recover pyrite from the washery refuse with high sulphate by gravity preparation method.

The pile fire can be eliminated by spraying lime sludge or the mixture of lime sludge and sodium hydroxide, a method of low cost and wide polularity.

The coal waste resources are abundant in China but utilized few. The comprehensive utilization of coal waste should be popularized in fields such as building materials, road construction, filling in mined area or surface subsidence area as well as reclamation in which most of the waste will be utilized until no more piles will be left and thus eliminate pollution caused by the waste pile. It is suggested that no more permanent waste piles will be piled up and the utilization of the waste must be taken into consideration when coal mine design and planning is under way to ensure the full utilization of coal waste. Since the waste from heading hold 70-80 per cent of the total coal mining waste, it is necessary to develop methods of filling them under-ground. China will conduct more research and testing works in this field.

REFERENCES

1   Proceedings of the First Conference of Environment Protection of China Coal Industry, Office of Environmental Protection, China's Ministry of Coal Industry, 1983.
2   Control Of Waste Piles Fire, West Fengcheng Coal Mines Administration, 1983,
3   A new Method of Extinguishing Waste Piles: Comprehensive Utilization of Coal Mining Wastes and Stone-like Coal, 1983, No2.
4   A New Way to Develop Energy Resource--Recover Coal from Coal Mining Wastes, Committee of Coal Technical Consultants, China Ministry of Coal Industry, September, 1984
5   Dr. of Technical Science M.P.Zborshchir et al, Application of New Methods to Extinguish Waste Pile Fire--An Urgent Task, Ukraine Coal, 1985, No11, pp 30-34
6   Surfactants Can Control Acid Mine Drainage, Coal Age, 1986, No7 p61
7   Candidate of Biological Science O.G.Borrov, Application of Sediments in Biological Treatment of Sewage for Spraying at Waste Pile, Ukraine Coal, 1985, No4 pp30-31

Reclamation, Treatment and Utilization of Coal Mining Wastes, edited by A.K.M. Rainbow   207
Elsevier Science Publishers B.V., Amsterdam, 1987 — Printed in The Netherlands

COAL WASTE IN CIVIL ENGINEERING WORKS
2 Case Histories from South Africa

F.W. SOLESBURY

Partner; Hawkins, Hawkins and Osborn; Consulting Engineers, Rivonia; Republic of South Africa.

SUMMARY

Many large heaps of coal mine waste in the form of shale and coal discards are present in the mining areas of the Eastern Transvaal and Northern Natal. Much of this material has been subject to spontaneous combustion which appears to have overcome some of the deleterious engineering properties associated with the natural Karoo aged sediments. Changes to the boiler feed in the modern power stations has robbed the civil engineer of a useful source of clinker and left him with millions of tonnes of relatively useless P.F.A. Two studies, one involving clinker and the other burnt shale are reviewed in this paper.

BACKGROUND GEOLOGY

The coal resources of Southern Africa are derived from the Karoo Supergroup which ranges in age from 100 to 160 million years, which covers the later Carboniferous, Permian, Triassic and early Jurassic periods of the Northern Hemisphere.

These rocks are part of the last depositional epoch of Gondwanland prior to its break up around 160 to 130 millions years ago. Similar suites of shallow water sediments are found in India, Antarctica, Australia and South Africa.

The Karoo period opened with the widespread Dwyka glaciation which was followed by the development of distinct basinal sedimentary facies, with only the northern one containing coal measures. Most of these carbonaceous deposits appear to be derived from drift vegatation associated with backswamps and deltaic conditions; hence the characteristic seat earths of the Carboniferous aged coals are missing. The coals are associated with siltstones and fine sandstones in upward coarsening cycles. The purity of the coal deposits, their thickness and laterial continuity are thus determined by contemporary sedimentalogical processes. This suite of rocks are the Ecca Group.

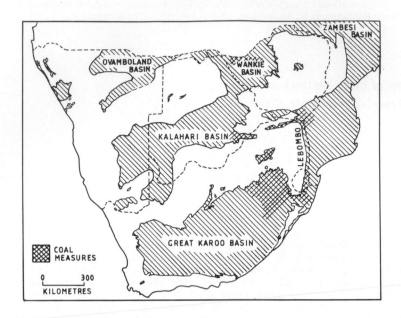

Figure 1. The distribution of Karoo Rocks in Southern Africa.

The argillaceous rocks in which the coal seams occur contain varying amounts of carbonaceous material and are also generally characterised by the presence of phyllo-silicate minerals such as micas and chlorites and by various clay minerals. All three of these mineralogical groups are associated with micro-laminations and incipient parting planes. This latter fabric only manifests itself when these rocks are subject to dessication resulting from by tunnelling, or mining or quarrying activities.

As a consequence of this relatively rapid decay brought about by exposure, there is a great reluctance to use the fine grained sediments in any form of engineering work.

Even the coarser arenaceous rocks still contain linear segregations of these minerals. In a major double-span rail tunnel recently completed for S.A. Transport Services, the RQD of a fine grained sandstone deminished from 100 to 0 over a period of 6 months.

The Ecca is followed by the Beaufort Group which is a similar group of argillaceous rocks which currently cover more than half of South Africa. The Karoo finished with the Molteno sandstone and the Stormberg volcanic outpourings. The total thickness of Karoo rocks is nearly 9 000 metres of shallow water or terrestial deposits.

Prior to the advent of coal pulverisation for powerstation boiler feed, raw coal was used and this tended to form into a hard gravel size clinker.

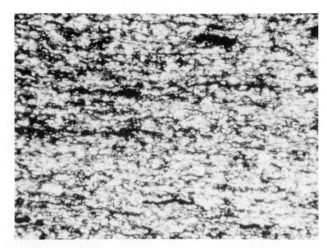

Figure 2. Thin Section Showing the Incipient Bedding in Ecca Siltstone.

The dumps of this material, associated with the older powerstations are much sought after for various civil engineering purposes.

The vast majority of coal waste and shale dumps have over the years caught fire and many still burn causing pollution problems in some of the older mining areas of the Eastern Transvaal and Northern Natal. In the more modern operations these waste tips are rolled to preclude spontaneous combustion and also to await the days of the fluidised bed boilers; when some operators think these dumps will be reused in the same way that the gold mine dumps are being recycled.

Two examples of the use of coal mine waste are covered by this paper, both of which have been constructed.

CASE HISTORY I : HEAVY DUTY COLLIERY HAUL ROAD

In order to supply $\pm$ 900 000 tonnes of coal per month to a new Escom power station, situated 20 kilometres to the south of Vereeniging on the Vaal River, a new opencast coal mine was brought into production.

To connect the opencast workings to the crushers and washing plant a haul road approximately 4 km long, has been constructed. At the time of design a final decision on the haulers to be purchased had not been made, other than the fact that they would be in the $\pm$ 180 tonne range.

A detailed centre line soil survey of the most suitable route was carried out. This revealed that there are two major subgrade soil types, a stiff potentially expansive, alluvial clay adjacent to the initial open cut, - this is called the lower terrace, and a deep, potentially collapsible aeolian soils at the processing end of the route or upper terrace. These occur over about 50% of the route respectively. Laboratory tests were carried out on these materials which confirmed their respective potential for movement, but also confirmed the initial view that the aeolian soils would be suitable for selected fill and subgrade material, as the road had to be constructed before mining operations. As no hard or even intermediate grade rock would be spoiled as a result of the mining operations, off site sources for the subbase and wearing course layers were examined. These included ash from the nearby SASOL plant, dolerite from a large dyke occuring 10 kilometre to the south, andesite scalpings, a previously non saleable material, from a local stone crusher and clinker from the adjacent old Vaal Power Station. All these materials were tested for strength, grading and durability.

MATERIAL	PERCENTAGE RANGE			PLASTICITY INDEX	MAX. DRY DENSITY	OPTIMUM MOISTURE CONTENT	CBR % AT 95% M.O.D
	GRAVEL	SAND	FINES				
AEOLIAN SAND	5	35	50	8 - 10	2 020 kg/m^3	8,2%	48
CLINKER	70	20	10	NON PLASTIC	1 495 kg/m^3	21,8%	16
SAND & CLINKER	42	40	19	5	1 880 kg/m^3	9,1%	41

Table 1. Major Soil Parameters for Engineering Design Purposes.

Pavement Design

The Mechanistic Design Procedure (MDP) was used for the design of the pavement layers. This involves a linear elastic analysis to determine the stress, strains and deflections for any number of pavement layers at any depth. Vehicular inputs include, wheel loads, axle loads, axle spacing, tyre pressures and frequency of vehicles. Soil parameters require the Elastic Modulus and the Poissons Ratio of each material for each proposed pavement layer. The reaction of each layer to the applied load can then be computed and compared against safe values. In this way the final design, the relative merits of different materials and layer thicknesses can be evaluated and costed.

The calculations where carried out for a Euclid, Komatsu, Titan 33 (rear dumper type haulers) and the Unit Rig BD180 (mid dumper) all with fairly similar loaded mass - namely 180 tonnes; the former type having 2 axles and a much higher wheel load than the 3 axle mid dumper type hauler. Each vehicle is 6,5 metres wide. It was estimated that a maximum of 22 loaded vehicles per hour would travel the haul road, with an average of 16 per hour running 24 hours per day, 6 days a week. Between 1 in 6 and 1 in 8 would return loaded with reject material.

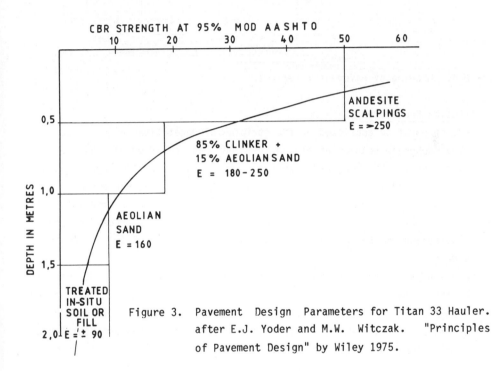

Figure 3. Pavement Design Parameters for Titan 33 Hauler. after E.J. Yoder and M.W. Witczak. "Principles of Pavement Design" by Wiley 1975.

The final design of the road prism depended on the expected traffic but generally consisted of a 500 mm thick subbase of Vaal Power Station clinker blended with aeolian sand, which in turn was overlaid by about 500 mm of quarry scalpings as a upper subbase and wearing course. Each of these two layers required approximately 10 000 m^3 per kilometre of 20 m wide road. A similar pavement was constructed in the tip area, but a lighter specification was designed for the workshop area which was not normally trafficked by loaded haulers.

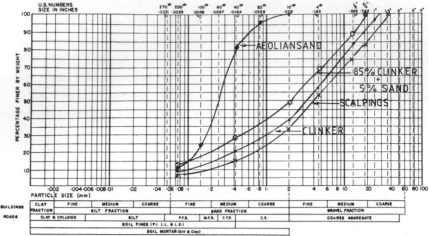

Figure 4.   Grading of pavement materials.

Construction Procedures

Site preparation consisted of the striping of vegetation and topsoil.   In the clay subgrade section no ground treatment was specified and these active soils were covered by compacted layers of aeolian sand from the upper terrace area.   These layers were further traversed by loaded haulers moving on predetermined routes to provide compaction.   Weak spots were repaired and additional layers placed and the compaction repeated.   This process preceeded the construction of the selected layers by 3 months.   This fill is 1 to 2 metres deep.

On the upper terrace the insitu loose aeolian soil was scarified, heavily watered and compacted by heavy vibrating rollers.   This process densified the highly voided natural material and achieved the required minimum selected fill CBR of 7%.   This area was similiarly traversed by loaded haulage vehicles on their way to the lower terraced roads.

Properties of Clinker

The clinker used on this project was obtained from the Vaal Power Station which is situated approximately 10 kilometres south of the project area.   The clinker had been stockpiled for many years and in places has become cemented into massive hard concretions.   The individual "pebbles" are hard but are typically cellular and hence this material has a low dry density and absorbs large amounts of compaction water. See Table 1.

This material has been used over the past few years on mine service roads and appears to be disintegrating due to wetting and drying.   This fact together with the relatively low compaction strength and the lack of fines, resulted in this material only being utilised in the lower subbase layer and

then mixed with 15% of the aeolian sand. This blending improved the CBR
strength considerably and the Dynamic Penetrometer Tests have recorded insitu
CBR values of more than 50 - See Figure 5.

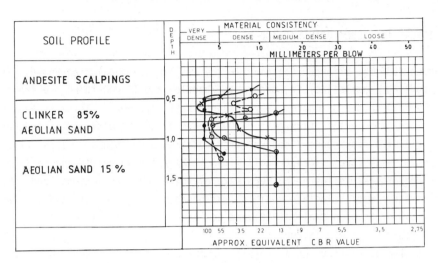

Figure 5. Dynamic Penetrometer Test Values.

Figure 6. End user of haul road
180 tonne Euclid.

Figure 7. Clinker dumps awaiting
blending with sand.

The 1983/1984 construction costs (£1 = R2,00) using these different
materials were:-

Scalpings	R9,80 m^3
Clinker/Sand	R7,80 m^3
Aeolian Sand	R2,20 m^3
Insitu Preparation	R1,80 m^3

This resulted in a price of R300 000 per kilometre or £150 000 which compares favourably with the new 20 metre wide runway at Jan Smuts Airport (Johannesburg) which cost R500 000 per kilometre when it is considered that the 180 tonne haulers have twice the axle load of the Boeing 747.

CASE HISTORY II : BULK COAL STORAGE BINS

During 1985 the Blinkpan Colliery in the Eastern Transvaal, expanded its coal handling facilities in order to handle the 10 000 tonne ore trains currently used on the Richards Bay Export rail system. This expansion necessitated the installation of a 34 000 tonne surge capacity and a load out bin capable of a continuous discharge of 4 000 tonnes per hour. This paper only deals with the storage facilities. The relative merits and costs of stacker/reclaimers, silos or bunkers were evaluated by the client and the latter system was selected. This was based on the costs of handling two products (Power Station Smalls (PSS) and Low Ash Coal(LAC)), site conditions, flexibility and economic factors.

The system as constructed is totally automated and interlocked throughout to ensure continuous loadout.

Three types of bunker support were evaluated:-

o   compacted fill, with a stabilised base to support interlocking blocks or concrete lining.

o   roller compacted concrete lining, supported by a fill embankment.

o   Reinforced Earth, with requisite support embankment.

The latter was selected as having the least severe geotechnical constraints, its speed of construction and for the control of hopper geometry and hence flow characteristics. Laboratory tests showed that the LAC ideally required 45° slopes and the PSS 46°. It was feasible to obtain this geometry with Reinforced Earth and a common slope for both bins of 46° was utilised.

Bunker Design

The site of the two 17 000 tonne bunkers was on loose sandy transported soils which overlay fine grained residual soils, with weathered sandstone bedrock at ± 3,0 metres below ground level. Ground water was present at this interface.

The mechanical and structural requirements necessitated that the bunkers be founded on the sandstone with the reclaimer tunnel being excavated a further 2 metres into bedrock.

The Reinforced Earth structure is composed of precast cladding elements 1,5 metres high and 3,0 metres long, which form two eliptical structures - 18 metres high and 45 metres across at the top. These have a face angle of 46° braced by 6 metre long corrugated galvanised tie strips. These strips are bedded into a 70 m wide laterite layer immediately behind the panels and then an embankment built of natural sandy hillwash and burnt shale / coal discards. Both the latter materials or common fill were obtained on the mine property. Altogether 20 000 m³ of laterite, 70 000 m³ of aeolian sandy hillwash and 80 000 m³ of burnt shale / discard were used.

Due to differential movements between the panels and the corrosive nature of drainage water from the stockpiled, a geofabric drainage layer was used behind the panel joints as opposed to a sealed joint.

Laboratory testing on the various embankment materials resulted in a design that forced the failure circle behind and beneath the reinforced section of the structure - Figure 8. The stability of the common fill is treated in the same way. Internal stability is calculated to ensure that a sufficient contact area of reinforcing strip is available in the resistant zone to mobilise the frictional forces and to provide anadequate factor of safety under all loading conditions. As a consequence a strict grading envelope is specified as well as compaction characteristics.

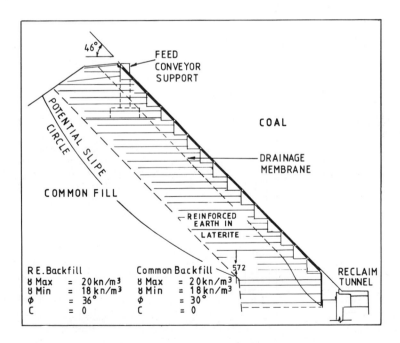

Figure 8.   Cross Section through Embankment.

Common Fill Embankment

Laboratory testing was carried out on the available fill materials in order to obtain the basic soil mechanics properties for embankment design purposes. In addition chemical tests were conducted on the burnt shale / discard materials.

In the burnt shale dumps, whilst there is a variation in the coarse gravel fraction, the fines content appeared to be relatively uniform - Figure 9.

The Atterberg constants showed the natural sandy soil to have a low to medium plasticity with ± 6% linear shrinkage. As the burnt shale / discard is non-plastic no Atterberg Limit tests where carried out.

A series of consolidated, soaked drained triaxial tests were carried out on both common fill materials which had previously been compacted to around 95% modified AASHTO dry density. Table 2 summarises fill material parameters. The main points are:-

o The burnt shale has a low compacted density 1 370 - 1 400 kg/m^3.

o The burnt shale has a high void ratio - 0,9 even at 95% modified AASHTO density.

o The burnt shale has a considerable intake of moisture.

o The individual deviator stress versus strain curves indicate that the compacted materials behave in a more or less pseudo-plastic manner at strain values in excess of 5%.

o The shear strength of the burnt shale increases in a curvilinear manner for increasing consolidation pressure.

From this data recommended drained shear strength parameters were proposed.

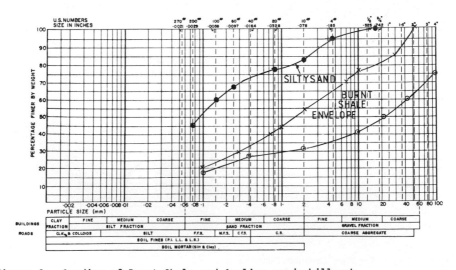

Figure 9. Grading of Burnt Shale and Aeolian sandy hillwash.

Material	Max. Dry Density (kg/m^3)	Optimum Moisture Content %	Shear Strength Parameters	
			Cohesion (c)	Angle of Friction
Burnt Shale	1471	18,5	150kPa	31°
Silty Sand	1903	11,3	30kPa	25°

Table 2. Summary of Main Material Parameters.

Routine compact tests were carried out to determine compaction specifications. The lower dry density and high optimum moisture content for the burnt shale is a reflection of the high porosity.

Chemical tests were carried out on samples of the burnt shale as this was suspected of possessing a high sulphate content and other corrosive properties; and this is confirmed - see Table 3.

pH Value	6,25
Conductivity mS/m at 25°C	332
Total Dissolved Solids	4228
Chloride, Cl	46
Sulphate, SO	2507
Saturation Index (Langelier)	-0,44

Note   :   Values expressed in mg/l where applicable.

Table 3. Summary of Chemical Testing on Burnt Shale

These results indicate that the burnt shale is indeed aggresive to concrete and steel particularly in the presence of moisture. Consequently this material was recommended only for use in the outer section of the embankment away from the reclaimer tunnel, Reinforced Earth panels and reinforcing strips.

Construction of the structure and its earthwork proceeded according to schedule - 9 months and within budget and is currently handling export coal.

218

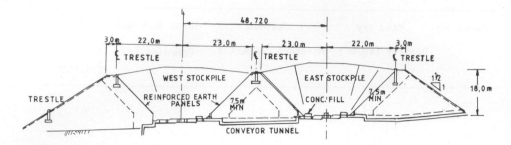

Figure 10. Cross Section through the Bunkers.

Figure 11. Aerial View of Bunkers (Note reinforcing strips in inner lateritic
layer).

ACKNOWLEDGEMENTS

The author is grateful to Andrew Smith of Reinforced Earth (S.A.) (Pty)
Limited for providing the data and photograph of the coal storage bunker; and
his colleaques Paul Taylor and Peter Hurry are thanked for their contributions
and critism of the Haul Road case history.

Reclamation, Treatment and Utilization of Coal Mining Wastes, edited by A.K.M. Rainbow   219
Elsevier Science Publishers B.V., Amsterdam, 1987—Printed in The Netherlands

MINESTONE IMPOUNDMENT DAMS FOR FLUID FLY ASH STORAGE

Dr A.K.M. RAINBOW[1] and Professor Dr. Hab. Inz, K.M. SKARZYNSKA[2]

[1]Head British Coal Corporation, Minestone Services, Philadelphia, Tyne and Wear, United Kingdom

[2]Professor of Soil Mechaics, University of Krakow, Poland

INTRODUCTION

The generation of electrical energy from coal-fired power stations create substantial quantities of residual waste materials. The principle source of waste originates during the underground coal mining operations where the volume of waste arisings can be as much as the volume of coal produced. The secondary source of wastes results from the burning of the coal where the yield may be of the order of 13-15 per cent of the volume of coal burnt. As a consequence both the coal mining and electrical generating industries are faced with a continual requirement to provide facilities for the disposal of the waste(s) produced.

The amount of material required for the construction of impounding dams for the storage of ash is substantial therefore, the utilization of the ash and/ or the mine waste (minestone) as substitutes for conventional waste retaining embankment material has a number of important benefits to the industries, (i) the conservation of conventional materials including a reduction in environmental disturbance which would have been caused in the winning and transport of conventional materials. (ii) the utilization of mine wastes which would otherwise have to be disposed of in some other way and (iii) substantial cost savings to both industries.

In the United Kingdom impounding dams are used extensively by British Coal to contain slurry and tailings from the coal preparation process. (Plate 1). These structures perform essentially the same function as impoundment dams for fluid ash storage. Tailings lagoon embankments are constructed using Minestone and the design criteria and construction methods can equally be applied to structures for ash storage.

The following factors are considered in the design of a lagoon bank by British Coal (National Coal Board 1970):

a. The purpose and size of the structure:

b. The stability of the foundations under the weight of the embankment on sloping ground, an investigation of its stability both under and downhill of the embankment will usually be necessary;

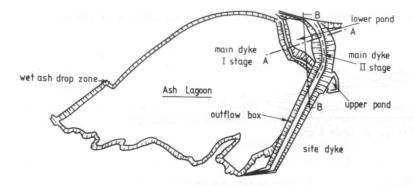

Fig. 6. Plan of Ash Lagoon and Dykes

The 61m high dam constructed near Moundsville, West Virginia, serves to impound fly ash in slurry form from the American Electric Power Company's Mitchell Power Plant. The dam is being constructed utilizing the coarse coal refuse (minestone) from the McElroy Mine (Saxena et al 1984). The cross sectio of the dam is shown in Figure 7.

The wastes, typical of the majority of coal mine operations in Eastern United States, consists of shales, siltstones, mudstones, stiff clays and include naturally contaminated coal. The gradation of the materials varied during the construction period and was dependent on the mine operation and the preparation plant operation. Various degrees of particle breakdown occurred during handling, conditioning and compacting of the fill.

Based on the findings of investigations carried out on several test pads, it was recommended that coal refuse should be compacted near the optimum moisture content to a density greater than 95% of the maximum dry density and that compacted lifts should not be greater than 0.3m in thickness. Types of equipment found suitable for this operation were the vibratory smooth drum and the sheepsfoot roller each having a total applied force of not less than 20,930kg.

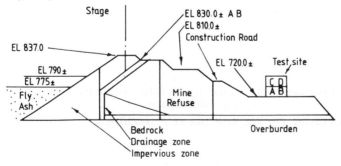

Fig. 7. Cross Section of Fly Ash Dam Near Moundsville U.S.A.

Theoretically, by the judicious selection from the enormous stocks of minestone material having the desired geotechnical properties can be found. However, in practice it has been found that selection is not normally required and that fresh wrought minestone delivered directly from the mine can be made to perform admirably for the construction of lagoon embankments.

The use of minestone, in the way that it is used in this particular application, relies on achieving the desired permeability and its ability to withstand the aggressive nature of the lagoon interface.

Minestone is an amalgam of mainly soft rock debris from the Coal Measures (Rainbow 1983) and may typically contain 60 per cent clay minerals (Collins 1976)

In this respect the main geotechnical characteristics which require consideration are therefore, those which effect the degree of particle packing which can occur; namely the particle size distribution (psd) and the relationship between achievable density and moisture content. Investigations suggest that minestone having a mean value in excess of 15 per cent passing the 75 micron sieve are capable of yielding permeabilities less than 10^{-6} m/sec.

By their very nature minestone can, with the appropriate degree of compaction, be made to breakdown to this level of fines content. An understanding of composition and original particle size distribution and moisture content is therefore invaluable. Figure 8 compares the particle size distribution for minestone used on the lagoon embankments at Gale Common (UK), Moundsville (USA) and Przezchlebie (Poland).

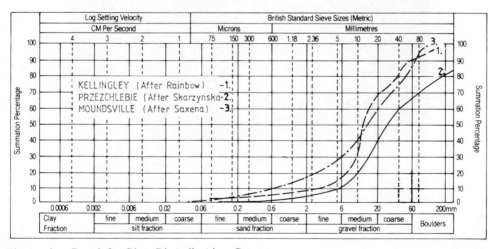

Figure 8. Particle Size Distribution Curves

It is however, the relationship between particle size distribution, fines content, moisture content and degree of compaction which ensures the final

228

permeability of the structure.

While much research has been conducted on the compaction characteristics and procedures for soils used in hydraulic structures such as embankments and earth dams a relatively small amount of information is available for utilizing minestone for similar structures. It is essential to identify insitu compaction characteristics for utilization of minestone in large scale engineering disposal facilities that later become reclaimed land. It is strongly recommended that investigations, prior to building minestone structures should be carried out on trial embankments with different thickness of layer and varying degrees of compaction energy (Saxena et al 1984, Michalski, Skarzynska 1984).

Plate 4.   Typical Placement and Compaction of Minestone Layer

Since it is impossible to control the quality of compaction of the minestone structure after construction by any kind of soundings/penetration test, etc., the emphasis has to be laid on control while construction is still in progress i.e. current control. For dams and embankments of reasonable height it will be essential to establish a field laboratory.

At Przezchlebie, for example, each compacted layer was monitored using the sand replacement method before the contractor was allowed to add a successive layer - the moisture content and maximum dry density was obtained (Figure 9).

From the tests carried out it can generally be said that in the majority of cases the required compaction was achieved. Of 400 control tests compaction was found not to be adequate in about 40 cases, and the layers were qualified for repeated compaction. The reason for inadequate compaction was usually too low a moisture content of material or too small a number of passes of the compacting plant. Accordingly, the systematic sprinkling of layers during compaction was introduced, or a repeat of compaction, which in effect gave the required results.

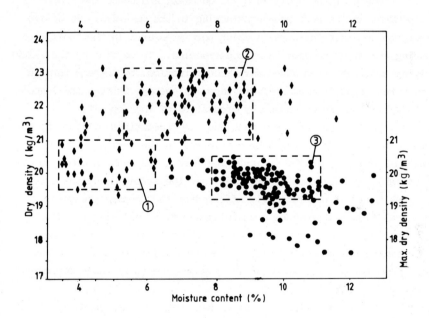

(1. Relative compaction >1.   2. Relative Compaction >1.
3. Data from Proctor Tests).

Figure 9 Distribution of Dry Density vs. Moisture Content for Minestone Used in the Embankment at Przezchlebie

The shear strength is the most important characteristic of the spoil material in the design of lagoon bank cross-section. Normally the appropriate shear strength parameters for stability analyses are those in terms of effective stress, i.e. the effective cohesion c', and the effective angle of shearing resistance $\emptyset$'. The test procedure for determining c' and $\emptyset$' in the laboratory should correspond as far as possible to the stress and drainage conditions in the lagoon bank. The consolidated undrained triaxial test with pore pressure measurements and the drained triaxial test are normally used by British Coal.

The specimens for the laboratory triaxial tests should preferably be taken either from an existing tip constructed similarly to that under investigation, or from the initial or trial construction of the tip. Where this is not practicable at the design stage, guidance for the design can be obtained from samples compacted in the laboratory at a moisture content and to a density corresponding to that expected in the actual construction. Where the data obtained from the laboratory compacted specimens are critical with respect to the stability of the proposed design, the validity of the design parameters should be checked by testing samples taken from the tip during the early stages of its construction.

The shear strength of the spoil in a lagoon bank, and hence the factor of safety, is dependent on the pore pressures acting in the material. In design it is necessary to consider the ways in which pore pressures in the spoil may be set up during and after completion of construction. These are; by the weight of the overlying spoil as it is added during construction, by seepage through the lagoon bank resulting from natural drainage, by seepage through the lagoon bank, by drawdown in the lagoon caused by excavation of the deposit.

Seepage of water through a lagoon bank and its foundations will set up pore pressures, the effect of which should be considered in the design. If an effective internal drainage system is installed within the bank in order to draw down the water table it will usually be found that the stability of the outer slope of the bank will not be critically affected by the seepage pressures.

A theoretical flow net provides a useful means of visualising the flow pattern of the water seepage through the bank. The net can be obtained by graphical, electrical analogue or finite-element analytical methods. In the stability analysis the pore pressures acting are assumed to be those determined from the relevant flow net.

The actual flow net produced in a lagoon bank will be influenced by the following factors.

    a.   The build-up of the deposit in the lagoon and the contact with the bank will have the effect of increasing the length of the seepage path. This will reduce the seepage pore pressures in the bank compared to those that would otherwise be acting if the lagoon was filled with water only. The presence of a layer of supernatant water on the surface of the deposit will superimpose additional gravity seepage through the bank.

    b.   The amount of water and also the time required to produce a flow net corresponding to steady-state seepage is dependent on the permeability of the bank material. In some cases the combination of these factors may result in the full flow net being developed, whereas in other cases it may not. For example, good draw-off arrangements reduce the amount of water available for seepage through the bank thereby affecting the development of the full flow net.

c.   For a lagoon bank constructed of relatively impermeable material the construction pore pressures developed within the bank by the weight of the fill may be more critical than the pressures developed by gravity seepage from the lagoon, particularly if the fill is placed at a high moisture content.

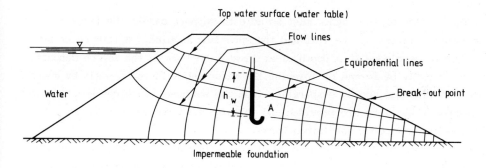

Flow line: a line indicating the direction of the flow of water through the section. Equipotential line: a line along which the level of the water is constant. (Water rises to the same level in piezometers installed anywhere along a given equipotential line.)

Figure 10.   Example of a Theoretical Flow Net in a Homogeneous Isotropic Bank Retaining Water

Once a lagoon is filled to design capacity the problem of placing fill over the lagoon deposits should be dealt with similarly to designing a tip to be placed over soft clay foundations.

Not more than 3m of coarse discard should be laid on top of a partially dried out lagoon deposit. If a thicker layer is considered necessary, then a comprehensive site investigation should be carried out to check on the engineering provisions required to ensure stability, especially in relation to the banks. These provisions should normally include special drainage measures to ensure:

a.   that no serious pore water pressures develop in the retaining banks;

b.   that adequate drainage paths are provided to take away water expelled during consolidation of the deposits; and

c.   that the shear strength of the deposits is adequate for supporting the overlying fill.

The control of the rate of placing the fill is an essential complementary factor to the above drainage measures to ensure that the pore pressures in the deposit, the retaining bank and the foundations do not develop to an extent which would reduce the factor of safety below the required value. The method of placing the fill must avoid overstressing the lagoon deposit, which would disturb and remould the stratified structure of the deposit. This would tend to weaken the deposit and also adversely affect its consolidation characteristics. The latter would in turn reduce the rate of gain in shear strength due to consolidation under the loading applied by the fill. The fill should be placed in successive horizontal layers, each layer covering as large an area as possible before beginning the next.

The surface of the lagoon must be able to support safely the first as well as subsequent layers of the fill and the spreading machine. A firm crust should therefore be developed to a depth of at least 1 to 1.5m in the lagoon deposit. This can usually be formed by lowering the water table in the deposit by means of gravity drainage and by evaporation from the surface, which can be effected if adequate provisions are provided and followed with regard to drainage arrangements in lagoons and lagoon banks.

It is desirable that the grading of the material in the bottom layer of the fill be selected so that it will form a drainage layer on the surface of the lagoon deposit, to aid the dissipation of excess pore pressures in the deposit.

Table 1 summarizes details of the examples cited where the waste from coal mining has been used in the construction of impounding dams for the storage of fluid fly ash and Table 2 summarizes geotechnical parameters obtained for the minestone used in those structures.

TABLE 1   Details of Minestone Dams used to Illustrate the Successful Use of Minestone

	Name of the Dam				
	Silesia	Valleyfield	Gale Common	Przezchlebie	Moundsville
Material Used in Dam Construction	Fly Ash	Fly Ash & Minestone	Minestone & Fly Ash	Minestone	Minestone
Volume of Material Used		325,000m³	18M m³	1M m³	
Length of Dam		7Km		940 m	
Layer Thickness		200mm	100mm	500mm	300mm
Moisture Content		28% (Ave.)			
Compacting Plant		Towed Vibratory Rollers		12 Tonnes Vibratory Roller	Vibratory Smooth Drum Roller
Lagoon Capacity		2 Mm³	43Mm³	7Mm³	

TABLE 2   Properties of Minestone Used in Cited Examples

Properties	Unit	Gale Common Dam	Przezchlebie Dam	Moundsville Dam
Fraction Content Cobbles Gravel Sand Silt, Clay	% 	16 65 16 3	17-33 49-59 11-14 7-10	
Uniformity Coefficient	-	-	87-267	
Specific Gravity	-	2.15-2.31	2.32-2.60	2.19-2.50
Moisture Content	%	6-9	4-6	-
Liquid Limit		30-38	-	29
Plastic Limit	%	16-22	-	16
Plasticity Index		14-16	-	13
Maximum Dry Density	kN/m^3	15.0-16.2	17.0-19.4	17.3-20.4
Optimum Moisture Content	%	8-9	9-11	7-9
Permeability Coefficient	m/s	10^{-4}-10^{-5}	2×10^{-4}	4×10^{-8}-10^{-4}
Angle of Internal Friction	deg	25-35	37-44	31-40
Moisture Content (as placed)	%	7-10	6-9	7-9
Permeability Coefficient (compacted)	m/s	10^{-7}-10^{-9}	3×10^{-5}	10^{-8}-8×10^{-4}
Dry Density (compacted)	kN/m^3	16.2-19.0	17.1-26.5	-

In addition to obtaining the geotechnical parameters of minestone a
comprehensive series of insitu tests should be undertaken in the lagoon material
as the properties of the fly ash are very important both in the design stage
and during construction of the minestone impoundment dams for fly ash (Table 3).

TABLE 3 Properties of Fuel Ash Deposited in Lagoons and Compacted in Dams

Properties	Unit	Silesia Lagoon	Gale Common Lagoon	Przezchlebie Lagoon	Valleyfield Lagoon
	−	Silt, Sandy-Silt	−	Silt, Sandy-Silt	−
Uniformity Coefficient	−	14	6.7	11	−
Specific Gravity	−	1.86–2.31	2.24	2.05–2.12	−
Moisture Content	%	62–70	18	40–50	>60
Dry Density	kN/m^3	7.84–8.66	−	8.72–10.0	−
Maximum Dry Density	kN/m^3	−	14.6	11.46	11.51
Optimum Moisture Content	%	−	17.9	25	33
Permeability Coefficient	m/s	2.10^{-6}	1.48×10^{-6}	10^{-7}–10^{-5}	−
Permeability Coefficient (insitu)	m/s	10^{-6}–10^{-5}	−	−	−
Angle of Internal Friction	deg	39	29	31	39
Cohesion	kN/m^2	12	−	0–5	5
Moisture Content (as placed)	%	−	−	−	30–40
Permeability Coefficient (compacted)	m/s	8.7×10^{-6}	−	−	5.3×10^{-7}
Dry Density (compacted)	kN/m^3	10.72	−	−	10.56–11.66

The geotechnical parameters collected and collated during the construction of the lagoons may prove invaluable since when the lagoon reaches its maximum potential capacity there may be a need for the development of the site for environment protection measures or for later utilization.

It may be that to fully develop the site surcharging of the lagoon may be necessary and an evaluation of the strength behaviour of the embankments and deposited ash will be essential. To achieve this additional investigations may be required and in this context it could be dangerous to carry out the investigations directly on the surface of fairly weak ash deposits. For this reason it may be appropriate to build experimental embankments on the lagoon surface in order to provide a working platform for site field studies. This procedure was adopted at Przezchlebie and the method and some results are described below.

Two experimental embankments were constructed of minestone on the ash lagoon surface. Boreholes were drilled from the surface of the embankments to a depth of 10m. Vane tests were also carried out, both in the vicinity of the boreholes and also from the ash surface at a distance of 3 to 20m from the base of the embankment. In order to gain some knowledge of deposit stratigraphy and to take samples for laboratory study, open pits were made in the vicinity of the embankments (Figure 11).

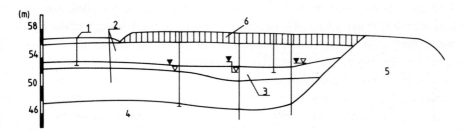

1. Loose   2. Moderately Compacted
3. Compacted and Highly Compacted 4. Substratum of Lagoon
5. Main Dyke   6. Experimental Embankment

Figure 11   Compaction Zones of Ash Deposits in the Sub-Base of
            Experimental Embankments Przezchlebie (Poland).

In the vertical section of ash deposits four basic observed zones were distinghished:

(i) a surface zone with a thickness of about 1-2m in a loose state.
       The shearing resistance within this zone was 15 to 30 kN/m^2

(ii) a transistional zone 1-2m thick, moderately compacted. Values of
       shearing resistance were from 40 to 150 kN/m^2

(iii) a highly compacted zone generally 0.5-1m thick but reaching 2m.
       In places the shearing resistance was greater than 250 kN/m^2.

(iv)   a zone lying below the highly compacted layer in which a large
       variability of shearing resistance in the range of 80-180 kN/m^2 was
       found. It may be inferred that particularly high compaction in ash
       deposits occurs in those regions in which the ground water level
       fluctuates.

Concluding Comments

On the basis of relatively long term observations of hydraulic structures
constructed using minestone or minestone-ash cited in this paper and
additionally observations of the use of minestone in hydraulically active
environments which are more aggressive that the passive lagoon embankments
(such as the use of minestone fill on the foreshore of the Pegwell Bay
International Hoverport and the considerable use on the Delta Project - the
scheme to close-off the Rhine - Maas - Schelde Delta along the North Sea) it is
not unreasonable to suggest that minestone can be successfully engineered for
use in impounding dams for fluid ash storage.

A further advantage is that engineering experience suggests also that the
cost differential in utilizing minestone in such a way compared to the cost of
building some heaps is not great. Due to the shortage of suitable land for
stockpiles of minestone more industrial complexes are turning to a "central
dumping" philosophy where fly ash and minestone from a number of power stations
and collieries deliver their wastes. Because the material for building lagoons
are an economic important factor it is sensible to utilize those materials more
widely but before this can be accomplished greater understanding of the
geotechnical properties of the material(s) is necessary.

As a result of the encouraging use of minestone as an alternative material
for constructing water retaining structures and from observations of other
types of structures perhaps with imagination we see minestone utilized in
more demanding hydraulically active applications - there is certainly no
shortage of relatively cheap minestone in many coal mining countries and if -
as in Britain one considers the possible application in tidal barrages we may
yet see structures like these utilizing considerable quantities of minestone.
Indeed an embankment as in Figure 12 may become a reality.

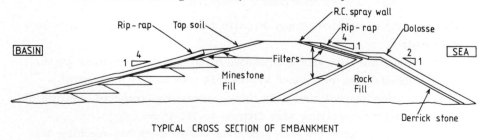

TYPICAL CROSS SECTION OF EMBANKMENT

Figure 12   Typical Cross Section of Embankment for Tidal Barrage

ACKNOWLEDGEMENTS

Dr A K M Rainbow wishes to thank British Coal Corporation for permission
to publish this paper and Professor K M Skarzynska wishes to thank Department
of Soil Mechanics and Earth Works, University of Agriculture in Krakow for
permission to publish this paper whilst undertaking the Talbot-Crosbie
Fellowship at Glasgow University.

Both Authors wish to thank the Central Electricity Generating Board,
Messrs, Rendel, Palmer and Tritton and Przedsiebiorstwa Materialow Podsadzkowych
for their assistance in providing data and figures where appropriate.

Both Authors wish it to be known that the views expressed in this paper
are their own and not necessarily those of their respective sponsoring
organisation.

REFERENCES

1. B. Bros, K, Parylak, "Seepage Control from Ash Lagoons", Proc., of the
    Eleventh International Conference on Soil Mechanics and Foundation
    Engineering, San Francisco, (1985), Vol. 3. p. 1183-1184.

2. R.J. Collins, "A Method of Measuring the Mineralogical Variation of Spoils
    from British Collieries, Clay Mineral". (1976), 11, p 31-50.

3. A.J. Cowan, "Valleyfield Ash Lagoons Building on Marginal and Derelict
    Land", Institution of Civil Engineers Conference, Glasgow (1986), p. 83-93.

4. P Michalski, K.M.Skarzynska, "Compactability of Coal Mining Wastes as a
    Fill Material", Symposium on the Reclamation, Treatment and Utilization of
    Coal Mining Wastes, Durham England (1984) p. 15.1-15.13.

5. National Coal Board - Technical Handbook, "Spoil Heaps and Lagoons",
    National Coal Board, Hobart House, London (1970).

6. A.D.M. Penman, "Tailings Dams and Lagoons, Building on Marginal and
    Derelict Land", Institution of Civil Engineers Conference, Glasgow (1986),
    p. 37-57.

7. A.K.M. Rainbow, "An Investigation of Some Factors Influencing the
    Suitability of Minestone as the Fill in Reinforced Earth Structures",
    National Coal Board, London   (1983), p. 562.

8. A.K.M. Rainbow, M Nutting, "Geotechnical Properties of British Minestone
    Considered for Landfill Projects", Symposium on Environmental Geotechnology
    Allentown, Pennsylvania, U.S.A. (1986).

9. A.K.M. Rainbow, K.M. Skarzynska, "Marginal and Derelict Land Fills and
    their Development", 8 Karjowa Kong. Mech., Gruntow i Fundamentowania,
    Wroclaw, Poland (1987).

10. S.K. Saxena, D.E. Lourie, J.S. Rao, "Compaction Criteria for Eastern Coal
    Waste Embankments", Journal of Geotechnical Engineering (ASCE) (1984),
    Vol. 110, No. 2, p. 262-284.

11. K.M. Skarzynska, P. Michalski, "The Use of Colliery Shales for Hydraulic
    Embankment Construction", 8th Danube-European Conference on Soil Mechanics
    and Foundation Engineering, Burnberg, (1986), p. 231-234.

238

12. K.M. Skarzynska, H. Burda, H. Kelpacz, "Hydraulic Structures Built of
     Unburnt Coal Mine Waste", (in Polish) Gospodarka Wodna (1987).

13. E. Zawisza, K.M. Skarzynska, "The Effect of the Ground Water Level on the
     Behaviour of an Ash Substratum", Proc., of Ninth European Conference on
     Soil Mechanics and Foundation Engineering, Dublin, Ireland (1987).

Reclamation, Treatment and Utilization of Coal Mining Wastes, edited by A.K.M. Rainbow   239
Elsevier Science Publishers B.V., Amsterdam, 1987 — Printed in The Netherlands

FLY ASH PONDS AND THE DAM LEGISLATION IN FINLAND

J. SAARELA[1]

[1]National Board of Waters, Finland Geotechnical design

SUMMARY

The new dam safety legislation in Finland came into force on
1 August 1984. It concerns also fly ash ponds with dams. The
Act applies to dams 3 m or more high.

INTRODUCTION

The new dam safety legislation in Finland came into force on
August 1984. It concerns also fly ash pond with dams. The Act
applies to dams 3 m or more high. The heigth is measured from
the lowest point, where the dam slope and earth surface meet to
the highest intended surface of the dammed substance. To reduce
the danger of potential damage, a safety surveillance programme
must be prepared for each dam falling within the scope of the
Act. The supervison of the observance of the Act and its regulations are the responsibility of the National Board of Waters and
its district administration with the exception of the rescue service.

In Finland there are 5 dams or the fly ash ponds at the power
plants, which are higher than 3 m.

FLY ASH PONDS IN FINLAND

In Finland the fly ash from power plants is generally stored
in ponds. Because power plants are situated on the coast, the
ponds are dammed in the sea. Generally they have earth dams,
and the fly ash sludge is pumped into the ponds. When the old
pond is full, one must buld the new pond. Mostly there are several ponds at one power plant, which are used. The height of
dams are 3-5 meters and the areas of greatest ponds are several
hectares. There has not been any serious problems with them,
for example collapses.

In Finland, there are 5 fly ash ponds, which are higher than
3 m.
In figures 1-2 there are some examples of fly ash ponds.

Fig. 1. An example of the fly ash pond dammed in the sea.

Fig. 2. An half-full fly ash pond.

DAM SAFETY LEGISLATION

The new dam safety legislation came into force on 1 August
1984. It conserns also the dams of the fly ash ponds, which are
3 m or higher.

The new legislation includes:

- the Dam Safety Act;
- the Dam Safety Statute;
- the amendments to the Water Act; and
- the amendments to the Water Statute.

In additon, the Dam Safety Codes of Practice were issued by the
Finnish National Board of Waters in March 1985.

The Dam Safety Act covers dams built for permanent use, together with related structures and equipment, regardless of the building material or method, or the quality of the substance impounded. It also covers temporary dams where applicable. Assembling
and keeping a safety file is however not required for the temporary structures.

The Act applies to dams 3 m or more high. The height is measured from the lowest point where the dam slope and earth surface
meet to the highest intended surface of the dammed substance.

However, it also applies to lower dams if the quantity of the
dammed substance is large, or if the substance is such that, in
the case of an accident, human life or health might be endangered,
or there is possibility of serious damage caused to the environment or property.

The Act specifies that the building of the a dam shall be carried out so that the structure, by its strength and design, poses
no threat to safety, and:

- the owner or responsible party must keep and maintain a file relevant to the dam's safety.
- the owner or the operator is required to assess the likely damage to the population or the property downstream and provide the
National Board of Waters, provincial Goverment, District Fire Service Chief and Municipal Fire Service Authority with his assessment; and

- where dams, by failing, could endanger life or property, the owner is obliged to prepare emergency plans in cooperation with the fire service authorities, to maintain the equipment and material referred to in the plans, to take other action to safeguard people and property and to participate in implementing these plans.

To reduce the danger of potential damage, a safety surveillance programme must be prepared for each dam falling within the scope of the Act.

The supervision of the observance of the Act and its regulations are the responsibility of the National Board of Waters and its district administration with the exception of the rescue service.

THE INSPECTIONS OF THE DAMS

All dams are classified in four categories in the following way:
- P dams are those which in the case of an accident will endanger life or health, or cause serious damage to the environment or property.

- N dams are those which do not belong to categories P, O or T
- O dams are those which in case of an accident cause only minimal danger.
- T dams are temporary structures as defined in the law.

The Dam Safety Codes of Practice include some technical recommendations concerning the stability, the drainage system, the heigth of the dam (including frost protection), the slope protection and the permitted and size of trees.

The safety surveillance programme includes:

- regular inspections every 5 years by a competent expert;
- yearly inspections in the intermediate years, by maintenance personnel; and
- surveillance between inspections according to the programme defined in the basic inspections.

The responsibilities of different parties for dam safety are as follows:
- the owner is responsible for the safety of a particular dam (the dam must be designed by a competent and experienced person); and,

The National Board of Waters, the State Power Companies and
some municipal and private dam owners have already begun to prepare potential damage assessments and emergency plans.

- the National Board of Waters must supervise the fulfilling of
all requirements of the dam safety legislation.

DAM BREACH HAZARD ANALYSIS

The National Board of Waters may order the constructor, owner
or operator of the dam to procede or prepare an analysis about
the danger to the population or property downstream. The results
of this Dam Breach Hazard Analysis (DBHA) have to be delivered to
the

- National Board of Waters;
- Provincial goverment;
- district head of the fire- and rescue organization;
- local fire- and rescue authorities; and,
- dam owner

The National Board may give more dltailed guidelines on the
preparation of DBHA.

The main approach for the preparation of the DBHA is to:
- elucidate facts pertinent to organizing operations in case of
a disaster;
- prepare basic information to prevent a potential hazard or to
limit it by the various necessary activities; and,
- confirm the hazard classification of the dam and, depending on
it, the level of technical requirements for the inspections.

The DBHA deals first with dangers threating human life or property in case of a dam disaster. In special cases the DBHA may
deal, for example, with cases where the hazardous substance causes threat to health and environment (such as tailing dams).
The first sten of the DBHA, also named Dam Breach Sensitivity
Analysis, is to simulate and compare the hazards of different possible dam disasters with reference to the magnitude of discharge
and its abruptness. As a second step, at least one breach hyd-
rograph, preferably the most dangerous, is chosen for each main
floodway to be routed through the entire valley. Different failure assumptions and flood condition in the area covered by the
DBHA, concidered under the Dam Safety Code of Practice, include:

- Failure under normal operational conditions. The dam breach
flood has to be routed until it has decreased to the magnitude
of a natural flood of 1 in 20 to 1 in 100 years yrequency, or
until it arrives at another P-dam.

- Failure related to an extraodinary flood.
- Investigation of the effect of different dam operations during
extraordinary flood events.

REFERENCE

Loukola,E. - Dam safety Legislation in Finland.
Communication No. 13 to the Congress of ICOLD,
Lausanne 1985

Loukola,E., Kuusisto,E. and Reiter,P. -
The Finnish Approach to Dam Safety.
Water Power and Dam Construction. November, 1985

Reclamation, Treatment and Utilization of Coal Mining Wastes, edited by A.K.M. Rainbow
Elsevier Science Publishers B.V., Amsterdam, 1987 — Printed in The Netherlands

THE UTILIZATION OF COAL ASH IN EARTH WORKS

J. HAVUKAINEN

Geotechnical Department of the City of Helsinki, Finland

SUMMARY
    During the work on the ash research project undertaken in 1979 by the City of
Helsinki, ash constructions have been done on more than 100 different sites. The
compacted ashes used on these earth works amount to a total of more than
200.000 cu.m. Ashes have been used in the building of roads and pavements,
parking lots, foundation fills of small houses, athletic fields, and beds of
pipe lines besides serving as backfill and excavation fill. In Helsinki, ashes
have proved to be a good substitute for gravel and sand as foundation-
reinforcing and construction-layer material thanks to their low price, light
weight, load bearing capacity, good thermal insulation properties and, in the
case of fly ash, hardening qualities. However, ashes corrode most metals
generally placed inside earth works. Corrosion of concrete is negligible or
slight and ashes do not corrode plastic. Plants should not be placed in coal ash
unless there is proof that they thrieve in ashy soil.

INTRODUCTION

    In Finland, coal-fired power plants anually produce about 600 000 tons of

ashes (1982). The three coal-fired power plants in Helsinki produce together

some 200 000 t/y of ashes, of which roughly half is fly ash and the rest either

dry bottom ash of dry bottom boiler slag. Part of the fly ash output is sold at

present to the cement industry, the rest being utilized by the City of Helsinki

for earth construction purposes.

    Coal ash can be used in earth works as a substitute for uncrushed natural

aggregates. In Helsinki, ashes can be classified by type as fly ash, dry bottom

ash and dry bottom slag. Structures made of coal ashes possess good strength

properties, and they are lighter of weight and afford better thermal insulation

than natural aggregates.

    Ashes can be handled using conventional equipment. Dry bottom ash and dry

bottom boiler slag are similar to gravel and sand from the standpoint of

handling. In the compacting of fly ash, particular care should be taken that its

moisture content and the thickness of each layer constructed at any one time

remain sufficiently small. Ashes absorb water in a utilized state, but when

compacted with care they are not susceptible to frost.

    The significance of ashes as a substitute for aggregate is great in Helsinki

where

- the distances involved in the transportation of gravel and sand are long

- coal fired power plants are located close to construction sites
- there is a lack of places for the disposal of ashes or the old method of disposal in, for instance, dumps or areas of fill involve disproportionate expense
- the need for aggregate is great

TYPES OF ASHES

Factors affecting the quality and type of ash are, for example, the composition and coarseness of the coal burned, the type of furnace used in the power plant and the firing temperature. Coal ashes can be classified into three main types:

I Fly ash, which is separated and recovered from flue gases

II Bottom ash, which is left at the bottom of the furnace in a pulverized-coal firing power plant

III Bottom slag, which is left at the bottom if the furnace in a crushed-coal firing power plant.

Ashes can be differentiated by their grain size as follows:

Type of ash	Grain-size category
Fly ash	coarse silt
Bottom ash	medium or coarse sand
Bottom slag	fine or medium gravel

TABLE 1

Variations in chemical constituents of fly ash and bottom ash in Helsinki

Property	Fly Ash Range of weight - %	Bottom Ash Range of weight - %
SiO_2	41.9 – 50.2	32.5 – 35.6
Al_2O_3	16.6 – 22.6	11.0 – 12.1
TiO_2	0.8 – 1.3	0.6 – 0.8
Fe_2O_3	10.0 – 13.6	6.8 – 8.5
CaO	6.7 – 9.4	3.5 – 5.2
MgO	4.3 – 5.9	2.1 – 3.1
Na_2O	0.8 – 1.1	0.4 – 0.6
K_2O	1.7 – 2.3	1.0 – 1.3
P_2O_5	0.2 – 0.4	0.1
SO_3	0.4 – 1.5	0.1 – 0.3
Water content	0.04 – 0.17	34.0 – 39.8
Loss of Ignition	1.0 – 12.6	0.2 – 0.8
Water soluble Salts	1.6 – 3.4	0.16 – 0.45
pH	12.0 – 12.6	9.5 – 10.0

The bituminous coal burned in Helsinki is mainly imported from Poland and the variations in chemical composition of fly ash and bottom ash are shown in table 1.

In addition to the constituents shown in table 1, coal ash contains many other elements in trace quantities.

MAIN GEOTECHNICAL CHARACTERISTICS

The laboratory experiments done on the ashes proved that besides granular composition several other geotechnical properties as well vary among the main types of ash. It was endeavored by means of in-situ experiments to obtain information specifically about the techniques of handling ashes and about properties the investigation of which can most reliably be done not in the laboratory but in practice.

One of the advantages of ashes in comparison with conventionally used aggregates is their light weight. The maximum dry unit weight of coal ashes varies from 10 to 15 kN/m^3 and the optimum water content $w_{opt} \approx 20 \pm 5$ %. Moreover, a compacted ash construction absorbs additional moisture as it might be present, upon which w is likely to rise as high as 35 % and the unit weight will be in practice about 14.5 - 16 kN/m^3 when the density of ash is 90 - 95 % from the maximum density determined by the modified Proctor compaction test.

The permeability to water of an ash construction depends on the type of ash as follows:

Type of ash	Coefficient of permeability (m/s)
Fly ash	$10^{-7} - 10^{-9}$
Bottom ash	10^{-6}
Bottom slag	$10^{-5} - 10^{-6}$

A compacted ash construction is compressed quite rapidly under loading, like friction soils. The maximum values of the total settlements observed in the oedometer tests at a loading of 400 kN/m^2 after the second loading were as follows:

Fly ash	$\varepsilon = 1.6$ % (0 days)
Bottom ash	$\varepsilon = 2.8$ %
Bottom slag	$\varepsilon = 5.3$ %
Drainage sand (control material)	$\varepsilon = 2.7$ %

The just built construction made of every type of coal ash possess good shear-strength properties ($\emptyset = 32$ -$42°$, c = 0 - 34 kN/m^2).

The load-bearing capacity of just constructed fly ash layers observed in the plate-loading test is on the same order as that of well-sorted natural gravel. Hardening has the effect of increasing strength properties and bearing capacity with the passing of time. The load-bearing capacity of the other types of ashes is not quite so good, but good enough, for practical purposes. Ranging of E_2-values in a plate-loading test is as follows:

Fly ash
$$E_2 \approx 100 - 200 \text{ MN/m}^2 \text{ (0 days)}$$
$$E_2 \approx 200 - 400 \text{ MN/m}^2 \text{ (after 28 days at 10 - 20}^\circ\text{C)}$$

Bottom ash
$$E_2 \approx 80 - 150 \text{ MN/m}^2$$

Bottom slag
$$E_2 \approx 50 - 80 \text{ MN/m}^2$$

In the lights of reading from thermoelements installed in experimental constructions, it was noted over a period of three winters that the depth of frost penetration in ash constructions is only about 40 ... 60 % of that in gravel constructions. Contributing to the good thermal insulation are such factors as the porous structure and high water content of ashes.

The observations and measurements made in the field show that ashes, when carefully compacted, are not susceptible to frost. This means that commonly applied criteria based on grain-size composition are applicable to ashes. Even so, thin layers of ice have been preceived in fly ash layers. Those have been formed in joints produced in connection with compacting operations. To prevent the appearance of joints before compaction of a new fly ash layer, the surface of the previous layer must always be roughened when fly ash is used. Hardening and low permeability contribute to diminish the frost susceptibility of compacted fly ash.

PRACTICAL APPLICATIONS

From ashes have been built, for example, foundation fill for small houses, construction layers for streets, roads, pavements and parking lots, athletic fields, the beds of rain-water culverts and sewers, as well as backfills and excavation fills. The compacted ashes used amount to a total of more than 200 000 cu.m. on more than 100 different sites. Ashes have proved to be a good substitute for gravel and sand. The types of ash constructions built in Helsinki are shown in Fig. 1.

EFFECTS ON THE ENVIRONMENT

The corrosive effects of coal ashes on other construction materials and their detrimental effects on the environment have been under investigation during the ash research project 1979 - 1984. In the light of the observations made, it would appear as if the detrimental effects are exeedingly slight when they are taken carefully into account in conjunction with design and construction.

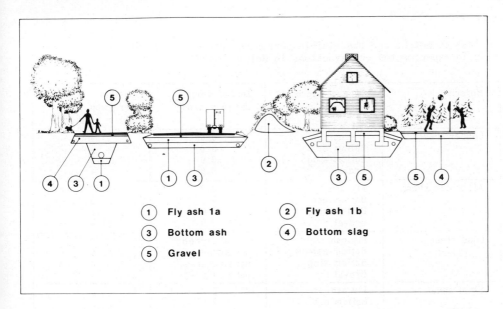

Fig. 1. The types of ash constructions built in Helsinki.

Most generally, the materials susceptible to corrosion are metals and concretes. About the corrosive effects of ashes, the world has fairly little knowledge. There are many factors that have a bearing on the phenomenon, and therefore the corrosiveness of substances can vary greatly.

The corrosion undergone by metals is electrochemical by nature and often reveals itself spottily. In general, protective measures against the corrosion of metal structures have to be taken when conventional materials are used in earth works, In a system composed of soil or ash, water, air and metal, factors contributing to the corrosiveness are, for instance, the conductiveness of the earth or ash construction, its moisture content, pH value, porousness and salinity. By studying any individual factor, one cannot, however, determine for sure the rate of corrosion and its magnitude. The best way to make sure of the matter is perform practical tests.

Practical tests conducted in Helsinki and reports received from other countries on the corrosive effects of ashes on metals prove that the metals commonly used, such as carbon steel, cast iron and aluminium are liable to corrode in coal ashes fairly rapidly. Before being placed in an ash construction, such metals should be treated protectively against corrosion. Galvanized and zinc-plated steels are also somewhat susceptible to corrosion in ashes. On the other hand, the corrosion of copper and lead is, in the light of present experience, trifling.

TABLE 2

The loss of weight and the visible corrosion rate of the metal samples after being in experimental constructions in Helsinki 1981-1982.

Material	Type of Aggregate	Loss of weight per year g/dm^2	Visible corrosion rate after 1 year
Copper sheet	Fly ash	0.0	+
	Bottom ash	0.0	+
	Bottom slag	0.0	+
	Gravel	0.0	+
Aluminium sheet	Fly ash	0.2	0
	Bottom ash	0.1	0
	Bottom slag	0.0	+
	Gravel	0.0	+
Lead sheet	Fly ash	not measured	0
	Bottom ash	not measured	0
	Bottom slag	not measured	0
	Gravel	not measured	+
Steel sheet	Fly ash	3.2	-
	Bottom ash	2.4	-
	Bottom slag	0.3	0
	Gravel	1.2	-
Reinforcing steel rod	Fly ash	4.0	-
	Bottom ash	3.6	-
	Bottom slag	1.2	0
	Gravel	not measured	-
Cast iron pipe	Fly ash	3.3	-
	Bottom ash	3.1	-
	Bottom slag	0.0	+
	Gravel	not measured	0
Hot zincified steel sheet	Fly ash	0.8	0
	Bottom ash	1.7	-
	Bottom slag	1.2	-
	Gravel	0.0	+
Galvanized steel sheet	Fly ash	not measured	-
	Bottom ash	not measured	-
	Bottom slag	not measured	-
	Gravel	not measured	0

Corrosion rate: - = high corrosion rate
                  0 = medium corrosion rate
                  + = low corrosion rate

The corrosion of concretes is by nature chemical and in type either one of dissolution or of expansion. Dissolving agents can be, for example, acid and soft waters, diffusive salts, oils and fatty substances. Substances causing expansion of concrete are sulphur compounds, among them sulphates. In practice, conctere structures have not been observed to corrode in ashes.

Plastics on the whole last well in earth constructions. Their good resistance to corrosion has likewise been observed in ashes. Therefore, they do not need protection against corrosion.

Whenever corrosion of material is feared, protective measures should be taken or the placing of such material in ashes should be avoided.

Leaching experiments on fly ash and bottom ash for Cd, Cr, Cu, Ni, Pb, Zn and Fe indicated a slight potential for contamination of ground - and surface - water supplies when comparing the results with the commonly used sanding material. However, the potential is much slighter than indicated on waste-water sludge utilized as component of a plant nutrient in field husbandry. The results of the leaching experiments made of the City of Helsinki are shown in Fig. 2. The corresponding quantities (mg/kg Ts) of trace elements in waste-water sludge are as follows:

	Cd	Cr	Cu	Ni	Pb	Zn	Fe
Total (mg/kg Ts)	9.5	100	450	95	460	1800	120000
Easily soluble fraction (%)	24	7	5	31	1	40	1
Acid soluble fraction (%)	44	55	44	44	86	5	56
Poorly soluble fraction (%)	32	38	51	25	13	55	43

With regard to the elements included in the composition of ashes, they also contain nutrients needed by plants. However, ashes contain constituents detrimental to plant life, too. In connection with the combustion of coal, the boron contained in it undergoes a change to become more soluble in water, whereupon the abundance of boron passed on to the plants growing in the ashes is harmful to some of them. The high alkalinity of ashes, with a pH value of 9...12, can likewise prove to be an unfavorable circumstance. A further negative factor is generally a lack of nitrogen and phosphate. At the planning stage, measures should be taken to prevent the roots of plants from being in direct contact with coal ashes, unless evidence has been obtained that the plants in question thrive in ashy soil.

CONCLUSIONS

The adjustment of the economic, technical and environmental factors involved in the utilization of coal ashes makes of a former waste material a significant aggregate for earthworks, liberating dumps and disposal areas to receive real waste and opening the gates to ever new geotechnical applications.

REFERENCES

1 The Utilization of Coal Ash in Earth Works. Technical Guidelines. The Geotechnical Department of the City of Helsinki, Bulletin 33, Helsinki 1983.
2 The Utilization of Ash from Coal-Burning Power Plants in Municipal Earth Constructions. The Final Report of the Ash Research Project in 1979-1984. The Geotechnical Department of the City of Helsinki, Bulletin 45, Helsinki 1986.

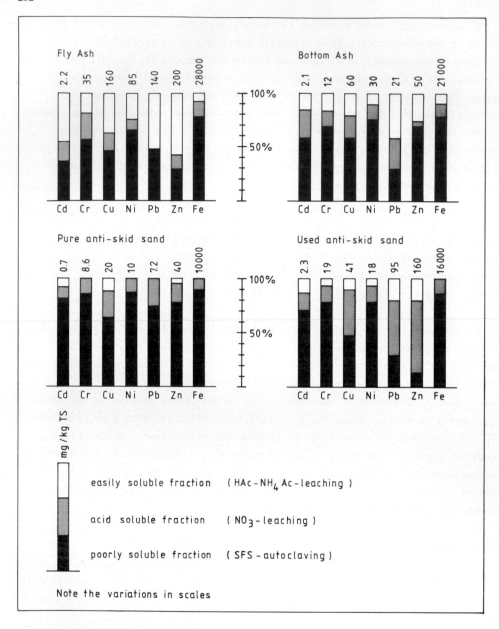

Fig. 2. The concentration of some trace elements in fly ash, bottom ash, pure sanding material and used sand in Helsinki. The total amount of an element is separated to the three fractions according to its solubility.

Reclamation, Treatment and Utilization of Coal Mining Wastes, edited by A.K.M. Rainbow   253
Elsevier Science Publishers B.V., Amsterdam, 1987 — Printed in The Netherlands

BIOTECHNICAL METHODS IN THE TREATMENT AND RESTORATION USE OF COAL MINING
WASTE.

P.J. NORTON B.Sc, Ph.D, C.Eng, F.I.M.M., M.I. Geol.
Research and Development Manager, British Coal Minestone Services,
Philadelphia, Houghton-le-Spring, Tyne and Wear (United Kingdom).

SUMMARY
    Large areas of dereliction which occur in the United Kingdom could be
effectively reclaimed using coal mining in conjunction with other waste
products, and at the same time giving the additional benefit of removing the
environmental eyesores of abandoned pit heaps in mining areas. By mixing
different waste products a suitable soil making material can be formed at a
considerable cost saving compared to previous methods using chemicals and
fertilizers. The restoration to areas of amenity use, including lakes, and
agricultural use by using materials such as sewage sludge, pelletized refuse
and power station fly-ash are discussed.

INTRODUCTION
    Some coal mining wastes, if not treated carefully, are potential sources of
pollution in that they may cause acid mine drainage or provide aggressive
environments for successful plant life. It is important that when mining
wastes are used in restoration that neither of these situations arise.
During the last decade more stringent planning and pollution regulations at
both National and International level have led to a much greater awareness of
these problems.
    The 1984 Survey of Derelict Land in England (ref 1 and 8) showed that there
were over 45,000 hectares of derelict land of which 34,000 justified
restoration. Colliery waste constituted the largest proportion (29%) of this
area and there is considerable benefit to be gained from its restoration and
also its use in reclaiming other areas of dereliction. In order to enhance
sales of colliery waste (Minestone) it is vital that such reclamation is to
environmentally acceptable standards. There is often an acute lack of topsoil
for final restoration for many of these schemes.
    Nowadays it is considered uneconomic to use large expensive and repetitive
applications of chemicals as water neutralizing agents or fertilizers.
Recent studies by British Coal Minestone Services have shown that the use of
other waste materials are a much cheaper alternative to chemical products
resulting in the possibility of once-only applications to solve longterm

problems with regression and unnecessary aftercare treatment.

In order to fully assess their benefits a full programme of field and laboratory research is being carried out under the auspices of British Coal Minestone Services. The results so far are extremely encouraging and it is hoped that the methods will provide for a greater use of Minestone in reclamation of derelict land and an improved environmental image for the mining industry.

Where a good cover of vegetation is required on land reclaimed by mining waste then the use of variable mixes of the following waste products is being investigated:

- Sewage Sludge, to provide organic material and some nutrients for both soil making material and in wet restoration.
- Flue-Gas Desulphurization end-products from coal-fired power stations. Some of these have the effect of raising the pH in the acidic colliery waste.
- Pelletized refuse from urban incinerators, again to raise the pH and provide some nutrients.

The use of other materials to aid plant growth may also be needed in specific cases. Where a rapid cover of dense vegetation is required then the use of mycorrhizal inoculants of fungi on the roots of young trees during transplantation has been proved to give a higher success rate of survival, especially for trees.

Many low lying surface mines, infilled quarries and other derelict reclamation areas have problems with a high groundwater table that issues at surface. (ref 2). In such cases it is important that the water does not become contaminated and issue as ferruginous and/or acid water. This will not only cause pollution of downstream watercourses but will also seriously affect plant growth and cause ugly scars on the reclaimed land surface.

Some mines and quarries cannot be completely infilled and where natural groundwater rebound occurs the resultant lake may well become polluted with acidic water. In these cases the use of activated sewage sludge to purify the water by the process of eutrophication and the additional construction of downstream wetland areas have proved successful in preventing long-term pollution problems. This latter system has been most successful in North America where restoration and pollution controls are strictly enforced. Wet restoration of abandoned surface mines or quarries stands a better chance of approval with the local planning authorities in the United Kingdom as water sport recreational areas are extremely popular with the public. With more money becoming available from government funds and the growth of the leisure industry then this type of reclamation is bound to become more prevalent in future.   British Coal Minestone Services are actively involved in seeking new ways of using colliery waste in such projects to the overall benefit of the environment in the old mining areas.

USE OF FERTILIZERS IN COLLIERY WASTE RECLAMATION

When colliery waste is being restored or being used to restore other derelict areas it is important that a vigorous growth of vegetation is produced.   In the past there has been a considerable dependance on chemical fertilizers to provide the necessary nutrients.   Unfortunately leaching of the chemicals and regression of vegetation due to the regeneration of acidity has resulted in some areas becoming unsightly with poor vegetation cover and ochrous deposits associated with the ferruginous acid leachate.   Only the repetitive application of further costly fertilizers could halt this process, especially in areas of high rainfall and on the more permeable of the colliery spoils.

A typical high dose application was at one large pit heap in a high rainfall area of the United Kingdom which suffered considerable regression after considerable amounts of fertilizer were applied.   The pit heap is on a steep valley side and some of the slopes to be restored were also steep - up to 1 in 2 in places.   Elsewhere slopes were around 1 in 10 and even horizontal in places.   The area to be restored was 2,000 square metres.   A preliminary soil survey revealed that most of the surface was extremely acid with an average pH of 2.   The sulphur content was around 1.2% which was mainly con-tained in iron pyrites (FeS_2) therefore still available for oxidation and hence acid production, viz:

$$4FeS_2 + 4H_2O + 14O_2 = 4FeSO_4 + 4H_2SO_4$$

This breakdown of iron pyrites is greatly increased in the presence of oxygenated water (rainfall) and usually occurs within a metre of the surface. Other reactions involving bacteria (Desulphovibrio and Thiobacillus) also attack the pyrites and after further oxidation the ferrous sulphate is con-verted to ferric hydroxide $Fe(OH)_3$ which gives rise to the familiar ochrous discharge at the surface issues of ground water.

As a result of the high amounts of acid in the colliery waste it was decided to add the following amounts of fertilizer to provide a neutral and nutritious soil;

0.4 tonnes/hectare of Compound Fertilizer 15:15:15

0.5 tonnes/hectare of triple super-phosphate

30 to 80 tonnes/hectare of 'Calcitic' limestone.

The whole area was thus treated and sown with grass and white clover (Trifolium retens) or lupins (Lupinus) with some areas of 10% gorse (Ulex

Europaeus) and other areas, especially the steeper slopes, with deciduous trees; Silver Birch (Betula pendula), Grey Alder (Alnus incana) and White willow (Salix alba) and also with conifers; Lodgepole Pine (Pinus contorta) and Larch (Larix eurolepis). The deciduous trees were individually protected small 'feathered whips' i.e. 1 to 2 metres high and the conifers 'sturdy transplants' about 0.6 metres high. The latter being planted in fenced areas to protect them from sheep and wild animal trespass. Most of the species chosen were capable of withstanding low pH with the broadleaved varieties able to tolerate a pH of 6, the pine trees a pH of 4.5 and the birch as low as 4. The alder trees are particularly beneficial on colliery waste as, like the clover, they will replace nitrogen in the soil. However it is essential that the acidity in the soil is reduced so that the goodness in the soil is released for absorption by the plant roots. The other essential requirement is that the moisture in the waste is maintained, and that it is not over-compacted during deposition. Unfortunately colliery waste is dark in colour and sometimes very permeable, and hence during summer months will overheat and cause drought conditions. Contour ploughing and a rapid formation of grass helps to alleviate this. In this particular case neither were successful and regression of the vegetation occurred and most of the trees eventually died after two years, the main reason being leaching of the chemicals and hence a lack of nutrients for the plants. Repeated doses of fertilizers are the only remedy here and obviously very costly in order to maintain vegetation growth.

This regression is quite common on colliery spoil (ref 4) and the answer would appear to lie in a greater use of organic material to retain the moisture and less use of expensive soluble fertilizers which can be easily leached out of the waste. Many projects in the past have suffered from the lack of necessary aftercare required by this latter method. What is needed is a one-off treatment during restoration which will last until a full vegetation cover has been established and is self-generating and resistant to drought. The use of waste and other products to provide this has many attractive possibilities and have been grouped under the general heading of biotechnical restoration.

BIOTECHNICAL METHODS IN RESTORATION

In order to make the vast resources of colliery waste more suitable for use in derelict land reclamation then it is considered essential that the above problems are not encountered. The use of other waste products mixed with the colliery waste is an attractive idea as it keeps costs to an absolute minimum and avoids the use of other products which can be better utilized elsewhere.

Sewage Sludge

The most abundant material is sewage sludge which can be mixed with the top layer of the colliery waste to provide nutrients, moisture retention, and raise the pH slightly. The usual amount of sewage sludge applied to the top layer of fill is around 100 tonnes of dry solids (tds) per hectare or per 2,000 tonnes of colliery waste. To date the main problems have been concerned with the logistics of matching sludge production with site and colliery waste availability (ref 5), and only 7% of the sludge disposed of on land is used in actual land reclamation; the majority being spread on agricultural land (40%) or wastefully disposed of at sea (30%) and creating further environmental problems. There are large regional variations in its disposal methods, of which some will contravene European directives in the near future; such as the 78% sea disposal in Scotland. Its use in reclamation may therefore be forced on local authorities. However there have been some notable successes, especially on abandoned pit heaps and restored surface mine sites, and with the correct support there is no reason why much more use should not be made of this valuable resource which totals some 1.2 million tds per annum.

One of the main problems with sites restored with conventional fertilizers is that the nutrients are soon used up or leached out and such sites require costly aftercare treatment. The benefit of sewage sludge and other waste products is that they give up their nutrients slowly thus allowing a well-established vegetation cover to form without regression. Colliery waste is, on average, an excellent soil-forming material; it has a granular structure, is well graded and has low permeability (usually around 10^{-5}cm/second). With the addition of sewage sludge an acceptable soil for most amenity and poor grazing agricultural use can be made.

Sewage sludge has an unfortunate reputation with the public and some local authorities. On most reclamation schemes dewatered sludge is used which can be either digested (undergone the effect of anaerobic bacterial treatment) or undigested (ref 6).

All the sludges contain organic matter, phosphorus or trace elements. The latter can be a problem if heavy metals are present as is the case in sewage arising from metallic industrial areas such as in Yorkshire and the Midlands. Sometimes lime is added which may have a pH of 12 and can be especially useful in combatting the acidity of some colliery spoils. The organic matter in sewage sludge is especially important here as it can inhibit the oxidation of pyrites in colliery spoil (ref 7).

The toxic elements and pathogens which have been the cause of many of the restrictions on the use of sewage sludge in the past are now in much lower concentrations as a result of more efficient treatment control (ref 8). If the Department of the Environment and EEC directives are followed and proper management of the reclamation is undertaken then these problems can be overcome especially for amenity use restoration (see Table 1).

Zootoxic Metals	Agricultural	Amenity
As	10	40
Cd	3	15
Cr	600	1,000
Pb	500	2,000
Mg	1	20
Sc	3	6

Table 1. Department of Environment guideline for some toxic metal 'Trigger' concentrations (mg/kg) (ref 6).

It is therefore important that the type of sludge to be used is fully understood. There have now been so many successful applications that its use with colliery spoil cannot be ignored as being one of the most economic reclamation fill materials available (ref 5 and 9).

The most efficient way of overcoming the logistics of the supply problem would be to create soil 'banks' in the areas of greatest need. Here, ready-mixed colliery spoil and sewage sludge would be stored for known future use, either at active collieries or on abandoned pit heaps where the material was already being rehandled.

Several trials have already been undertaken in the U.K. by the Water Research Centre. Most have been on Opencast sites and abandoned colliery tip heaps and it is proposed to do trials on the use of colliery waste/sewage sludge mixes for land reclamation in the near future. In the meantime, lysimeter or column experiments in the laboratory with various mixes and newer types of material mentioned in this paper are being undertaken.

Mycorrhizal Inoculants

In order to further guarantee the successful growth of plants and especially trees on colliery waste, the use of new biotechnical products is also being considered. It has been noticed (ref 10, 11 and 12) that certain trees have particular fungi associated with their roots (see Figure 1). Ectomycorrhizae are a symbiotic association between the fungi and plant root. The fungus forms a sheath around the roots of both deciduous and coniferous trees and penetrates between the calls of the root. Shoots (hyphae) from the sheath penetrate the colliery waste, thereby extending the root system.

Fig. 1. Symbiotic association of fungi on roots of birch trees growing
       naturally on colliery waste.

The promotion of the growth of these mycorrhizae on tree roots has the
following beneficial effects:

- They increase plant nutrient and water absorption from the colliery waste
- They increase the plant's tolerance to drought conditions and to low pH
- They protect the plant from pathogens and toxic metals

Tree seedlings and nursery transplants which are inoculated with
mycorrhizae are thus better able to withstand the aggressive environment of
the colliery waste and other waste mixtures of soil-making material.

It is hoped that with these developments in the use of other waste materials
to combine with the colliery waste, the use of expensive fertilizers as
recommended in the past (ref 13, 14, 15) can be reduced or even eliminated.

Power Station Flue Gas Desulphurization (FGD) End-products

It is envisaged that in order to reduce air pollution and the 'acid rain'
problem, all coal-fired power stations in the U.K. will need to be equipped
with flue gas desulphurization equipment. The end product of this process is
variable depending on which method is used.

The methods are:

(1) Wet Scrubbing using limestone and producing a gypsum ($CaSO_4$) end product.

(2) Regenerative systems which produce sulphuric acid as an end product.

(3) Spray Drying absorption (SDA) producing sulphides, hydroxides and lime.

(4) Sea water washing; not yet developed fully.

The wet scrubbing and SDA end products are neutral or alkaline and vary
between pH 6.5-9. The geographical location of these products close to the
coalfields makes them suitable for mixing with colliery waste in order to

reduce its overall acidity when used in land reclamation. The fly ash
content should also help reduce permeability of some of the more granular
colliery spoils. As there is a similar disposal problem with these materials
then it would seem logical to combine them to provide a cheap bulk landfill
or reclamation product.

The wet scrubbing process produces commercial gypsum, most of which would
be used in the plasterboard industry. A typical 2,000 Megawatt powerstation
would produce 500,000 tonnes of gypsum per annum, of which a rough estimate
considers about 100,000 tonnes may be available for land reclamation.

The SDA process (ref 16) on a similar sized power station would produce
700,000 tonnes per annum of about 45% $CaSO_3$, 10% $CaSO_4$ and 10% $Ca(OH)_2$ as
well as small amounts of CaO and $CaCO_3$, with the remainder composed of
fly-ash. This material would appear to be more suitable for soil making than
the wet scrubbing product. Almost all of it would be available as it has no
other suitable commercial value.

It appears that the wet scrubbing method is preferable on economic,
environmental and commercial grounds. The first 'retro-fit' to an existing
power station is expected to be working by 1993. There is therefore
sufficient time available to find suitable reclamation projects and carry out
further research into the combination of these problematic waste products.

Pelletized Refuse

In recent years it has proved economical to incinerate domestic refuse in
urban areas. The pelletized end product has a high pH around 8 as a result
of its high $CaCO_3$ content. It is also high in organic material and nitrogen.
It is therefore an ideal material for blending with colliery waste in the top
layers of land reclamation to provide a base for good vegetative cover.
Although there are only a few incinerators at present it is thought that many
more may be built and hence provide another cheap waste product for use with
colliery waste in landfill and reclamation projects.

WET RESTORATION

In the context of this paper the term is used to describe restoration of
abandoned surface mines or quarries or derelict land sites where there is a
high natural groundwater table. In the case of surface mines and quarries
there may be a final void which cannot be filled for a variety of reasons.
In this case it is often expedient and economical to allow the void to fill
with water to form an amenity value lake. However, where mining waste is
involved it is important to prevent the water becoming polluted with acid and
ferruginous water resulting from the oxidation products of iron pyrites which

is ubiquitous in coal mining wastes. The percentage of total sulphur in
unburnt minestone can vary between about 0.2% and 7.5% depending upon its
origin.

Where sites cannot be used for profitable landfill projects, usually
because of the high water table, then the lake amenity use is often the only
recourse left to the planners and mining companies involved. Until recently
the most common method of curing acid lakes was simply to apply lime to
neutralize the water. Unfortunately, where external influences maintain the
inflow of acidic water or the promotion of acids within the lake, then the
water would soon revert to acid and further applications of lime would be
necessary to maintain neutrality. This is extremely costly, especially where
the inflow is constant. At one surface mine reclamation project about
50 tonnes of lime per week were needed to treat an inflow of only 160 l/s of
water which had a pH of less than 4. The methods were inefficient and
probably caused as much environmental problems as the already existing water.
Acid lakes caused by air pollution from coal-fired power stations have also
undergone the same treatment. Sometimes regression to acidic water occurs
after the applications of limestone or lime and the costly treatment must
be repeated.

Fig. 2. Acid quarry lake restored to amenity use using activated sewage sludge
(Courtesy of British Industrial Sands Ltd.).

In 1984 the Freshwater Biological Association launched a pilot scheme at
British Industrial Sands quarries in Norfolk to reclaim acid lakes with a pH
of 3. (ref 18 and 19). A small amount of lime and activated sewage sludge

was added to the lake, and through a process of eutrophication a self-regulating system has been achieved which requires no further work and maintains a neutral pH. The scheme was successful and is planned for more projects, in particular a large abandoned surface mine.

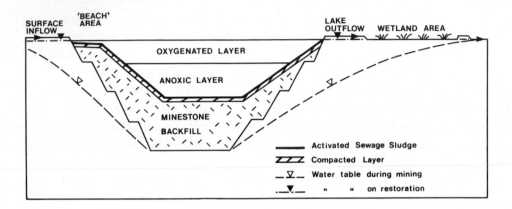

Fig. 3. Diagrammatic cross-section of typical wet restoration of a surface mine or quarry using minestone.

The preferred method for restoring abandoned mines and quarries, as suggested by the Freshwater Biological Association, is as follows:

Backfilling

Minestone is an eminently suitable material for this purpose as it is freely available in a 'stock pile' form in the pit heaps and often its removal improves the environment and releases land for development. Only transport costs are normally incurred and it is therefore usually the cheapest source of fill material.

The backfill must be placed to a suitable geometry for the lake bed. An absolute maximum slope angle is around 14° and preferably as large an area of the lake bed as possible should be horizontal. The deeper the lake the better so that the backfilling is only used to provide a suitable geometry and shallow beach areas for fishing, bathing or sporting access etc. A relatively impermeable lake bed should be provided by simple compaction or lining with a suitable material if available (e.g. FGD end product). This is to reduce as much as possible the ingress of any potentially acid groundwater passing through the pyritic minestone fill (see Figure 3).

Inundation

Next, all pumping systems should be withdrawn and the lake should be filled as rapidly as possible, without causing scour of the lining. In some cases the diversion or re-location of surface watercourses can be used to assist the groundwater rise once pumping has ceased. The addition of un-polluted surface water should also help to raise the pH. Once the lake is full, the groundwater recharge through the fill should be reduced to a minimum.

Neutralization

If the pH of the water is still less than about 4 then some lime or caustic soda will need to be added to the water to allow the process of eutrophication to work. Caustic soda is preferred as often the waters associated with coal mining in particular will have a high calcium content. A rate of about 100 tonnes of caustic soda per 10 million cubic metres of acidic water is roughly needed if the pH is around 3 to 4.

Eutrophication

The addition of activated sewage sludge provides organic material on which the anaerobic bacteria feed and convert sulphates in the polluted water to sulphides by eutrophication in a reducing environment. The deepest part of the lake is anoxic and allows the bacteria to thrive and maintain the water free of acids and precipitate the hydroxides which cause the ochrous colouring. Eventually the floor of the lake is composed of organic sediment and layers of harmless metal sulphides in a similar way to that found in natural deep lakes (see ref 2).

The upper oxygenated warm layer of the lake is fed by nitrates and phosphates from the sludge which provide food for the growth of algae which in time support higher forms of life and eventually fish stock can be introduced.

One of the main criticisms of the scheme has concerned the possible problems with the presence of pathogens and heavy metals in the sewage sludge. Modern methods of sewage treatment and the fact that the water is acidic when the sludge is introduced will prevent these problems occurring as proved in the schemes so far undertaken. Some shallow lakes (5m deep) have shown a slight regression to acidity and it is thought to be due to the absence of a deep and stable reducing environment.

The process of eutrophication appears to have great potential as a cheap and effective way of utilizing waste materials to provide suitable lake restoration in aggressive conditions. However, it must be stressed that further research is continuing to improve this relatively new technique.

WETLAND RESTORATION

Many abandoned colliery pit heaps and restored areas suffer from ground-water leachates which are potential pollutants to the receiving surface watercourses. Most of these problems relate to pit heaps which may be over 100 years old or pre-date the installation of proper drainage works during construction. The leachate is predominantly ferruginous and sometimes may be acid or contain toxic metals. Very small volumes of water are usually involved and the amounts vary seasonally with little rainfall recharge. Because the volumes are so small there is little to be gained from the installation of full drainage and treatment works to remedy the situation.

A method which has gained considerable success in North America (ref 22) is the construction of a wetland or bog downstream of the issue. The idea is to provide a 'filter' for any noxious materials in the water by passing through a 'mini' anoxic environment where sulphides are produced in the deeper layers of the bog. About 200 sq. ft. of wetland is needed per gallon per minute of polluted water.

All that is required is to construct a flat area which is impounded so that the water flow is almost static. Peat and marshland plants are then provided and usually the plants will colonise the area successfully if a marshy environment is maintained. An added benefit of this type of restoration is that it will provide an 'ecological niche' for endangered marshland wildlife and plants. This meets with general public interest and assists in winning planning and environmental approval for the overall restoration scheme.

CONCLUSION

Various combinations of the above methods should provide for all requirements in land restoration using colliery spoil. Minestone Services are continuing their involvement with funding for research projects into the ideal combination of methods and materials to make the best use of waste products in land reclamation. In this way other more important strategic minerals and greenfield quarry sites for other bulk fill materials are saved for future use. The dual benefit of simultaneously cleaning up abandoned pit heaps whilst using the material for land dereliction cannot be ignored and Minestone Services forsee considerably more use being made of this readily available and valuable resource.

ACKNOWLEDGEMENTS

   The author would like to thank his colleagues in the coal mining industry
and in particular the Water Research Centre, Freshwater Biological Association
and the Environmental Advisory Unit for their assistance in the preparation
of this paper. The views expressed are those of the author and do not
necessarily represent those of British Coal.

REFERENCES

1. Department of the Environment, Survey of Derelict Land in England,
   Department of the Environment Report, London, U.K., 1984.
2. P.J. Norton, Groundwater rebound effects on surface mining in the U.K.,
   Second International Symposium on Surface Mining, Bristol, U.K., 1983.
3. P.J. Norton & R. Henderson, A method of predicting the pollution potential
   of mining backfill, International Mine Water Journal, Vol. 3, No. 1,
   Spain, 1983.
4. H.E. Bloomfield, The aftercare of reclaimed derelict land, PhD Thesis,
   University of Liverpool, U.K., 1978.
5. J.E. Hall, A.P. Dow & C.D. Bayes, The use of sewage sludge in land
   reclamation, Water Research Centre, U.K. Report No ER1346-M, U.K., 1986.
6. Department of the Environment, Sewage sludge survey (1980 data), Report of
   the Standing Committee on Disposal of Sewage Sludge, London, U.K., 1983.
7. C.A. Backes, The oxidation of pyrites and its environmental consequences,
   PhD. Thesis, Glasgow, U.K., 1984.
8. Environmental Data Services, New question mark against land disposal
   guidelines for metals in sewage sludge, Report 142, U.K., November 1986.
9. J.E. Hall & E. Vigerust, The use of sewage sludge in restoring disturbed
   and derelict land to agriculture, Conference on European Concerted Action
   on the treatment and use of sewage sludge, Sweden, 1983.
10. G.W. Thomas & R.M. Jackson, Growth responses of sitka spruce seedlings to
    mycorrhizal inoculants, New Phytologist, Vol. 95, U.K., 1983.
11. S. Lanning & S.T. Williams, Nitrogen in reclaimed land, Environmental
    Pollution, (Series B), No. 2, U.K., 1981.
12. G. Glatzel & J. Fuchs, The application of special organic fertilizers with
    difficult afforestations, Allegmeine Forst Zeitschrift, Germany, 1984.
13. R.E. Ridley, Afforestation Scheme for New Brancepeth Colliery, National
    Coal Board publication, London, U.K., 1970.
14. K. Wilson, A Guide to the Reclamation of Mineral Workings for Forestry,
    Forestry Commission Research and Development Paper No. 141, U.K., 1985.
15. J. Jobling & F.R.W. Stevens, The establishment of trees on regraded
    colliery spoil heaps, Forestry Commission Paper No. 7, U.K., 1980.
16. J. Donelly, Disposal and Utilization of Spray Dryer FGD end-products,
    Canadian Electrical Association Seminar on Sulphur Dioxide removal,
    Ottawa, Canada, 1981.
17. G.D.R. Parry & R.M. Bell, The use of colliery spoil as a blinding for
    toxic wastes, Symposium in the reclamation, treatment and utilization of
    Coal mining wastes, U.K., 1984.
18. W. Davison, Sewage sludge as an acidity filter for groundwater fed lakes,
    Nature Vol. 322, No. 6082, U.K., August 1986.
19. R. Needham, Wet Restoration, Mineral Planning No. 26, U.K., March 1986.
20. R. Fernendez-Rubio (Ed.), Abandono de Minas Impacto Hidrologico,
    Instituto Geologico y Minero de Espana, Spain, 1986.
21. L.G. Love, Sulphides of metals in recent sediments, Proceedings of the
    15th Inter-University Geological Congress, U.K., 1967.
22. R.C. Severson & L.P. Gough, Rehabilitation materials from surface coal
    mines in Western U.S.A., Reclamation and Revegetation Research, Elsevier,
    Amsterdam, 1984.

Reclamation, Treatment and Utilization of Coal Mining Wastes, edited by A.K.M. Rainbow
Elsevier Science Publishers B.V., Amsterdam, 1987 — Printed in The Netherlands

COAL MINE SPOIL TIPS AS A LARGE AREA SOURCE OF WATER CONTAMINATION

J. SZCZEPANSKA [1] and I. TWARDOWSKA [2]

1 University of Mining and Metallurgy, Inst. of Hydrogeology and Engineering Geology, 30-059 Cracow, Al. Mickiewicza 30 / The Poland/

2 Polish Academy of Sciences, Inst. of Environmental Engineering 41-800 Zabrze, M.Skłodowskiej- Curie st.34 / The Poland/

SUMMARY
    The scale of negative effects coal mine spoil tips on the water environment in the vicinity of the tip has been exemplified on the result of long-term investigations of the active tip Smolnica situated in the area of the Upper Silesian Coal Basin / Poland /.
The problem have been presented on the basis investigations of surface and underground water quality in the vicinity of the tip against the background of dynamics of soluble solids leaching from coal mine spoil deposited on the tip and chemical composition of pore solutions in vertical cross - section of the tip. The results of chemical analyses of pore solutions and effluents from the tip as well as investigations of underground and surface water have proved the negative changes of water quality in consequance of tip impact.

INTRODUCTION

    Coal mine spoils have a negative effect upon the aquatic environment in the vicinity of the tips in consequence of the soluble substances leaching. The problem of water contamination as a result of coal mine spoil deposition is at moment importance, as these spoils are the biggest by quantity group of industrial solid wastes in Poland. About 60 mln m^3 of spoil are produced annualy from underground coal mines in the Upper Silesian Coal Basin / USCB / in Poland.

    The deposition of so great quantitys of spoils on the limited area and frequently on a permeable bedrock, will cause the increasing threat of water environment contamination by the soluble compounds leached out from the spoil. Among these substances, the particularly negative effect is caused by the chloride leaching as well as by the oxidation of iron sulfide minerals and the subsequent hydrolisis of the reaction products to form acidic, high sulphate and high iron and heavy metal drainages.

Taking into consideration the annual production of coal mine spoil in the USCB and the mean sulphate production estimated at the rate of 20,8 g/tonne of spoil/day the total annual sulphate production in the spoil deposited within the tips has been found to amount some 740 000 tonnes SO_4/ year.

The annual chloride load deposited on the tips has been estimated to the be about 65 000 tonnes/year.

Also the heavy metals which present in spoil in trace concentrations, are the essential factor of water contamination in the vicinity of tips.

The scale of the water contamination caused by coal mine spoil tips has been exemplified on the result of long – term investigations of water quality in the vicinity of the Smolnica-tip /USCB, Poland /, against the background of dynamics of soluble compounds leaching from coal mine spoils deposited on the tip.

MATERIAL AND METHODS

Object of investigations

The active spoil tip Smolnica covers at present some 60 hectares and has situated in the valley of the Bierawka -river / Fig.1/.

The construction of the tip proceeds towards the river bed according to the direction of the surface run off and ground water flow. In the foundation of the tip occur the permeable sand.

Spoils have been deposited since 1965 in two layers, each about 10 m thick. They have been discharged from the carboniferous strata of group 300 and 400, those being in Westphalien A and the upper Namurian C series. The spoils consists mainly of claystones / 79,5 %/, mostly of kaolinite – illite – sericite type.

The permeability to water of the spoil is generally high and ranges from $2.47 \cdot 10^{-6}$ m/s to $1.28 \cdot 10^{-3}$ m/s depending on the position in the cross-section of the tip.

Chlorides in freshly produced spoils are present in mean concentration $0,05\%$ Cl^-. The mean total sulphur content is $0,90\%$ S_t. The sulphur is distributed as iron sulphide /$0,81\% S_p$/ and in small amount as organic sulphur and in the form of sulphates / $0,009\%$ S_{SO_4} /.

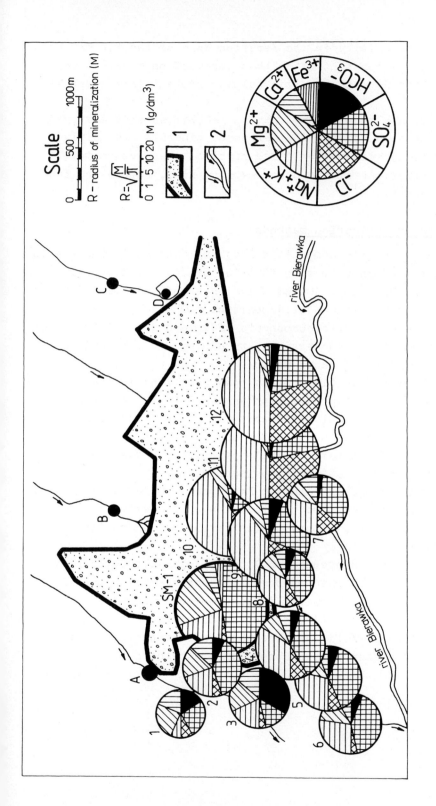

Fig.1.Hydrochemical map of the investigated area.Smolnica coal mine spoil tip-Upper Silesian Coal Basin,Poland. 1-Smolnica tip,2 - river and streams.Sites of waters sampling: A,B,C,D - streams /N/ above the tip; 1,2,3 - ponds /W/ below the tip; 4,5,6,7 - streams /S/ below the tip; 8,9,10,11,12- effluents from the tip; SM-1 - prospecting - shaft /pore solutions/.

The sulphide reactivity in the aeration zone of the tip has been
described by the constant of sulphate production $r_s = 0,210$ meq/eq
G_s . day, where G_s — the pyrytic sulphur content in the spoils
/eq/t/.

The spoils are low-buffered /$\widetilde{s}_{buf.}$ =0,71/, so they are pre-
disposited to turn acidic. The carbonate minerals are mainly
sideroplesite and minor quantities of calcite and dolomite.
Trace concentrations of the heavy metals: zinc, copper and lead
are also present.

Sampling and investigation methods

Dynamics of soluble components leaching from coal mine spoils
has been observed in situ as a function of the age of spoils.
The spoils in the outer, 100 cm thick layer of this tip of various age since the moment of deposition / 0 - 12 years/ have
been sampled. The soluble components have been derived from the
spoil by means of the successive 3-time repeated standard water
extractions of 1 : 5 / rock: water/.

The chemical composition of the pore solutions in the vertical
cross-section of the tip have been analysed too. The samples for
investigation of pore solutions were taking from the prospecting
shaft / Fig.2/ about 11 m deep. The upper parts of profile up to
the depth of 7,5 m contained spoils deposited 10 years ago. The
spoils in the bottom part were 15 years old.
Spoil of natural moisture content have been sampled every 0,5 m.
From these samples the pore solutions have been extracted using
pressure method / ref.1 / and concentrations of major ions have
been determined.

The changes in hydrogeochemical conditions in the vicinity of
the tip have been observed by comparing the chemical composition
of the adjacent natural waters flowing towards the tip with the
water quality below the tip; drainages ditches, effluents from the
tip, ponds, streams / Fig.1/. The chemical composition of the water samples have been analysed by standard methods.

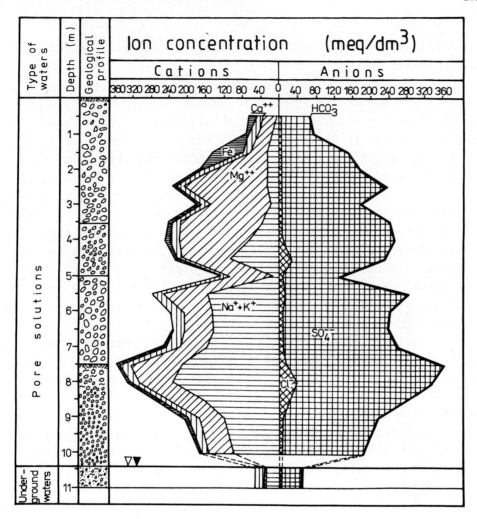

Fig.2. Hydrochemical profile of the pore solutions extracted from the spoil deposited on the Smolnica tip. The prospecting – shaft SM-1.

RESULTS

Dynamics of soluble components leaching from the outer layer of the tip.

Spoils of various age / from 0 to 12 years/ sampled from the outer layer of the tip / 0 - 1 m / reveal a very high dynamics of chloride leaching.The rate of chloride removal was found to range 80 % just in the first year after deposition / Table 1/. In the following years chloride concentrations show only negligible changes. High susceptibility of chlorides to leaching off is the result of their occurence in the liquid phase i.e. in pore solution infilling the carboniferous rocks / ref.2 / as well as the consequence of the lack of equillibrium limitations.

The dynamics of sulphate leaching shows the diffrent character.The sulphate concentration in the spoils is a resultant of two processes: the sulphate generation by the action of iron sulphide oxidation and the sulphate leaching by infiltration water.
On the analysed tip during the 5-6 years long period, after spoil deposition, constant increase of sulphate concentration in spoils have been observed / Table 1 /. It indicates the predominance of production over the leaching off process. In the following years the tendency of sulphate concentration decrease in spoil is marked. It evidences a gradual sulphide depletion in the spoils.
At the same time, the spoils, become acidic in reaction just in the first year after deposition / pH.3.3 - 3.7/ in consequence of their low buffering capacity.

The major component of the freshly produced spoils are chlorides / Table 1 /. After a lapse of a year, the dominant anions become sulphates. In the period from 1 to 6 years after spoil deposition, the sulphate concentration in the outer layer of the tip ranges from 160 to almost 500 % of the initial content in the freshly produced spoils.

The results of field investigations of the dynamics of the spoil leaching imply that the risk of contamination of the aquatic environment by chlorides is determined by the annual load of chlorides deposited on the tip together with coal mine spoils.

The hazard of water contamination by sulphates is defined by the total quantity of spoils filed in the aeration zone of the

Table 1

Retention of chloride, sulphate and total dissolved solids in the spoil deposited on the Smolnica tip

Approx. age of tip at sampling point /years/	Depth of sample from surface /cm/	pH	Chloride as Cl /dry weight/		Sulphate as SO$_4$ /dry weight/		Total dissolved solids TDS /dry weight/	
			g/t	% chloride content of fresh wrought spoil	g/t	% sulphate content of fresh wrought spoil	g/t	% TDS content of fresh wrought spoil
Fresh wrought spoil		7,90	531,1	100,0	281,0	100,0	1462,0	100,0
1 - 2	0 - 20	3,33	70,5	13,3	450,0	160,2	725,8	49,6
	70 - 100	6,06	101,0	19,0	722,9	257,3	1214,5	83,1
3	0 - 20	3,59	70,5	13,3	452,5	165,4	775,6	53,0
4 - 6	0 - 20	3,75	88,2	15,5	658,6	234,3	1004,9	68,7
	70 - 100	3,14	94,0	17,7	1396,3	496,9	2020,0	138,2
10 - 12	0 - 20	3,38	72,8	18,7	278,9	99,1	489,6	33,5
	70 - 100	6,20	86,9	15,3	1143,2	406,8	1700,8	116,3

tip in the condition of a relatively free acces of oxygen.The total surface area of the tip is great importance too.It decides
of the quantity of leaching water, and of the size of the outer
layer of tips where the conditions for generation of soluble
components and their leaching off are the best.

Chemical composition of pore solutions in the aeration zone of the tip.

In the vertical profile of the prospecting-shaft SM-1 /Fig.2/
the distinct differentiation of the chemical composition of pore
solutions has been marked. It has been caused of the two-layer
construction of the tip body. The highest concentration of sulphates and total dissolved solids in the pore solutions occurs
on the contact of the upper 10 years old with the lower 15 years
old spoil layer i.e. on the depth of about 7,5 m.
Total dissolved solids concentration is there as high as 24 300
mg/dm^3. Sulphates have been present in concentrations 16 500 mg/
dm^3. Pore solutions are of SO_4 -Na-Mg type.Up the profile of the
upper layer, the concentrations of dissolved components gradually
decreased. Chemical type of pore solutions changes from SO_4- Mg
in the layer 5,0-1,5 m to SO_4 - Mg- Fe in the most washed out
outer layer 1,5 - 0 m. In this layer also the highest acid potential of spoils and the lowest the buffering capacity has been
recorded.

The impact of the different period at spoil deposition has
been marked in the lower 15 years old layer of spoils by the
decrease of the total dissolved solids to the level of 13 000 mg/
dm^3. It has been caused by the partial, substantial washing out
of this layer before the deposition of the upper spoil level.On
the other hand the chemical type of water SO_4 - Na - Mg remains
the same.

In the whole profile a very low chloride concentrations amounting to 70 mg/dm^3 Cl^- have been observed.
The concentration of this ion in the pore solutions extracted
from freshly produced spoils have been found to be about 7000 mg
/dm^3.Reduction of chloride content which accounts for about 99 %
has been the result of the multiple washing out of the spoils in
the whole analysed profile and the high dynamics of chloride

leaching.

The observed increase of dissolved components in the vertical profile of the tip is a consequence of the influence of two factors, i.e. a vertical redistribution of salt loads / ref.3 / and a various degree of washing out of the each particular spoil layer in the profile of the tip body. The extent of vertical redistribution of dissolved component loads is limited by the geochemical constrains imposed by equillibrium conditions.Chlorides redistribute the most intensive because of the lack of equillibrium limitations within the observed range of concentrations. The rate of vertical redistribution of sulphates depends on the buffering capacity of spoils contributed by calcium carbonates, which limits a sulphathe solubility to the level of equillibrium with gypsum.

Because of the low buffering capacity of spoils analysed, the equillibrium limitations has a little effect thus the vertical redistribution of sulphate loads is significant.

Besides of the vertical redistribution loads process,the chemical composition of pore waters in the vertical profile of the tip is influenced also by the different degree of washing out of spoil layers in the vertical cross-sections of the tip body, determined by flow rate of infiltration water. For the analysed tip, the period of one-fold washing out including transport of contaminants leached off from the outer layer of the tip by percolating water to the waters beneath the tip toe, lasts some 5 years. In this period the outer layer 1 m thick will be washed out about 20 times. The changes of chemical composition of pore solutions in the vertical profile of the tip have reflected the different age of infiltration waters. In the lower part of profile waters of the chemical composition corresponding to the erlier stages of leaching process are present.It is testified to the occurence of sodium ions. In the upper, outer layer sodium is replaced by iron and hydrogen ions.

Chemical composition of underground and surface waters in the vicinity of tip

The results of analyses of drainages from the spoil tip as well as investigations of the adjacent under ground and surface

Table 2

Chemical composition of pore waters extracted from the spoil deposited on the Smolnica tip and surface and underground waters in the vicinity of this tip

Type of waters		Location of sampling points	pH	Total dissolved solids content TDS /mg/dm³/	Chloride content Cl⁻ /mg/dm³/	Sulphate content SO₄²⁻ /mg/dm³/
Pore solutions		Prospecting-shaft SM-1 0 - 7,5 m /10-years old spoil/	4,0 - 5,2 / 5,0	4162,4 - 24287,5 / 13553,9	84,8 - 931,8 / 247,6	2829,3 - 16510,8 / 9559,0
Pore solutions		Prospecting-shaft SM-1 7,5 - 10 m /15-years old spoil/	4,6 - 5,2 / 5,0	12249,2 - 22854,0 / 16275,9	129,5 - 1276,2 / 494,1	8490,6 - 14511,6 / 10880,9
Under-ground waters		Prospecting-shaft SM-1 /10,43 - 11 m/	7,0 - 7,9 / 7,5	3610,5 - 3637,4 / 3624,0	148,4 - 168,2 / 158,3	1787,3 - 1799,2 / 1793,3
Surface waters	Fresh waters	Streams /N/ above the tip	7,4	123,6 - 363,3 / 228,8	23,1 - 34,4 / 29,2	50,0 - 131,7 / 100,4
Surface waters	Polluted waters	Ponds /W/ below the tip	7,6 - 9,3 / 8,7	3693,3 - 4929,8 / 4354,1	34,4 - 172,0 / 124,9	133,9 - 2708,9 / 1340,2
Surface waters	Polluted waters	Streams /S/ below the tip	5,8 - 7,9 / 7,1	5755,9 - 8128,2 / 7230,8	153,7 - 845,1 / 591,2	3045,1 - 4450,4 / 4064,9
Surface waters	Polluted waters	Effluents from the tip	7,4 - 8,2 / 7,7	6457,8 - 19236,8 / 13919,4	742,0 - 5886,5 / 3776,8	3856,6 - 6341,2 / 4866,7

277

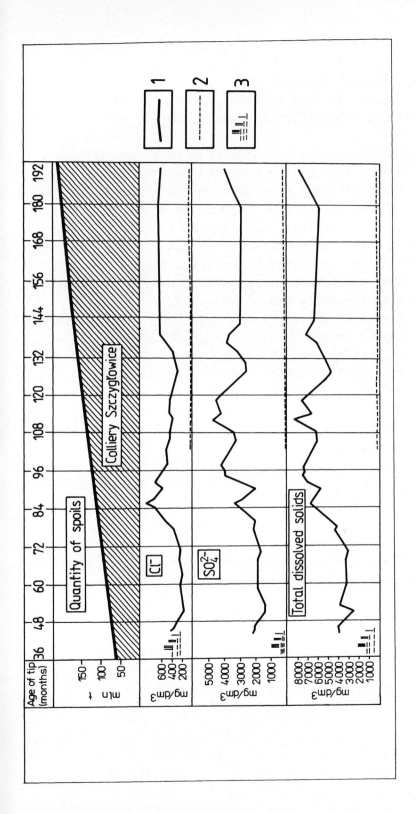

Fig.3. Variability of chloride /Cl⁻/, sulphate /SO_4^{2-}/ and total dissolved solids in surface waters caused by coal mine spoil deposition. Smolnica spoil tip. 1 - drainage ditch from the tip /sampling point—4/; 2 - stream above the tip /sampling point - A/; 3 - classes of water quality / I,II,III/.

waters have proved the negative changes of water quality in consequence of tip impact. / Table 2, Fig. 1 - 3/.

Total dissolved solids concentrations in waters beneath the tip have increased from 6 to over 14 times, in comparison with those in adjacent natural waters flowing towards the tip on its uphill side.

Sulphate concentrations increase from 11 to more than 33 times, and chloride content from 2,5 to over 25 times. The highest fluctuations of chloride content have been observed in adjacent waters beneath the tip. It corresponds with the degree of washing out the spoil on the particular parts of the tip of different age according the dynamics of chloride leaching.

In the adjacent waters beneath the tip /on its the downhill side/ the distinct increase of heavy metals, mainly copper, zinc and lead have been observed, too. / Table 3 /.

The chemical type of waters changes from HCO_3 - Ca- Mg for natural waters flowing toward the tip to SO_4 - Na or SO_4 - Cl - Na in waters beneath the tip.

Natural waters quality flowing towards the tip fulfill the standards for pot water. Water quality in drainages, watercourses and ponds below the tip exceeds not only the standard limits for pot water, but also the standards for the lowest III quality class for the inland waters.

C O N C L U S I O N

On the basis of the studies and long-term investigations of coal mine spoil tip Smolnica / Upper Silesian Coal Basin / has been concluded:

Coal mine spoil tip are found to be the serious source of contamination of natural underground and surface waters in the vicinity of the tip.

In the studied area the waters, yielding over normative amount of dissolved matters, sulphates, chlorides and heavy metals. The pollution is due to leaching of soluble matters / chloride, sulphate and heavy metal / in the process of percolation of precipitation water through the spoil deposited on the tip.

Table 3

Heavy metals content in the spoil deposited on the Smolnica tip and in the waters in the vicinity of this tip /Approx. age of tip 7 - 16 years/

Parameter	SPOILS /Content - g/t/	WATERS /Content - mg/dm³/					
		Fresh waters		Polluted waters /drainage ditch from the tip/			
		Stream /N/ above the tip		Ditch W		Ditch S	
		min	max	min	max	min	max
Chromium — Cr	50	0,00	0,09	0,00	0,10	0,00	0,10
Nickel — Ni	60	0,00	0,78	0,11	0,45	0,11	0,80
Copper — Cu	30	0,00	0,001	0,00	0,99	0,001	1,03
Zinc — Zn	50	0,00	0,11	0,10	2,70	0,30	1,89
Cadmium — Cd	—	0,00	0,001	0,00	0,02	0,00	0,05
Lead — Pb	14	0,00	0,10	0,00	1,07	0,00	0,80
Cobalt — Co	80	0,00	0,25	0,06	0,12	0,05	0,08

On the tip with high permeability of spoil generation and leaching of soluble solids have been observed in the full profile of aeration zone.

The waters percolate through the spoil tip becaming saturated with soluble components. Chemical composition of infiltration waters in the vertical profile in the aeration zone of the tip is a consequence of influence of two factor:
vertical redistribution of salt loads and a various degree of washing out spoil layer in the profile of the tip body determined by rate flow of infiltration water.

According to dynamics of chloride leaching from the spoils the risk of contamination of the water environment by chlorides is determined by the annual load of chlorides deposited on the tip together with coal mine spoil.

The water pollution by sulphates is defined by the total quantity of spoil in the aeration zone of the tip in the conditions of air penetration, to the spoil, which is a factor limiting sulphide oxidation.

R E F E R E N C E S

1. J. Szczepańska, Chemistry of pore waters from marine and continental clays of Tertiary formation, Geologia, t.8, z.3, Z. Nauk. AGH, Kraków, 1982, pp. 105 / in Polish, sum. English/
2. J. Herzig, J. Szczepańska, St. Witczak, I. Twardowska, Chlorides in the Carboniferous rocks of Upper Silesian Coal Basin, Fuel, 0172 / 1986 /.
3. I. Twardowska, Mechanism and dynamics of Carboniferous spoils leaching on the tips, Prace i Studia IPIS – PAN No 25, Wrocław, 1981, pp.206 / in Polish, sum. English /.

Reclamation, Treatment and Utilization of Coal Mining Wastes, edited by A.K.M. Rainbow   281
Elsevier Science Publishers B.V., Amsterdam, 1987 — Printed in The Netherlands

MINESTONE AND POLLUTION CONTROL

M Nutting, B.Sc., M.Phil, C.Eng., M.I.M.M., M.I.Geol., A.M.I.W.M., F.G.S.

Land Development Manager (South), Cleanaway Ltd

ABSTRACT
     An overview of the utilisation of minestone during the controlled disposal
of solid wastes is presented via three considerations:-

1)   a three-fold definition of minestone

2)   a review of the classification of potential landfill sites together with
      methods for their upgrading, and

3)   a study of some physical and chemical characteristics of each of the
      three minestone types, together with an examination of how these types,
      together with an examination of how these may be best exploited during
      utilisation.

INTRODUCTION

     The controlled landfill disposal of many solid wastes remains to this day
the most cost effective, and therefore most widely practised method of disposal.
Their disposal in an environmentally acceptable manner, according to modern
standards, requires the use of often extensive pollution control measures.

     These measures are chiefly associated with the preparation, operation and
successful completion and restoration of any site.

     Historically sites that have been used have either required little or no
treatment to render them suitable for such an operation or have relied upon
the presence of naturally occurring materials on or close to the site for their
treatment. Increasingly, sites that require such limited treatment are
becoming less common, and those that do exist are to be found at greater
distances from the centres of waste generation. The use of hitherto less
suitable sites through the provision of more extensive, engineered, pollution
control measures has been the most recent trend.

     Localised depletion of the more conventional, readily available, materials
such as Clays and sub-soils, most commonly cohesive in classification, and more
commonly retained for their soil-making properties, has led to the use of
other materials for the provision of such measures. These other materials
include manufactured polymeric materials, processed and enhanced naturally
occurring materials such as bentonite, and materials which require little or
no processing, such as minestone.

Minestones, however, are both limited in their application for such uses and in their geographical availability, and it is not intended that they be considered in competition with the other materials mentioned above, rather that there are applications in which they may be more cost effective in achieving the desired and designed end-product. It is for this market that minestones have been considered.

MINESTONE – A GENERAL DESCRIPTION

A summary description of minestone may be proposed as follows:- it is an aggregation of fragments of , typically cyclothemic, Coal Measures rocks and associated minerals, notably (in order of decreasing grain size) sandstone, siltstone, and mudstone (shale), together with lesser amounts of ironstone, pyrite, coal, and, more rarely, limestone. Varying quantities of weathering or secondary minerals may also be present.

No detailed study of the petrologies of minestone from around the country has been prepared. Individual petrological profiles are controlled by the precise nature of the strata being worked at the time, and may therefore vary from one coalfield region to another, and within this broad-scale variation from one colliery to another, depending upon the (range of) seams worked.

Similarly, with the exception of specific clays (see below) only localised mineralogical studies have been carried out. A range of typical mineral compositions is presented in Table 1.

TABLE 1

MINERAL	WEIGHT %	MINERAL	WEIGHT %	MINERAL	WEIGHT %
Quartz	15 to 23	Siderite		Limonite	
Illite	7 to 36	Calcite	3 to 8	Hematite	trace
Mixed Layer Clay	7 to 15	Ankerite		Magnesite	
Kaolinite	3 to 38	Feldspar	1	Gypsum	
Chlorite	2 to 7	Jarosite	2		
Allophane	1 to 2	Alunite	2		
		Pyrite	1 to 5		

Typical range of mineralogical analyses of colliery spoil (after Collins, 1976; Parry, 1982).

MINESTONE – NOMENCLATURE

The term 'minestone' so far used in this paper as a general term may be further sub-divided into the three more specific forms detailed below.

Fresh-wrought Minestone

Minestone utilised directly after coal preparation has ceased may be termed as 'fresh-wrought', and normally accounts for the coarse fraction of the discard, i.e. the plus 0.5mm fraction, the fine fraction being disposed of separately, termed 'fines'. However, more frequently, as disposal methods change and are restricted by available space at many collieries, the 'dried' finer materials are remixed with the coarse fraction prior to combined disposal. The physical characteristics of the material may therefore be affected together with the physical requirements it is capable of meeting.

It may therefore be proposed that 'fresh-wrought' minestone is likely to be lower in 'fines' than the associated 'as-dug' minestone will be, allowing where necessary for the re-addition of the finer fraction prior to disposal.

A further product that is being used in increasing amounts is the coarse fraction produced by the re-processing, i.e. washing, of the older heaps for their coal content. In terms of material type, and therefore engineering capability, although such materials are undoubtably re-processed 'as-dug' minestone they show many of the characteristics of 'fresh-wrought' minestone, in that they are chiefly composed of the plus 0.5mm fraction alone and many have not been re-subject to physical degradation brought about by their re-placement during replacement.

'As-dug' Minestone

Once placed on a spoil heap, and therefore subjected to a degree of both physical handling and weathering, and to some degree of chemical weathering, the spoil, whether composed of the coarse fraction alone or of the re-combined colliery waste, may be termed 'as-dug' minestone.

The residence time on the heap is not considered to be the single most relevant criterion to have effected the material, as most physical degradation is caused by the process of handling, placing and compacting the 'fresh-wrought' minestone in accordance with present methods of tip construction. The degree of such degradation experienced depends partially on the petrological composition of the material constituents, i.e. its resistance to such breakdown, and partially upon the age of the heap concerned, and therefore its method of construction.

Although some weathering, both physical and chemical, takes place it is again dependent upon petrological composition (Rainbow, 1983) in that the finer grained rocks are the more susceptible to such degradation, and the duration of any surface exposure prior to burial. Indeed such weathering effects have been shown to be limited to less than one metre below any exposed surface (Spears etal, 1971) and to be selectively reduced in the absence of freely available oxygen (Glover, 1975).

In line with the use of re-washed minestone, any proposal to use any such replaced and restored material would again necessitate its consideration as 'as-dug' minestone.

'Burnt' Minestone

'Burnt' minestone is that which has been subjected to raised temperatures through the combustion of any inherent coal content.

Typical physical changes are variations in particle size distribution and rock strength, brought about by the fusing together or breakdown of the clay-based rock fragments. Chemical variations centre on the loss of minerals through oxidation or heat induced decomposition.

A CLASSIFICATION OF POTENTIAL LANDFILL SITES AND METHODS FOR THEIR UPGRADING - A REVIEW

The classification of site proposed for landfill operations is derived directly from the D.O.E. Waste Management Paper No. 4 (D.O.E., 1976):-

i)   those providing a significant element of containment for wastes or leachates

ii)   those allowing slow leachate migration and significant attenuation

iii)   those allowing rapid leachate migration and insignificant attenuation

Disposal into Classification 1 sites, or those upgraded to comply with its specification, may be more simply termed (John and Davies, 1976) as 'concentrate and contain'. In contrast, the disposal into Classification 2 sites, or those upgraded to comply with its specification, may be more simply termed 'op cit' as dilute and disperse.

Through the use of selected materials and their correct engineering, in this case one or more of the specific types of minestone already detailed, it may be possible to upgrade sites into the classification(s) above the one under which it is naturally defined: to provide suitable engineered additions to sites of any classification to affect a reduction in hydraulic loading on those natural or engineered features providing the required degree of leachate containment or regulated migration: or construct disposal facilities where previously none existed in any complete form.

Current site designs related to the control of pollution for any of the above general scenarios chiefly centre on the provision of a site lining to ensure that liquids are either retained within the site or are released at a rate compatible with their attenuation, either wholly or partially by the material itself, within the groundwater regime or by the bedrock; the provision of a site cap to restrict the ingress of water, and therefore reduce hydraulic loading on the base of the site once it has been filled; and the operation of many sites on what may be termed the small cell principle, through the con-struction of internal, or secondary, control bunds.

Day-to-day pollution control in the successful operation of sites hinges on effective site management. This is likely to be largely dependent on the provision of daily or more general binding layers to prevent wind blow together with vermin and bird access.

Methods of Site Upgrading

Site Lining

The selection of a material for use as a site-lining is dependent upon its ability to prevent pollution of ground or surface water resources. It must therefore be capable of providing either an 'impermeable' layer for 'containment' sites or a layer of low permeability for 'dispersal' sites. For this aspect, its most important parameter is therefore the level of impermeability it is capable of producing when used in an engineering environment.

Despite the fact that the control of landfill operations is through accepted acts, notably the Control of Pollution Act, 1974, precise specified controls for acceptable site operation in individual locations would appear to largely depend upon the type of aquifer protection policy operated by the Water Authority concerned. The author has been unable to verify a nationally acceptable coefficient of permeability value which could be referred to as a 'maximum acceptable limit' for either 'containment' or 'dispersal' sites.

However, a working value of 10-7 cm/sec, corresponding to the "practically impermeable" classification of Terzaghe and Peck (1967) - (figure 1), would appear to be acceptable to selected Authorities for 'containment' sites where the disposal of liquids is not proposed and where the hydraulic head is expected to be minimal. This figure would appear to be supported by independent research carried out by the Harwell Laboratories (Stevens, 1984).

Higher values are therefore likely to be necessary where these conditions are exceeded in any way.

Conversely, lower values may be suitable for 'dispersal' sites, although the precise value may ultimately depend on the amelioration potential of the bedrock or ground-water regime below each individual site.

General Binding Materials

For general and public health reasons, each layer of waste tipped typically required a daily or weekly covering of inert materials, the main aim of which has been outlined above.

Materials should therefore have a reasonably high density, should not give rise to dust and should provide for a reasonable degree of trafficking. Furthermore, their use should not exacerbate the quality of any potential leachate, and they should be relatively permeable so as to prevent perched leachate tables or the build-up, and possible lateral migration, of landfill-gas.

Site Capping

The purpose of a capping layer over a landfill site is essentially to reduce rainfall infiltration, hence the material, in this case minestone, will be required to have similar properties to that used for a 'containment' site lining, i.e. it should be "impermeable".

The permeability is again the most important feature of materials used for this purpose. Again, a general working value accepted by selected Water Authorities of 10-7 cm/sec. can be considered, although a higher figure of 10-8 cm/sec. is considered to be more suitable in Stevens (1984).

A further important aspect of a capping layer is that it should be capable of being compacted to maintain a high degree of integrity at seasonally fluctuating moisture contents, i.e. it should be 'crack resistant'. It may be said that the ability of a material to resist this effect of weathering depends on its 'soil strength', i.e. a combination of shear strength and permeability, together with its ability to retain moisture at depth. Two important physical parameters are therefore the relationship between moisture content and placed density and the "aggregate" nature of the material with respect to its crack resistance and susceptibility to loss of moisture through capillary action.

Methods of Site Construction

As well as the use of one or more of the above detailed methods of general site engineering and management, it is possible to construct sites for the controlled disposal of wastes. Their construction typically consists of the provision of one or more bund walls behind or within which tipping operations proceed above the general ground level. Typical purpose built structures facilitate tipping in blind valleys, against raised ground or may provide for additional tipping space above that occurring naturally, i.e. by way of raising land, most frequently to bring it into beneficial use.

It is frequently the case that at least one bund, normally the major one, is constructed at the low point of whatever feature is being exploited, and therefore in addition to being "impermeable", the bund may have to withstand both lateral and hydraulic pressures.

The most important geotechnical characteristics to be considered are there fore permeability, O.M.C./maximum dry density relationship and shear-strength.

SELECTED PHYSICAL AND CHEMICAL CHARACTERISTICS - A REVIEW
Physical Characteristics

From the above review of those parameters which control whether or not any material, specifically minestone in this instance, is used for the operations covered, it may be seen that the most important ones are those that affect the way in which the material performs as an engineering material during the control of any pollution which could otherwise occur during a site's operation.

It is proposed to review those parameters below.

Permeability

A summary of the coefficients of permeability of those exclusively 'as-dug' minestones so tested is presented in Figure 1, which has the classification proposed by Terzaghi and Peck (1967) superimposed for clarity.

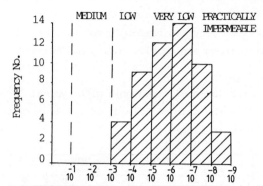

Figure 1   Laboratory Constant Head Coefficient of Permeability
of Minestone (after Rainbow, 1983), with Terzaghi
and Peck (1967) Classification.

It may be seen that a significant proportion of those samples so tested have coefficients of permeability equal to or less than 10-7 cm/sec. previously discussed.

It may therefore be proposed that 'as-dug' minestones may be readily selected to produce the necessary permeability requirements for their use as linings or caps in conjunction with 'containment' or 'dispersal' type sites.

It should be noted that the bulk of these 'as-dug' minestones originate from the coalfields in the centre of the country.

It should also be noted that a coefficient of permeability of as low as 10-10 cm/sec. has been obtained in field trials for one 'fresh-wrought' minestone, admittedly one with a high percentage of re-mixed 'fines'. 'Fresh-wrought' minestone is beginning to be considered for use in increasing quantities for its low permeability characteristics, although detailed data concerning coefficients of permeability are not generally available to date.

It may also be proposed that 'fresh-wrought' minestones may be selected where it is recognised that there has been some addition of 'fines' prior to disposal, for the provision of linings or caps for 'containment' or 'dispersal' sites.

The coefficient of permeability of 'burnt' minestones are consistently in excess of those required for its consideration for use as either a lining or a cap.

Bearing in mind that minestone is an aggregation of soft rock fragments, and although it may be composed of in excess of 60% clay minerals (Collins 1976), its mixed-layer clay content differs in each coalfield (Taylor, 1984) and may be relatively low (Collins, 1976). The effect of these clays on individual permeabilities is likely to be, locally at least, of secondary importance. It may therefore be deduced that the permeability of the placed minestone constructions depends mainly on the degree of particle packing which may be achieved during the compaction of the material. Of importance to individual coefficients of permeability are therefore the particle size distribution and relationship between achievable density and moisture content.

Particle Size Distribution (P.S.D.)

A summary of P.S.D. data for 'as-dug' minestones with typical 'fresh-wrought' curves superimposed, for the main coalfields of the country and for the U.K. as a whole, is presented in Figure 2.

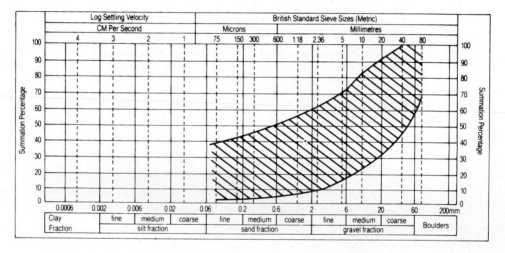

Figure 2. P.S.D. Ranges for as-dug Minestone (after Rainbow, 1983).

A comparison of individual P.S.D. and coefficient of permeability data reveals that a value of 10-7 cm/sec. or less equates with 'as-dug' minestones having in the order of 20% or more of material passing the 75u sieve, and that a value of 10-5 to 10-7 cm/sec. can generally be obtained from minestones having in the order of 15 to 20% of material passing the 75u sieve. In conjunction with the statements made earlier it may be said that 'as-dug' minestones suitable for use as linings or caps without pre-treatment are likely to be readily available from Scotland and Central England, and that with limited pre-treatment, such as screening to lower the relative percentage of the coarser fraction or the addition of 'fines', may be found in South Wales and Kent.

Furthermore, it may also be proposed that 'fresh-wrought' minestones suitable for use for the provision of linings or caps may also be available from Central England without pre-treatment, and from Scotland, South Wales and Kent with pre-treatment. This statement must be taken in the context that the suitability of 'fresh-wrought' minestones will depend very much on the precise method of operation of individual collieries.

The P.S.D. is also important in relation to minestone's resistance to cracking. Minestone is classed as a 'well graded, dry cohesive soil', (D.o.T., 1976). It may be said that the extensive size range of particle sizes found in any one, typically 'as-dug' or 'fresh-wrought' minestone means that it is likely to display a resistance to cracking not found in caps of the more frequently used clays or cohesive soils which have a reduced range of particle sizes, and which may show signs of stress after repeated cycles of desiccation and re-wetting.

Again, it may be proposed that all of the likely sources of either 'as-dug' or 'fresh-wrought' minestone considered above for use as linings or caps show a sufficiently wide range of particle sizes to meet this requirement.

A further consideration concerning the P.S.D. of minestone is in its use for daily or general cover. It is typically for this use where the greatest demand for 'fresh-wrought', including re-processed and 'burnt' minestones occurs.

Minestone selected for such use are ormally chosen because of their low price and ready availability as much as the fact that limited amounts of the less than 0.5mm fraction gives improved trafficking qualities and does not give rise to dust during trafficking. Similarly, a reduced percentage of the finer fractions may typically eliminate any potential problems with leachate or gas build-up within the tip's structure.

Optimum Moisture Content (O.M.C.) and Maximum Density

A survey of O.M.C. values, Table 2, reveals that the 'natural' content lies within plus or minus 2% of that required to give the maximum compacted density necessary for an acceptable engineering placement whether it be for a lining, a cap or for the construction of internal or major control bunds. In general, therefore no conditioning of the minestone is required prior to their use.

Experience from field trials has shown that if minestone is placed according to D.O.T. specifications (D.O.T., 1976) then the maximum density is readily achieved. Such densities are considered essential if the desired coefficients of permeability are to be maintained. A reduction of such density to, say 90% of maximum, may result in a reduction in the coefficient of permeability by as much as 20%. Through the use of these specifications the construction of any such structures remains a straight-forward engineering project, using standard equipment and techniques. It is thus possible to

290

equate the provision of any number of structures from any coalfield on an
equal footing.

TABLE 2

Analysis	Leachate from Domestic Waste Sites (Robinson and Maris)	Leachate from Domestic/ Colliery Spoil Sites Site 1	Site 2
p.H.	6.2 - 7.45	7.2	7.7
Total iron (mg/1)	0.09-380	35	43
C O D	66-11600	3705	280
B O D	2-8000	393	67
Ammoniacal Nitrogen	5-730	84	-
Chlorides	70-2777	318	-
Sulphides		<1	-
Sulphates	55-456	322	450

Comparison of Leachate Characteristics (after Small, 1983).

Shear Strength

The most comprehensive reviews of shear strength are to be found in
Taylor and Garrard (1984) and Rainbow (1983) as reviewed in Figure 3.

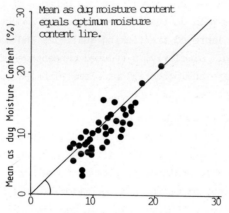

Figure 3   Relationship between As-Dug Moisture Content and
Optimum Moisutre Content for Maximum Dry Density
(after Rainbow, 1983).

Experience shows that, assuming all minestones are tested at maximum
placed density, the shear strength is dependent upon P.S.D. and individual rock
petrologies. Both of these may be equated ultimately to their original depth
of burial, and hence geologic compaction (Taylor and Garrard, 1984).

The values of shear strength obtained from minestone indicate that suitable retaining embankments can be constructed at economically acceptable rates.

Chemical Characteristics

The importance of the chemical characteristics on the potential use of minestones in pollution control works centres on their likely pollution potential and on any potential for the natural amelioration of other sources of pollution and on their behaviour as an engineering material.

As minestones are generally composed mainly of non-water soluble minerals (Collins, 1976) the main pollution potential is likely to be related to the release of chloride ions contained in connate waters held within rock fragments (Taylor, 1984); the formation of low pH, high water soluble sulphate and water soluble metal ion concentrations through the oxidation of inherent pyrite, coupled with the release of sulphates and associated leaching of clays (Smith and Ward, 1984); and with their possible release of water soluble sulphides or hydrogen sulphide gas through the reduction of the oxidation released sulphates above.

Experience has shown that the oxidation of pyrite is controlled by oxygen availability (Glover, 1975) and therefore depth of burial on the heap to be worked.

The vast majority of minestones tested have pH values of between 6 and 8 although extremes of 2 and 10 have been noted, most frequently with respect to 'burnt' minestone.

The use of fresh, i.e. unweathered minestones, typically 'as-dug' or 'fresh-wrought', of pH between 6 and 8 is therefore unlikely to increase leachate strength from an acidity aspect, especially as most sites are strongly reducing in nature once filling has been established.

Should high pH material be used, for example 'burnt'minestone, the dilution factor and strength of many site leachates, Figure 4,means that any possible contribution made by minestone is likely to be far outweighed.

To date no study on the possible release of hydrogen sulphide or water soluble sulphides from any of the three types of minestone has been carried out, although any such study must consider the amounts of pyrite in any 'fresh-wrought' or 'as-dug' minestones or the soluble sulphate content of any 'burnt' minestone proposed for use, together with any likely dilution effects or the influence of any appreciable amounts of sulphate-rich wastes likely to be found in the tip. Apart from the brief mention in Small (1983), no such polluting incidents have been brought to the attention of the author.

It may be proposed therefore that any of the above three categories of minestone may be used for pollution control works, although, from the point of view of chemical parameters, unweathered materials may be best suited.

292

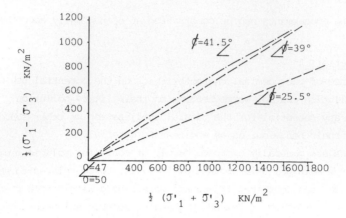

Figure 4.   Undisturbed Drained Shear Strength at As-Dug Moisture
            Content (after Rainbow, 1983).

To date minestones have been used in conjunction with 'containment' sites
alone, although the possibility of its use for 'dispersal' sites, relying to
some extent upon its own amelioration potential, is beginning to attract
attention.

The preliminary work on the cation exchange capacity of the expanding
layer clays found in 'as-dug' minestone reported in (Taylor, 1984) will
require further investigation in order to pursue this possibility.

Preliminary studies concerning the gas attenuation potential of 'as-dug'
minestone have been carried out by the Harwell Laboratories (Emberton and
Young, 1985). It is understood that any attenuation is dependent upon the
P.S.D. of the material together with the presence of clays, finely particulate
carbonaceous matter, freely available iron oxides. Moisture content is also
an important factor.

Of the materials tested, the 'as-dug' minestone gave some of the best
results, both treated as a single material and blended with others tested,
although it should be noted that the best materials only had attenuation
potentials of one sixth that of ground charcoal.

It may be therefore, that a new demand for minestones for such a use could
develop over the next few years.

CONCLUSION

It may be concluded that, from the extensive stocks of all three
categories of minestone available for consideration throughout the U.K.,
materials may be selected with or without prior treatment, or with limited
pre-treatment, to perform a variety of pollution control measures during the
development, operation and completion of waste disposal sites.

ACKNOWLEDGEMENT

The author is indebted to British Coal, during whose employ all work
necessary for the production of this work was undertaken, and for permission
to present this paper. The views expressed herein are those of the author and
not necessarily those of British Coal.

REFERENCES

1. Collins R.J., A method of measuring the mineralogical variation of spoils
    from British Collieries: Clay Minerals. 11. pp. 31-50.
2. Department of Environment. The licencing of waste disposal sites,
    Waste Management Paper No. 4, 1976.
3. Department of Transport. Specification for Road and Bridge Works, Fifth
    Edition, 1976.
4. Emberton J.R. and Young P.J., The use of cover materials to attenuate
    landfill gas at Ugley Landfill, Essex, AERE Harwell, Environmental and
    Medical Sciences Division, 1985.
5. Glover H.L., Coal Mining, water, and the environment, in Mining and the
    surface environment, a short industrial course, Nottingham University,
    wp. 16-23, 1975.
6. John D.G., and Davies D.R., Some hydrogeological aspects of waste
    disposal at landfill sites, at: Land Reclamation Conference, 1976,
    Paper 24.
7. Nutting M., The use of minestone in conjunction with the controlled
    disposal of domestic and commercial waste, in: Symposium on the
    Reclamation, Treatment and Utilisation of Coal Mining Wastes, Ed.
    A.K.M. Rainbow, Durham, 1984. Paper 48.
8. Rainbow A.K.M., An investigation of some factors influencing the
    suitability of minestone as the fill in Reinforced Earth structures,
    N.C.B. Publication, 1983.
9. Small G.D., The effect of coal mining waste on domestic landfill sites,
    in: Landfill Completion, Proceedings, Harwell, 1983, pp. 73-81.
10. Smith A.C.S. and Ward P., Pollution potential from the reclamation of
    coal washery dumps, in: see Ref. 7, 1984, Paper 28.
11. Spears D.A., etal, A mineralogical investigation of a spoil heap at
    Yorkshire Main Colliery, Q.Jl., Eng, Geol., 3,4, pp. 239-252.
12. Stevens C., Landfill lining and capping, AERE Harwell, Environmental and
    Medical Sciences Division, 1984.
13. Taylor R.K., Composition and engineering properties of British Colliery
    discards, N.C.B. Publication, 1984.
14. Taylor R.K. and Garrard F.G.F., Design strengths for United Kingdom coarse
    colliery discards, in: see Ref. 7, 1984, Paper 34.
15. Terzaghi K. and Peck R.B., Soil mechanics in engineering practice,
    2nd Edition, John Wiley and Son, 1967.

Reclamation, Treatment and Utilization of Coal Mining Wastes, edited by A.K.M. Rainbow
Elsevier Science Publishers B.V., Amsterdam, 1987 — Printed in The Netherlands

THE STUDY OF SATURATED COAL MINING WASTES, UNDER THE INFLUENCE
OF LONG-TERM LOADING

K.M. SKARŻYŃSKA and E. ZAWISZA
Soil Mechanics and Earthworks Department, University of Agriculture,
Kraków /Poland/

SUMMARY
The study was carried out using large-scale laboratory apparatus
which enabled actual sized minestone to be studied. The equipment
also allowed long-term loading conditions to be applied. The paper
shows that minestone, when fully saturated and compacted and subjected to a loading intensity of 300 kPa for 13 months, displayed
variations in particle size, moisture content and bulk density.
Under the conditions obtained in the test it was possible to measure the extent of the settlement. It is anticipated that the data
may enable more acurate predictions of settlement to be made.

INTRODUCTION
Coal mine waste is used for the construction of hydraulic structures and for infilling surface voids created by the extraction of
minerals. Geotechnical parameters of the waste material from a number of coal mines in Upper Silesia where determined over a period
of many years /refs.1, 2/. Methods of determining its suitability
together with specification requirements where prepared /ref. 3/.
It was considered important that the behaviour under full saturation was determined. Because little information on this aspect is
currently available, a programme was initiated to determine the
effect of saturation on the long-term behaviour of minestone when
incorporated into hydraulic structures. These studies were also
conducted with the aim of elaborating methods of large-scale model
studies of coal mining wastes.

EXPERIMENTAL ARRANGEMENT
The experiment was carried out on large-scale model apparatus,
which made it possible to study material with a natural grain size
distribution, and to maintain loading for a longer period of time
/refs. 4, 5/. The equipment for model studies consists of a box, measuring 100 x 50 x 90 cm /see Fig. 1/. On the front, transparent
panel, there is a 4 cm square grid, making it possible to measure

Fig. 1. General view of testing equipment

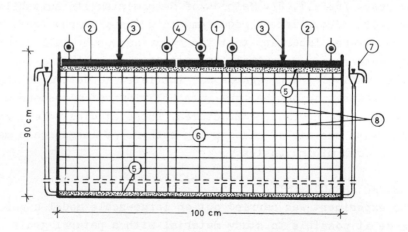

Fig. 2. Diagram of model apparatus. 1. centre plate, 2. side plates, 3. load, 4. dial gauges, 5. filtration layers, 6. minestone, 7. water level, 8. square grid.

any deformation arising in particular layers of material. The box is equipped with 4 plates: a large plate having the same area as the upper surface of the box, and 3 smaller ones; the center plate measuring 50 x 15 cm, and the two side plates 50 x 42 cm each. The loading system enabled a force of 300 kPa to be applied. The apparatus also included dial gauges to record surface movements, and a water control facility.

METHODS OF INVESTIGATION

The material used in the study was fresh; 78 % of it came from the washery plant, and 22 % from collieries. Ninety percent of the material was claystone and silt-stone, the remaining 10 percent was coal shale and sandstone. The grain size composition is illustrated in Fig. 3.

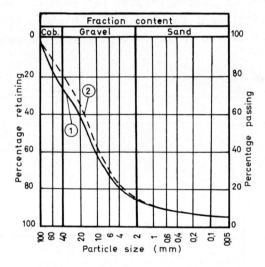

Fig. 3. Particle size distribution of minestone. 1. before test, 2. after test.

Material with a 7.6 % moisture content was compacted in the model apparatus in 12 cm layers up to a total depth of 70 cm. The compaction index was 0.91. Two layers of sand, one above the sample and one below, were incorporated to aid of minestone saturation during the test.

The experiment was conducted in two stages. In the first stage, a load of 49 kPa was applied to the whole surface by means of the

large plate, and maintained until the settlement ceased. The system of loading was changed in the second stage, first of all a load of 49 kPa was applied to each of the 3 plates /see Fig. 2/.When
no further settlement was recorded the minestone was saturated
/the water level being maintained at the level of the upper filtration layer/. The load on the centre plate was then increased
by 24.5 kPa every 24 hours /at the rate of 1 kPa per hour/ up to
300 kPa, the loading on the side plates being maintained at 49
kPa. The loadings were maintained until all settlement ceased,
then the water was drained off the sample and the quantity was
measured, the moisture content and grain size composition was also determined.

RESULTS AND ANALYSIS

The first stage of loading lasted for 46 days. During this period the final settlement was 4 mm, fig. 4 shows the actual shape
of the settlement curve. To avoid disturbing the model, moisture

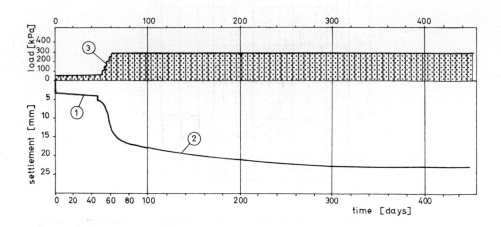

Fig. 4. Settlement of minestone vs time. 1. big plate, 2. centre
plate, 3. rate of loading.

content in this phase was only taken near the surface and fell
from an initial value of 7.6 % to a final value of 5.9 % during
the test sequence. In phase two when a constant load of 49 kPa
was applied to the three plates, settlement continued for five
days. This settlement amounted to 2 mm. The further gradual loading

of the centre plate, up to 300 kPa, lasted for 10 days, and caused
a settlement of 7.8 mm. It also had the effect of making the side
plates tilt towards the centre plate. The total mean settlement
of the two side plates was about 7 mm. Settlement under the centre
plate continued for 349 days, with a settlement of 17.6 mm being
recorded. The total settlement including the settlement induced by
the earlier loading was 22 mm. From the above results it may be
seen that settlement increased nearly double during the period
when subjected to the 300 kPa load.

By interpreting the results of the investigations carried out,
the compressibility of the material can be calculated. At a load
of up to 49 kPa, the oedometric modulus of primary compression is
about 8600 kPa and the modulus of secondary compression is about
17000 kPa for the minestone used in the model test. It is likely
that the increase in the modulus of secondary compression is also
the result of the modified loading conditions /i.e. the exchange
of one large plate for 3 smaller ones/. Changing the load on the
middle plate /49 to 300 kPa/ yielded a modulus of primary deformation of about 9400 kPa, 44 % of the settlement was observed to
have taken place rapidly whilst the remaining 56 % took over 13
months.

TABLE 1

Characteristics of the material tested

Denotation	unit	before test	after test	difference
Fraction content				
cobbles		27	18	-9
gravel	%	59	67	+8
sand		8	9	+1
silt and clay		6	6	0
Uniformity coefficient		29	24	-5
Moisture content	-	7.59	7.57	-0.02
Maximum bulk density	g/cm^3	1.92	1.99	+0.07
Dry density	g/cm^3	1.78	1.85	+0.07
Compaction index	-	0.91	0.95	+0.04

Following the investigation, after the water had been drained away,
the moisture content returned to 7.6 % /see Tab. 1/. The water
drained off from the model amounted to 14 l, thus the mean moisture content of the material was about 10 % during stage two of the

a

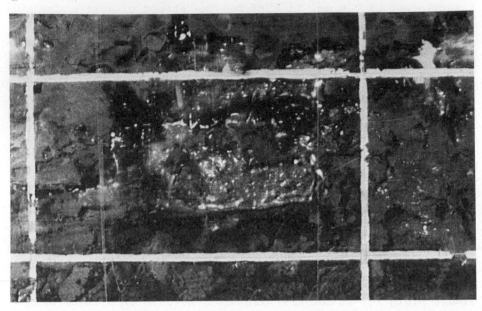

b

Fig. 5. Partial disintegration of rock fragments during the test period. a. initial state, b. state after 13 months.

study. It was also found that the grain size distribution was slightly altered, expressed as a partial disintegration of larger rock fragments /see Fig. 5/, and showed a decrease of 9 % in the cobbles fraction, an increase of 8 % for gravel, and an increase of sand content by 1 %.

SUMMARY

To resume the results, it may be stated as follows:

1. When saturated with water and subjected to loading of 300 kPa for over a year, the compacted minestone studied /I_s = 0.91/, does not show significant changes in the grain size distribution, moisture content and bulk density /see Tab. 1/. This is evidence of the considerable slowing down of the weathering process, as compared with the course of similar processes, taking place under field conditions, on the surface of dumps and earth structures. The checking of weathering processes is caused by the compaction and saturation of minestone, and elimination of atmospheric influences.

2. The course of settlement under conditions of long-term loading, which is unusual for coarse-grained soil, may be accounted for by the partial weathering of the edges of rock fragments, this bringing about a slightly better compaction of the material.

3. It ought to be stressed that at fairly low moduli of compressibility, the angles of internal friction and cohesion are relatively high /refs. 2, 3/, this leading to a fairly high stability of embankments composed of minestone.

4. The method of model studies adopted is correct, and makes it possible to predict changes in the behaviour of minestone with time. Its weak point is that it is time-consuming, hence it is difficult to carry out studies for various kinds of waste material, and various conditions under which structures are used.

REFERENCES

1 K. Skarżyńska, P. Michalski, H. Burda, E. Kozielska-Sroka, Determining the possibility of applying minestone of the "Julian", "Andaluzja", "Grodziec" mines for the construction of embankments for the River Brynica, Soil Mech. a. Earthworks Dep., Univ. of Agriculture, Kraków 1981.

2 K. Skarżyńska, P. Michalski, H. Burda, E. Kozielska-Sroka, Investigating the geotechnical properties of minestone from the "Ziemowit" coal mine, in order to assess the possibility of utilizing them for the construction of embankments of Goławiecki

Stream. Soil Mech. a. Earthworks Dep., Univ. of Agriculture, Kraków 1983.

3　K. Skarżyńska, J. Setmajer, P. Michalski, H. Burda, E. Koziels-ka-Sroka, Guidelines for designing, construction and acceptance of hydrotechnical structures made from unburnt coal mining was-tes. Soil Mech. a. Earthworks Dep., Univ. of Agriculture, Kraków 1985.

4　E. Zawisza, Model studies of the course of deformation in a silt substratum under loading. PhD thesis. Univ. of Agriculture, Kraków 1983.

5　E. Zawisza, Method of model studies of the deformation of cohe-sive soils under loading. Conference on Theoretical and expe-rimental problems of soil consolidation. Kraków-Janowice 1985.

SUCTION PRESSURES IN COLLIERY EMBANKMENT SURFACES

R.K. TAYLOR[1], S.J. BILLING[2], F.T. BICK[1] and R.J. SIMONDS[2]

[1] School of Engineering & Applied Science, University of Durham,
South Road, Durham DH1 3LE, United Kingdom

[2] Department of Geological Sciences, University of Durham, South
Road, Durham DH1 3LE, United Kingdom.

SUMMARY
    Suction pressure measurements were carried out in two
experimental tips constructed of fresh and highly weathered coarse
colliery discard, respectively, from different sources.
Continuous data logging was achieved by means of pressure
transducers attached to tensiometers set up at three different
depths below the surface of the heaps.

    The results are considered in conjunction with chloride ion
trends established in an earlier field research project. Results
suggest that saline discards are best seeded in the autumn when
suctions are negligible and chloride ions concentrate downwards
into an embankment.

    Limited measurements at a colliery in the Nottinghamshire
coalfield were found to show higher maximum values and greater
fluctuations in suction pressures than those in the experimental
tips. Differences between model tips and real structures are
believed to be due to the greater exposure to the weather of the
latter structures. The partly saturated zone was shown to be
only about 0.5m thick in the experimental heaps and a lagoon
embankment.

INTRODUCTION

    An investigation of the concentration and movement of chloride

ions in a colliery lagoon embankment in the South Yorkshire

coalfield was reported at the 1st Symposium in 1984 (ref. 1). Down

the flank of this embankment, chloride ions in the coarse discard

(mudrock) construction materials tended to be leached downwards

into the embankment in sustained wet periods. In contrast, during

dry summer months a more even distribution of chlorides was

recorded in the top 1m of discard. At the extreme, the surface desiccation of seepages revealed local 'hot spots' with concentrations of up to 2453mg/1. (as Cl).

The effect of chlorides is to promote physiological drought, particularly in young vegetation (ref. 2). Consequently, it was deemed important to determine more precisely the movement of water (and electrolytes) in the partly saturated zone of an embankment in order to try and recognize an optimum window for seeding embankments - that is, when downwards leaching is operative but weather conditions still favour restoration operations.

SOIL SUCTION

Water held in soil can be thought of in terms of gravitational water in the larger pores, capillary water in intermediate sized pores and adsorbed or hygroscopic water in the ca. 400Å wide zone of strongly bonded water around clay-size particles. The model of Buckman and Brady (ref. 3) is useful for illustrating the relationship between soil water and plants. Figure 1 illustrates the three zones: adsorbed water, capillary water and gravitational water. Also shown is the range of root suctions which determines the water available to plants.

The attraction of soil water to mineral particles is customarily described in terms of water tension or suctiton pressure. Conventional units of pressure may be used for quantifying the negative water pressures involved, or the pF scale may be adopted (Fig. 1). In the latter scale a pF of unity is the pressure equivalent to a 10cm high column of water, pF3 (the negative equivalent of atmospheric pressure) is equal to 1000cm of water and so on. Oven drying is approximately pF 7 (that is, 10^7cm of water) and full saturation is pF0.

In Figure 1 it should be noted that water held at a tension of less than 34kN/m^2 (pF = 2.54) will be subject to movement by any 'free' gravitational water in the larger pores and capillaries.

There are a number of methods by which suction pressures can be measured in the laboratory and in the field. The suction pressure-water content curves for material of less than 2mm size (Fig. 2), separated from coarse discard from Maltby Colliery, were determined by laboratory suction plate apparatus. Pressures up to pF3 can be measured with this equipment (Ref. 4).

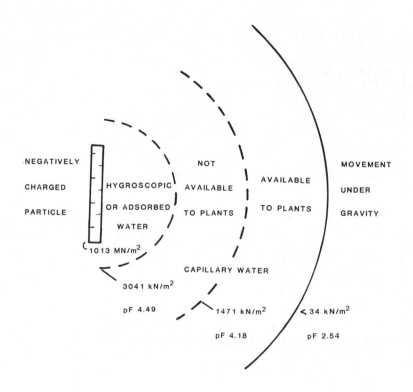

Fig. 1.   Illustrating relationship between water in soil and mineral particles. Suction requirements of vegetation shown (modified after Buckman and Brady, ref. 2).

The drying curves shown in Figure 2 for the full particle size ranges of Maltby and Bilsthorpe coarse colliery discards were determined when tensiometers embedded in discard samples were

306

removed from the embankments concerned and allowed to dry out in the laboratory. These latter curves should only be regarded as indicative but it can be seen that they span a much more restricted water content range than the Maltby fines. Two of the bulk discard curves show a sharp drop in water content (with air entry) at about pF 2.5 or greater.

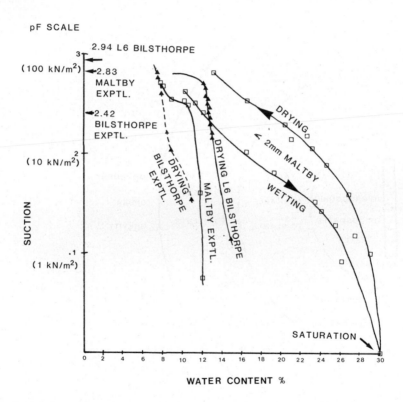

Fig. 2.   Suction pressure (pF) vs. moisture content relationships for coarse discards from Maltby and Bilsthorpe collieries and less than 2mm size fraction (Maltby).

The less than 2mm size material referred to above displays the more characteristic shaped curves of silty and argillaceous soils. The marked hysteresis of the wetting and drying curves is related to the disparity in shape of void spaces and entry channels in the material. Plants generally obtain their moisture and nutrient requirements from this smaller size fraction (ref. 5) so it might be argued that the ideal size/water content range in the partly saturated zone of an emabnkment is greatly different from that of the bulk discard in Figure 2. However, scarifying followed by soiling is commonly used in restoration and thus vegetation requirements can be more nearly satisfied in terms of water content.

DESIGN OF PROJECT

The project commenced in June 1984 with the construction of an experimental tip of coarse discard believed to be about 7 years in age from Malty Colliery (S. Yorkshire Area). The particle size distribution of the material is shown in Figure 3. About 2.2 tonnes of discard was used for this tip which was constructed in the laboratory court-yard at Durham University. The structure (Fig. 4) was compacted in 3 layers to a nominal bulk density of 2.0Mg/m^3 with the exposed face being inclined at 18 degrees.

During the spring of 1985 a similar experimental tip was constructed using fresh coarse discard from Bilsthorpe Colliery (N. Nottinghamshire Area). This material was considerably coarser-grained than Maltby (see Fig. 3), but over a period of about 15 months it degraded to a size range not greatly different from the Maltby material. However, during this period it was subjected (amongst other experiments) to leaching by water, simulating 5 years rainfall.

The only difference in construction details relating to the Bilsthorpe mini-embankment was that 3 tonnes of discard were used and it was contained within breeze-block rather than timber walls.

There were two reasons for constructing the experimental tips,

308

namely, that continuous monitoring of suction pressures could readily achieved, and that research would not be impeded by the miner's strike. Subsequently, some measurements were made in 1985 at two locations at Bilsthorpe Colliery; in the 3 years old embankment of Lagoon 6, and below the 3 years old soil cover at a location denuded of grass on the restored part of Tip No. 1. The discard in the latter tip was about 15-20 years old.

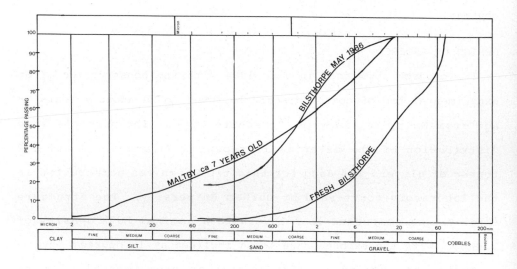

Fig. 3. Particle size distributions of fresh and experimentally degraded coarse discard from Bilsthorpe Colliery and 5-7 years-old discard from Maltby Colliery.

Tensiometers

High-air-entry ceramic piezometers (Figs. 4 and 5) with a pore diameter of approximately $1\mu m$, an air entry value of approximately 1 atmosphere (ca. $100kN/m^2$) and a permeability of 2 x $10^{-8}m/s$ were modified as tensiometers. In essence the porous ceramic cylinder, sealed at one end, is connected at the other end to a sealed water reservoir and a suction measuring device (Fig. 5). In the trial stages of the Maltby experimental tip,

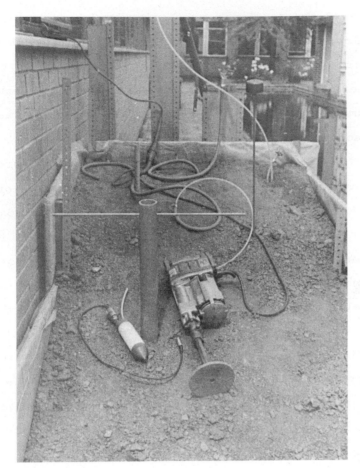

Fig. 4. Construction of Maltby experimental tip, showing tensiometer ready for installing in hole made by cutter-tube.

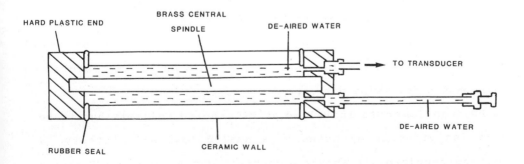

Fig. 5. Cross section of tensiometer (piezometer).

standard mercury manometers were used for measurement, but these were replaced for continous monitoring by Schaevitz pressure transducers (no P721/001) calibrated in both tension and compression. The transducers minimised the time lag in tensiometer response since practically no water flow occurs as the instrument adjusts. The water inside the tensiometer assumes the same solute composition and concentration as the soil water and therefore it is the matric potential which it measures and not the osmotic suction. Measurements by a tensiometer of this type are limited to suction pressures of below 1 atmosphere ($101kN/m^2$). Indeed, the highest measurement recorded in the present work (see Fig. 8) was $86.7kN/m^2$ (pF 2.94) in Bilsthorpe tip No. 1. It was suspected that in dry weather higher suctions would apply. Prior to installation of a tensiometer it is necessary to de-air the ceramic and to ensure that it is not leaking after it has been filled by submergence in de-aired water and the top section assembled. Installation at the depths indicated in Figures 6, 7 and 8 was accomplished by centering each instrument in a narrow hole filled with water following the removal of any large gravel or cobble-sized fragments. The discard was then tamped down around the tensiometer, the excess water helping to provide a good seal. Periodically air was drawn into the tensiometers, particularly after protacted intervals of high suctions. Flushing and replacement with de-aired water was therefore necessary on a regular basis. Some breaks in the record of individual tensiometer readings (Figs. 6 and 7) are related to maintenance following air entry.

The tensiometers installed at Bilsthorpe Colliery had to be read individually by means of a sealed lead acid battery, a voltage regulator to ensure a constant excitation of 10V, and a battery powered multimeter to display the output voltage.

Considerable problems were encountered with these tensiometers because of leakage. The fault was a consequence of this series of piezometer tips being home-produced.

SUCTION MEASUREMENTS

Output for the two experimental embankments is shown in Figures 6 and 7, whilst field measurements at Bilsthorpe Colliery are shown in Fig. 8. Rainfall figures, maximum daily temperatures and wind speeds for the experimental tips were taken from the Durham University Observatory records, whilst those shown in Fig. 8 were supplied by the Nottingham Weather Station at Watnall.

The Maltby record commences in July 1984 and runs through to the end of August 1985, Bilsthorpe experimental tip was monitored from early in May to early August 1985 after which other experiments were carried out in the material.

Maltby and Bilsthorpe records (Figs. 6 and 7) indicate that throughout the monitoring period, only small negative or positive pore pressures were measured in the deepest tensiometers (430mm Maltby, 500mm Bilsthorpe). Although the early Maltby measurements (Fig. 6, July 1984) were affected by air entrainment, the technique of refilling and stabilising tensiometers was mastered within a few weeks. Bilsthorpe discard, which was about 6 months old when instrumented, caused air entrainment difficulties throughout the monitoring period because of its inherent coarseness.

The highest suction pressure reading in the experimental tips was recorded in July 1984 ($65kN/m^2$, Maltby discard). In the following year (July 1985) the tensiometer at 100mm depth in the Maltby experimental tip registered a maximum of $59kN/m^2$. This compares with a maximum of only $26kN/m^2$ for the shallowest tensiometer in the much coarser Bilsthorpe material. A comparison of Figure 7 with the requisite time period in Figure 6 illustrates the much lower suctions measured in Bilsthorpe material. Maximum suctions in terms of the pF scale are shown in Figure 2.

The shallowest tensiometers (ca. 100mm depth) customarily recorded the highest suction pressure, although in more continuous wet periods there was a tendency for the middle tensiometer to record the highest measurements. Reference to Figure 6 shows that in August 1985 the middle tensiometer is measuring about $20kN/m^2$, whilst readings from the top instrument drop below $5kN/m^2$.

Examination of the records for both discards suggests that rainfall is probably the main control on suction pressure in the experimental tips. It was not possible to assess the influence of wind and temperature on these heaps because they were to a large extent protected by buildings.

The main findings of the monitoring programme are illustrated by the Maltby record (Fig. 6). In mid-September the suction pressures dropped dramatically in all tensiometers and remained at negligible levels throughout the winter months. By April 1985 they had started to rise, although it was not until June that (fluctuating) sharp increases occur (see also Fig. 7). The highest suctions were measured in July and August in both 1984 and 1985. This pattern suggests that the sowing and planting of vegetation is best initiated during the autumn window, starting in mid-September. Previous work (ref. 1) has indicated that the downwards concentration of chlorides into an embankment commences as a late autumn event. In contrast, spring sowing would mean that young vegetation might well be subject to the higher summer suction pressures and thus to higher chloride levels in new embankments constructed of saline discards.

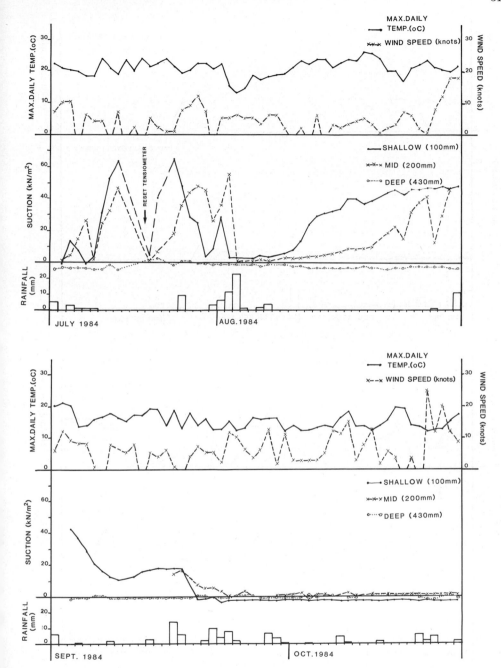

Fig. 6.   Record of suction pressure measurements from 3 tensiometers in <u>Maltby</u> experimental tip (1984-85). Weather data after Durham University Observatory.

314

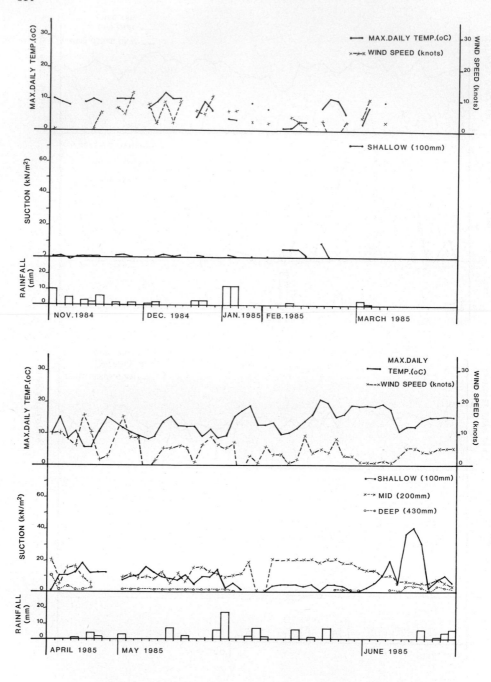

Fig. 6. continued.

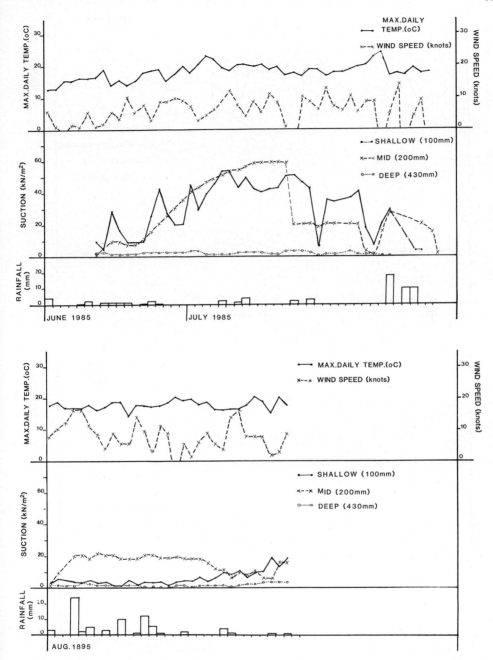

Fig. 6. continued.

316

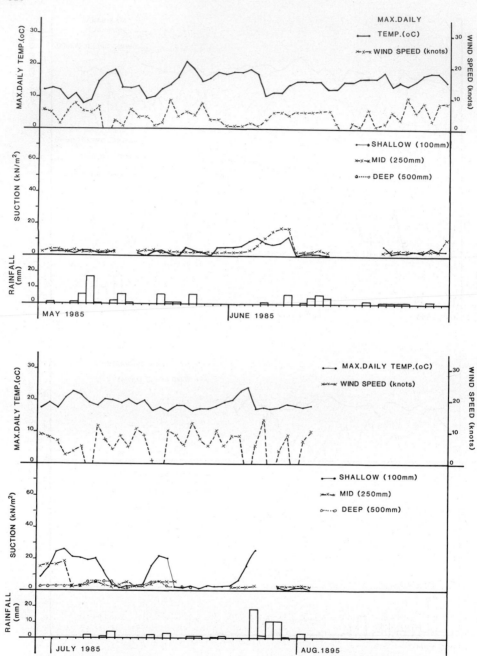

Fig. 7.   Record of suction pressure measurements from 3 tensiometers in <u>Bilsthorpe</u> experimental tip (1985). Weather data after Durham University Observatory.

Field measurements

Figure 8 is a record of suction pressure measurements taken in lagoon embankment L6 and within the grassed part of Tip No. 1 at Bilsthorpe Colliery. The 3-year-old exposed coarse discard of embankment L6 forms the best comparison with the experimental tips, in that Tip No. 1 has been top-soiled and seeded. Indeed, the pattern of measurements in L6 embankment also favours this interpretation.

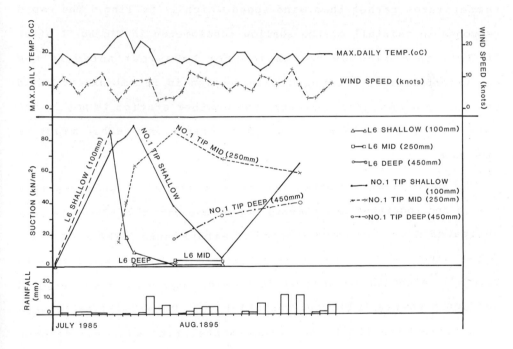

Fig. 8. Suction pressure measurements in Lagoon 6 embankment and Tip No. 1, Bilsthorpe (1985). Weather data after Nottingham Weather Station, Watnall.

The maximum readings of both tip and lagoon embankment are notable at 89 and 86kN/m^2, respectively. These are more than 20kN/m^2 higher than the maximum value measured in the Maltby experimental tip. However, both field locations at Bilsthorpe Colliery are in open areas with greater exposure to the elements than those experienced by the experimental tips. Lagoon 6 embankment is 11m high, Tip No. 1 is 17m high, which should be compared to a height of under 1m for the experimental structures. Fluctuations in suction measurements from shallow tensiometers in both embankments at Bilsthorpe (Fig. 8) are very marked. It would appear from Figure 8 that the high suctions on the 23 July (L6) and 26 July (Tip No. 1) tend to correspond to increasing daily temperatures rather than wind speed which is falling. The rapid response to rainfall of the shallow tensiometer in Tip No. 1 after 26 July is analogous to tensiometer behaviour noted in the experimental tips, but the drop in suction in L6 embankment would seem to be premature. However, the weather station is not in the immediate site vicinity so some discrepency in weather might be expected.

The overall tensiometer response in Tip No. 1 suggests that soiled and vegetated embankments may behave rather differently from untreated coarse discard exposed in embankments. The middle and deepest tensiometers are apparently unaffected in the short-term by rainfall, although the deepest instrument showed a gradual rise in suction pressure following installation. Embankment L6 is more comparable with the Durham experimental tips with the deepest tensiometer at 450mm registering negligible suction pressures in rainy periods. The response to rainfall for all 3 tensiometers over the period from 26 July to 6 August 1985 suggests that the partly saturated zone of the embankment may be little more than a skin some 450mm in depth.

CONCLUSIONS

Soil suction pressures developed in the partly saturated zone of a waste embankment are of particular interest with respect to the early (progressive) restoration of saline colliery discards. Results to date suggest that 5 to 7 years natural leaching may be necessary to reduce contained chlorides to low levels. Because chlorides can cause physiological drought in young vegetation it is advantageous to assess their movement in colliery embankments as precisely as possible. The monitoring of two experimental spoil banks by means of tensiometers connected to pressure transducers and a data logger has identified a number of important suction pressure characteristics.

The highest suctions were measured in the months of July and August when previous work (ref. 1) has shown desiccated chloride/sulphate 'hot spot' measurements to be at a maximum. Chloride ion levels in general are more evenly distributed through the partly saturated zone during this period. In mid-September suction pressures dropped significantly to minimal levels and they fluctuated around pF0 until the following April when they started to rise again. It is suggested from this pattern of soil suctions that autumn is the optimum sowing period for vegetation, since suction pressures are then minimal and chloride ions are beginning show a downwards concentration into the embankment. In contrast spring sowing would tend to subject young vegetation to enhanced suction pressures and the generally higher chloride ion levels of the summer months.

Suction measurements made in a 3-year-old lagoon embankment and a restored colliery tip at Bilsthorpe Colliery indicate a much higher maximum (x1.5) and greater flucturations in suction pressures than in the experimental tips. These differences are believed to be a function of the greater exposure to weather – particularly drying winds - in the case of the Bilsthorpe structures.

The pattern of suctions developed in the restored tip is rather different from that of the unrestored lagoon embankment, which shows certain similarities with the experimental tips. One similarity is the apparent depth of the partly saturated zone which suction measurements suggests is only about 0.5m in thickness. Chemical oxidation is likely to be almost entirely restricted to the partly saturated zone in a 'shale' embankment. In the older generation of unrestored, loose tips intense weathering was restricted to a depth of about 1m, and in the 50-year-old Yorkshire Main tip a low degree of weathering was detected even below this to a maximum depth of 3.81m. Similarly, in discards placed in loose 1.5m layers at Gedling Colliery soil suctions of between 6-17kN/m^2 were measured to a depth of 4.5m (Section 5 of ref. 6). The present work suggests that a much thinner partly saturated zone may well obtain in modern colliery embankments.

ACKNOWLEDGEMENTS

We are grateful to British Coal for the grant which financed this research. The views expressed are the authors and not necessarily those of the Board.

REFERENCES

1   R.K. Taylor, S.J. Billing and P.K. Spencer, Chlorides in coarse colliery discards. Symposium on the Reclamation, Treatment and Utilization of Coal Mining Wastes, Durham, England. (1984) 31.1-31.15.

2   J. Sutcliffe, Plants and Water, London, Edward Arnold Ltd., 1976.

3   H.O. Buckman and N.C. Brady, The Nature and Properties of Soils, 1969, Macmillan, 653p.

4   J.D. Coleman, An Investigation of the Pressure Membrane Method for Measuring Suction Properties of Soil. Road Research Note RN/3464, Road Research Laboratory, Crowthorne, 1959.

5   E.W. Russell, Soil Conditions and Plant Growth. Longman Group Ltd. London, 1973.

6   National Coal Board, Review of Research on Properties of Spoil Tip Materials, Lab Ref. No. S/7307. National Coal Board, Hobart House, London, 1972.

Reclamation, Treatment and Utilization of Coal Mining Wastes, edited by A.K.M. Rainbow
Elsevier Science Publishers B.V., Amsterdam, 1987 — Printed in The Netherlands

"DEEP RIPPING: A MORE EFFECTIVE AND FLEXIBLE METHOD FOR ACHIEVING LOOSE SOIL
PROFILES"

A.R. BACON[1], BSc, MSc(Eng), CEng, MICE and
R.N. HUMPHRIES[2], BSc, MA, PhD, C Biol, MI Biol, AMIQ

[1]British Coal, Headquarters Technical Department, Ashby Road, Stanhope Bretby,
Burton-on-Trent, Staffs
[2]Humphries Rowell Associates Ltd, Holts House, 25 Wymeswold Road, Hoton, Leic.
(formerly - Midland Research Unit, University of Nottingham)

SUMMARY
    The paper reports the results of investigations carried out to determine
the physical characteristics of soils replaced by motorscraper in colliery
tip restoration, reviews current recommendations on soil handling and
suggests a new approach.
    Using a recording cone penetrometer it is shown that (a) the simple 'single
row' placing technique of placing soil results in less compaction that more
complex operation procedures involving turning on to and off soil layers;
(b) deep ripping on completion of soiling leads to less compact lower horizons
than conventional sequential ripping of individual soil layers.
    The application of these findings will lead to economy in operations and to
the required standards of restoration being more readily achieved.

INTRODUCTION

    The massive and compacted nature of many reinstated soil profiles has been

blamed on the use of motorscrapers or towed scraper boxes for top and sub-soil

stripping and replacement. In most respects however the scraper is ideal for

soil handling, this being the role in civil engineering earthworks for which

it was designed and developed.

    The results of investigations carried out to determine the physical

characteristics of soils replaced by motorcraper are given, current

recommendations for soil replacement are reviewed and a new approach suggested.

The work described formed part of a three year programme of research by

British Coal into the agricultural restoration of colliery spoil heaps with a

deep soil cover. It was undertaken by the Midland Research Unit, School of

Agriculture, University of Nottingham and included a wide range of field

studies of equipment, techniques, crops and soil conditions.

SCRAPER OPERATIONS

    It was apparent from a review of the available literature that much current

advice appeared to be based firstly on 'casual' observation rather than

physical measurements and secondly on general conclusions drawn from

experience at a single site, for one soil type and depth and using only one

322

soil handling technique. It is not surprising that there has been confusion
and disagreement as to the best practice and a failure to fully recognise that
different approaches may be appropriate depending upon the scale and type of
operations and nature of site conditions.

Where scrapers have been observed it is generally agreed that compaction
occurs. The wheels cause most of the compaction with only a minor
contribution from the action of the box during soil lifting and placing.
The response has been either to abandon the use of scrapers or to control
traffic movements to minimise the compaction induced.

In the latter case two general scraper operating patterns have been
advocated:-

I 'Single Row' technique

In this method there is no turning; the scrapers following common straight
tracks in laying each strip of soil (Fig 1a). This approach is based on the
assumption that most compaction occurs during the first passage of the scraper
and therefore relatively little further compaction occurs from repeated
trafficking. Also an uncompacted strip of soil between the wheelings remains.

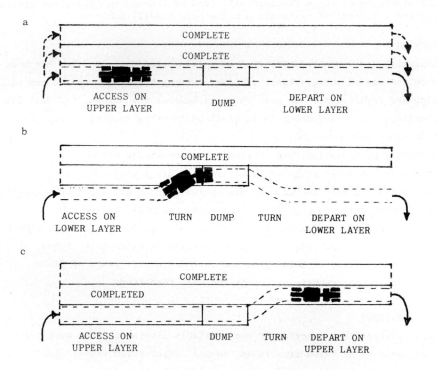

Fig. 1 Scraper Operating Patterns. (a) Single Row. (b) Dump and turn - run on
lowest layer. (c) Dump and turn - run on upper layer.

II Dump and Turn Techniques

A technique recommended for a shallow cover of soil involves the scraper running on the lowest layer of soil exposed and turns on to the layer being placed only when actually depositing the soil (Fig 1b). The assumption with this method is that total trafficking of the layer being placed should be kept to a minimum.

An alternative technique is recommended for a deep soil cover of several layers. This involves the scraper running as far as possible on the upper most layer and then turning onto the adjacent and completed strip as soon as the load is discharged. This approach contrasts with that above being based on the assumption that the greatest compaction will occur in the uppermost layer where it can be most readily be removed.

Both of these techniques require the scraper to turn and cross from one set of wheelings to the adjacent wheelings. This results in compaction of the soil between wheelings and potential for greater soil damage due to smearing and shearing during the turn.

INVESTIGATION OF COMPACTION CAUSED BY SCRAPERS

Measurements of soil strength were made during the various stages of reinstatement of a soil profiles at two sites in Nottinghamshire, details of which are given in Table 1. The effects of single row and dump and turn and random traffic patterns on a trial area of 0.8 ha were monitored.

TABLE 1

Details of the site, soil and specification for soil replacement

	SITE B	SITE S
Soil type	Sandy loam	Clay loam
consistency	Dry, loose	Dry, friable
depth		
top soil	250 mm	250 mm
subsoil	500 mm (2 layers x 250)	500 mm (2 layers x 250 mm)
Equipment	Terex TS14 motor scraper	Terex TS14 motor scraper
Slope	1 in 12	1 in 12

A recording penetrometer was used to determine the soil strength at fifteen 35 mm intervals in the upper 525 mm of the reinstated profile (including colliery shale where applicable) (ref. 1). It is assumed that the resistance to cone penetration is a measure of relative compaction or looseness of soil

324

"aggregates". The theoretical shortcomings of this type of instrument are recognised but the advantages of convenience and speed of use in an operational situation were of greater importance.

EFFECT OF TRAFFICKING ON SOIL STRENGTH

In Figure 2 are shown typical results for the immediate effect on penetration resistance of varying numbers of passes of an unladen motorscraper over clay loam sub-soil. It is apparent that, as one might expect from experience of earthworks engineering and indeed from observation of wheeled plant operating on loose soil, most compaction occurs during the first pass. In the context of agricultural soil handling, subsequent passes have little further effect.

In Figure 3 the average cone resistance of soils placed by scrapers operating to different traffic patterns are reported. It can be seen that there is little to choose in terms of degree of compaction between an uncontrolled random traffic pattern and a strictly controlled 'Dump and Turn' pattern. The average resistance for the 'Single Row' pattern is much lower.

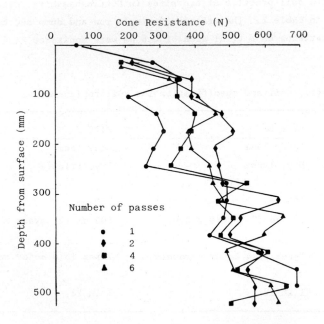

Fig.2. Mean cone resistance in wheelings after passes by motorscraper over a clay loam subsoil.

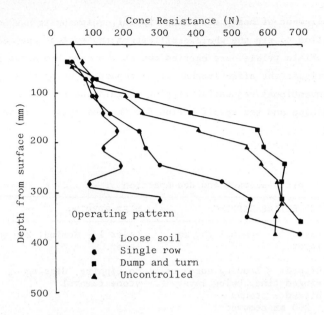

Fig.3. Mean cone resistance of sandy loam subsoil placed by motorscraper operating to different patterns.

Thus there is more to commend the 'Single Row' ' system of scraper operations with which about 40% of the soil lies between wheel tracks and remains uncompacted.

These investigations confirm that the use of scrapers causes compaction of replaced soil and quantifies its extent. The average compaction is much less with the 'Single Row' method than with the 'Dump and Turn' method though heavy compaction in the wheelings still occurs.

DECOMPACTION OF RESTORED SOIL PROFILES

The conventional practice to overcome compaction in restored soil profiles is to rip each soil layer in turn after it is placed referred to as 'Sequential Ripping'. From site observation it was obvious that this procedure was not achieving decompacted soil profiles. To place a new layer equipment must operate on the previously decompacted layer below and this causes the re-compaction of the loose soil. In addition, the passage of plant on loose soil causes deep ruts in the trafficked layer which make it impossible to place a layer of uniform thickness. In the event of heavy rainfall the loose layer is inundated forcing cessation of operations for an extended period; in some cases a whole season has been lost.

With the development of heavy duty deep ripping equipment it has become a feasible proposition to rip the whole profile from the surface upon completion of replacement. Field trials were carried out at the same two sites as in Table 1 to investigate the effectiveness of decompaction using this approach compared with conventional sequential ripping. Again the recording penetrometer was used and the specification for the soil placement was as shown in Table 2.

TABLE 2

Specification for soil placement and decompaction

	SEQUENTIAL RIPPING	DEEP RIPPING
Sub-soil	Layer 1 - Nominal 250 mm layer	Layer 1 - Nominal 250 mm layer
	Ripped* - leading edges of winged tines below layer. Disced - stones > 200 mm removed	No ripping, discing or stone removal
	Layer 2 - Nominal 250 mm layer	Layer 2 - Nominal 250 mm layer
	Ripped* - leading edges of winged tines below base of layer. Disced - stones > 200 mm removed	No ripping, discing or stone removal
Top-soil	Laid in single layer 250 mm nominal	Laid in single layer 250 mm nominal
	Ripped* to 350 mm Disced and stones > 150 mm removed	Ripped* to 650 mm minimum Disced - stone removed > 150 mm
Cultivation	Fertiliser, seeding and harrowing	

* Equipment D 8 K Caterpillar tractor with parallelogram linkage with 3 shank ripper equipped with winged boots.

On adjacent areas 150 m x 25 m the two decompaction treatments were applied On one of the areas the conventional sequential ripping approach was adopted with each layer being ripped through into the underlying layer before the next was placed. On the other the soil was placed in the same manner but no ripping was done until the profile was complete when deep ripping to a minimum of 650 mm was employed. Each area was then subject to normal agricultural cultivations and sown to winter barley.

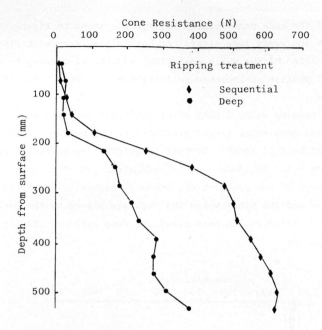

Fig.4. Mean cone resistances of clay loam soil profile when ripped sequentially and with a final deep rip.

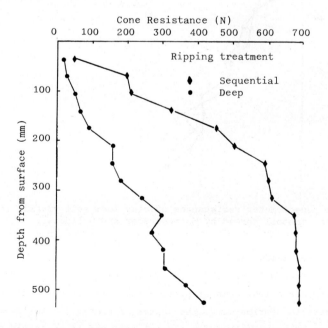

Fig.5. Mean cone resistances of sandy loam soil profile when ripped sequentially and with a final deep rip.

The results of the cone penetrometer surveys are shown in Figures 4 and 5 for the clay loam and sandy loam sites. At both sites, even with close and continuous supervision of the whole operation, sequential ripping resulted in a shallower final profile of loosened soil compared to the profile ripped only on completion.

The method of placing using a back acter excavator to avoid any trafficking of the soil was not considered either practical or appropriate in this case where the soil had been in stock. However an alternative method of spreading the top-soil using a LGP bulldozer was investigated. The typical results of penetrometer surveys of two profiles are shown in Figure 6, one where the bulldozer was used and the other where the soil was placed by the single row motorscraper method followed, in both cases, by deep ripping. No significant difference is apparent.

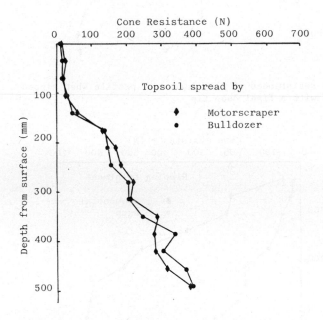

Fig.6. Mean cone penetrometer resistances of clay loam soil profile following deep ripping after topsoil spread by motorscraper and bulldozer.

IMPLICATIONS FOR TIP RESTORATION

Deep ripping of the whole profile upon completion of soil replacement by motor scraper produces a less compact profile than the conventional sequential decompaction practice. Furthermore, the degree of compaction following deep ripping appears to be no more than for 'loose' soil and is certainly as loose as can be achieved using the alternative technique of soil spreading using a low ground pressure bulldozer. It is therefore considered that the use of the

motorised scraper for soil replacement is justified where it can be followed
by the use of heavy duty deep ripping equipment to relieve compaction.

The deep ripping approach also has a number of other significant benefits
both operationally and in terms of restoration performance.

These are:-

(i)   Full advantage can be taken of the availability, efficiency and
economy of the civil engineering mobile plant which is present on call at
most colliery waste disposal sites.

(ii)   As the method of operation is simple, specification and control of
the work is made easier.

(iii) Because the plant is not running on a very loose layer, soil layers
can be placed more easily to accurate uniform thicknesses avoiding the
corrugated effect which is inherent in the sequential ripping method. This
is particularly important in the case of top-soil.

(iv) With the sequential ripping method almost the whole area is loose
throughout the operations. Any significant rainfall inundates the surface
layer making it totally untraffickable for many days at least with risk of
severe soil damage if operations recommence too soon. By not ripping until
the whole profile is replaced these delays are reduced as water cannot
readily enter the placed layer. The minimisation of delays is important not
only from the point of view of cost but also because a whole planting season
may be lost.

(v)   It follows from the above point that very valuable operational
flexibility is introduced into the restoration programme. For example, soil
laying can be carried out very quickly when soil and weather conditions are
suitable and the scraper fleet is available. The ripping operation is
independent of this and can be carried out at any time after soil placing.
Thus both activities can be carried out with maximum efficiency.

The application of deep ripping does not preclude the implementation of
measures such as stone removal and liming, which may be needed as the soil
profile is being replaced.

Flexibility is also introduced into the cultivation and aftercare
programme. The decompaction operation can be undertaken to suit the
cultivation programme rather than the soil replacement activity.
Decompaction is only effective if carried out under dry soil conditions and
is likely to be maintained longer if vegetated immediately as this helps to
develop and stabilise the structure of the disturbed soil. It is therefore
considered that deep ripping is best carried out as part of the normal soil
cultivation programme when it is anticipated that weather conditions will be
such as to allow the whole sequence from deep ripping through to the final

330

harrowing of the planted seed bed to be completed in one operation. If a decompacted profile is left unplanted, much of the benefit is rapidly lost due to slaking and settlement if heavy rain follows and access may then be impossible for months.

CONCLUSIONS

It has been shown by penetrometer surveys that decompaction of complete soil profiles can be more effectively carried out by deep ripping than by conventional sequential ripping. In fact, after decompaction, motorscraper laid profiles are as loose as those placed by LGP bulldozer.

With the powerful caterpillar tractors and heavy duty deep ripping equipment now available, profiles can be ripped to depths of up to 1 m. Thus it is possible to decompact either the whole soil profile or the depth appropriate to the after use from the finished surface and obviously this process can be carried out to suit the agricultural programme.

The application of these findings will, it is believed, lead to considerable economy in tip restoration operations and to the required standards of restoration being more readily achieved.

ACKNOWLEDGEMENTS

The authors wish to thank British Coal for permission to publish this paper. Any opinions expressed are those of the authors and not necessarily of British Coal.

REFERENCE

1. G Anderson, J D Pidgeon, H B Spencer and R Parks. A new hand-held recording penetrometer for soil studies, Journal of Soil Science, 31 (1980) 279-296.

Reclamation, Treatment and Utilization of Coal Mining Wastes, edited by A.K.M. Rainbow
Elsevier Science Publishers B.V., Amsterdam, 1987 — Printed in The Netherlands

COMBINED TIPPING AND OPENCAST COAL SCHEME AT BENTINCK COLLIERY

D. M. BROWN, C.Eng., M.I.C.E., M.I.H.T.

British Coal, Nottinghamshire Area

ABSTRACT

In order for Bentinck Colliery to continue producing coal, additional
tipping capacity was required to cater for the disposal of discard.
Opencast coal production and colliery discard disposal schemes like other
schemes of this nature have to take account of safety, financial
considerations and be environmentally acceptable. This joint Deep Mines
and Opencast coaling scheme satisfied these criteria by allowing viable
opencast coal reserves to be extracted from an area designated for Deep
mines discard disposal, at a site local to the colliery and at the same
time produce additional tipping capacity by taking advantage of the
Opencast coal operation to create a new land form. Environmental gains
were achieved by the rapid formation and restoration of embankments using
opencast overburden that screened subsequent Deep Mines tipping. The
retrieval of suitable soil making materials from within the opencast
overburden enabled the standard of restoration of the colliery disposal
site to be improved. As it was necessary for colliery discard to be tipped
onto the opencast site whilst it was being worked, close liaison between
the various parties was essential. Careful planning and rigorous
implementation of phasing of works and machine movements were required to
ensure security of the site and safety of the operations.

SUMMARY

This combined tipping and opencast coal scheme was carried out in a
satisfactory manner with its objectives being attained. The site was fully
exploited for coal extraction amounting to 268,000 tonnes and in such a
manner as to minimise the cost of recovery. Advantage was taken of using
the tipping site for overburden placing thus avoiding double handling. The
existing tip site settlement ponds were utilised for both opencast mining
and tipping requirements in order to provide suitable treatment of surface
water run off. The new land form created by the opencast coal operations
produced an additional three million cubic metres tipping capacity for
Bentinck Colliery with early restoration completed round the site
perimeter. Some 220,000 cubic metres of soil making materials were
recovered from the opencast overburden. The use of this material will
enable a higher standard of restoration to be achieved on the tipping site
than would otherwise have appertained.

INTRODUCTION

Bentinck Colliery is within the Deep Mines Nottinghamshire Area of
British Coal and 18Km north west of the City of Nottingham. Annesley
Colliery was connected underground to Bentinck Colliery in 1982 with the
run of mine coal amounting to 3 million tonnes per annum being raised at
Bentinck for treatment, and the discarded material of 1.4 million tonnes
per annum being disposed on the Bentinck tip site.

The coal seams recently worked are the Blackshale, Waterloo,
Tupton, High Hazles and Deep Hard at depths below the surface of 200 to
570 metres.

The Collieries have been operating before the turn of this century
and employed a total workforce of 2940 men.

The Bentinck Colliery tipping site is situated within the British Coal
Central East Region of the Opencast Executive who investigated the site in
1979 and 1980 and decided that they wished to exploit the site prior to
tipping of spoil by Bentinck Colliery, which would have practically
sterilised the Opencast Coal reserves. The coal seams worked were the
Swinton Potteries laying between 3m and 21m below the surface. The
thickness of the coal seams ranged from 90mm to 280mm yielding 268,000
tonnes of coal that were to be transported by road to the Opencast disposal
points at Oxcroft some 29Km North and Denby some 22Km south west of the
site in the County of Derbyshire and Bennerley some 27Km south of the site
in the County of Nottinghamshire.

BRIEF HISTORY OF TIPPING SITE

Tipping of spoil on the 108Ha tipping site commenced in 1952, and
contained some 9 million cu.m. of material, 7 million cu.m. being coarse
spoil laid by earthmoving equipment and 2 million cu.m. of a fine material,
pumped onto the site in suspension with water and contained in lagoons with
banks formed by the coarse spoil.

Progressive restoration of the site to agriculture has been carried out
with 25 Ha completed by 1982. Part has been used as permanent pasture for
cattle and part with a rotated use for pasture, hay making and arable
crops. A shortfall in the amount of soil required to totally restore the
site to present day standards had occurred due to (a) part of the site
having been worked as a clay quarry by a private operator and soils lost,
(b) part of the site being low lying marshland and soil was not fully
recovered, and (c) present day standards requiring greater depths of soil
for agricultural uses.

PERMISSION PROCEDURES AND CONDITIONS

The Mineral Planning Authority responsible for assessing and determining permission for the Deep Mines tipping scheme was Nottinghamshire County Council. A Planning Application was duly submitted in October 1981 and a consent granted in November 1982.

Opencast Executive were required to make an application (under the Opencast Coal Act 1958) to the Department of Energy for authorisation by the Secretary of State. An application was made in October 1981 and the Authorisation granted in August 1982.

Changes to the permission procedure for opencast coal schemes were made in 1984 requiring a Planning Consent from the Mineral Planning Authority and an Authorisation from the Secretary of State for the Environment. It is expected that legislation will be introduced shortly to make the Mineral Planning Authority the sole assessing and determining body.

As opencast coaling and spoil tipping were to be carried out on the same site in sequence, close liaison was established between the various interested parties at the outstart of the planning stage. Both the opencast coal Authorisation and spoil tipping Planning Consent reflected the needs of both operations and advantage was taken of the mutual benefits that could be gained.

Conditions were placed on both Opencast Executive and Deep Mines and included the following:-

(i) The Authorisation required the Opencast Executive to carry out operations in accordance with a programme of phases agreed with the County Planning Authority. (As the County Planning Authority were responsible for determining the conditions in the Deep Mines Planning Consent this enabled the Authority to ensure that the two schemes were knitted together to their satisfaction.)

(ii) The Authorisation called for steps to be taken to ensure that all vehicles leaving the site did not emit dust or deposit materials upon the highway and that a wheel cleaning device shall be installed and used.

(iii) The boundary hedges and walls, and trees within the site shall be preserved where-ever possible and the site secured by additional fencing as agreed.

(iv) Phasing of works and final contours to be in accordance with the plans submitted.

(v) Stripping of topsoil and subsoil to be carried out in sequence and stored separately in a manner and place to be agreed with the Mineral Planning Authority and used for the restoration of the completed areas. Operations to be carried out only when conditions were suitable.

(vi)   The tipped spoil shall be spread and compacted in layers to ensure stability and to avoid spontaneous combustion.

(vii) Existing watercourses be safeguarded and diversions provided, as agreed, with precautions taken to avoid pollution.

(viii) If suitable soil making materials are found during the stripping of the overburden, they shall be recovered for use during the restoration of the final surface of the site.

(ix) Noise from vehicles and machinery be minimised and dust controlled

(x)   Restoration and landscaping shall be in accordance with the details shown on the plan submitted. The works shall be carried out to the satisfaction of the Mineral Authority with respect to soil depths, methods of placement, suitability of site conditions, preparation of the surface of the tipped material and soils, types of seeds and trees. Samples of tipped material and soils shall be analysed to assess fertiliser, lime and other ameliorants required to promote normal growth. A five year aftercare scheme shall be submitted for agreement by the Mineral Planning Authority, specifying the steps required to bring the land to the required standards for use for agriculture and woodland.

METHOD OF TIPPING COLLIERY DISCARD

As mentioned above, two types of colliery discard are produced, i.e. (i) Coarse and (ii) Fine.

(i)   Coarse discard. Coarse discard is transported from the coal preparation plant to the tip on a 1600m long conveyor belt which end discharges to form a heap approximately 7m maximum height. The spoil is then loaded into motorised tractor scrapers, transported and placed and compacted in 300mm maximum layers. This is a good quality general fill material comprising mainly shales and mudstones having a particle size distribution between 100mm and Zero. When compacted it is suitable for forming banks used to contain the fine discard.

(ii)   Fine discard. Fine discard is pumped in suspension with water from the coal preparation plant in a twin 100mm diameter 1700m long pipeline into lagoons formed from coarse discard and located on the tipping site. The solids settle out and clarified supernatant water is decanted from the lagoons and either recirculated to the coal preparation plant or discharged to the River Erewash.

Tipping operations are required to be carried out over a 20 hour working day, five days a week without fail in order to maintain the output of coal through the colliery coal preparation plant.

METHOD OF OPENCAST COAL WORKING

Following competitive tendering, the Opencast Executive awarded a contract to a company to carry out the site operations. The contract documents reflected the requirements of the terms of the Authorisation, the deep mines needs relating to timing, phasing and design and the Executive's requirements with regard to timing, phasing, safety, coal production, coal quality, and delivery.

Stripping and storage of soil was carried out when conditions were suitable using motorised tractor scrapers. Overburden comprising clays and mudstone was removed by a combination of machinery consisting of motorised tractor scrapers, back actors and dump trucks in a series of cuts with material from these cuts being deposited in its final position, as far as possible, in order to avoid double handling. Soil making materials contained within the overburden were placed on the adjacent tipping site for future use in restoration of the deep mines site. The coal was excavated by a tracked forward loading shovel loaded into coal lorries, weighed and transported to the Opencast Disposal Points. These vehicles passed through a wheelwash before entering the highway. The route taken by onsite vehicles was agreed with the colliery personnel in order to avoid any conflict, the route by offsite vehicles being agreed with the highway authorities. Restoration of the opencast site was carried out by replacing top soil and sub-soil and seeding with a grass mix under the guidance and supervision of the Ministry of Agriculture Foods and Fisheries.

Due to the interaction between the Colliery and Contractor's activities, strict control and co-ordination of all operations were necessary. This was effected by day to day liaison between the contractor's site foreman and the colliery Tip Foreman. Monthly site meetings between the Opencast Executive Resident Engineer, the Contractor's Agent and Deep Mines Engineers decided policies and strategies.

As the Opencast operations progressed and Deep Mines began tipping discard into those areas that had been worked for coal it became necessary to have clear demarcation boundaries so that the Contractor, Colliery Manager and their staff knew what they were responsible for. This was achieved by erecting large marker posts and their positions being translated onto plans.

Drainage of the Opencast site and the Deep Mines site was the responsibility of the Contractor and Colliery Manager respectively, each being accountable to the Severn Trent Water Authority for the quality of their effluent. Effective control was maintained by close co-operation and

planning which took account of Deep Mines coarse discard tipping and
changing lagoon system together with the effects of opencast workings on
the existing land drainage system and the construction of a new final
drainage system.

DETAILS OF SCHEME

The scheme is described in four stages as detailed below, each stage
being illustrated by plans and sections as at Figures 1 to 4.

Stage 1.   (Fig.1) - Site plan and section showing final contours with
and without Opencast coal workings.

Stage 2.   (Fig.2) - Site plan and section of Opencast Coal workings.

Stage 3.   (Fig.3) - Site plan and section at completion of Opencast Coal
workings.

Stage 4.   (Fig.4) - Final site restoration plan and section.

Stage 1.

Figure 1 is a plan and section of the site showing the final contours of
tipping schemes with and without opencast coal workings. A Planning
Consent for the scheme without opencast coal workings was granted in 1973.
The Planning Consent for the scheme with opencast coal workings was granted
in 1982. The limits of the western boundary of the 1973 scheme were
determined by the ground levels required to allow free drainage of surface
water from the tipping site and the area of land bounded by the toe of the
tip, the M.1 motorway and Park Lane, account also being taken of the
surface watercourse that flows through a culvert beneath the motorway
draining the land to the west of the motorway. The opencast coal workings
necessitated removal of the overburden in the shaded area bounded by the
motorway, Park Lane and the proposed line of the watercourse under the 1973
Planning Consent. This facilitated the diversion of the watercourse, as
shown on the plan and section, which in turn allowed the 1973 site boundary
to be moved westward and hence increased the tip site capacity by some
three million cubic metres without increasing the maximum height of the
tip.

Stage 2.

Figure 2 is a plan and section of the site showing the extent of the
opencast coal workings. The site was work in three phases to suit the
requirements of Deep Mines spoil disposal by allowing tipping of both
coarse and fine discard to occur progressively over the site following on
opencast coal working, but before opencast activities were completed.
Lagoons L.10 and L.11 on the existing tipped area were nearly full and new

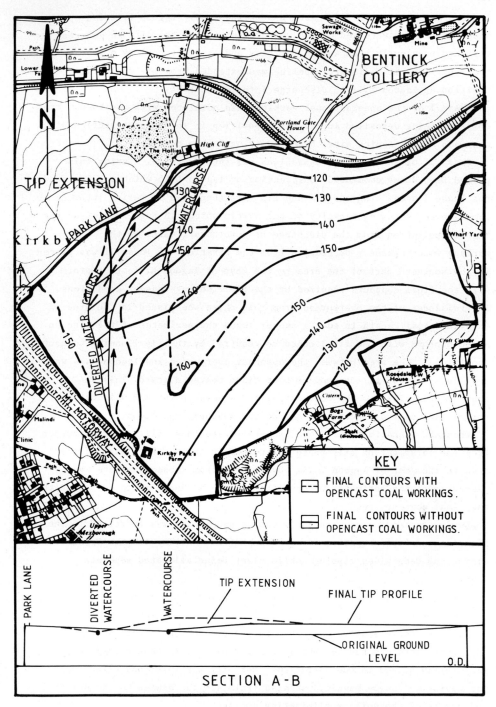

Fig.1   Site plan and Section showing final contours with and without
Opencast coal workings.

lagoons to contain the fine discard were required. Lagoon L.14 was progressively constructed on the western section of the site in an area that had previously been worked for clay and opencast coal by a private operator. The fine discard pipeline was extended from its discharge position at lagoon L.11 to discharge into lagoon L.14.

Phase I of the opencast coal workings was completed with overburden from this cut being deposited in its final position in an adjacent section of the Deep Mines tipping site, hence avoiding double handling.

Phase II of the opencast operations was then executed. During this period Deep Mines commenced construction of lagoon L.15 which incorporated the Phase I area. Lagoons L14 and L15 were then used in series in order to increase the settling capacity of the system which, in turn, minimised the risk of pollution from the discharge of supernatant water to the surface water coarse. Overburden from the first cut of Phase II was placed in the southern part of the area to the west of lagoon L.14, to conform with the final contours required by the tip Planning Consent conditions. The remainder of the overburden from this phase was placed in the area of excavated coal, again in such a manner round the perimeter of the site so as to conform with the final contours required by the tip Planning Consent conditions. The area between the opencast overburden and lagoon L.14 was left low in order to accommodate Deep Mines coarse discard. This was conveniently close to the conveyor belt discharge point and was most useful for placing coarse discard during adverse weather conditions.

Opencast operations were then transferred to Phase III. The overburden from the first cut of this phase was placed in the northern part of the area to the west of lagoon L.14. Overburden from subsequent cuts was placed round the perimeter of the site in such a manner as to conform with the final contours required.

During the period of opencast workings great care was exercised in the planning and implementation of on-site traffic control with the Opencast mining and Deep mines tipping mobile plant being allocated separate routes.

Whilst opencast operations were being carried out Deep Mines completed tipping to final contours in the area to the north of lagoons L.10 and L.11. Opencast Executive were able to restore this area with soils from the site. This had the advantage of allowing soil to be stripped and immediately placed in one operation. Not only was this cost effective by not having to double handle, but also minimised the damage to the structure and nature of the soil by eliminating storage.

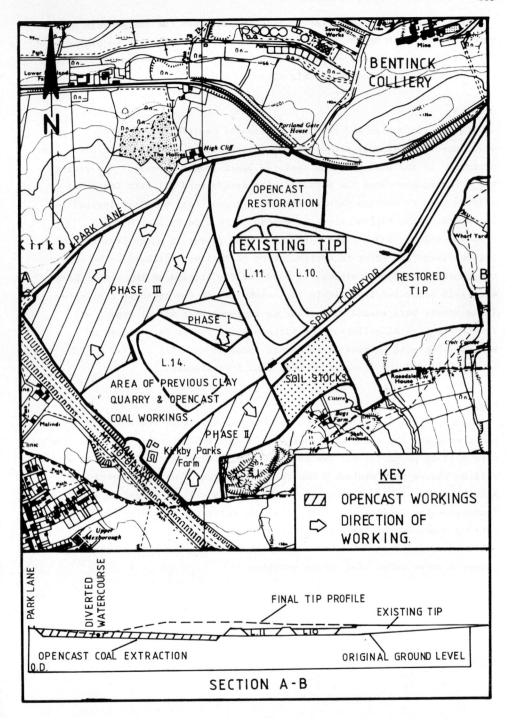

Fig.2 Site plan and sections of opencast coal workings.

The area to the south of Lagoon L.11 was utilised for the storage of some 220,000 cubic metres of soil making materials which were recovered from within the opencast overburden. This material will be used to improve the overall restoration of the tip site.

Stage 3

Figure 3 is a plan and section showing the position of the site development immediately after the completion of the Opencast workings. The perimeter of the site had been restored apart from two areas. The area in the north accommodates the metalled site access road off Park Lane that will be used to service future proposed Opencast Coal workings located to the south of the tipping site, an access track used by Deep Mines tipping vehicles and a clean water pond used to retain water from the lagoon discharges prior to recirculation to Bentinck Colliery. The area to the south, marked soil stocks, is used for the storage of the soil making materials recovered from within the overburden of the opencast coal site. These stocks were seeded with grass to improve their visual aspect and reduce the risk of polluting the surface water course by washdown of solids. The void created by the final opencast cut will be used to form Lagoon L.16 for the disposal of Deep Mine fine discard.

Stage 4

Figure 4 is a plan and section of the proposed final restored site as required under the terms and conditions of the 1982 Planning Consent. The site will be progressively restored for agricultural use. The better soils will be placed to a minimum 900mm thickness in the areas designated for arable crops. Soils of 250mm minimum thickness will be placed in areas designated for pasture and those areas set aside for tree planting will receive a minimum of 150mm thickness of soil.

It is proposed to establish a footpath following the line of the diverted watercourse as a public amenity.

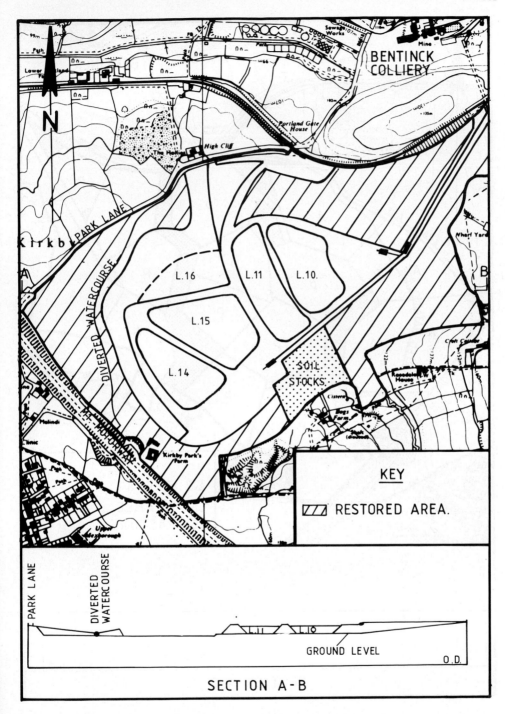

Fig.3 Site plan and section at completion of opencast coal workings.

342

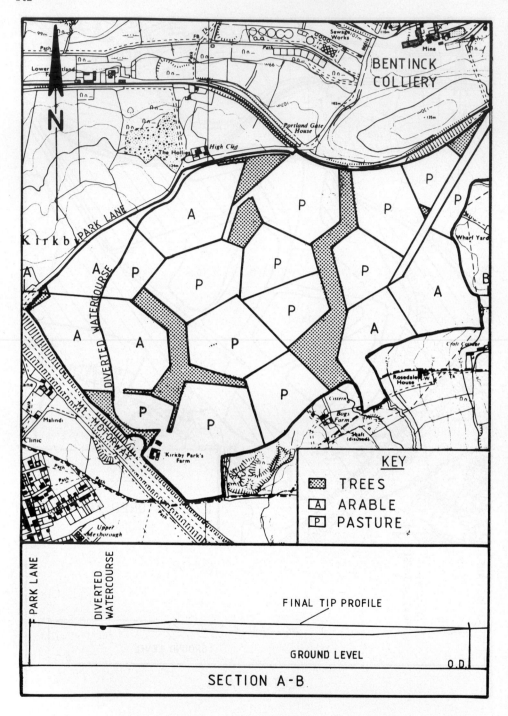

Fig.4 Final site restoration plan and section

Reclamation, Treatment and Utilization of Coal Mining Wastes, edited by A.K.M. Rainbow   343
Elsevier Science Publishers B.V., Amsterdam, 1987 — Printed in The Netherlands

THE DESIGN OF COLLIERY SPOIL TIPS : OBJECTIVES AND TECHNIQUES

J R Talbot, Landscape Architect
British Coal

INTRODUCTION

Over the past two decades there has been a great deal of good work done
to help solve the technical problems of reclaiming derelict land and in
particular in restoring disused colliery spoil tips with grass or trees.
However, there is a danger in confusing the problem of past dereliction with
the new generation of modern spoil tips, and, over emphasis on the technical
and agricultural aspects can be to the detriment of the overall objectives
and landscape design. This paper considers some of the design problems
associated with the tips planned to accommodate the current and future spoil
production.

MODERN TIPS

It is generally recognised that despite many laudable efforts to find
alternative disposal methods, local tipping will be required to accommodate
most of the colliery spoil produced well into the future.   Currently over
200 hectares of land is required annually to accommodate over 50 million
tonnes which is disposed of in surface tips close to mines.   This represents
the equivalent of two of British Coal's largest tips being planned, tipped
and restored every year.

These modern tips may be extensions to existing tips, they may be
combined with reclamation or opencast sites or they may be literally on green
field sites.   They are well engineered structures and can be tipped to any
shape with progressive restoration using generous soil depths.   They are
very different from the reclamation schemes of old tips where soil is scarce,
slopes are steep, and dereliction plentiful.   They are not derelict sites.
They are the last stage of the coal mining process and are an essential and
integral part of this industry.

The number of tips that are not subject to planning controls is becoming
fewer and fewer as they are completed and either restored by British Coal or

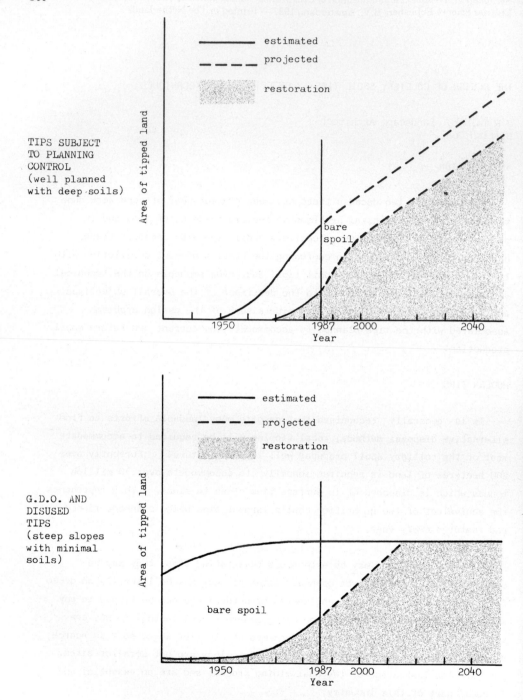

Figure 1 Diagramatic representation of spoil tipped since 1930 showing areas of bare spoil and restoration

by the local authorities with the aid of central government grants.
Conversely, an increasing proportion of active tips are subject to planning
controls and all new tips now designed, must satisfy the requirements of the
Minerals Planning Act. Thus, as shown in Figure 1, the problems associated
with old tips of steep slopes and minimal soil will become a thing of the past
and the only bare spoil will be the operational areas of active tips where
subsequent restoration with deep soils will take place progressively.

With good depths of soil the technical aspects of restoration require
care, but present few problems.   The success of a tipping scheme depends more
on a careful selection of the site, it's appearance during construction and
it's landform and landscaping; in other works, it's design.

DESIGN OBJECTIVES AND IMPLICATIONS

Before a spoil tip can be designed the objectives need to be determined
and these are listed below in Table 1.

TABLE 1

Design Objectives

```
┌─────────────────────────────────────────────────────────┐
│                                                         │
│   1.    ECONOMIC WASTE DISPOSAL                         │
│                                                         │
│   2.    COMPLIANCE WITH SAFETY STANDARDS                │
│                                                         │
│   3.    ENVIRONMENTAL CONSIDERATIONS                    │
│                                                         │
│         Minimum Loss of Landscape                       │
│         Minimum Disturbance to Farmland                 │
│         Minimum Disturbance to Wildlife                 │
│         Minimum Disturbance to Recreation               │
│         Not too much noise                              │
│         Not too much dust                               │
│         Not too visible                                 │
│         Not to marr the Landscape                       │
│         No loss of Topsoil                              │
│         Beneficial Afteruse                             │
│         Removal of Dereliction                          │
│                                                         │
└─────────────────────────────────────────────────────────┘
```

Economic Objectives

The main objective is the economic disposal of colliery spoil. This obvious fact often appears to be forgotten in the protracted negotiations that lead up to a planning permission being granted or refused. This main objective must immediately be qualified by the two controlling factors of safety and environment.

Safety

Safety and engineering standards are controlled by the Mines and Quarries (Tips) Act 1969 and are strictly observed in all British Coal's tip designs. In practice, the main limitations these standards impose on the design is on that of slope. However, generally other considerations such as agricultural restoration or landscape appearance become the limiting factor on the angle of the finally restored slopes.

Environmental Objectives

The second controlling factor, that of environment, becomes the main constraint on tip-design and it is this aspect that is now considered.

The first and main objective of economic disposal is a positive objective but nearly all of the environmental objectives are negative, and herein lies the real design problem. Apart from the reclamation of derelict land (which may be possible if it happens to be in the right place), the only environmental objective which is positive is the restoration to beneficial after use. It is very difficult to design for negative reasons, important though they might be, and thus the objective of "beneficial after use" often assumes undue importance and can result in "the tail wagging the dog".

The issue of restoring the tip to beneficial after use is not important in itself as the land has been used very beneficially for tipping (thus enabling many millions of tonnes of coal to be mined), and has in economic terms 'paid its way'. Likewise, the loss of any farmland is not important economically as the loss of income is extremely small compared to the income from the coal. The loss of food production is totally insignificant compared with the national total of food production. Loss of farmland and returning it to beneficial use are only important in so far as they affect the appearance of the landscape and its management thus satisfying the other environmental objectives.

There is a soil conservation argument for returning land to good use and there is certainly a landscape argument but there is little economic or agricultural argument.

The choice of what the land can be used for once the tip is complete should therefore be subordinate to the other environmental objectives. However, this choice of after use is very important since once made it quite obviously and quite properly imposes limitations on the tip design, i.e., site selection and appearance during construction as well as the landform and landscaping of the finally restored tip.Thus we are presented with the real design problem. We know that the main objective is economic spoil disposal, and we know that there must be no adverse environmental impact. In achieving these the only positive requirement is to restore to a beneficial after use, but this is primarily a means of restoring and managing the landscape and not an end in itself.

DESIGN VARIABLES FOR LOCAL TIPS

Faced with such vague and at the same time, very limiting guide lines, what design choices are there?

Again, contrasting the old and the new, reclamation schemes of old derelict spoil tips are restricted in their design by the high cost of moving spoil to change their shape, by the limited amounts of soil available for revegetation and by chemical problems peculiar to certain spoils such as high acidity and salinity.

When designing a modern tip, there are fewer restrictions and a larger number of design choises or variables and these are listed in Table 2.

Site Location

The extent to which the design objectives are going to be achieved is often determined at the site selection stage. The location of the site almost entirely dictates the economics of the scheme, it can minimise or preclude any noise and dust problems by being away from dwellings and it can minimise the conflict with other land uses. Site selection is a relatively straight forward planning process which requires the economic and environmental evaluation of alternative schemes. Comparing one site with another requires outline schemes to be prepared for each site and, for true comparison to be possible and where appropriate these schemes should be similar in size, appearance and after use. The number of suitable alternative sites is usually small as economics and often environmental considerations requires a site close to the mine and parcels of land of sufficient size free of houses, industry, roads, railways, rivers, overhead power lines etc. are usually scarce.   The best way of increasing   tipping capacity at existing mines is almost always to extend the existing tip. This is often not only the cheapest solution, but keeps all the possible environmental problems in one place without making them any worse.   It also requires far less additional land for any given additional capacity.   The diagrams in Figure 2 illustrate two alternative methods of extending the tipping capacity at existing mines. Table 3 shows the effect of these alternatives on the amount of land required.

TABLE 2

Design Variables

SITE LOCATION
APPEARANCE DURING CONSTRUCTION Tipping Methods Tipping Sequence Screening
FINAL APPEARANCE AND RESTORATION   Shape of Tip Depths of Soil Stripped Afteruse/Restoration

FIGURE 2   Two alternative methods of extending the tipping capacity at
            existing mines

SEPARATE SITES

EXTENSION OF EXISTING TIP

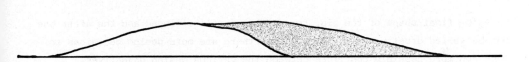

TABLE 3
Effect of separate and extended tips on landtake.

EXAMPLE 1

	ADDITIONAL CAPACITY (Cubic Metres)	ADDITIONAL AREA (Hectares)	AVERAGE DEPTH (Metres)
SEPARATE SITE	14.5 M	120	12
EXTENSION	14.5 M	72	20

EXAMPLE 2

	ADDITIONAL CAPACITY (Cubic Metres)	ADDITIONAL AREA (Hectares)	AVERAGE DEPTH (Metres)
SEPARATE SITE	25 M	164	15
EXTENSION	25 M	87	29

Appearance During Construction

The decisions on tipping methods, tipping sequence and screening are all made bearing in mind the possible environmental problems caused during the construction period of the tip.   The tipping methods are usually by conveyor to the site with spreading by scrapers, or dump trucks and dozers or in two cases with stacker spreaders.   With all of these methods any reasonable tip shape can be constructed.   The tip may or may not include lagoons for settling the fine wastes pumped up in a slurry pipeline from the coal preparation plant.   These impose constraints on the construction sequence and restoration phasing, but with careful design, they need not adversely affect the appearance of the tip during construction or once tipped.

Final Appearance and Restoration

The final shape of the tip, the depths of soil stripped and the after use can be varied greatly and it is here that there are more design decisions to make and more possible conflicts between the various design objectives.

As previously argued, the after use will of necessity affect the whole of the tip design and in particular the shape of the tip, and yet it is not the main object of the exercise.   The main object is to make the tipping process and the final tip environmentally acceptable.   How do we therefore decide on the most suitable after use?

RESTORATION OPTIONS

There are many alternative uses to which a colliery tip can be restored and the main options are listed in Table 4.

Return to previous agricultural grade

The after use that most often emerges out of the planning process is a restoration to the previous agricultural grade.   This can impose the most severe restrictions on the whole design and is often in conflict with the important design objectives such as minimal land take and restoration of the landscape.   This is most clearly illustrated by the limitations imposed on the slopes suitable for various afteruses as shown in Table 5 below and by the effects of shallow slopes on landtake as shown in Figure 3 and Table 6.

TABLE 4

Restoration Options

```
RETURN TO PREVIOUS AGRICULTURAL GRADES

VIABLE AGRICULTURE

COMMERCIAL FORESTRY

NATIVE WOODLAND/SCRUB/GRASSLAND

PUBLIC ACCESS (INFORMAL)

SPECIFIC LEISURE OR SPORTS FACILITIES

INDUSTRIAL OR HOUSING DEVELOPMENT
```

The effect of the slopes on the total landtake of a colliery tip is shown in Figure 3 and Table 6. The example illustrated shows three alternative landforms using different gradients on the outer slopes but giving the same capacity of 15 million cubic metres. For comparison the shape of each tip is a truncated cone on a flat site with a height limited to 35 metres. The Figure shows plans and sections of each of the alternatives. Table 6 gives, in each case, the areas of land taken up by the tip and by the slopes. By insisting on slopes of less than 1 in 8 rather than 1 in 3, up to 40% more land can be required and the area of sloping land created can be 160% greater.

Restoration to previous grades of grade 3 and above can therefore be in conflict with landuse and landscape objectives.

Viable agriculture

Accepting restoration to lower grades of agriculture allows for a greater flexibility in design, the site can more easily be varied in size and location there are fewer restraints during the construction phase & the final restoration can be designed more imaginatively to fit into the surrounding landscape.

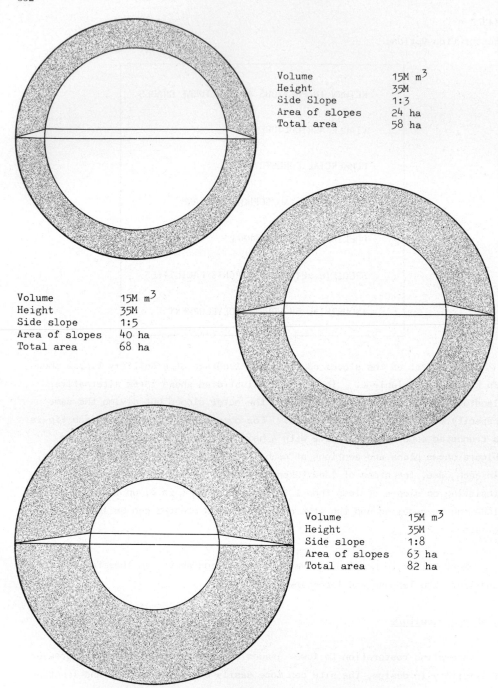

Volume   15M m^3
Height   35M
Side Slope   1:3
Area of slopes 24 ha
Total area   58 ha

Volume   15M m^3
Height   35M
Side slope   1:5
Area of slopes 40 ha
Total area   68 ha

Volume   15M m^3
Height   35M
Side slope   1:8
Area of slopes 63 ha
Total area   82 ha

Figure 3   Effect of slopes on landtake illustrated by plans and sections of
3 alternative tips giving the same volume.

TABLE 5

Limitation of Landuse on Slopes

USE	MINIMUM SLOPE
Agricultural field drainage	1:200
Natural drainage	1:40

USE	MAXIMUM SLOPE
Agricultural land classification grade 3a	1:7
Agricultural land classification grade 3b	1:6
Combine harvester (up and down only)	1:5
Two way ploughing	1:5
Agricultural land classification grade 4	1:3
Permanent pasture	1:3
Preferable maximum for trees	1:3

TABLE 6

Effect of side slope on landtake

	SIDE SLOPE	AREA OF SLOPES ha	TOTAL AREA ha
A	1:3	24	58
B	1:5	40	68
C	1:8	63	82

The main advantage in restoration to agriculture is that it takes care of the management of the land and rehabilitates the soil. Providing the price or rent is low enough, farmland can nearly always be sold or let and this maintains the tip in a manner which is visually attractive and acceptable in the surrounding countryside.

Commercial forestry

Restoration to commercial forestry imposes few limitations on the shape of the tip, but whilst trees can be established in thin soils or even on bare spoil, growing trees on a truly commercial basis would require at least as much soil as for agriculture. Their initial establishment is also more costly than grassland and very definite arrangements for their adoption and future management need to be made prior to the design stage.

Natural vegetation

Native woodland, scrubland or grassland could be established on spoil tips, and would impose few limitations on the design of the tip shape or slopes. It would fit the tip into the landscape (even though it might contrast with the adjacent monocultures of modern farming), and it would undoubtedly provide many wildlife habitats for native flora and fauna. The usual argument against this type of restoration is that the management is difficult and costly. I am not convinced that this is so, or that any management at all is required after the initial establishment. Natural vegetation managed quite satisfactorily on its own for a few years before man appeared on the scene, but certainly in the early stages its appearance could give an impression of neglect.

Public access or Sports Facilities

Restoration to public access or a specific leisure or sports activity is not often attempted simply because of the long time scale over wich these tips are planned. Tips restored to agriculture or forestry can of course be subsequently used for leisure purposes, but the tip cannot be specifically designed for leisure facilities unless very definite arrangements have been made for their adoption and subsequent management at the planning stage.

Industry or Housing Development

Use of tipped land by industrial or housing development is possible but the location and demand make this an unlikely use for newly planned tips.

CONCLUSION

The main objective of the tip design is to dispose of spoil in an economical and environmentally acceptable manner.   The two main purposes of a tip restoration scheme are to conserve the topsoil and restore the landscape. In the absence of viable proposals from third parties for projects such as commercial forestry, nature reserves, public open space, sports and leisure facilities or industrial and housing development it is inevitable that tips will generally be planned for restoration to farmland.   If agricultural considerations alone are not allowed to dictate the design, more imaginative solutions can be created with a greater variety in siting, landform and planting thus fitting the tipped land back into the surrounding landscape.

The tipping will only have been a temporary use of the land, the soils will have been conserved and the potential for alternative uses will remain.

ACKNOWLEDGEMENTS

I thank Mr. C.T. Massey, Head of Mining, H.Q. Technical Department for his permission to present this paper and also all those who have helped in its preparation.   The views expressed are those of the author and not necessarily those of British Coal.

Reclamation, Treatment and Utilization of Coal Mining Wastes, edited by A.K.M. Rainbow
Elsevier Science Publishers B.V., Amsterdam, 1987 — Printed in The Netherlands

THE UTILIZATION OF DIRT FROM COAL MINES AND LAND RECLAMATION

GAO YOULEI, HAN GUANGXU and SUN SHAOXIAN
[1]Director, Administrative Office, Technical Development Department, Ministry of Coal Industry
[2]Director, Environmental Protection Office, Ministry of Coal Industry
[3]Deputy Chief Engineer, Tangshan Coal Research Institute, CCMRI

SUMMARY

In the process of coal mining and preparation operations, a large quantity of dirt are rejected. At present, about 100 million tons of dirt are produced per year, 1,300 million tons are deposited, occupying an area of more than 6,700 hectares. China pays more attention to the comprehensive utilization of dirt and land reclamation. The utilization of dirt includes the recovery of low quality coal, increasing the width and height of the foundations of roadway and railway lines subjected to subsidence arising from coal mining, making bricks with dirt, the recovery of pyrite and the production of coal needed by fluid-bed furnaces. Land reclamation with dirt mainly refers to filling subsidence pits nearby the shaft mouth, which is, after being treated and strengthened, used as construction site or foundations for villages movement. For this reason, tests on the loads of foundations made of dirt in part of the areas have been carried out to determine their bearing loads. While the subsidence pits in other areas, after being filled with dirt, were reclaimed as farm land and planted with trees, or the subsidence pits accumulated with water have been built into artificial lakes.

INTRODUCTION

At present, the coal production in China is 850 million tons per year, 15% of which is dirt. The large anount of newly-deposited dirt and the old dirt piles not only occupy an area but also cause pollution problems. According to statistics, there are 850 dirt piles occupying an area of more than 6,700 hectares, 18% of which experienced spontaneous combustion. Therefore,the development of the work in the utilization of dirt and land reclamation can offer economical benefits and permenantly environmental benefits as well. The chinese government and the Ministry of Coal Industry have paid more attention to it and ascked every mine district to work in this field actively and arrange it as a special subject to carry out studies.

COMPREHENSIVE UTILIZATION OF COAL MINING DIRT

Power generation using dirt as fuel

Since its first dirt-fired power plant provided with fluid-bed furnace was built in Yongrong Mine District, Sichuan Province in 1975, China has set up 7 dirt-fired power plants with a total installed capacity of 89,500 kw and an annual power production of 400 million KWh. 1.2 million tons of dirt were consumed resulting in a raw coal save of 400,000 tons. Better economical and environmental benefits were achieved. In Pingxiang Gaokeng dirt-fired power plant, for instance, the installed capacity is 18,000 kw and the heat efficiency of the three 35-ton (vaporization rate) fluid-bed furnaces is up to 71% after long-term operation, showing a low power generation cost and offering an average annual profit of more than 2 million Yuan. Didao dirt-fired power plant, Jixi, provided with two 130-ton (vaporization rate) fluid-bed furnaces combined with two 25,000 kw power generation units is currently the largest dirt-fired power plant in China.

Being used as construction materials

(i) Making bricks. To make bricks by using mining dirt as material can realize several aims including save of energy and soil, the utilization of solid wastes as well as environmental protection. Since 1964 when China began to make bricks by using mining dirt, more than 190 dirt-brick factories have been set up and spreaded in every mine district, offering an annual production of more than 1,600 million pieces of bricks. Most of the industrial and civil buildings in mine districts are made of bricks.

In Jiaozuo Coal Mine Administration, Henan Province, it was started to make dirt-bricks in 1970 and the profit earned was used to expand the production scale. Up-to-now, 9 dirt-brick factories have been set up and provided with 6 tunnal-shaped kilns, 1 close-type kiln and 6 open-type kilns operating in turn. In 1985, 72.85 million pieces of dirt-bricks were produced, offering a profit of 580,000 Yuan.

(ii) Making cement. In 1976, it was firstly and successfully tested to make bricks by using dirt instead of clay and a part of coal in Zhangdian Cement Plant, Shandong Province. In 1978, dirt was used to produce #400 cement in Handan Coal Mine Administration. For recent 20 years, the Ministry of Coal Industry has owned 22 dirt-cement plants, 3.38 million tons of cement was produced from

1981 to 1985, and 1.014 million tons of dirt was consumed as raw material resulting in a coal save of more than 0.6 million tons.

(iii) <u>Making other construction materials.</u> According to the specific local conditions and the requirements, the following construction materials have been made by different mine districts:

    a. Light aggregate made of dirt or ash from fluid-bed furnaces;

    b. Cellular concrete made of ash from fluid-bed furnaces;

    c. Cement containing no clinker and steam-cured ash bricks;

    d. Small-sized hollow blocks by using ash from fluid-bed furnaces as semi-light aggregate instead of common sands and stone, etc..

Filling road way and railway dams in mine districts

In China, nearly all the coal under the road ways of mine districts and most of the coal under railway lines have been mined. Both the road ways and railways are in a normal dynamic subsidence condition, with the maximum subsidence value being more than 10 m. In order to ensure the smooth operation of the road way and railway in subsided areas, mining dirt has been used as the most convenient and cheapest material to widen and heighten their foundations. This is the main application of the mining dirt in China.

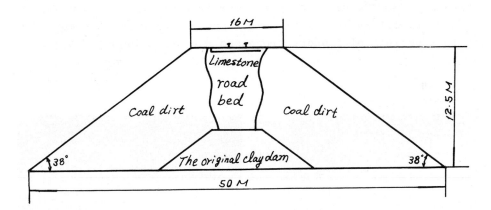

Fig. 1. Schematic drawing of the structure of the dam made of dirt.

Hegang, Fengfeng, Kailan and Huainan Coal Mine Administrations have accumulated rich experiences, For example, in Liyi Coal Mine, Huainan Coal Mine Administration, where a total of 12.16 million tons of coal is lying under the railway line, with the help of

Tangshan Coal Research Institute, more than 3 million tons of coal has been mined since 1975 causing a railway foundation subsidence on the order of more than 7 m. To maintain the reasonable profile of the railway line, 300,000 m^3 of mining dirt was filled. Now the railway dam is 12.5 m high, the structure of which is shown in Fig.1. Its final height will be up to 19 m. At present, long-term observation and study of the spontaneous combustion feature of the dirt and the stability of the railway line dam, etc. are being carried out by Huainan Coal Mine Administration and Hefei Coal Mining Research Institute.

Other applications

(i) The recovery of pyrite. Now, 8 plants have been built for recovery of pyrite from dirt with an annual capacity of 195,000 tons. The sulphur content of the feed dirt is 4-11%. Coal is also recovered simultaneously with the recovery of sulphur. This is a comprehensive approach to change wastes into treasure, reduce pollution, sufficiently make use of the resources and effect environmental protection in mine districts.

(ii) The production of mullite sands and powder. In the Precision Casting Material Factory, Nanpiao Coal Mine Administration, the finely-selected clinker dirt was, after high-temperature calcining at 1,300°C, used to make a new type of high-grade aluminosilicon refractory — mullite sands and powder which are applicable to a variety of foundry processes such as molten-casting and

Fig. 2. Mullite sands and powder.

suction casting, etc.. This kind of material is high in quality, wide in application and perfectly effective in environmental protection and prevention of foundry workers from silicosis, therefore, it has been sold to abroad. Shown in Fig.2 are the mullite sands and powder produced by this plant.

(iii) The production of Aluminium chloride and vanadium pentoxide. Nanpiao Chemical Plant is the main enterprise engaging in the comprehensive utilization of dirt, where dirt is used as raw material to produce crystal aluminium chloride and basic aluminium chloride both of which can be used as high-quality coagulants for drinking water purification.

Additionally, vanadium-contained coal or ash from fluid-bed furnace are also used in China to produce vanadium pentoxide the purity of which is higher than 98%.

LAND RECLAMATION BY USING DIRT IN COAL MINES

Underground mining is the main mining method in China's coal production, the majority of which utilize long-wall coal face mining where fully roof caving in is employed, thus leading to serious ground subsidence. Since 1956, more than 2,000 ground subsidence observation lines and 2 grid-form observation stations have been set up. According to practical measurements, in case where the dipping angle of the coal seam is less than 45°, the maximum ground subsidence is about 0.8 times the thickness of the coal seam mined. However, due to the supporting effect of various kinds of coal pillars, the subsidence volume is around 60% of that of the coal seam mined. The land in mine districts was destroyed by ground subsidence in a large scale. Statistics shows that, about 20 hectares of land subsided when every one million tons of coal was mined. In some districts, water accumulation occured in a large area due to high underground water level. This problem is especially conspicuous in mid-east China, which has constituted one of the main environmental problems in coal mines. Therefore, the Ministry of Coal Industry entrusted Tangshan Coal Research Institute, CCMRI, to conduct research work on this problem and take it as a major subject. The test site is at Huaibei Coal Mine Administration, Anhui Province. Land reclamation by filling dirt in subsided areas is one of the main research contents.

At present, apart from farm land restoration, the purpose of land reclamation by using dirt in China is to change the subsided areas into land applicable to industrial and civil construction, used as foundations needed by moving the valliges where coal is lying underneath, for harness of pollution arising from coal mining dirt, as well as for afforestation and beautification of the environment in mine areas.

The method of land reclamation making use of mining dirt

In Daihe Colliery, the use of the original dirt pile has been brought into a stop, while the nearby subsidence pits accumulated with water has been used for waste-dumping the elevation and leveling degree of which are strictly controlled according to the requirement of land reclamation. Now, 12 hectares of land has been reclaimed. Shown in Fig.3 is the land reclamation method where small mine cars and trolley locomotive are used to dump dirt into the subsicence pit accumulated with water.

Fig. 3. Land reclamation by dumping dirt into the subsidence pit accumulated with water.

The coal seams mined and the subsidence occured in this mine are shown in Fig.4. Three coal seams have been mined in underground with the total mining thickness being 6.2 m. The ground subsidence was 4.7 m deep and accumulated with water, the water level being 2.5-3.8 m. The thickness of the backfilled dirt is 5.1 m. The

reclaimed land is mainly used as construction sites, where the
culture and sports activity center of Huaibei Mine District were
planned to set up (see Fig.5). Now, the culture building (Fig.6)
and a swimming pool (Fig.7), a floodlit court (Fig.8) and 2 tourist
lakes making use of the subsidence pits (Fig.9) have been set up.

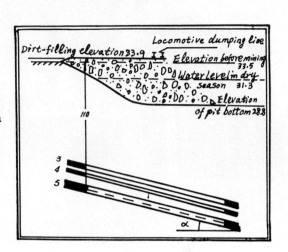

Fig. 4. The profile of
underground mining beneath
land-reclaiming area and
dirt-filling operation in
subsided area.

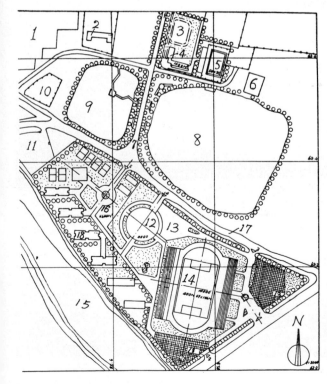

1 - workers' village
2 - kindergarten
3 - floodlit court
4 - culture building
5 - swimming pool
6 - settling pond
7 - cement road way
8 - east lake
9 - west lake
10- Coal Mine Cons-
    truction Company
11- garage and oil
    store
12- gymnasium
13- car parking yards
14- stadium
15- Daihe River
16- middle school
17- Xu-Huai Highway
18- classrooms and
    office, etc.

Fig. 5. The culture and
sports activity center
planned to be built in
the dirt-reclaimed
area.

Fig. 6. The culture building.

Fig. 7. The swimming pool.

Fig. 8. The floodlit court.

Fig. 9. One of the 2 tourist lakes.

The middle school, a gymnasium, a stadium of the Mine, etc. are to
be built and integrated with the afforestation of the banks of
Daihe River. After the completion of these engineering works, the
original dirty and untidy subsidence pits and dirt piles will become culture and sports sites of the mine district with comforta-
ble environmental conditions.

Tests on load-bearing capacity and reinforcement of the dirt-filled foundations

The foundations newly filled with dirt are very loose. Static
load-bearing tests carried out at field indicate that the permissible load-bearing capacity is only 7.25tons/m^2. In order to build
multi-story buildings on this kind of foundations, strong compaction was carried out. The weight of the ram of the compactor is
9.2 tons,which can be raised 9 m high. The ram can automatically
disengage from the hook through a automatically disengaging device
so as to compact the dirt-filled foundation. Shown in Fig.10 is
the equipment used to effect strong compaction of the dirt-filled
foundations, with which 10 times of compaction are carried on each
point on average and a 0.85 m subsidence of the foundation is achieved after strong compaction. Based on measurements, the density
of the foundation was increased by 10%. The load-bearing capacity
of the foundation was increased to 40 tons/m^2 based on the load-
bearing test carried out after strong compaction the cost of which
was only 4% of the construction engineering cost of the building.

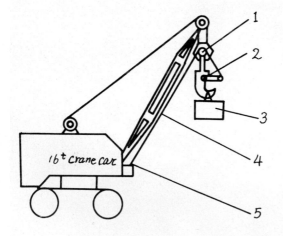

1 - slewing block
2- automatically disen-
  gaging device
3 - 9.2ton compactor ram
4 - limit steel rope for
  disengagement
5 - fixing position of
  limit steel rope

Fig. 10. Schematic drawing of the strong compaction equipment.

The strongly compacted dirt-filled foundations were superior to the previous soil foundation which existed before subsidence occurred, therefore, the construction cost was 10% less than mormal case.

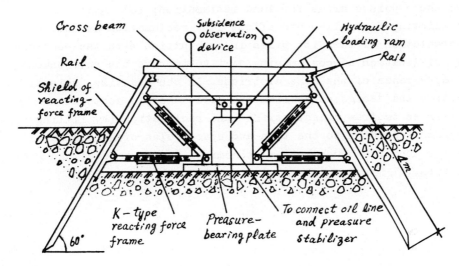

Fig. 11. Schematic drawing of load-bearing test on the dirt-filled foundation.

Soil-overlay and planting of the dirt-reclaimed land

Apart from the use of the dirt reclaimed land as construction sites, farming was effected by using hydraulic dredging unit to take soil from the edge of the subsided pit and pipeline to transport the soil to the land filled with dirt. The overlaid soil was 0.5 m thick. Many kinds of grains, vegetables and trees have been planted on the reclaimed land to test the suitability of the dirt-reclaimed land for farming and afforestation.

Land reclamation in subsidence area by means of hydraulic dredging unit is a harness technology high in efficiency, less in cost consumption and better in operation conditions. The operation of which is shown in Fig. 12. This method has gained warm welcome from the peasants and its use has been widely popularizing in the localities.

CONCLUSIONS

Its relatively late in China to start land reclamation making use of mining dirt and the task is very heavy. But, the Ministry of Coal Industry pays more attention to this work. The regulations concerning land reclamation is now being drawn up. The Ministry of Coal Industry asked not to build permanent dirt poles in the newly-built collieries; while in operative mines, to dump mining dirt into the subsided areas for land reclamation; to stop the use of the existing dirt piles and effect land reclamation as well as afforestation of the dirt piles in combination with the comprehensive utilization of coal mining dirt to improve the environment and appearance of the mine districts. In the meantime, scientific research and design establishments were asked to cooperate with collieries to draw up scientific land reclamation (by filling dirt) programs according to the time and space rules of ground subsidences for achieving much better economical, environmental and social benefits.

Fig. 12. Soil-overlay operation by means of hydraulic dredging unit.

Reclamation, Treatment and Utilization of Coal Mining Wastes, edited by A.K.M. Rainbow   369
Elsevier Science Publishers B.V., Amsterdam, 1987 — Printed in The Netherlands

AN INVESTIGATION INTO CHEAPER MONOLITHIC PACKING MATERIALS UTILIZING COLLIERY

TAILINGS

A.H. Zadeh[1], A. Barkhordarian[2], P.S. Mills[3], A.S. Atkins[2] and R.N. Singh[1]

[1]Department of Mining Engineering, University of Nottingham, University Park, Nottingham, NG7 2RD, England.

[2]Department of Mining Engineering, North Staffordshire Polytechnic, College Road, Stoke on Trent, Staffordshire, ST4 2DE, England.

[3]Headquarter Technical Department, British Coal, Ashby Road, Stanhope Bretby, Burton on Trent, Staffs.

SUMMARY
    This research is aimed at developing cost effective and environmentally acceptable methods of disposing liquid tailings at existing and new coal mines. Previous laboratory tests by the authors have indicated that tailings from a coal preparation plant when mixed with a suitable percentage of pulverized fuel ash and bentonite/cement mixture can produce a cheaper monolithic material offering acceptable initial pack strength as compared with many established packing systems. This paper aims to develop a cheaper filling material to reduce the cost of the tailing pack system and improve the initial strength of the mixture by using different accelerator and additives. The results indicate that by using Aluminium Sulphate as an accelerator the resulting mixture provides a high initial pack strength exceeding 0.15 MPa at one hour curing time, which is an acceptable pack strength design criterion. The paper presents results of a comprehensive laboratory testing programme for the optimization of constituents at various moisture contents. Strength and deformational properties of the selected mixtures are also presented. These results are very promising and a surface trial utilising this system is envisaged in the near future.

INTRODUCTION

    Recent developments in the energy market have highlighted an ever increasing need for the production of deep mined coal at competitive prices. In a previous paper (ref 1) the authors have proposed a method of utilising fine colliery waste to replace imported materials for monolithic pack construction in longwall advance mining. The aim of this scheme was to reduce the unit cost of the monolithic packing operation together with reducing the environmental impact of the liquid tailings disposal on the surface. It has been shown that a mixture of colliery tailings with a suitable percentage of cementicious or cheaper pozzolanic material can form a monolithic pack of adequate designed strength to meet pack design requirements. However, the cost of this packing material is only marginally lower than the alternative systems available.

The aim of this work is to further reduce the cost of tailing systems by using different accelerators and additives. Studies of the tailings from two selected coal mines have indicated that moisture contents of tailings vary between 52 and 80%. Tailings containing more than 70% moisture have to be recycled as the cost of improving the strength of the monolithic pack would be prohibitive. The paper describes the results of the time dependent uniaxial compressive strength tests for various tailings mixtures using different types of additives along with their unit costs. The results of various strength and deformational properties of the processed liquid tailings for various selected moisture contents for different curing times are also presented.

RESEARCH PROGRAMME

In order to improve the geotechnical characteristics of tailings pack system the following experimental program was carried out.

(a) Selection of ideal accelerators to formulate the optimum constituents of tailings mixture,

(b) Optimization of constituents with respect to moisture content,

(c) Evaluation of the spontaneous combustion risk potential for the selected tailings mixture,

(d) Assessment of the geotechnical properties of the selected tailings mixtures by conducting the following tests:

(i) Direct shear test,

(ii) Uniaxial compressive strength,

(iii) Deformational properties test,

(iv) Triaxial strength test.

TABLE 1

Various tailings mixtures used for optimization of constituted at 70% moisture content

Mix.	Tailings %	PFA %	OPC %	SB %	Accelerator %			Mean Uniaxial Compressive Strength (MPa)							Price
					Na_2Co_3	K_2Co_3	TEA	2 Hours	4 Hours	8 Hours	24 Hours	48 Hours	1 Week	4 Weeks	(£) $1m^3$
A	58.43	14.76	22.50	2	1.13	0.23	0.33	Weak	Weak	Weak	0.60	1.88	-	-	29.70
B	56.90	14.22	25.00	2	1.25	0.25	0.37	Weak	Weak	Weak	1.17	2.73	-	-	32.80
C	56.40	14.10	25.00	2	2.50	-	-	Weak	0.17	1.13	-	-	-	-	24.80
D	54.40	13.60	25.00	2	5.00	-	-	0.10	0.34	0.73	1.56	1.92	2.88	3.74	33.60

SB: Sodium bentonite OPC: Ordinary portland cement
TEA: Triethanolamine

SELECTION OF ACCELERATOR AND FORMULATION OF THE OPTIMUM CONSTITUENTS

The preliminary trial was conducted on tailings containing 70% moisture with all commercially available cement accelerators which were likely to improve the pack characteristics. Table 1 shows the variations of Uniaxial Compressive Strength (UCS) with curing time for 4 different mixtures using sodium and potasium carbonate and trietheloamine as accelerators. The results indicate 5% sodium carbonate gives an inital strength of 0.1 MPa. at two hours curing time (mixture D). This strength is substantially less than other commercially available monolithic packing materials (Figure 1).

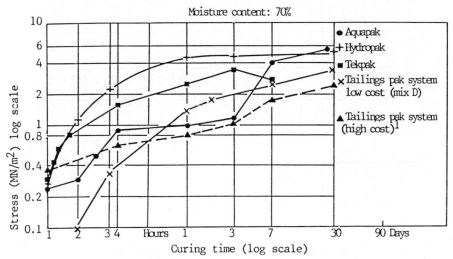

Fig.1 Comparative Strength of monolithic packs (mix D)
(Tailings/PFA: 4/1   OPC: 25%   B: 2%   Na₂Co₃: 5%   Cost: £33.60   1m³)

Optimization of constituents with respect to moisture content

The moisture content of tailings from two collieries showed that it can vary from 50 to 80%. This high water content can be stabilized by addition of PFA with insignificant cost increase. Tailings with moisture content of 50%, 60%, 70% and 80% were obtained and mixed with PFA, Cement, Bentonite and aliminium Sulphate. The results are shown in Table 2 (M to R). The ratios of cement , Bentonite, and Aluminium Sulphate were kept constant throughout the test. The results indicated that where the moisture content exceeds 70% the initial setting time is reduced although in the long term it gives the desired strength (mix O 80% mosture and R70%moisture). The strength of mix M (52% moisture) in Figure 2 reaches 0.33 MPa after an hour curing time. This rapid early strength is attributed to low moisture content and the use of Aluminium Sulphate. The Strength of mix M reaches 4.63 MPa after 30 days. The strength of mix P (62% moisture) in Figure 3 is lower than mix M due to a 10% higher moisture content. The final strength is considered to be adequate for mine packing requirements.

372

TABLE 2
Tailings mixtures used for optimization of constituents at various moisture contents

Mix	MC %	Tailings %	PFA %	OPC %	SB %	Na₂Co₃	K₂Co₃	Fe₂So₄	Al₂So₄	1 Hour	2 Hours	4 Hours	24 Hours	48 Hours	1 Week	2 Weeks	4 Weeks	Price £ 1m³
						Accelerator %				U.C.S. (MPa)								
E	62	60.40	15.10	20.00	1	2.00	1.50	-	-	Weak	Weak	Weak	1.19	1.91	-	4.10	-	32.90
F	62	58.40	14.60	22.50	1	2.00	1.50	-	-	"	"	"	1.51	1.98	-	4.03	-	34.79
A	68+10	34.00	34.00	20.00	-	-	-	2.00	-	Weak	Weak	Weak	Weak	0.09	1.89	-	-	19.74
B	68+10	30.00	30.00	25.00	-	-	-	2.50	-	"	"	"	"	0.14	2.38	-	-	23.80
G	62	60.40	15.10	20.00	1	-	-	-	3.50	"	"	0.11	0.61	0.76	-	-	-	24.50
H	62	59.20	14.80	20.00	1	-	-	-	5.00	"	"	0.13	0.61	-	-	-	-	26.00
I	62	59.20	14.80	20.00	1	3.50	1.50	-	-	Weak	Weak	Weak	0.91	1.53	-	3.70	-	36.00
J	62	56.80	14.20	20.00	1	8.00	-	-	-	"	0.14	-	0.90	0.95	-	-	-	34.80
K	62	55.20	13.80	25.00	1	-	-	-	5.00	"	0.28	0.33	-	-	-	-	-	32.50
L	62	55.20	13.80	22.50	1	10.00	-	-	-	"	-	0.31	1.67	-	-	3.31	-	41.00
M	52	55.20	13.80	25.00	1	-	-	-	5.00	0.33	-	0.84	1.50	1.73	2.62	3.24	4.63	32.50
N	62	51.75	17.25	25.00	1	-	-	-	5.00	Weak	0.14	-	0.40	-	0.45	0.92	-	32.70
O	80	55.20	13.80	25.00	1	-	-	-	5.00	Weak	Weak	Weak	0.30	-	0.87	1.21	-	32.50
P	62	48.30	20.70	25.00	1	-	-	-	5.00	0.19	⌐	0.44	1.30	-	1.45	-	3.80	33.00
Q	80	41.40	27.60	25.00	1	-	-	-	5.00	Weak	-	0.20	-	-	-	-	-	33.50
R	70	54.40	13.60	25.00	1	-	-	-	5.00	Weak	0.13	0.34	1.56	1.92	2.88	-	3.74	32.90

MC: Moisture content   OPC: Ordinary portland cement   SB: Sodium bentonite

At this stage it was decided to evaluate the spontaneous combustion risk potential of the selected mixtures.

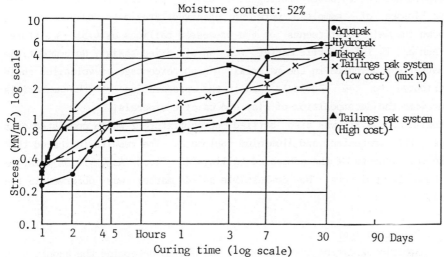

Fig.**2.** Comparative strength of monolithic packs (mix M)
(Tailings/PFA: 4/1   OPC: 25%   B: 1% Al_2So_4: 5%   Cost: £32.50 lm^3)

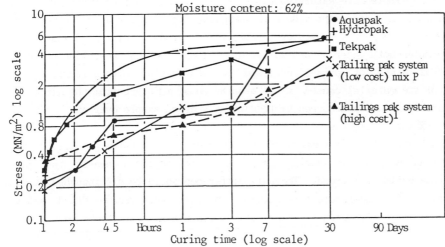

Fig.3. Comparative strength of monolithic packs (mix P)
(Tailings/PFA: 3.5/1.5   OPC: 25% Al_2So_4: 5%   Cost: £33.00 lm^3)

EVALUATIONS OF SPONTANEOUS COMBUSTION RISK POTENTIAL

The processed tailings were tested to evaluate the risk of spontaneous combustion for following mixtures:

(a) Tailings,PFA, Bentonite, Ferrous Sulphate,

(b) Tailings,PFA, Bentonite, Aluminium Sulphate.

Proximate Analysis

A proximate analysis was carried out on the mixture 1 which was dried at room temperature for 10 days. During the drying period, the time dependent moisture losses of samples were not observed. Proximate analysis results have shown the main constituents of the processed tailings as 4.51% moisture, 83.36% Ash and 12.13% volatile matter (on an air dried basis). No combustible solid matter was revealed in the material. A high percentage of volatiles could be attributed to the chemicals used in the preparation of the materials together with the decomposition of clay and calcite minerals.

A proximate analysis was also carried out on mixture 2 which consisted of a Tailings , PFA, Bentonite, and Aluminium Sulphate. The results indicated that the main constituents of the processed tailings as 84.5% Ash, 10.5% Volatile matter and 5% moisture. No combustible solid matter was observed in the material.

Adiabatic Oxidation Test

Two adiabatic oxidation tests were carried out to determine the spontaneous combustion potential of the pack material. The first tests were run by using an air dried sample with saturated air. The second tests were run by using a vacuum dried sample with saturated air conditions. The samples were ground to - 210 microns in order to provide a maximum reactive surface area between the sample and air. Figure 4 indicates the oxidation trends of air dried and vacuum dried samples both under saturated air conditions. The high heat liberation at the initial stage of oxidation could be considered to be a result of the wetting of the material caused by the absorbtion of moisture from the air flow. The amount of moisture absorbed by 100gms of air dried sample over an

8 hour oxidation test was 0.7%.

As indicated by the temperature-time curves, both samples produced nearly the same amount of heat during the tests and followed nearly the same temperature build up trend. The only difference could be distingushed by the examination of temperature time curves, in that the vacuum dried sample indicated a slightly higher rate of temperature rise at the beginnig than that of the air drier sample and started cooling much earlier (curves I and II). This material has a very poor chance of being crushed to the particle size of -210 microns and dried to the condition of the tested samples in an underground environment. Bearing this in mind, it can be concluded that although the occurence of fissurations to some extent do allow the humid air to penetrate into the pack material underground, the heat of liberation due to the oxidation and wetting will not significantly promote the conditions required for spontaneous combustion to occur (Figure 4).

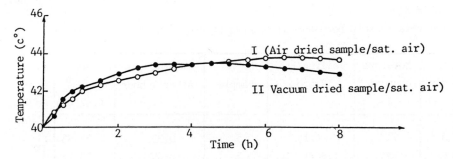

Fig.4. Spontaneous combustion test for mix 1.

The results of adiabatic oxidation tests on mixture 2 are shown in Figure 5, both for air dried saturated air and vacuum dry/saturated air test environments. As can be seen from figure 5 an increase in temperature is due to the heat from wetting. After a period the sample loses the capacity to absorb more moisture and there is no heat emissions and therefore no increase of temperature. On the contrary, the material starts cooling down due to the convection of heat by the air flow (Figure 5).

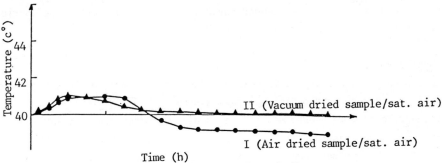

Fig.5. Spontaneous combustion test for mix 2.

GEOTECHNICAL TESTING PROGRAMME

It was stated earlier that the moisture content of tailings can vary from 50 to 80%. This change in the moisture content has a direct effect on the strength of tailing pack. The addition of PFA, which is the least expensive compound in the pack, can absorb water and control the compressive strength.

A series of short term strength tests were carried out to establish the amount of PFA required to achieve high early strength (Table 3). The percentage of ordinary portland cement, bentonite and aluminium sulphate were kept constant throughout the test. The only variable in the system was the ratio of tailings to PFA in accordance to moisture content.

TABLE 3

The relationship between Compressive Strength and moisture content

Mix.	Moisture content %	Tailing/PFA Ratio	No of Sample	σ MPa After 1 hour	Cost 1m^3 Pack (£)
A	52	4:1	5	0.33	32.53
B	58	1.84:1	5	0.60	33.26
C	58	4:1	5	0.17	32.53
D	62	2.35:1	5	0.19	33.00
E	62	1.50:1	5	0.69	33.50
F	69	2:1	5	0.15	33.18
G	73	1.50:1	5	0.39	33.50
H	80	1.50:1	5	0.17	33.18
I	70	1.50:1	5	0.30	34.46

Testing programme for tailings pack system with 70% moisture content

The object of the test was to examine the various strength properties of the tailings pack system over an extended period. The following tests were carried out with time intervals of 1 hour to 1 month.

(i) Shear strength

Time dependent laboratory tests were carried out in order to study the shear and cohesive strength of the tailings pack system.

All shear tests were performed in a small commercially available shear testing rig. The maximum size of specimen accepted by this machine was (60x60x10) mm. The normal load is applied by a dead weight system through a lever arrangment with an arm ratio of 10:1. The rate of shearing was controlled by an electronic motor, gearbox and a feed arrangement. The shear rate could be varied from a rate of strain of 0.33x10 mm/min to 4.23 mm/min. The shear force developed by the constant rate of strain was monitored through a proving ring and load cells. The shear box was interfaced with a microcomputer which permitted complete monitoring of shear tests at various normal load values until a residual stress value was obtained. The normal loads used for the tests were 226.5, 326,5, 426,5 Kgs. The results of 1 hour to 1 month tests are shown in Table 4.

377

TABLE 4
Summary of strength and deformation properties of tailings pack at 70% moisture content
(Tailings/PFA: 1.5/1) mix 1'

Curing	Uniaxial Compressive Strength MPA	GPa E	Poissons ratio ν	Direct Shear Test		Triaxial Test	
				Cohesion KPa	Friction Angle φ	Cohesion KPa	Friction Angle ø
1 hour	0.35	0.025	—*	317.13	11.85	275	20
2 hours	0.39	0.06	—*	422.13	16.17	275	22
4 hours	0.43	0.066	—*	476.56	18.26	375	28
24 hours	1.30	0.76	—*	520.60	21.60	400	29
48 hours	1.60	1.06	—*	569.22	22.00	700	42
1 week	2.81	1.66	—*	640.75	24.00	700	45
2 weeks	3.91	2.27	0.18	779.93	31.79	900	47
4 weeks	4.43	2.00	0.22	1120.43	41.66	1001	49
3 months	6.19	2.50	0.16	1086.00	40.00	1100	49

The results from Poisson's ratio from 1 hour to 2 weeks showed they are obscure. The samples are not strong enough to measure this value less than 2 weeks curing.

378

(ii) Determination of deformational properties of the tailings pack

The most common method for studying the deformational properties of the tailings specimen is by axial compression of a cylindrical specimen under a uniaxial stress state. For any level of stress applied to the specimen, the axial and lateral strain can be measured and used to determine the deformation relationship of stress to the strain for the specimen. The prefailure deformational behaviour was characterised by two constants. The first is Young's modulus of elasticity, "E", which is the ratio of axial stress to axial strain The second is Poisson's ratio, "v", which is the ratio of lateral strain to axial strain. Young's modulus and Poisson's ratio of the tailings pack are shown in Table 4.

(iii) Triaxial test

The triaxial tests were employed not only to determine the shear strength of rock, but also to investigate the behaviour of the tailings pack under three dimensional stress fields. The triaxial test was carried out using an RDP testing machine.

The test is effected by applying a constant hydraulic pressure to the curved surface of a cylindrical specimen, prior to causing its failure by increasing the normal stress (σ_1) applied to the specimen plane ends. A number of tests were carried out with confining pressures between 0.17, 0.34 and 0.69 MPa. The shear strength of the pack has been determined from the graphical presentation of Mohr's circles and the intercept of the failure envelope with the shear strength axis. The internal agle of friction was evaluated from the slope of σ_1 and σ_3. The results are shown in Table 4.

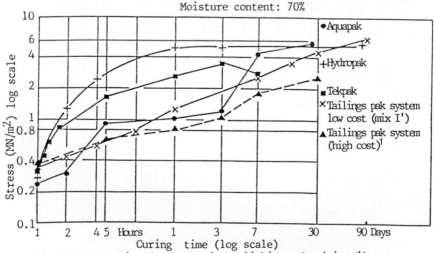

Fig.6. Comparative Strength of monolithic packs (mix I')
(Tailings/PFA: 1.5/1 OPC: 25% B: 1% Al_2So_4: 5% Cost: £34.46 1m^3)

(iv) Uniaxial compressive strength

The tests were carried out in a Denison 500 KN testing machine. The samples were loaded and failure strength was recorded. The results are shown in Table 4. The final strength after 3 months reached up to 6.19 MPa. Figure 6 (Mix Í) shows the strength of a tailings pack system containing 70% moisture in comparison to Tekpack, Aquapack and the high cost tailings pack system, described in ref 1. The results indicate that the selected low cost tailings pack systems achieves a high initial strength of 0.35 MPa and a strength of 6.19 .in 3 months. This strength is higher than most established monolithic packing systems.

CONCLUSIONS

Results from extensive laboratory tests have shown that a mixture of colliery tailings with a suitable percentage of cementicious or cheap pozzolanic material can form a monolithic pack of adequate design strength to meet pack design requirements. The unit cost of the developed tailings pack is lower than many other established packing systems.

The moisture content of the tailings is the most important factor in gaining desirable early strength. Continuous monitoring of the moisture content of tailings is necessery for field installation utilising tailings.

There are other factors governing the pumpability of the-run-of mine tailings such as the efficiency of the pump, transportation and mixing arrangements. A number of logical methods have been established and discussed in a paper by the authors (ref 3).

Results from two different mixtures showed that the tailings pack system has no risk potential for spontaneous combustion.

ACKNOWLEDGEMENTS

The authors wish to express their gratitude to the Science and Engineering Council and British Coal for jointly financing the project. Thanks are also due to various managers and engineers within British Coal and specially to the Transport and Coalface branch of the Headquarters Technical Department for their co-operation and assistance with the project.

REFERENCES

1 A.S. Atkins, D. Hughes, D. Parkin and R.N.Singh, Utilization of colliery tailing in mining activities,Symposium on the Reclamation , Treatment and Utilization of Coal Mining Wastes, Durham, England, 1984, pp. 18.1-18.11.
2 A.S. Atkins, J.C. Atkin, R.N. Singh and A.M.H. Zadeh, An alternative method of surface disposal and stabilization of coa mine tailings, Geotechnical and Geohydrologicl Aspects of Waste Manegements, Colorado State University, 1986, Fourth Collins, pp.227-286

3   A.S. Atkins, R.N. Singh, A Barkhordarian and A.M.H. Zadeh, Pumpability of coal mine tailings for underground disposal and for regional support, Second International Symposium on the Reclamation, Treatment and Utilization of Coal Mining wastes, 7–11 September 1987 Nottingham University (to be published)

Reclamation, Treatment and Utilization of Coal Mining Wastes, edited by A.K.M. Rainbow
Elsevier Science Publishers B.V., Amsterdam, 1987 — Printed in The Netherlands

GEOTECHNICAL ASPECTS OF FINE COAL WASTE DISPOSAL IN
LOWER SILESIA , POLAND

B. BROŚ

Professor of Civil Engineering , Agriculture University of Wrocław,
Wrocław / Poland /

SUMMARY
    The physical and mechanical properties of coarse and fine coal
wastes from colliery spoil tips and lagoons of the Lower Silesia
coalfield in Poland are given. The paper considers the various types
of coal waste embankments for tailings ponds in this region. Some
examples of piping failures of tailings lagoons banks are discussed.

INTRODUCTION

    Lower Silesia coalfield in Poland covering an area of about
500 km^2 requires special emphasis because of the difficult geological
and hydrological environment and the local topography, typical for
mountainous region. The area is subject to the rainfall of the range
between 800 and 1200 mm a year, when an average for Poland
amounts to about 600 mm a year.

    Usually there are 10 to 26 coal seams found in the Carboniferous
rocks of the Lower Silesia mountains, suitable for mining. Coal seams
are thin and steep, often faulted, difficult and costly for mining. The
importance and the value of the Lower Silesia coalfield lies in the
fact, that it is a source of very good high - quality coking coal.
Annual output of this coal exceeds 3 million tons and is accompanied
by the output of coal mine wastes of about 7 millions tons a year.
Coal seams occur in the sedimentary rock which consists of sandstone,
siltstone, mudstone and shale. Coal mining involves the driving of
shafts and roadways, mining of the coal and processing of the raw
mined coal in separation plants where impurities as coal mining wastes
are removed.

    The mining rock waste comes from the development of shafts and
mine roadways and consists mainly of sandstone, siltstone and
mudstone, with particles in the gravel to about 500 mm size range.
It constitutes about 20 to 40% of the whole amount of the coal mining
wastes. Part of this rock waste as the stone is disposed of
underground. The rest of the stone goes out on the belt mixed up with

the coal and is removed in the washery. Two types of coal wastes are produced in this process, respectively: coarse coal waste, which is also known as coarse discard, consisting predominantly of shale and fine coal waste known also as fine discard. Coarse discard mostly consists of the coarse material, and with particles in 20 mm to 200 mm size range, which is separated in the washery. Fine discard consists of lighter particles in the silt to sand size range, there is below 20 mm in size. It is the fine material, called often the tailings, usually fine particles of coal and shale remaining in suspension in water after the washing process and rejects from flotation process.

The coal mining wastes from the two separation stages are usually disposed of separately. The coarse discard is mostly transported by conveyor belts or aerial ropeways to a nearby disposal site and tipped in the form of a cone or ridge and is generally remained at the angle of repose. The fine grained coal wastes, called also tailings, are disposed of by pumping in a slurry form in tailings ponds, known also as lagoons, created often by the coarse discard embankments.

In Lower Silesia exist 33 spoil tips and the area of tipping land is about 210 ha. Typical for this region are steep hillsides on which there are large mounds of even steeper mine spoil.

PROPERTIES OF COAL MINING WASTES FROM LOWER SILESIA COALFIELD

Properties of coarse discard

The shale and mudstone are normally the largest components of spoil tips in Lower Silesia. The most stable rock type is the sandstone, but it is the smallest component and therefore has a small influence on the engineering properties of coarse discard.

Physical and chemical weathering of the shale and mudstone and the oxidation of pyrite found in the shale change the coarse coal waste properties with time. Physical weathering occurs predominantly during the handling and disposal of coarse discard. The mechanical breakdown of the coarse particles of coal waste may be observed on truck pathways where it is subjected to repeated loading. Chemical weathering of the coarse discard is predominatly caused by the oxidation of the pyrite present in the shale and it is mostly surficial and extends only about 1.5 m below the spoil tip surface "/ ref. 1 /" .

Long – term tests of coarse discard from the old and new spoil tips of WAŁBRZYCH – colliery were conducted. Tests consisted of soaking test in the period of 33 days, in which coarse discard was placed in a beaker and was covered with distilled water, followed by successive freezing and thawing of tested material every 5 days during an arbitrary period of 21 days. After 54 days of testing it was found that relatively small degradation of the larger particles into gravel and sand-sized particles occured, but without the formation of silt and clay-sized particles.

Consolidated drained direct shear tests were conducted on coarse washery discard,from old and new spoil tips, before soaking and successive freezing and thawing tests and after these tests. The results were of the same order and corresponded to values of about $\phi' = 39°$.

The moisture content of the coarse washery discard varied in the range:

    for old tip   −   from 5.7% to 8.1%
    for new tip   −   from 0.8% to 4.1%

The moisture content in existing old tip is significantly higher than the values of fresh delivered coarse discard from coal preparation plant in the new tip. The same correlation has been observed in the case of old and new tips in Wales " / ref. 1 / " .

The values of maximum dry density for coarse washery discard were as follows:

	Optimum moisture content	Maximum dry density
for old tip −	10.7 %	1.81 tonne/m^3
for new tip −	13.5 %	1.69 tonne/m^3

Specific gravity of coarse coal waste is significantly lower than common soils and rocks and depends on the amount of coal carbonaceous material, shale and sandstone.

The shear strength parameters of the coarse coal waste vary with gradation, unit weight, petrographic composition and with the confining pressure because of breakage of the weak, plateshaped particles. According to " / ref. 1 / " the overal ranges of these parameters for England and Wales are of the same order. The lower limit represents a value of ϕ' of about 25° and the upper limit about 40° . The values of these parameters of the coarse coal waste from Lower Silesia coalfield are nearly the same.

Some typical ranges of the engineering properties of the coarse washery waste of Lower Silesia coalfield are given in Table 1.

TABLE 1

Coarse washery coal waste typical range of properties

Water content / in tips /, %	0.8 to 18.9
Specific gravity, tonne/m^3	2.29 to 2.34
In embankment:	
Dry unit weight, tonne/m^3	1.19 to 2.29
Total unit weight, tonne/m^3	1.81 to 2.47
Standard Proctor test:	
Maximum dry unit weight, tonne/m^3	1.69 to 2.02
Optimum water content, %	8.4 to 13.5
Coefficient of permeability, cm/s	
in tips	0.5×10^{-1} to 3.0×10^{-1}
after Standard Proctor test	2.06×10^{-6} to 1.35×10^{-5}
Direct shear test parameters:	
c' – cohesion, kN/m^2	$c' = 0$
ϕ' – degrees	$23°20'$ to $44°$

Properties of fine discard

Fine coal waste often called tailings is pumped in a slurry form into tailings lagoons where sorting of the particles occurs, there is heavier and coarser particles settle out near the discharge area and fine particles at the opposite side of the lagoons near the outlet end. The mineralogical composition of lagoon deposits is predominantly coal and carbonaceous shale. Typical for this material is a low specific gravity in the range of 1.4 to 2.0 tonne/m^3 , what indicates on the high coal content. The specific gravity of coal lies between 1.3 and 1.5 tonne/m^3 " / ref. 2 / " .

The main difference between fine coal waste and natural soils is their low total unit weight due to their low specific gravity. The fine coal waste is predominantly cohesionless material composed of angular particles, which varies from silt to medium sand.

The values of the angle of internal friction ϕ' of fine coal waste from Lower Silesia lagoons were little higher than those obtained for coarse discard at the corresponding colliery. Some observations in England indicate the same tendency " / ref. 1 / " .

The engineering properties of tailings from WAŁBRZYCH-colliery lagoons are summarized in Table 2.

TABLE 2

Typical range of properties of fine coal waste of Lower Silesia

Water content / in lagoons / , %	48.4	to	75.2
Specific gravity, tonne/m^3	1.56	to	1.82
In lagoons:			
Dry unit weight, tonne/m^3	0.75	to	0.91
Total unit weight, tonne/m^3	1.27	to	1.41
Organic content, %	24.7	to	45.3
Coefficient of permeability, cm/s	2.66×10^{-7}	to	9.7×10^{-8}
Consolidated, drained direct shear test:			
c' - cohesion, $kN/_m 2$	$c' = 0$		
ϕ' - degrees	27°	to	$39^\circ 30'$

DISPOSAL PROBLEMS OF COARSE COAL WASTE

Fresh coarse coal waste is always dumped in Lower Silesia coal-field on sloping ground because of the topographical conditions. Sometimes it is used as a building material for dams and banks of tailings lagoons. The coarse coal waste embankments are enddumped in large lifts with no compaction and slopes of embankments are at the angle of repose due to dumping of the waste from the crest. The cases of instability of existing waste embankments are very rare. Sometimes the toe material moves outwards producing a bulge or occur small local slipes.

Physical disintegration of the coarse coal waste in the waste embankments due climate effects, erosion by wind and rain and because of chemical weathering promote deterioration and gradual flattening of slopes.

The observation of existing spoil tips in the region of Lower Silesia coalfield indicates that the outer slopes of tips remain at the angle of repose, but for old tips this angle is about 30° and for new formed tips of about 38° to 42°.

Landscaping and vegetating of coal waste embankments is desirable for aesthetic, control of erosion and reduction the possibility of spontaneous combustion. In Lower Silesia coalfield vegetation on the slopes of the old tips is partly due to planting of trees, and partly due to self-seeding of grass and trees.

There are some projects of utilization of coarse coal waste for other purposes, i.e. as building material for flood protection embankments and for railway earthworks.

FINE COAL WASTE DISPOSAL

In Lower Silesia coalfield the fine coal waste is disposed of in a slurry form into lagoons, which embankments are usually constructed from the coarse coal waste or from the mining rock waste. The most common forms of the lagoons are dams from compacted coarse washery discard across valley tributaries, which create upstream ponds.

Cross-section of this type of lagoon with homogeneous dam from compacted coarse coal waste, mainly shale, 22 m high and 695 m in length, is shown in Fig.1.

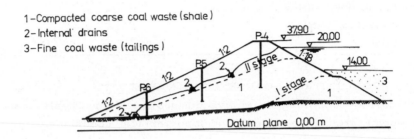

1-Compacted coarse coal waste (shale)
2-Internal drains
3-Fine coal waste (tailings)

Fig. 1. Cross-section of the homogeneous lagoon dam from compacted coarse coal waste.

Some properties of the coarse coal waste, mainly shale, placed and compacted in the dam:

Total unit weight, tonne/m^3	1.8	to	2.0
Water content, %	6	to	10
Optimum water content, %	8	to	9
Maximum dry unit weight, tonne/m^3	1.85	to	1.90
Coefficient of permeability, cm/s	1.3×10^{-3}		

Direct shear test parameters:

c' - cohesion, kN/m^2 $c' = 0$

ϕ' - degrees $36°$ to $41°$

The measurements of the crest dam vertical movements, carried out every month on several benchmarks distributed along the crest, showed that the dam expanded as the result of penetration of seepage water from the reservoir into the embankment, causing the swelling of shale. The crest heaved from 1 mm to 13 mm.

Cross-section of another dam formed for large tailings pond is shown in Fig. 2.

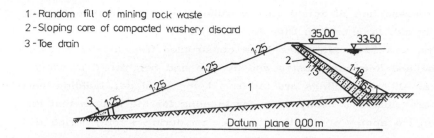

1 - Random fill of mining rock waste
2 - Sloping core of compacted washery discard
3 - Toe drain

Fig. 2. Sloping core lagoon dam from mining rock waste and washery spoil

The dam is 30.5 m high and 1285 m in length, mostly from mining rock waste and coarse coal waste, tipped from belt conveyors as random fill but in the way to form the dam of the shape of an arc in plane. The main portion of the embankment consists of a large downstream zone mostly of mining rock waste which was dumped in high lifts with upstream semi-impervious sloping core of placed in thin layers and compacted washery spoil.

Some properties of compacted washery spoil in core:

Total unit weight, tonne/m^3	1.81 to	2.47
Dry unit weight, tonne/m^3	1.62 to	2.29

Standard Proctor:

Maximum dry unit weight, tonne/m^3	1.78 to	1.88
Optimum water content, %	7.7 to	12.1

Coefficient of permeability at maximum dry unit weight, cm/s	1.35×10^{-5} to	2.06×10^{-6}

Direct shear test parameters:

c' – cohesion, kN/m^2	$c' = 0$	
ϕ' – degrees	$33°$ to	$43°50'$

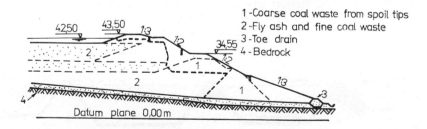

1 - Coarse coal waste from spoil tips
2 - Fly ash and fine coal waste
3 - Toe drain
4 - Bedrock

Fig. 3. Zoned dam of tailings pond for fine coal waste and fly ash

Cross-section of zoned dam of tailings pond for fine coal waste and fly ash is shown in Fig. 3.

The dam 31 m high has been constructed from locally available coarse and fine coal wastes and fly ash and heightened in many stages. Although tailings are far from being ideal dam-building material, they are utilized in many dams for lagoons for the reason that they are on the spot. Construction of this dam during a long period of many years and according to different projects caused the very great heterogeneity of the dam. However the principle of heightening of banks was always the same: it was done in upstream method, i.e. an

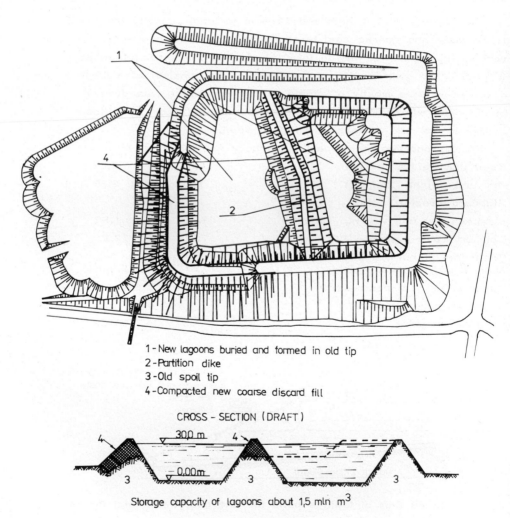

1 - New lagoons buried and formed in old tip
2 - Partition dike
3 - Old spoil tip
4 - Compacted new coarse discard fill

CROSS - SECTION (DRAFT)

300 m

0.00 m

Storage capacity of lagoons about 1,5 mln m³

Fig. 4. New lagoons formed in old spoil tip

additional bank of coal waste was placed with its base partly on the crest of the existing bank and partly on the lagoon deposit, placing fine waste material in upstream zone and coarse coal waste in downstream zone.

New lagoons for fine coal waste disposal buried and formed in old spoil tip are shown in Fig. 4. In the immediate vicinity of a colliery is situated an old spoil tip 15 m to 45 m high and of the area of about 15 ha. It creates on a flatter valley slope a flat topped rectangular embankment of mining rock and coarse coal waste. The outer slopes of the tip remain at the angle of repose about 30° . Two lagoons divided by partition dike, about 30 m deep, were excavated and formed in the tip. New parts of lagoon banks and partition dike were formed from fresh coarse discard, placed in thin compacted layers and creating outer slopes 1 : 1.8. The minimum width of the outer crest is 15 m.

At the toe of the reconstructed tip a pipe drain and a ditch were installed for colecting the run-off from the slopes and the seepage water from lagoons.

Storage capacity of both lagoons is about 1.5 mln m^3.

A very low submerged unit weight of fine coal waste is responsible for low shear strength and high liquefaction potential of this fine waste material.

The serious trouble from piping in fine coal waste disposal results from progressive backward erosion of concentrated leaks which develop mostly through the lagoon banks.

The piping failure shown in Fig. 5. occured in 1969 on the side dam of the large lagoon for fine coal waste and fly ash " / ref. 3 / " .

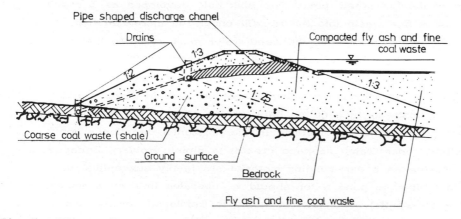

Fig. 5.   Fine coal waste and fly ash lagoon piping failure

The initial bank was constructed from coarse coal waste, mainly
shale, and the second-stage bank from compasted fly ash and fine coal
waste, covered with a layer of waste rock. The sudden piping failure
developed as a result of loss of fine material into the pipe drains
under the increased hydraulic head that developed when free pond
water came close to the upstream face of the bank. The large amount
of a slurry deposits poured over a distance of about 500 m
covering the valley to a depth of about 1 m.

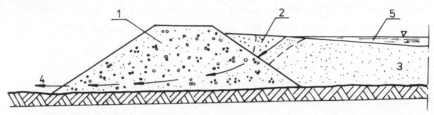

1 - Mining rock fill
2 - Sink hole caused by piping failure
3 - Fine coal waste
4 - Water and fine coal waste flowed through rock fill
5 - Pond

Fig. 6. Piping failure in waste rock dam

Another example of a piping failure is illustrated in Fig. 6.
In this case the main body of the fine discard lagoon bank was
constructed from mining rock waste. During construction of the rockfill
bank the upstream filter layer was omitted. During the period of
operation, when free pond water was allowed to come close to the
upstream face of the bank, a piping failure suddenly occured. The
cause of the sudden failure was a sink hole which breached the
edge of the fine waste pond. The sink hole developed as a result
of loss of fine waste into the rockfill under the increased hydraulic
head that developed when the pond water came close to the upstream
face of the fill.

The piping failure can develop when the water surface in the pond
rises above the level of the beach, so that it is in direct contact
with the upstream face of the bank, which often is not homogeneous
and exhibits higher permeability in the horizontal direction. The
resulting concentrated seepage creates a piping problem. Because
fine wastes as a dam-building material are highly susceptible to
internal piping a pond water should be operated in such a manner
that a substantial beach of fine waste is maintened between the
upstream face of the bank and the free water surface in the pond.

The method of calculation of the minimum width of the beach is illustrated in Fig. 7. , where i_{kr} = 0,15 is a critical gradient for fine coal waste and fly ash.

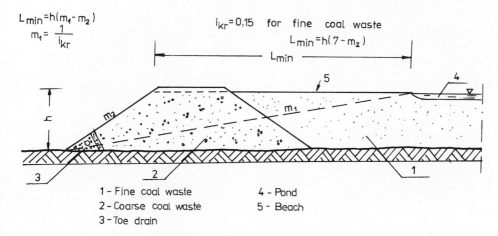

$$L_{min} = h(m_1 - m_2)$$
$$m_1 = \frac{1}{i_{kr}}$$

$i_{kr} = 0,15$ for fine coal waste
$$L_{min} = h(7 - m_2)$$

1 - Fine coal waste   4 - Pond
2 - Coarse coal waste   5 - Beach
3 - Toe drain

Fig. 7. Method of calculation of the minimum width of the beach

CONCLUSIONS

Despite of the heterogeneity of coarse coal waste and its associated density and shear strength it do not significantly affect stability of spoil tips and dams from this coarse discard. Several authors are of the same opinion / refs. 4 - 5 / .

Among the types of failure most common to fine coal waste lagoons deposits is piping. The fine discard lagoons with the pond water immediately behind the dike are much more susceptible to piping failure than these in which pond water is situated as far as possible from the retaining bank.

REFERENCES

1   G.M. Thomson, S. Rodin, Colliery spoil tips after Aberfan, Pap. 7522, Inst. Civ. Eng. , London, 1972.
2   I. Holubec, Geotechnical aspects of coal waste embankments, Can. Geotechn. J. Vol. 13, 1976, pp. 27 - 39.
3   B. Broś, Pollution from ash lagoons and use of ash for embankments, Proc. X ICSMFE , Vol. 2, Session 6/3, Stockholm 1981, pp. 309 - 312.
4   G. Annen, V. Stalmann, Waschberge im Deich- und Dammbau, Glückauf, nr. 26 , 1969 , pp. 1336 - 1343.
5   C.O. Okagbue, The geotechnical characteristics and stability of a spoil heap at a Southwestern Pensylvania coal mine, Engineering Geology Nr 20 , Elsevier , Amsterdam, 1984 , pp. 325 - 341 .

Reclamation, Treatment and Utilization of Coal Mining Wastes, edited by A.K.M. Rainbow
Elsevier Science Publishers B.V., Amsterdam, 1987 — Printed in The Netherlands

"THE APPLICATION OF THE MULTI-ROLL BELT FILTER TO DE-WATERING OF FINE COAL
WASTE SLURRIES"

H G King, Mineral Processing Division, Mitchell Cotts Mining Equipment Ltd

SUMMARY
During the past eight years the Andritz continuous pressure filter, (multi
roll filter), has been used to de-water fine coal tailings arising from coal
washing plants throughout the world. This paper describes the method of
operation of the continuous pressure filter and tabulates a range of
parameters obtained from operating units.

INTRODUCTION

It is common practice to remove all particles below 1.0 or 0.5 mm from a
run-of-mine coal prior to treatment of the coarser material by jigging or
heavy medium separation. This fine fraction may be re-combined with the
coarse coal product, treated by flotation or spirals, or discarded without
treatment. Whether fine fraction is treated or not it will inevitably be
present at the end of the process as a suspension in water. In addition, the
water used for processing the coarse fraction will accummulate fine solids.
The outcome of the coal washing process is the presence of considerable
quantities of fine waste slurries. The disposal of the fine waste slurries
has become a major problem in recent years.

Disposal in ponds (or lagoons) which allows the liquid phase, water in the
case of all conventional coal preparation techniques, to drain into the
ground, is becoming limited for reasons of technology, economics and
legislation. The practical objections may be:

a)   Increasing environmental opposition
b)   More stringent safety regulations
c)   Increasing surface space requirements - often in or near built-up
        areas
d)   Cost of transportation by road or pumping to a distant site

Demands for a device which will provide a disposable cake product, (and
often a filtrate capable of re-use in the washery system), and which is of
reasonably low capital cost and not labour intensive gave rise to the
development of the Andritz continuous pressure (belt) filter referred to
hereinafter as CPF. This filter, also known as the multi-roll filter in the
United Kingdom, was developed in Austria and first applied to the mining
industry in the mid-seventies.

The first commercial production units began de-watering coal tailings in 1979, in the Federal Republic of Germany and the United States of America. Since that time the CPF has also been used for the de-watering of waste slurries in Peoples Republic of China, Canada, United Kingdom and Japan. Between thirty and forty units have been installed, usually with an operating belt width of 2.2 or 3.5 metres.

The Continuous Pressure Filter

The design and operation of the CPF unit is arranged so that pressure is placed on the slurry in stages, increasing only so much as the stability of the mixture allows. If the rate of increase of pressure exceeds the rate of increase of cake stability the following deleterious effects are caused:

    i)   Extrusion of solids from side of machine
    ii)   Extrusion of solids into and through filter medium
    iii) Production of wet cake
    iv)   Dirty filtrate

The solid-liquid separation process is carried out on and by means of endless filter belt moving over rollers. Initially water is allowed to drain by gravity from the slurry on a single belt. Maximum pressure here is only due to the depth of the slurry, rarely more than 5 cm at a specific gravity of 1.4.

The second stage sees a slow increase in pressure as the second belt joins on the thicker slurry in the wedge zone. At this stage pressure on the slurry becomes greater than atmospheric (0.1 to 0.2 bar) as the available volume decreases as the belts converge at an angle of four degrees.

The third, and (in the case of waste coal slurries), final stage is the S or press zone. Here the belts, now parallel, and with a fairly stable material between them, pass round the S rolls. The considerable belt tension, being constant throughout the machine, leads to an increase of intensity of pressure as each of the seven succeeding S rolls is of decreasing diameter. In simple terms:

$$\text{Pressure} = \frac{\text{Belt tension}}{\text{Diameter of roll}}$$

Pressures of 1.2 to 1.3 bar are achieved here. In addition the cake undergoes repeated shearing as the belts move round the rolls in alternating senses. This shearing effect forces the solid particles to pack more efficiently thus aiding the de-watering effect.

Figure 1 shows a side view of the CPF indicating the gravity drainage, wedge and S pressure zones and major mechanical components.

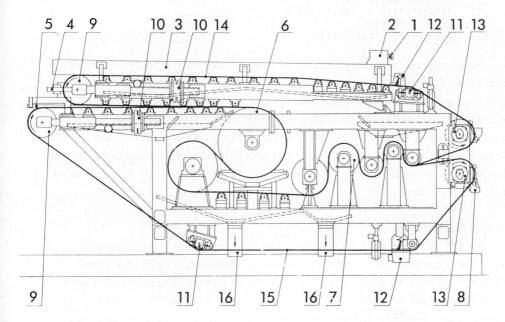

Figure 1 - Side view of CPF S7 Unit

1. Slurry inlet
2. Slurry head box
3. Gravity dewatering zone (2 - 4)
4. Baffle
5. Wedge zone (5 - 6)
6. Perforated drum roll
7. S Pressure Zone
8. Press cake discharge (Doctor blade)
9. Belt tensioning rolls
10. Belt tensioning device
11. Belt tracking device
12. Belt cleaning device
13. Machine drive
14. Upper belt
15. Lower belt
16. Filtrate drains

Direction of Belt Travel:

Upper Belt: from 2 to 4   Lower Belt: from 13 to 9   Wedge zone: from 5 to 6

Waste slurries are difficult to de-water. The intermediate sludge has poor mechanical stability while economic and technical considerations call for a machine with reliable operation and high throughput. The requirements are therefore for a long gravity zone, a long wedge zone and a multiple S zone.

The resultant CPF, designated the S7 is large with 2.3 and 3.6 m actual filter belt widths and belt lengths of 20 m and   m respectively.

Water

The water in the waste slurry or suspension may be present in several forms, the main ones being:

i) adsorbed water
ii) interstitial or cavity water
iii) capillary water
iv) water of crystallisation

The mechanical dewatering system of the CPF can only effectively remove the interstitial water. The removal of fine capillary water would require a great expenditure of mechanical energy (and a heavier stronger machine) whilst adsorbed moisture constitutes a minor part of the total water. Similarly water of crystallisation is not present in large quantities in the minerals which constitute coal tailings and thus may be disregarded with the consequent saving of energy required, possibly in the form of heat, to break this chemical bond.

Slurry Conditioning

The CPF S7 may remove 30 tonnes of water from a feed slurry every hour. In order to effect this high rate of removal the open area of the woven filter belt must be large and each aperture will have dimensions of several tens or hundreds of microns whilst the solids in the feed may contain large proportions, possibly 50 per cent, of particles below this order of size. The high feed rate and belt speed do not allow for the forming of a pre-coat and thus much of the solid phase would pass through the belt at the operating conditions and speed required.

The slurry must therefore be conditioned prior to filtration. A dilute solution of organic conditioner, often with a molecular weight of several million, is added to the slurry. These solutions, which may be anionic, cationic or non-ionic according to the state of the slurry, assist conditioning in the following ways:

i) by neutralising the electrostatic repulsive forces present with the fine particles
ii) by removing the water adsorbed to the surface of the particles
iii) by producing large stable agglomerations of particles ("floccs") by binding individual particles together

The conditioning agent (or flocculant), often a solid polyacrylamide, is usually made into a 0.5 per cent w/w solution on site and often diluted to 0.1 per cent just prior to addition to the feed slurry.

Typical flocculant addition rates are between 0.25 and 0.50 kg per tonne of dry solids.

Effective mixing of slurry and polymer solution must then take place.
Sufficient energy must be expended to ensure good dispersion and mixing
without damaging the newly formed floccs. Slow speed mixing drums or paddles
may be used although in-line mixers, which have no moving parts are popular.
The mixing procedures, although difficult to quantify, must be of a defined
nature in order that the procedure may be duplicated or altered in a
controlled manner.

Figure 2 shows an Andritz CPF S7 operating a tip-washing waste in North-
East England. Of particular note is the efforts which are made to assist the
gravity drainage stage. Two rows of vertical tubes can be seen. These
gently plough the flocculated solids, allowing free water, trapped on the
surface, access to the belt.

Ropes trailing across the belt disturb the floccs lying on the belt, again
permitting free water to pass through the belt.

Good operating practice will ensure that no free water is present when the
material reaches the breast roll at the end of the drainage zone and enters
the wedge zone.

Figure 2. Andritz CPF in operation in North-East England.

Further Process Requirements

On discharge from the CPF the resultant cake is removed by a conveyor placed below the doctor blade. In cases where extra dump stability is necessary careful mixing of the filter cake with coarse tailings is practised. Where greater stability is required, or coarse material is not available, cement may be added to the cake. The filter cake is fed to a mixer where two to four per cent (w/w) is added.

Normal British practice is to operate the CPF in closed circuit with a thickener. In most cases the thickener already exists to increase the solids content of the slurry prior to conventional methods of disposal. The returned filtrate contains a small proportion of carried over flocculant which assists the settling effect of the thickener. The thickener overflow may be used as a source of high pressure belt wash water and, on occasions, to make up or dilute the polymer solution.

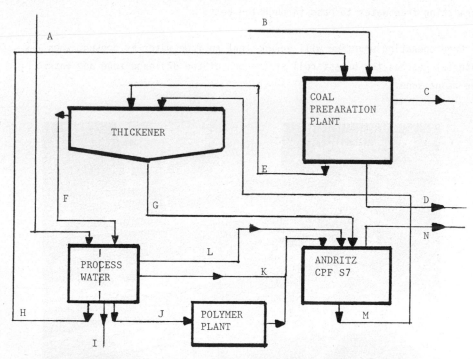

Figure 3 - Circuit with CPF, Thickener and Coal Preparation Plant.

A	Water supply	H	Process water
B	Run of Mine	I	Effluent system
C	Washed coal product	J	Polymer make-up water
D	Coarse tailings	K	Polymer dilution water
E	Fine tailing slurry	L	Belt wash water
F	Thickener overflow	M	CPF effluent (filtrate & belt wash)
G	Thickener underflow	N	Filter cake

A conventional sump pump is adequate for the effluent, which will contain 0.5 to 1.0 per cent solids.

Figure 3 illustrates diagrammatically the combination of coal preparation plant, thickener and filtration system.

Operating Data

Table 1 below summarises major CPF machine and operating data typical of industrial applications in the United States, Canada and the United Kingdom.

TABLE 1

DESIGNATION		CPF S7
Actual belt width	m	2.30
Operating belt width	m	2.20
Typical belt length	m	19
Weight	t	18
Drive motor	kw	11.5 - 15.0
System Power Consumption	kw	50 - 75
Belt type		Polyester monofilament
Belt Speed	m min^{-1}	8 - 15
Belt tension	N mm^{-1}	6 - 12
Feed solids		Minus half millimetre washing or flotation tailings
Feed solids concentration	% wt w/w	30 - 40
Solids ash content (dry)	% wt w/w	60 - 80
Slurry Feed rate	m^3/h	50 - 80
Solids below 63 microns	% wt	60 - 85
Solids in effluent (filtrate)	% wt	1.0 - 2.0
Flocculant consumption	kg/t	0.25 - 0.50
Cake thickness	mm	10 - 20
Cake moisture	% wt w/w	30 - 40
Throughput rate, dry solids	t/h	20 - 35

Conclusion

Eight years' operational experience in three continents have confirmed the suitability of the continuous pressure filter for the de-watering of fine waste suspensions.

The advantages of the continuous pressure filter are:-

    Continuous operation
    Low capital cost (compared with filter press)
    Low power consumption
    Flexibility
    Low manpower requirements

400

The disadvantages are:-

Cake product may require additions before dumping
Conditioning (polymer) costs may be high

The Author wishes to acknowledge the help given by Messrs Andritz GmbH of Graz, Austria and its overseas associates.

Reclamation, Treatment and Utilization of Coal Mining Wastes, edited by A.K.M. Rainbow   401
Elsevier Science Publishers B.V., Amsterdam, 1987 — Printed in The Netherlands

PUMPABILITY OF COAL MINE TAILINGS FOR UNDERGROUND DISPOSAL AND FOR REGIONAL SUPPORT.

A.S. Atkins[1], R.N. Singh[2], A. Barkhordarian[1] and A.H. Zadeh[2]

[1]Department of Mining Engineering, North Staffordshire Polytechnic, College Road, Stoke on Trent, Staffordshire, ST4 2DE, England.

[2]Department of Mining Engineering, University of Nottingham, University Park, Nottingham, NG7 2RD, England.

SUMMARY

The coal mining industry in the U.K poses great environmental concern over the disposal of extraneous dirt which is projected over the next two decades to exceed 1100 million tonnes. This size of operation is expected to present problems in obtaining planning permission for both existing and new mines. Currently 12 - 14 percent of the Run of Mine (R.O.M) output consists of fine liquid tailings containing 45 - 80 % moisture, which may present considerable environmental and geotechnical disposal problems. In some countries this proportion of fines can be as high as 38 percent of the ROM output through intensive mechanization and materials handling system, presenting both technical and financial problems of waste disposal.

The paper describes a laboratory investigation on liquid tailings from two large mines in the United Kingdom over complete weekly mining and coal preparation cycle. The results of physical, chemical and geotechnical analysis of tailings are presented together with their implications to the pumpability characteristics for underground disposal.

The paper considers the engineering characteristics of tailings for the design of tailings transport and utilization systems and presents three alternative schemes. Further considerations are given to the pressure/quantity characteristics and to the selection of cost effective pumping systems for tailings disposal to underground workings for backfilling operation.

INTRODUCTION

In the United Kingdom, the production of waste is an inevitable consequence of coal mining of thin and poor quality coal seams by mechanised techniques and waste support by caving methods (ref. 1). In 1920, the amount of dirt produced was some 5% of the saleable production which gradually increased to 61% in 1980. This trend is now continuing and over the past 6 years, some 95% of dirt produced is brought to the surface and disposed of in surface tips and lagoons presenting severe surface environmental problems. In order to reverse the trend of increasing dirt disposal at the surface, it is logical to utilize colliery dirt for ground support purpose. This paper examines some Aspects of underground transportation of liquid tailings in pipelines.

WASTE MANAGEMENT IN THE BRITISH COAL MINES

Current Techniques:

In 1985/86 British Coal (BC) produced 87.6 million tonnes of coal from deep mines, of which 18.8 million tonnes were discarded as waste. The dirt produced from coal preparation plant is divided into two categories, coarse and fine discards. Fine discards commonly known as flotation tailings, liquid tailings or simply as tailings are mainly the by-product of the cleaning of fine coal in froth flotation cells. The maximum particle size of tailings is normally 1.00 mm and about 30-60% is below 63 μm suspended in 50%-80% or more water.

The traditional method of colliery waste disposal at inland collieries is the formation of local tips with coarse dirt. However, liquid tailings are customarily disposed off in lagoons where coarse dirt is used to construct lagoon walls to contain pumped fine tailings. Mines located close to the sea sometimes dispose of the waste material into the sea (North East Area). In 1978/79, 3.5 million tonnes of spoil were tipped onto beaches and 1.5 million tonnes into inshores waters (ref. 2), (ref. 3). The visual impact of the present lagoon system for disposal of tailings has caused vociferous opposition from the general public. In spite of stringent factor of safety governing the stability of these lagoons it is very difficult to obtain planning permission in cases where there could be a remote possibility of loss of life or property. The growing awareness of environmental matters on disposal of tailings by the general public has been reflected in the increasing number of conditions attached to planning permissions for such methods of disposal. Planning permision for tailings disposal in a lagoon took almost eight years recently at a modern mine in Staffordshire, after a number of public enquires (ref. 4).

Although in some cases it is still possible to obtain planning permision, the current policy of tailings disposal, to comply with environmental concern, is by dewatering tailings using deep cone thickeners, solid bowl centrifuges followed by cement stabilisation or filteration methods. The stabilised tailings are disposed of at the surface at a cost of approximately £5.00 to £6.00/tonne of dry solids treated (ref. 5).

Alternative Proposals

An alternative method is to utilise tailings in mining activites for the formation of face end packs, solid stowing and backfilling of arches and roadway supports. The method of achieving this objective is to mix tailings with pulverised fuel ash (P.F.A), a by product from power stations burning coal, ordinary portland cement (O.P.C) and bentonite to form monolithic packs. The long term objective is to dispose the tailings for solid stowing and also to use it as a carrying media for the underground disposal of coarse discards (ref. 1).

RESEARCH OBJECTIVES

In order to design a tailings disposal system, the significant physical characteristics of tailings which will actively influence the handling, disposal and utilisation of this materials need to be examined. The range of properties together with their limits of variability should be determined with the following objectives:-

o The degree of flexibility which a disposal or utilisation system will
   need to be used in order to cope with the waste as it arises.

o The extremes of the range of variability may indicate where materials could
   rise which require special treatment.

The purpose of this paper is to investigate the physical properties of tailings from two long term, representative sites which will influence the selection of pumping equipment for the underground transportation of liquid tailings.

RESEARCH TECHNIQUES AND PROGRAMME

Testing Programme

Tailings samples were collected from two large collieries located in Staffordshire and Nottinghamshire coalfields over a one week period, from Monday to Friday. Wednesday, being mid week, was also chosen to collect samples on three shifts basis in the case of site 1.

The samples of tailings from site 2 were collected over a week from Monday to Thursday, and on Friday the samples were collected both on the morning and the afternoon shift respectively.

The methods of physical testing were conducted in accordance with British Standards 1377:1975 (ref. 6), methods of Tests For Soils For Civil Engineering Purposes, and according to B.C. application of British Standards to the testing of colliery spoil (ref.7).

Selection of samples

The samples of tailings from the under flow of the thickener were collected so as to be as representative as possible and a valve situated under the thickener was used for collecting all samples used in the test work for comparative purposes.

Particle size distribution

Particle size distribution of tailings investigated are shown in Figures 1-4. The determination of particle size distribution was made by conventional dry and wet screening, and the Malvern 2600/3600 particle size analyser.

Atterburge limit

Liquid, plastic limit and plasticity index were conducted in accordance to British Standards 1377:1975, methods of Test For Soils For Civil Engineering Purposes (ref. 6).

Void Ratio (e)

Void ratio of tailings was calculated using the following formulae (ref. 8)

$$Sc = 10*(100-M) / M + (100 \quad M)/SGs \tag{1}$$

$$Sc = (100)/(1+e) \tag{2}$$

$$e = (100/Sc) - 1 \tag{3}$$

Where Sc : Solids Concentration; M : Percentage of Moisture; SGs : Specific Gravity of Solids; e : Voids Ratio

Proximate Analysis of Tailings

Proximate analyser Mac 400 was used for analysing the tailings samples in terms of moisture, volatile matter, ash and fixed carbon. Mineral matter of tailings was calculated using Parr formula (ref. 9) as follows:-

$$M : 1.08 A + 0.55 S \tag{4}$$

Where

A : Ash percentage; S : Sulphur percentage.

The percentage of sulphur in the sample was determined by using LECO sulphur determinator SC 132, in which SO emmision is measured by a solid state infrared detector.

RESULTS AND DISCUSSIONS
Physical and Chemical Analysis

Figures 1 and 2 show the variation of particle size distribution of tailings from both collieries over a one week period and Figures 3 and 4 illustrate the variation of particle size distribution of tailings over a one day cycle. The particle size distribution curves Figures 1 to 4 for each sample show a relatively well graded material varying between 0.001 - 1.00 mm in size distribution, with the exception of samples A and D from site 1 which showed a much lower percentage of fine particles.

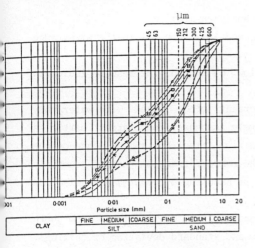

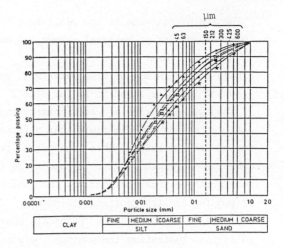

Fig.1. Particle size distribution of
tailings from site 1 on a week basis.
Samples A, B, C, D, E, F and G.

Fig.2. Particle size distribution of
tailings from site 2 on a week basis.
Samples B, C, D, E, F and G.

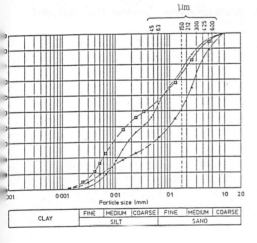

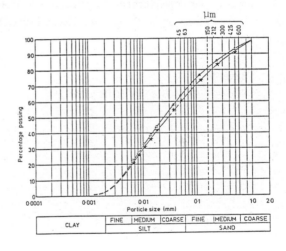

Fig.3. Particle size distribution of
tailings from site 1 on a three shift
basis. Samples C, D and E.

Fig.4. Particle size distribution of
tailings from site 2 on a two shift basis.
Samples F and G.

SYMBOLS

✳ A	▲ D	△ F	
■ B	● E	★ G	
☐ C			

The main features of the particle size analyses are:
o Minor amounts of material above 1.0 mm, and less than 17% in the size range 0.5mm to 1.0mm.
o High concentration of fine materials below 45 micron, the results indicating 30-55% for site 1 and 58-72 for site 2.

The fine particles are particularly important, since mixed with water they act as a heavy media fluid that effectively suspends the larger particles during transportation. This fine size fraction also has an influence on the cohesive strength (ref. 10) and it is the coarser fraction that is generally more abrasive thus influencing the choice of handling procedures and selection of pumping and transporting equipment.

Table 1 summarizes the physical characteristics of tailings from both collieries indicating the maximum , minimum and standard deviation of tailings. The results illustrate that there is only a small variation in the physical characteristics of tailings from different collieries. This is important in the design of the underground transportation system. Specific gravity values shown in Table 1 are relatively uniform for each sample. The lower values of specific gravity indicated in the table can be attributed to the presence of coal in the samples as demonstrated by a carbon content varying from 11% to 23%.

Equations 5 to 8 indicate the empirical relationships between SGs, SGt and moisture content.

Site 1

$$SGt = 1.74 - 8.17 * 10^{-3}M \qquad\qquad r = 0.84 \qquad\qquad\qquad (5)$$

$$SGt = 1.026 + 0.141\ SGs \qquad\qquad r = 0.46 \qquad\qquad\qquad (6)$$

Site 2

$$SGt = 1.59 - 5.47 * 10^{-3}M \qquad\qquad r = 0.97 \qquad\qquad\qquad (7)$$

$$SGt = 1.86 - 0.29\ SGs \qquad\qquad r = 0.82 \qquad\qquad\qquad (8)$$

Where SGt : specific gravity of tailing; M : moisture content; SGs : specific gravity of solids; r : Correlation coefficient.

The correlation coefficient between specific gravity of tailings and moisture contents for tailings from site 1 indicates that there is limited correlation between these two variables. In the case of site 2 the variables have a closer correlation. Also, the correlation coefficient between SGt and SGs for site 1 indicates that there is a limited inter relationship between these variables as compared to the same variables from site 2. This variation

407

TABLE 1.
Maximum, minimum and standart deviation values of physical characteristics of tailings from sites 1 and 2.

Physical Properties	Symbol	Site 1			Site 2		
		Maximum	Minimum	S.D.	Maximum	Minimum	S.D.
Percentage of moisture	M	58.10	47.42	3.42	65.87	54.90	4.13
Specific gravity of solids	SGs	2.14	1.85	0.11	2.18	2.03	0.07
Specific gravity of tailings	SGt	1.35	1.26	0.03	1.29	1.23	0.02
Solids fraction by mass	Sfm	0.53	0.42	–	0.45	0.34	–
Liquid fraction by mass	Lfm	0.58	0.47	–	0.66	0.55	–
Solids fraction by volume	Sfv	0.36	0.26	–	0.28	0.19	–
Liquid fraction by volume	Lfv	0.74	0.64	–	0.81	0.72	–
Specific gravity of liquid	SGl	1.00	1.00	–	1.00	1.00	–
Percentage of mineral matter	Mm	78.62	60.62	6.51	75.97	71.60	1.65
Percentage of moisture of air dried sample	Ma	1.45	1.28	0.76	1.45	1.28	0.06
Percentage of volatile	V	19.03	14.97	1.71	16.34	14.83	0.55
Percentage of ash	A	72.44	55.61	6.42	69.58	65.41	1.58
Percentage of fixed carbon	Cf	23.44	11.61	4.41	17.54	12.86	1.63
Percentage of sulphur	S	1.67	0.63	0.41	1.75	1.23	0.19
Plastic limit %	PL	33.00	29.13	1.33	31.46	28.83	0.92
Liquid limit %	LL	37.00	33.00	1.63	39.00	36.50	0.97
Plasticity index %	PI	7.51	1.50	2.29	10.17	5.04	1.66
Void ratio	e	1.39	0.90	0.16	1.93	1.22	0.27

could be attributed to the high moisture content of tailings from site 2 as compared to site 1. The degree of saturation (moisture content) is one of the most important factors governing the pumpability of mine tailings. The analysis presented in Table 1 indicates that moisture content of the liquid tailings varied between 46 to 58% for tailings from site 1 and between 54 to 66% for site 2. The variation in moisture will control the quantity of material to be pumped in relation to the pre-determined compressive strength. All samples from both collieries had moisture contents well above their liquid limit values given in Table 1. The mineral matter from the analysis indicates minimal variation and consequentely has little effect on the pumping characteristics of liquid tailings. However the chemical properties of the tailings are important for stablization.

The percentage of solids in tailings varies from 34% - 53% and the ease of pumping of tailings is determined from solids concentration and void ratios. Solid concentration is used as an inverse measure of ease of pumping of tailings. Voids ratio, however, reflects directly on the ease of pumping; as voids ratio increases so the ease of pumping increases.

DESIGN CONSIDERATION FOR TAILINGS TRANSPORT AND UTILIZATION SYSTEM
Selection of a disposal system

To select a system for preparation and transportation of tailings materials for the fill from the surface of the mine to the underground disposal point, a number of logical alternatives methods were considered. It has been established that the following materials are required to form a monolithic pack of pre-determined compressive strength:

o Tailings

o Pulverised fuel ash

o Bentonite

o Ordinary Portland Cement

o Accelerators

Mixing Arrangements

The major constraints for the design of the mixing and transporting systems are as follows:

1 Slurry should remain in suspension throughout in the pipe line between the coal preparation plant and the underground disposal point.

2 Pipe blockages must be eliminated.

3 Wear to the system kept to a minimum.

4 Automatically respond to the change in the properties of the tailings, (moisture content and pulp density)

5 Burden to the colliery transport system is minimal.

Fifteen alternative systems from different combinations of material were studied, six were shortlisted and consequently three systems were chosen on the basis of practicability and cost effectiveness. These systems are as follows:

System 1: Tailings and bentonite mixed at the surface. Cement, P.F.A and accelerators transported separately underground, and mixed.

System 2: Tailings and P.F.A mixed at the surface. Cement, bentonite and accelerators transported separately underground, and mixed.

System 3: Tailings, bentonite and P.F.A mixed at the surface. Cement and accelerators transported separately underground, and mixed.

Transportation Arrangements

Methods of transportation to underground for the above three systems were considered and 44 sub systems were generated from which six were chosen on the basis of practicability, cost effectiveness and system constraints, as illustrated in Figure 5. Further examination of the sub systems revealed that sub systems B,F and D would not fully benefit from utilization of hydraulic transportation. For a system to operate economically, maximum solid concentration should be transported hydraulically. For this reason systems A,C and E were choosen for further studies as illustrated in Figures 6 and 7.

In order for the pack materials to achieve maximum compressive strength in the desired time, the accelerators should be mixed thoroughly with the pack material. Since the pack material will set quickly in the presence of accelerators, in order to eliminate pipe blockages and difficulty in operating the system it is essential to have accelerators mixed at the discharge point via a spray gun which will give the maximum circulation of the material in the spray gun.

Pump selection

The following five steps are the basis for the selection of pumps (ref.11, ref.12, ref.13, ref.14).

o Plan of the pump and piping layout
o Determining the capacity
o Determining total head
o Study of liquid condition
o Choosing class and type of pump

From the results of calculations for a typical simulated mine layout the following data was identified:

Required Capacity	9.0-10.0 m/hr
Available head	15 - 20 m
Maximum particle size	1.0 mm
Solid fraction by mass	0.35-0.53

410

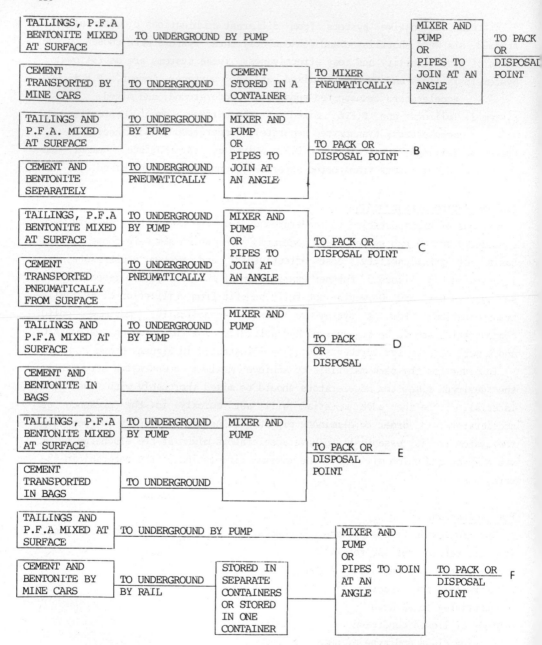

Fig. 5. Preliminary selection of alternative transportation arrangements.

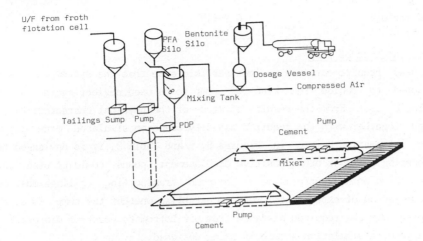

Fig. 6. Proposed delivery system using tailings, P.F.A. and bentonite mixed at surface, cement transported in bags.

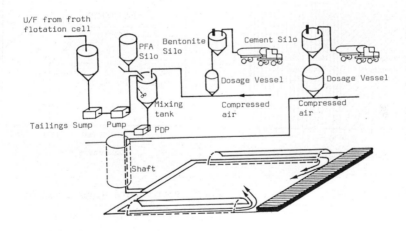

Fig. 7. Proposed delivery system using tailings, P.F.A. and bentonite mixed at surface, cement transported pneumatically.

Void ratio   0.9-1.9

Choosing the type of pump

The head requirements and maximum particle size that the system must handle may be used to decide the kind of pump to be used neglecting the pump and installation cost. From the results of study of the physical characteristics of tailings together with the required capacity based on simulated mine layout, and the information given in Figures 8, 9 and Table 2, it is envisaged that reciprocating pumps or other positive displacement pumps could be used, but a final decision on the performance of the pump transporting tailings must come after completion of experimental work to establish whether the pump is capable of pumping over the required distance from pit bottom to point of disposal. For this purpose, a simulation scheme is being designed.

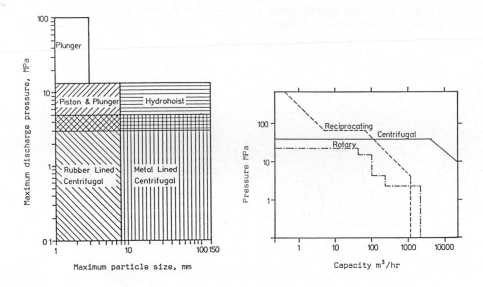

Fig.8. Pump capacity.
(after Gandhi, 1980)

Fig.9. Approximate upper limit of pressure and capacity by pump type.
(after Karassik et al, 1986)

Figure 8 and 9 show the general types of pump which are suitable. As the figure show, there are conditions for which only one type of pump is applicable, and there are areas where more than one pump type can be used.

TABLE 2
Flow and pressure capacities of reciprocating and centrifugal pumps.

pump type	maximum flow m/hr	maximum pressure MPa	maximum particle mm	mechanical efficiency %
plunger	225	41	8	85-90
piston	890	15	8	85-90
centrifugal	5700	55	150	40-75
diaphragm	325	20.5	4	45-60

CONCLUSION

The most important factors affecting pumpability of coal mine tailings are:

o Particle size distribution

o Moisture content

o Solids concentration

The results of experimental work indicate that there are only small variations in the physical characteristics of the tailings from the two collieries.

The most economical system for transportation of the packing material to underground is to mix pulverised fuel ash and tailings on the surface of mine and hydraulic transport the mixture to the underground disposal area. It is expected that the fine particles of tailings will keep the mixture in suspension and will prevent undue abrasive wear of the pipes and the pumping equipment.

It is envisaged that reciprocating pumps or positive displacement pumps are the obvious choice, but further experimental work is needed to establish the type of pump which will be able to handle tailings and deliver the required quantity over the required distance.

ACKNOWLEDGEMENTS

The authors wish to record their thanks to the Science and Engineering Research Council and British Coal for jointly financing the project. Thanks are also due to various managers, engineers, mineral processors and surveyers within British Coal for their co operation and assistance with the project.

Thanks are also extended to Mr P.Proudlove of North Staffordshire Polytechnic for his assistance in the illustration of this paper.

414

REFERENCES

1 A.S.Atkins, D.Hughes, D.Parkin, R.N.Singh, Utilization of colliery tailings in mining activities, Symposium on The Reclamation, Treatment and Utilization of Coal Mining Wastes, Durham, England September 1984, Paper 18 pp. 18.1 - 18.11 Published by British Coal.

2 Commision on Energy and Environment (1981), Coal and the environment, U.K. Her Majesty Stationary Office, 1981, pp. 58-69.

3 J.D. Blelloch, Waste disposal and environment, Colliery Guardian, August 1983, pp. 392-401.

4 Local Public Inquiry, Appeal by N.C.B in respect of land at waste farm Silverdale colliery Newcastle, Staffordshire County Council, 23rd July 1985, pp.1-33.

5 D.W. Brown, Disposal of colliery tailings, Colliery Guardian, February 1986, pp.68-70.

6 British Standards 1377:1975 , Methods of tests for soils for civil engineering purposes, pp. 13-46.

7 National Coal Board, Application of British standarts 1377:1967 to the testing of colliery spoil, Technical Memorandum Issued on Soil Mechanics Testing by The Joint Working Party, May 1971, pp 2-63.

8 National Coal Board, Spoil heaps and lagoons, Technical Handbook, Second Draft, September 1970, pp. 1-232.

9 Analysts' Handbook, Ultimate analysis of coal, National Coal Board, April 1952, pp.1-13.

10 L. Keren, S. Kainian, Influence of tailings particles on physical and mechanical properties of fill, Proceedings of International Symposium on Mining With Backfill, Lulea, 7 - June 1983, pp. 21-30.

11 R.Vanderpan, Proper pump selection for coal preparation plants, World Coal May, June 1982, pp. 54-56.

12 T.L Thompson, R.J Frey, N.T Cowper, Slurry pumps a survey, Hydrotransport 2, BHRA Fluid Engineering Paper H1, September 1972, pp 1 24.

13 J.W Crisswell, Selection and operation of slurry pumps, World coal, April 1983, pp. 47-53.

14 R. Steffek, Centrifugal pump selection, Plant Engineering November 25, 1981, pp. 109-111.

15 R.L. Gandhi, Evaluation of slurry pumps, Proceeding of 5th International Technical Conference on Slurry Transportation, Washington, March 1980, ISBN 0932066054, pp. 267-275 .

16 I.J. Karassik, W.C Krutzsch, W.H Fraser, J.P Messina, Pump handbook, Second Edition, ISBN 0070333025 McGraw Hill Book Company, 1986, pp. 1-5, 3-126.

Reclamation, Treatment and Utilization of Coal Mining Wastes, edited by A.K.M. Rainbow
Elsevier Science Publishers B.V., Amsterdam, 1987 — Printed in The Netherlands

IMPROVED ROCK PASTE: A SLOW HARDENING BULK FILL BASED ON COLLIERY SPOIL,
PULVERISED FUEL ASH AND LIME

K W COLE AND J FIGG
Ove Arup & Partners, 13 Fitzroy Street, London, W1P 6BQ

INTRODUCTION

In the West Midlands, in an area known as the Black Country, there
have in the past been many limestone mines, in addition to mines for
coal and ironstone. The underground cavities that are the remains of
limestone mines are often large. They may 5 to 8m high, with the roof
supported by pillars 5 to 15m square and 5 to 25m apart. The
majority of the mines that are still "open" (not collapsed) lie within
100m of the ground surface. The volume of "open" mines is believed to
exceed 5 million cubic metres. Many of the mines are below the
natural groundwater surface, and they are consequently waterfilled.

Collapse of the cavities, either through roof or pillar failure,
can produce very large subsidences of the ground surface. The
processes of degradation, leading towards collapse, of the rocks within
and surrounding the mines are not completely understood. It is however
certain that collapse "events" will continue to cause surface
subsidences above the remaining "open" mines. It is also likely that
each event will be small in extent compared to the extent of the mine
involved. It is worth noting that although the majority of mines
deeper than 100m appear to be fully collapsed, most collapses would
seem to have taken place without record of surface effects.

The particular time and location (the absolute likelihood) of a
collapse "event" causing surface subsidence above an "open" mine cannot
be predicted. However, the relative likelihood has been predicted in a
general way, from an assessment of mine age and depth, and the rock type
above the mine, that sporadic "events" are more likely above some mines
than others, Ref 1. The predicted relative likelihood has been found
upon intensive investigation of the mines to have good correlation with
the disclosed condition, particularly with the degree of deterioration
of the roof rocks and the slenderness of the pillars in the cavities.
The order of priority for treating the mines to remove the threat of
subsidence can thus be based upon technical justification.

Notwithstanding the merit of the technical justification, the
decision to undertake treatment of any mine has to take into account the
social and financial benefits, and weigh these against the risks
(particularly of not treating the mine) and the costs. These factors
inevitably affect the order of priority for treatment. The final
decision on treatment is a balance of social pressures and technical
need against resources.

The engineer can help tip the balance in favour of the decision to
undertake treatment, if the cost of treatment can be kept low. It was
the need to devise a cheap and effective bulk filling material, suitable
for filling large submerged cavities in a manner that would ensure that
surface subsidence was eliminated, that lead to the development of rock
paste.

ROCK PASTE

The original proposal for rock paste was made by Dr W H Ward,
(Consultant to Ove Arup & Partners) on the limestone mine problems,
Ref 2. He envisaged that colliery waste materials (colliery spoil),
abundant within economical travelling distances of the abandoned
limestone mines in the West Midlands, could be mixed with water to
make a paste-like material which he called "Rock Paste". The Rock
Paste should be capable of being pumped through a pipeline, and in the
mine cavities it should flow for large distances. By arranging
"injection holes" into the mine cavities to be at suitable spacings
the cavities would be filled.

Colliery spoil is the only material cheaply available in sufficient
quantities to fill the known "open" limestone mines, some of which have
a volume exceeding 0.5 million cubic metres. However, as colliery
spoil is a material which is discarded in the process of winning coal,
its nature is dependent upon the precise sequence of rocks (sandstones,
siltstones and mudstones) adjacent to the coal, the manner in which the
coal is separated out, and the way in which the spoil is treated and
dumped.

Fortunately, the manner in which the spoil is handled tends towards
ensuring a degree of uniformity. By the time it has been separated
from the coal, transported to the tip, dumped, spread and subjected to
a period of weathering the spoil degrades, and the resulting material
contains substantial amounts of sand, silt and clay sized particles.
The more resistant particles remain as gravel and cobble sized
particles randomly distributed in the spoil.

In some parts of spoil tips there are quite large amounts of coal
in the spoil, and the spoil may be contaminated with natural pollutants
such as chlorides and sulphates. There are also likely to be other
pollutants in the form of discarded materials from the construction
and running of the mine.

Some whole tips are burnt, while in others burning may have
affected only small pockets of spoil. The burnt colliery waste is
generally fused and abrasive, and not suitable for making Rock Paste.

It has been shown that the majority of colliery spoils, obtained
from tips that contain unburnt spoil, have gradings which fall within
the envelope shown in Figure 1. By trial it has been found that the
processes of handling the spoil, mixing it with water and pumping it
through a pipeline, tend to ensure that, even though the freshly dug
colliery spoil may have a "harsh" grading, lying below the lower limit
of the grading envelope, the resulting Rock Paste has a grading that
falls within the grading envelope.

FEATURES OF ROCK PASTE

The essential features of Rock Paste are that it shall

(i)   behave as a cohesive body

(ii)   spread as a plastic mass

(iii)   not segregate out and mix with water already in mines

417

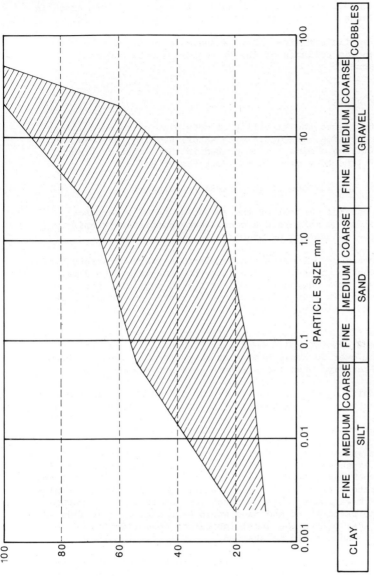

Figure 1. Partical size distribution "envelope" for suitable colliery spoil.

(iv)   not consolidate to such an extent that the gap left above is too large to ensure that proper roof support is given.

(v)   have sufficient strength to ensure stability of mines by preventing collapse of the surrounding rocks and mine pillars.

PLACEMENT OF ROCK PASTE

The trials to assess the viability of obtaining suitable colliery spoil in large quantities, mixing it with water, pumping the resulting Rock Paste, and filling a mine are described in the companion paper "The infilling of limestone mines using Rock Paste" Ref 3.

The essential conclusions of the trials were that:

(i)   With careful control of the amount of added water, Rock Paste can be made with a consistency sufficiently uniform for the concept to be practical.

(ii)   For practical purposes the permissible range of strength of the paste before injection into the mine is 1.5 to 3kPa. Below 1.5kPa segregation of the coarse particles from the fines matrix occurs, and the paste behaviour becomes more akin to a slurry.

Above 3kPa the paste is too stiff to be pumped continuously; a large concrete placement pump "laboured" to maintain an output over 50 cubic metres an hours through 200m of 200mm diameter pipe when the strength rose above 3kPa.

In addition, strong paste does not spread as far in the mine as weaker paste, as the distance it spreads is proportional to the paste strength. Injection holes would have to be at closer centres for the stronger paste.

(iii)   The measured amount of self weight consolidation was larger than expected, amounting to nearly 20% of the thickness of the Rock Paste deposited. The main reason was that all the laboratory tests were made on small samples. These samples would have exuded water through very short drainage paths during preparation and thus would have already consolidated 5 to 10%, before the usual laboratory consolidation measurements could be started.

In the mine cavities, the build up in thickness of the Rock Paste takes place at the point of injection, and the paste spreads outwards as a plastic mass; the opportunity to exude water rapidly through short drainage paths therefore does not occur. As the permeability of Rock Paste is low, most of the full amount of consolidation takes place after the cavities are filled.

ROCK PASTE STRENGTH

There are several ways in which Rock Paste may provide support to mine cavities and prevent collapse:

(i)   by completely filling the cavities and acting as a "cushion" to directly support the mines roof. This is unlikely to be possible, as it has been shown that the paste may undergo quite large self weight consolidation and thus eventually not completely fill the cavities.

(ii)   by supporting the layers of rock as they separate from the roof, and thus preventing further layers from separating. In theory this method would succeed with quite low paste strength of around 2 to 5kPa, provided that the fallen rock "bulked" sufficiently to fill the gap between the fallen rocks and the roof. In reality, with well-bedded rocks such as the Ludlow Shales and Nodular Beds (which overlie the two main limestone seams mined) this might not occur. Some of the layers could "pancake" with low bulking, or could disintegrate into a gravel-like mass, and "choking" therefore might not occur.

Further consideration of the likely mechanisms of behaviour lead to the concept that fallen roof rocks would aggregate to form columns in the rock paste, similar to stone columns used to support buildings on weak ground. It can be shown that such columns cannot expand and admit more fallen rock once they have formed to the full depth of the paste, provided the strength of the surrounding Rock Paste has a certain minimum value. This value depends upon the depth of the mine cavity; for a typical cavity 5m deep it is 12kPa, and for a typical cavity 8m deep it is 20kPa.

It should be noted that this "stone column" theoretical approach is regarded as very conservative. A gap up to 10% of the height of the mine cavity above the top of the paste has little adverse effect on the support provided.

(iii)   by providing lateral support to pillars, and by encapsulating them retard deterioration. It is not possible to quantify this effect, but it is evident that the stronger the paste, the greater is the support provided.

Laboratory and field tests on "basic" Rock Paste, made with colliery spoil and water only, have shown that some gain in strength will be obtained as the paste consolidates, amounting to a doubling or trebling of the placement strength to 5 to 8kPa.   For mines with cavities 5 to 8m deep for which the "stone column" theoretical approach to long term stability indicates the minimum strength should be 12kPa and 20kPa respectively, the possible maximum strength of "basic" Rock Paste is therefore too low.

If time permitted, one way of treating a mine would be to fill it with "basic" Rock Paste, allow the paste to consolidate and then inject a grout with a strength greater than (say) 50kPa into the cavity above the paste.   Some further consolidation would take place, but the essential elements of the support mechanism would have been provided. However, the delay of several years to allow consolidation to take place, followed by an injection of grout, would both add considerably to the cost.

Development of an "improved" Rock Paste, having the characteristics of higher strength and reduced consolidation, and requiring only a single stage of injection with no further treatment is thus seen to be an attractive goal.

IMPROVED ROCK PASTE

The characteristics of "improved" Rock Paste were readily defined from the results of the field treats and theoretical work. They are that it shall:

(i)   behave as cohesive body with a strength not greater than 3kPa during placement. This will allow it to spread as a plastic mass on a level floor to about 35m in a 5m high cavity, and about 85m in an 8m high cavity. If the cavity floor slopes, the distances the paste will flow downhill will be larger, depending upon the angle of slope; "infinite flow" (sheet flow, theoretically to infinity) will be achieved at angles of slope as small as 4°.

Because the mine cavities are large, placement from an injection hole may continue for many days; a time period of 20 days (the "delay period") before the strength exceeds 3kPa should be available, and a "delay period" of up to 30 days is desirable.

(ii) consolidate much less than "basic" Rock Paste, preferably not more than 5% of the height of the cavity filled.

(iii) gain in strength with time after reaching its final placement position, so that, preferably within a year, but certainly within 3 years, it has strength exceeding the minimum strength determined from the "stone column" theoretical approach.

It was envisaged that the necessary increase in strength could be obtained in two possible ways

(i)   by addition of a small proportion of Portland cement with a suitable long-acting retarder, or

(ii) from a slow hydrated lime-clay (pozzolanic)reaction. A pozzolana is defined as "a material which is capable of reacting with lime in the presence of water at ordinary temperature to produce cementitious products" (Ref. 4).

The Portland cement idea was rejected because of the exceptionally long time of set-retardation required to achieve the desired thirty day unchanged workability, and also because of the long term risk of deleterious sulphation of the hardened rock paste from sulphates derived from the colliery spoil.

For the lime-pozzolana reaction it is necessary that a suitable finely-divided reactive material (aluminosilicate) shall be present in the rock paste. The colliery waste naturally contains clay minerals, but the actual clay mineralogy and amount varies considerably. Therefore it was considered prudent to add a proportion of a reliable pozzolana to the rock paste to ensure that a long-term strength gain would always be obtained. The obvious choice material was pulverised fuel ash (pfa), since this has a good record of satisfactory use as a pozzolana, and is readily obtainable; its inclusion in improved rock paste has the merit of providing an additional use for an industrial waste product.

MIX PROPORTIONS

The opportunity was taken during the Dudley infilling trial (Ref 3) of experimenting with hydrated lime-pfa-colliery spoil mixes to determine suitable mix proportions and evaluate workability.

A problem was encountered in that even the smallest proportion of lime addition, as little as 0.5% by weight of solids, produced rapid loss of workability. A series of researches and laboratory investigations were commenced with the aim of solving the problem quickly, without having to delay the next mine infilling project for which it was anticipated grant aid would be available in the next financial year.

RESEARCHES

Consultations with lime specialists in industry, and literature studies identified three possible mechanisms for the rapid loss of workability:

(i)   chemical reaction between water soluble materials in the lime and colliery spoil leading to formation of hydrated compounds with consequent loss of free water in the rock paste.

(ii)   base exchange between alkali metal clay minerals and calcium ions from the lime with reduction in the lubrication/slip properties of the clay component of the rock paste.

(iii)   removal of free water from the rock paste by the extremely hydrophilic lime leading to poor workability/pumpability

The mechanisms were investigated by concurrent experiments. Reactions of the type considered in (i) have previously caused problems with impure (semi-hydraulic) limes where sulphates have reacted with aluminate phases to form ettringite-like minerals. Ettringite itself includes 32 molecules of water of crystallisation in each molecule. In the case of rock paste the colliery spoil contains both alluminates and sulphates and if lime is added the possibility of ettringite formation is feasible.

Type (ii) base-exchange reactions can be inhibited or delayed by chemicals which are absorbed preferentially by the clay minerals, thus blocking access to calcium ions derived from the lime. A number of types of chemical were identified from literature searches, viz:

(a)   surfactants (anionic, cationic, neutral)
(b)   phenols and phenolic compounds
(c)   tannin
(d)   sodium chloride (common salt)
(e)   quaternary ammonium compounds
(f)   urea
(g)   lignosulphonates
(h)   hexametaphosphate

With regard to the attraction of water by lime (iii) it ought to be possible to nullify this effect by ensuring that the hydrated lime is equilibrated with adequate water before use (ie use lime putty rather than dry hydrated lime) and/or to adjust the mixing procedure to

ensure there is always sufficient free water to "lubricate"the paste
particles and maintain workability.

LABORATORY EXPERIMENTS

Chemical interactions

Samples of the colliery spoil, pulverised fuel ash and lime were
obtained and various mixtures were prepared and stored for periods up to
7 days before being dried at low temperature, sampled and ground for
X-ray diffraction analysis.

This procedure was adopted since only secondary crystalline salts
were to be determined and because comparison samples of all materials
were available. In such circumstances a semi-quantitative analysis was
adequate.

Inhibitor chemicals

These experiments were made by mixing small quantities of material
which were then stored in sealed polyethylene containers to simulate the
condition of the interior of a large volume of rock paste.

To provide maximum opportunity for the inhibitor chemical to react
preferentially with the clay minerals, the colliery spoil and pfa were
first mixed together, followed by addition of the inhibitor and then
finally the lime slurry was added (with additional water if necessary
to maintain workability).

As far as possible commercially available substances were used
rather than pure chemicals in order to avoid a second series of
experiments to assess the effects of impurities. Therefore the actual
chemicals tested were:

    (a)   "Teepol L" - anionic surfactant
           "Hyamine 1622" - cationic surfactant
           "Tween 80" - non-ionic surfactant
    (b)   "Jeyes fluid" - phenol/cresol
    (c)   Tannin - used as chemical from BDH Ltd
    (d)   Sodium chloride (common salt) - water-softener grade (iodide
           free)
    (e)   "Comfort" - quaternary ammonium salt
    (f)   Urea - bought as chemical from BDH Ltd
    (g)   "Conplast 211" - lignosulphonate-based water-reducing
           admixture (FOSROC Ltd)
    (h)   "Calgon" - sodium hexametaphosphate

The surfactants, quaternary ammonium compound, lignosulphonate and
tannin were used at concentrations of 0.01, 0.1, 0.2, 0.4 and 0.8 per
cent of the total mixing water. The other chemicals were used at
concentrations of 0.5, 1.0, 2.0 and 4.0 per cent of the mixing water.

Earlier work with rock paste (Ref. 1) had assessed the strength
properties using a plate penetrometer. However, the relatively low
strength values required a rather large plate (125mm diameter) which
was not suitable for the small scale laboratory tests. Since the
plate penetrometer actually measures the stress required to cause
indentation of the paste, it seemed logical to use other
indentation-type tests and preliminary experiments were made using an
(unweighted) cone from a bitumen viscosity penetrometer (Ref.5). Cone

indentation appeared to have a reasonable relationship to the plate penetrometer values as shown on Figure 2. Subsequent tests utilised a Geoner (Swedish) fall-cone penetrometer (Ref. 6)

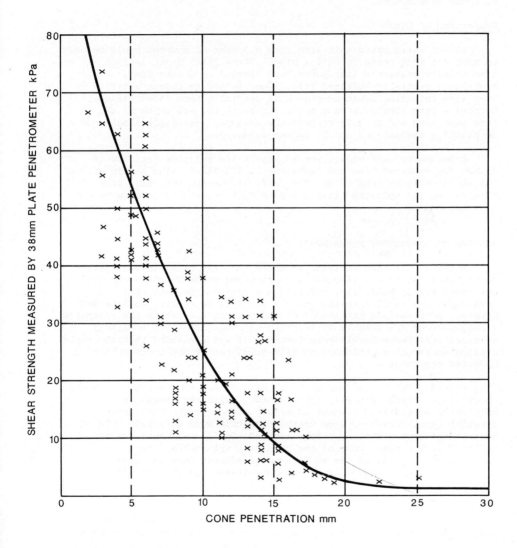

NB. Cone used was a bitumen viscosity measuring cone (Ref.5)

Figure 2.   Correlation of cone penetration with shear strength measured by 38mm diameter plate penetrometer.

Many of the experimental modified rock pastes mixes suffered from bleeding and settlement of the material, and therefore the settlement of the surface of each sample was also monitored.

To mimic the effects of flow likely to be experienced by the paste during actual mine infilling, some mixes were stirred each day for 20 days, before being allowed to stand for monitoring of long term strength development.

Comparison of Limes

It was anticipated that lime from a number of sources would be used to make the rock paste to fill a mine. These limes might include .commercially available lime putty, dry (bagged) hydrated lime, by-product partially-hydrated lime, lime treated to reduce activity or even lime releasing substances such as specially formulated glasses. Economics were likely to be a major factor in the selection of the lime component and in this respect the activity (available lime) must be balanced against the actual amount required.

Experiments were undertaken to assess the relative activity of Limbux dry hydrated lime and Peakstone "mixed lime" using Kaolin BP and the small-cylinder test of BS 1924 (Ref. 7). Tests were also made with alternative ICI hydrated limes and some Pilkington slow calcium release glasses.

RESULTS OF LABORATORY EXPERIMENTS

X-ray diffraction analyses revealed that the clay minerals present in the spoil materials from the Baddesley and Grove Colliery sources were essentially kaolinite, chlorite and mica/illite, and that Ironbridge pfa, whilst mostly glassy, did include quartz, mullite and gypsum. After mixing the amount of the mullite present in the rockpaste slowly reduced (presumably by reaction with the lime) but no likely acicular hydrated minerals were formed. It was concluded that the rapid stiffening effect experienced on site could not be due to formation of hydrated compounds.

Figures 3 to 5 show the results obtained for the various inhibitor chemicals. Several minerals appeared to provide the necessary short-term workability without adversely affecting the long term strength gain. However, some of the chemicals were effective only at quite large concentrations. This implied significant costs for the modified rock paste. Some of the chemicals were in any case unacceptable because of the known adverse effects on underground aquifers eg. Jeyes Fluid, urea, common salt and tannin. Of the remaining inhibitor chemicals the most effective at the lowest concentration appeared to be Comfort (quarternary ammonium compound).

The results of the comparison tests between Limbux and Peakstone limes showed that strength gains within 28 days were similar for both limes.

Longer-term tests were also commenced to provide additional data on the strength gain of the modified rock paste. The results of some of these long-term tests are presented in Figures 6 and 7. The strength developed and the rate of strength gain was found to depend upon both the origin of the colliery spoil, the method of preparation of the samples and the amount of lime added. Lime additions of 0.7% and 1% gave results within the desired range.

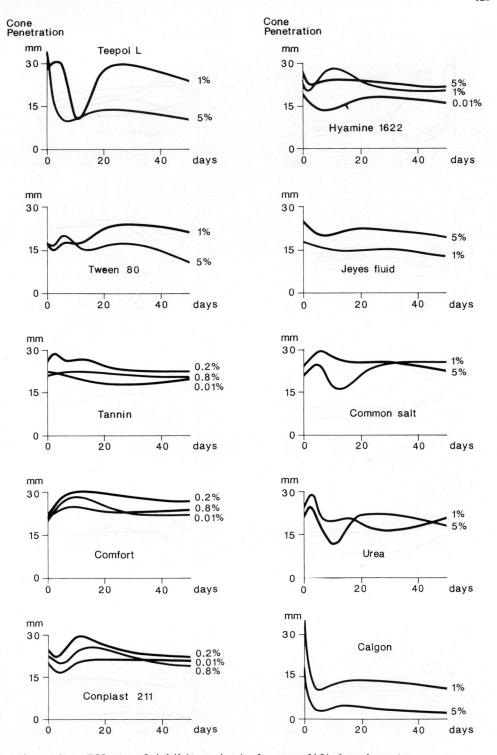

Figure 3. Effects of inhibitor chemicals on modified rock paste. Grove Colliery spoil with 4% pfa and 1% Peakstone mixed lime.

426

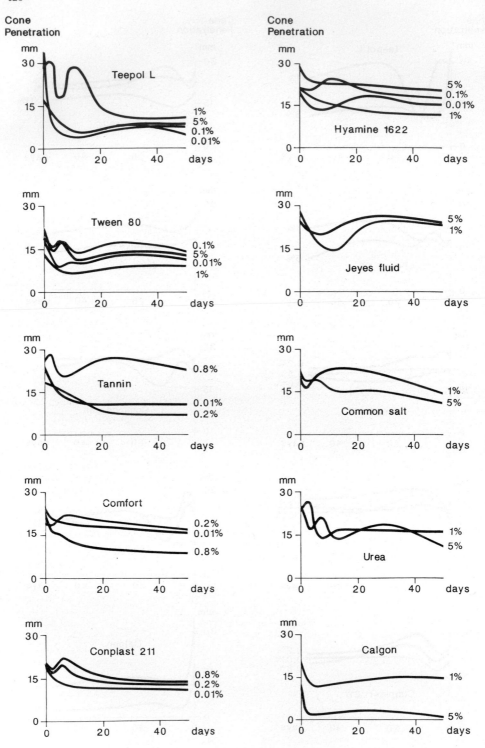

Figure 4.   Effects of inhibitor chemcials on modified rock paste.
Grove Colliery spoil with 4% pfa and 0.25% Peakstone mixed lime.

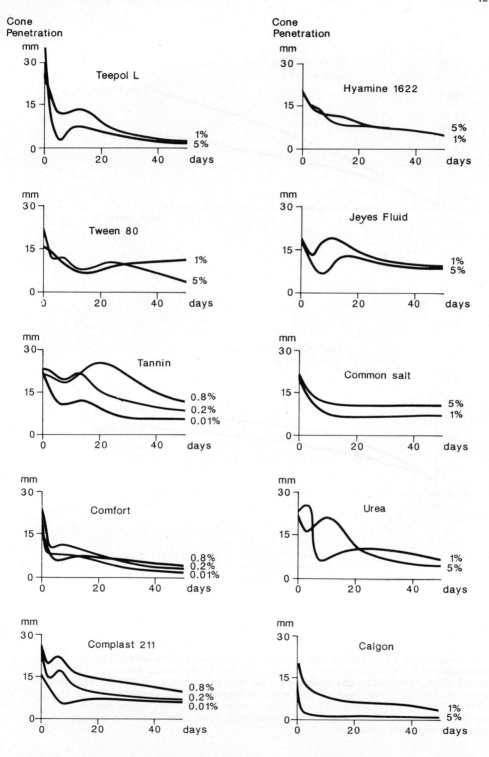

Figure 5.   Effects of inhibitor chemicals on modified rock paste.
Grove Colliery spoil with 4% pfa and 4% Peakstone mixed lime.

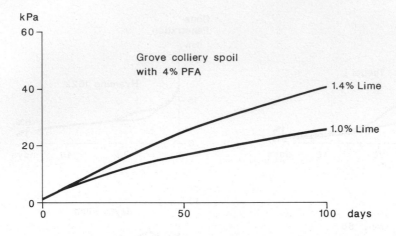

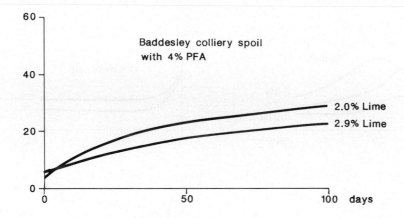

Figure 6.   Results of long-term strength tests on spoil samples with
            all particles greater than 2mm removed. Strengths measured
            with Geonor cone penetrometer.

Meanwhile experiments had also been in progress to establish the
optimum mixing procedure for the lime-pfa-colliery spoil system. It was
found that, providing the hydrated lime was prepared as a dilute slurry
with at least 75 per cent of the mixing water, before being added to the
already mixed colliery spoil and pfa, the modified rock paste would
retain its workability for an acceptability long term before a steady
gain in strength occurred.

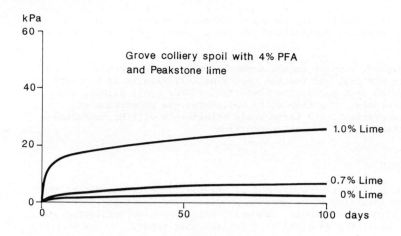

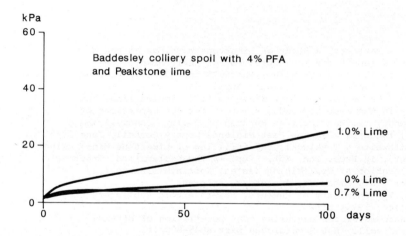

Figure 7.   Results of long-term strength tests on spoil with material
             greater than 40mm size removed. Strengths measured with
             Geonor cone penetrometer.

DISCUSSION OF RESULTS

    It appears that the incorporation of a small proportion of hydrated
lime together with pulverised fuel ash in suitable colliery spoil can
produce an "improved rock paste" which will remain workable for up to
30 days and which will eventually attain a strength well in excess of
the required 20 kPa at age greater than one year.

Only Comfort (quarternary ammonium compound) appears to be effective in sustaining workability, whilst allowing long-term strength to develop.

It is however, simpler to use a three-component rather than a four-component mixture, and for the next mine infilling it is proposed to use a modified rock paste of a lime-pfa-colliery spoil without any additional chemicals. For this to be successful the proportion of lime in mix is critical, and large scale experiments will be conducted on site to establish the correct proportion.

ACKNOWLEDGEMENTS

The authors thank the Department of the Environment, who direct the funding of the infilling works through Derelict Land Grant aid, and the Metropolitan Boroughs of Dudley, Sandwell, Walsall and Wolverhampton who promoted the works, for permission to publish this paper.

The authors also thank all of their collegues at Ove Arup & Partners who have been concerned in the project, both on site and in the office, for their help and encouragement.

REFERENCES

1. Ove Arup & Partners, a Study of Limestone Workings in the West Midlands, a report for the Department of Environment, the Metropolitan Borough of Dudley, Sandwell and Walsall, together with the West Midlands County Council, April 1983.
2. W H Ward, the use of colliery waste to fill flooded limestone workings in the West Midlands, a report for the Department of Environment, the Metropolitan Borough of Dudley, Sandwell and Walsall, together with the West Midlands County Council, June 1982.
3. P A Braithwaite & T Sklucki, the infilling of Limestone Mines using Rock Paste, in Proc. 2nd. Int. Symp. on the Reclamation, Treatment and Utilisation of Coal Mining Wastes, Nottingham University, 7-11 September 1987, Elsevier, Amsterdam, in press.
4. F M Lea, The Chemistry of Cements, Edward Arnold Publishers Ltd, 3rd Edition, 1970, p16.
5. Cone penetrometer for measuring cone penetration of bitumen stabilised soil. E L E catalogue part EL45-800/11.
6. Geonor A S Laboratory cone penetration apparatus 1974.
7. British Standards Institution methods of test for stabilised soils, London, BS 1926.

Reclamation, Treatment and Utilization of Coal Mining Wastes, edited by A.K.M. Rainbow
Elsevier Science Publishers B.V., Amsterdam, 1987 — Printed in The Netherlands   431

PRESENT SITUATION, PROBLEMS AND DEVELOPMENT OF UTILIZATION OF
COAL MINING WASTES IN CHINA

SUN MAO YUAN

Institute of Scientific and Technical Information Ministry of Coal
Industry, 21 Hepinglibeijie,P O BOX 1419 , Beijing China

ABSTRACT
    This paper detailed the resources, properties and utilization
of coal mining wastes in China.
    There are 10 types of fluidised bed boiler in use and many
achievements in research associated with such boiler have been
acquired. China will focus attention om the development of 35t/h
and 130t/h fluidised bed boilers in the near future. The two typical
boiler are introduced about their operation , problems and techni-
cal& economic indexes, one of them is at Guokeng power station,
other is at Di Dao power plant of Ji Xi mine area.
    Utilization of wastes in constructional material industry
include producing 1) cement, 2)brick, 3) lightweight aggregate,
4) concrete hollow block, 5) prottery & porcelain products and fire
resistant material, 6) others.
    Some people may be interested in the absorption and regeneration
method for refining sulphur from refuse in this article.
    Some results of scientic research in utilization of refuse are
presented: a) manufacturing new industrial stuffings (known asSAC)
b) 425# regular portland cement was produced in batches by lower
temperature baked technique, here clay is fully substituted by slag
of fluidised bed boiler, c) a solution of preventing wear of the
wall in boiler, d)"Hua Yi" efficient de-sulphuriring agent.
    Ministry of Coal Industry has made out a plan for utilization of
refuse from 1985 to 1990 in north-east, north China and east China
district, where is short of electric power.

PRODUCTION, PROPERTIES AND RESERVES OF REFUSE IN CHINA

    Coal is China,main energy resource,which occupies an important
place in our national economics. It is estimated that,by the end
of the century the situation that coal make up 70% or more of the
total energy consumption composition will not change. In 1985
China's output of raw coal was 870 million tons, ranking second
place of the world. In general, displacementof refuse in underground
mine makes up 10-15% of its raw coal production, individual mine
reach 30% . Refuse from coal proparation plant occupies about 15%
of raw coal to be processed. Up till now total amount of
accumulated rufuse over years have reached 1.2 billion tons or more
 , which occupied 50 thousand mu.At present the displacement of

refuse overs 100 million tons every years.

In 1980 the eight provinces or cities, including Hei Long Jiang
, Shan Xi, He Bei, Liao Ning, Shan Deng, An Hui, Bei Jing, Jiang
Su, possessed 660 coal mining waste heaps, amountingto 600 million
tons of refuse deposits . Of these heaps the refuse of Q-500kcal
/kg, Q= 500_1000 kcal/kg, Q= 1000_2000 kcal/kg and Q +2000kcal/kg
is respectively 33.9%, 35.1%, 26.5%, and 4.5%. Many of the 61
refuse heaps with higher Q value centralized in Xei Long Jiang,
Shan Xi, Gan Su and Guei Zhou provinces. At present the refuse
deposit in Liao Ning, He Bei and Shan Duang provinces make up 50
per cent of total refuse deposit in China.

The ash content in China's coal mining wastes is between 50 to
90%, an average of 70 - 80%. Calorific value (Q) is from dozens
kcal/kg to 4000 Kcal/kg, an average of 800 - 1500 Kcal/kg.
Calorific value in Ji Xi, Xi Shan and Yiao Jie areas is relatively
high. The sulphur content of refuse in state controlled collieries
is less than 1% in general, but in some areas, such as Nan Tong,
Liu Zhi and Lin Dong, the figure is at 8 - 16%. Aluminium content
in general is about 30%, but in some areas and collieries, e.g. in
Hai Buo Wan Area, Wan Zhung 1# Mine in Xin Long Colliery, Xiao Yu
Colliery in Shan Xi province, Nan Piao Area the value is more than
46%. The refuse in some regions contain rare elements, e.g. there
are 27 rare elements ascertained in the coal mining wastes from
Fu Shun Areas and AL_2Q, TiQ_2, Ga content exceed or approach the
industrial minigg boundary grade issued by the country, only above

50 m deep in the west of Xi Open Pit the reserve of above-
mentioned three elements in partings is respectivily 12.63, 2.26
and 2.33 million tons.

In 1985 the total displacement of washery tailings, that
suitable for fuel to fluidised bed combustion boiler , is about
30 million tons, Had used it to generate electricity, we should
have generate 8 billion KWHsof electricity , that corresponds to
70% of the total power consumption of state controlled collieries
and coal preparation plants in China.

The resource of refuse in China is rich , but its utilization
only occupied about 1/5 of annual displacement. The fact shows
that our country have great potentialities in the utilization of
refuse and that refuse heaps would be more and more like rolling
snow ball if we shouldn't do our better to solve the problem of
its utilization.

FLUIDISED BED COMBUSTION TECHNOLOGY FOR BURNING REFUSE

The research and spreaded usage of fluidised bed combustion
boiler was started from 1964 in China and in 1965 the first boiler
put into production at Shu Min Oil Company of Guang Dong Province.
At present there are about 2200 fluidised bed combustion boilers
in China and total vaporizing capacity more than 132000
vaporizang capacity of tons/hour. There are ten types of this
boiler, including 2, 4, 6, 6.5, 8, 10, 15, 20, 35, and
1 30 T/H. Wefocus our attention on the development of the 35T/H
and 130T/H fluidised bed combustion boilers.

The power plant owned by Gaokeng Colliery of Pingxing Area in
Jiang Xi Province has been equiped with three fluidised bed
combustion boilers (35 T/H) and three 6000 KW gas turbine
generators, with total installed capacity is 18000 KW. N.1 and
N.2 power generating sets were combined with the net work for power
generation in April and October in 1982 respectively. By the end
of 1985 electric power generated by the power plant reached
244.52 million KWHs with refuse consumpted of 73.56 million tons.
Main operation parameters of this power plant are as follows:

    Average calorific value of feeding fuel,   1550Kcal/kg.
    Pressure of over heat vapour,   39 kg/cm square.
    Temperature of over heat vapour,   450°C.
    Temperature of supply water,   150°C.
    Heat efficiency of the boiler,   67%.
    After 8000 hours of operation all the components and buried
    tubes of the boiler wore out,   0.3_0.4 mm.
    Cost of generating power,   0.049yuan/KWH.
    Annual net profit of the power plant,   2.9 million yuans.

    Di_Dao power plant owned by Ji Xi Area in Xei Long Jiang
province is equiped with two 130 T/H fluidised bed combustion
boilers and two 25 thousand KW gas turbine generator sets.
The boiler in the power plant is the largest one in China. By the
June of 1984 the two boiler operated accumulatively for 26949
hours with generating electric power 440 million KWHs and
burning refuse from coal preparation plant - 1.6 million tons
(calorific value 1800 Kcal/kg). Heat efficiency of the boiler is

more than 70%. The boiler of this power plant basically operated
in continuity, stability and full load, with operation was over
5000 hours a year on an average and one overhaul every years.

China have achieved some major successes on the research and
operation of fluidised bed combustion boiler as follows:

1) Various methods of ignition and starting operation for 2-130
T /H aforesaid boilers, including some critical techniques, e.g.
high-speed starting operation by thin fuel layer of high charcoal
fire and ignited by oil of fluid state; ignition of single/
multiple bed; distribution of fuel layer for ignition; increasing
temperature after ignition and distribution of gas for burning.

2) Stable burning and operation for a long time of various;
influence of the quality of the fuel; adjustment of operation.

3) Various well-distributingair systems for the boilers.

4) Reasonable arrangement and its thermal conductive
characteristic of immersed heated layer in fluidised bed
combustion.

5) Various methods and ways to improve combustion efficiency
of fluidised bed combustion, such as adding complete combustion
bed for fly ash, whirling gas tube in boiler, high temperature
suspended section and flue of V type, secondary blowing wind
for sowing coal, together with amplying velocity of flow in low
section onfluidised bed, low fuel layer and fine size fuel etc.

6) Effective methods to prevent wear of immersed heated
surface in fluidised bed.

7) Fluidised characteristics in fluidised layer, together
with influences of fuel size and sizing charater.

8) Design and operation experience of feed and removing slag
systems, equiped with efficient and low power-consumped fan.

9) Calculation methods of thermal of the boiler have been
preliminary studied.

It is worth while discussing the experience of the operation
of the boilers is thah , we should use processed refuse as far as
possible to increase the thermal efficiency of fluidised bed
combustion. Main ways are as follows: a) reprocessing
from washed or hand-adopted refuse. b) removing harder and low
calorific value refuse by applying the selective crush principle.
c) fully utilize the weathering refuse in refuse heaps. At Zhao

Ge Zhuong colliery in Kai Luan Area the caiorific value of the coal andrefuse blending fuel screened from the old refuse heap, that accumulated for several dozens years, is 2500-3000 Kcal/kg.

However, there have been many technical problems don't to completely solve, such as: Up to now we haven't systematically studied basic theories on the mechanism of fluidised bed combustion and transmitting heat, fluidised charaters of wide screened size and mechanism of desulphurization. Design of fluid bed combustion boilers have been depending on the data from experience. Thermal efficiency of the majority of the boilers isn't ideal. There were serious problems on the wear of the wall in boiler and the heated surface of buried tube.THEstudy on the steam-gas combined circulation system of the pressurized fluidised bed combustion boiler have just started. There weren't systematical consideration on the comprehensive utilization of slag and so on.

UTILIZATION OF REFUSE IN BUILDING MATERIALS INDUSTRY AND OTHERS

Building materials industry is the largest consumption on the comprehensive utilization of refuse, in 1978 - 7.12 million tons , in 1979 - 11.22 million tons and in 1980 - 12.50 million tons. By the end of 1981 , in coal mining system throughout China, 481 factories had been set up, which produce brick, tile and cement respectively by using refuse, stone-like coal and slag as material . The production capacity for brick and tile amounted to 2500 million pieces and for cement over 1 million tons.

PRODUCING CEMENT

By using refuse as burden Tua Zi cement plant in Liao Ning province produced 1.16 million tons of N.400 cement from 1966 to 1977, with cost of raw materials and fuel saved 1.89 million yans. By using refuse of low silicon and low iron content Zhang Dian cement plant produced rapid hardening cement, which compression strength is more than 600kg/cm square.

PRODUCING BRICK

Refuse consuption of producing brick is the largest and development is also rapid. Production of refuse and slag was 12.6

billion pieces in 1978 throughout China. 188 factories had been set up by 1984 in coal mining system with prodution capacity for brick amounted 1.6 billion pieces. Brick for building in Jiao Zuo, Yuong Ruong, Kai Luan, Bei Piao, Piang Zhuang, Piang Xiang and Shi Gu areas have been mostly or totally refuse brick.

PRODUCING LIGHTWEIGHT AGGREGATE

Refuse lightweight aggregate isproduced in many areas. FU Xin Area produced 250 thousand m^3 from 1978 to 1984.

PRODUCING CONCRETE HOLLOW BLOCK

On the basis of China's practice, building wall by using refuse concrete hollow block compared with using clay brick:

Save cement mortar	30 %
Increase construction efficiency	70 %
Wall weight itself reduce	40 %
Cut the time for a project down	25 %

At present refuse concrete hollow block is produced in major districts stored refuse throughout our contry. In 1977 a production line , which produces 75 thousand m^3 of the block a year , had been set up in Jiao Zuo City with proceessing refuse 190 thousand tons/year.

PRODUCING POTTERY PORCELAIN PRODUCTS AND FIRE*RESISTANT MATERIAL

Jin Zhou clay ware factory produces electricity porcelain element, by using refuse hand- proccessed in Nan Piao Area as major raw material, with consumption of the refuse is about 1100 tons/year. IN 1977 Chuong Qiang coal research institute manufactured steel container lining brick which is clinker manufactured by the parting in Jia Yang Colliery and 30 % of Jiang Jin Clay is blended in it. This lining brick achived great success in Chuong Qiang steel plant. Da Tuong Colliery in Huan Nan Area produces refractory brick, which is made up of 70 % of refuse and 30 %of white clay produced locally. Its refractory degree is 1610 - 1650°C , compression strength 271 kg/cm square.

OTHERS

According to our exprience ceramsite produced by refuse should

be developed from a long-term point of view. The refuse for
producing ceramsite contains 7 - 14 % of carbon. Comparing with
producing cement it have advantage of easily manufacturing of the
equipment, less investment and rapid effect, changing production
flexibly etc.

Various chemical products directly from refuse such as aluminum
chloride, poly-alumium and divanadium pentoxide from refuse should
be studied. The development of these chemical products should be in
line with the market demands at home and abroad in a planned way.
At present the technique to produce chemical products has been improved, with higher product quality and lower costs. Divanadium
pentoxide has already entered the internatiomal market. In 1981
two chemical plants, which extrat aluminum chloride and polyalumi-
num from refuse that content 38 % aluminum, had been set up in Nan
Piao Area, with production is respectively 10000 tons per year and
the value is above 8 million Yuans.

Utilization way of sulphur resource in high sulphur content
refuse is as follows: One is dressing C concentration method, another is chemical method. Ganbazi coal preparation plant in Nantong
has been put into operation, which adopts jig-table combination
process to recover pyrite from high sulphur content refuse. the
plant recovered about 20000 tons of sulphur concentrated from
1980 to 1981.

Currently, from the point of view of chemical methods sulphite
of absorption and regeneration method have been ripe technologically.
Here is its chemical equation:

$$SO_2 + H_2O + Na_2SO_3 \xrightarrow{\text{Normal Temp.}} 2NaHSO_3 \qquad \text{(absorption)}$$

$$2NaHSO_3 \xrightarrow{-150C} Na_2SO_3 + SO_2 + H_2O \qquad \text{(regeneration)}$$

It can notably improve the performance of absorption and
regeneration and reduce oxidizing rate. By using this method
Chang Zhou Second Chemical Plant refines SO_2 of liquid state,
with remarked technical & ecnomic results.

SOME SUCCESSFUL RESULTS ON RESEARCH AND DEVELOPMENT OF
COMPREHENSIVE UTILIZATION OF REFUSE

Research and development on comprehensive utilization of refuse
have been going on for a long-term time and much headway has been

made in it. Here illustrate with examples:

1. A new industrial stuffings (known as SAC) had been developed by Institute of comprehensive utilization of refuse in Jiang Su Province. Major raw material is washery discard produced in seven collieries, such as Xin He colliery etc. characteristics of the refuse is as follows: Q, 2260-2560 Kcal/ Kg; Fe_2O_3 content 2.61 - 3.05 %; ash about 6.5 %; quality is stable. There are three types of SAC-1,SAC-2 and SAC-3. Technology characteristics of SAC: less specific gravity; low absorbing oil nature, it can save plasticizer when use it in products added plasticizer; black colour; It has better mutual solubility with chemical compound of organic high polymer with enery consumption reduced and time saved in process. Good economic and technology results had been attained in dozens utilizations of SAC, such as: in manufacturing conveyor belt we substitute high wear-resisting soot carbon by SAC.(its content 50-70 %)with strength target up to standard,wear-resisting capacity and elasticity increased and 2000 Yuans / ton (SAC) of profit obtained. We use 10 % of SAC in PVC film as a substitute for soot carbon and light calcium of the prescription. The end product possesses 175-197 Kg/cm^2 of tensile strength and 227.5-245.1 % of extend rate and its quality surpass the standard issued by the ministry.

In rubber industrial products SAC is substituted for soft soot carbon with the same performance and 800-1000 Yuans/ton of cost reduced.

20 % of SAC is filled up in ABS plastic with 1917 Yuans / ton saved.

2. 425$^{\#}$ regular portland cement was produced by low temperature baked technique by Chuong Qing coal Institute cooperated with the cement plant in Yuong Ruong Area. Here clay is fully substituted by slag of fluidised bed boiler and the cement produced in batches from April,1985. The products can reduce 150-200 C of baked temperature,save 30 % of coal,increase 30 % of the consumption of slag, raise 50 % of production and reduce 25 % of cost compared with the traditional technology. It is the first of all that slag is fully substituted for clay with slag is mainly used as blending material in China. This possesses importance significance on comprehensive utilization of refuse,the development of fluidised

bed combustion technology and opening up the new ways of cement
production with washery slag displacement of fluidised bed boiler
is about 8 million tons evera year in China.

3. In some factories phosphoric acid aluminum-vitriol chamotte
fire-resistance daub replaces ordinary refractory clay in order to
solve the problem on wear of the chamber of the boiler. It achived
good performance and result in preventing wear.

4. " Hua Yi " high efficient de-sulphuriring agent was successfully developed by the Heat Energy Department in Qing Hua university.
It applys in fluidised bed boiler with low calcium/sulphur rate
(0.64-0.67) and de-sulphuriring rate 70-87 %. This agent have
the advantage of low cost, simple process and don't refit the
boiler when using it.

DEVELOPMENT IN FUTURE

To sum up above it is seen that there are tremendous potentia-
lity and good prospect in utilization of refuse in China. The
government and the ministry of coal industry pay great attention
to the environmental protection of colliery and utilization of
refuse. We considered that further study should be made to effect
a break through in the burning technique for large fluid-bed
boiler so as to realize standardization,universalization and
serialization in utilizing refuse.

The existing power stations burning refuse should be stren-
thened,developed and equiped with complete sets of equipment.
Efforts should be made to complete the project for comprehensive
utilization of slag and fly ash to increase economic benefit.

Ministry of Coal Industry has made out a plan for construction
of 45 power plants burning refuse from 1985 to 1990. These power
plants will locate near coal preparation plants in North-East,
North and East China district, where is short of electric power.
Total installed capacity of these power plants will be 1.003
million Kw with complete sets of 5 cement plants, 7 brick plants
and 3 aerocrete plants.

By estimate if above power plants and other plants will be put
into operation, it can generate power of 5.5 billion Kwh per year
and produce 250 thousand tons of cement,140 million pieces of brick
150 thousand m^3 of aerocrete, with profit of 140 million Yuans/year.

Reclamation, Treatment and Utilization of Coal Mining Wastes, edited by A.K.M. Rainbow   441
Elsevier Science Publishers B.V., Amsterdam, 1987 — Printed in The Netherlands

DIRECT TREE SEEDING ON COAL MINE WASTES IN BRITAIN - A TECHNIQUE FOR THE
FUTURE

A.G.R. LUKE

Cambridge Direct Tree Seeding Limited, Linden House, 40 Wilburton Road,
Ely, Cambs CB6 3SX (England)

SUMMARY
    Many observations have been made on the successful colonisation of
wastes from the coal mining industry by our native trees and shrubs.
Nature frequently overcomes even the most hostile site conditions, but the
timescale is often unacceptably long for the public, engineers and
planners.   In the last five years direct seeding trials and large scale
sowings in Britain have proven that man can dramatically speed up the
process of re-vegetation using woody species to a period of three to five
years.   Furthermore data from these projects illustrates that high
densities of seedlings and healthy rates of growth can be achieved directly
on coal shales and similar materials in the absence of top-soils.   When
compared to conventional techniques of grass establishment followed by
planting it appears that direct tree seeding offers biological, ecological
and economic advantages.   Vegetation establishment by direct seeding can
frequently meet several objectives at once by visually improving the site,
by providing erosion control and by creating a dynamic woodland or scrub.
For those involved in the re-vegetation of Britain's mining wastes direct
tree seeding offers a new approach for the future.

OBSERVATIONS ON NATURAL COLONISATION OF TREES AND SHRUBS

    The most extensive study of natural colonisation of coal shale wastes in

Britain was undertaken in 1952 (ref. 1). Of all the species recorded on

waste heaps 17.9 percent were tree or shrubs the most important species

being Betula pendula, B. pubescens, Fraxinus excelsior, Quercus robur,

Crataegus monogyna, Rosa canina, Rubus fruticosus agg., Sambucus nigra,

Cytisus scoparius, Ulex europaeus and Hedera helix. These eleven species

represented 4.4 percent of all the species recorded (Table 1) which

reflects more typically the importance of woody species in the vegetation

communities assessed. The relative proportion of trees and shrubs in

natural vegetation communities (and their prescence at all stages of

succession from bare spoil to closed woodland) was thought to be reflective

of the open, patchy and imature nature of the vegetation. Much lower

numbers of woody species usually occur in closed vegetational communities.

TABLE 1
List of tree and shrubs which naturally colonise or have been direct sown on coal shale

SPECIES	SPECIES RECORDED AS COLONISING BURNT AND UNBURNT SHALE		SPECIES DIRECT SEEDED ON COAL SHALE 1980-1986
	A	B	C
TREES			
Acer pseudoplatanus	+	+	+
Alnus glutinosa	–	+	+
Betula pendula	*	*	+
Betula pubescens	*	*	+
Fraxinus excelsior	*	–	+
Ilex aquifolium	+	–	–
Malus sylvestris	+	–	–
Pinus sylvestris	+	+	+
Populus nigra	–	+	–
Pyrus aria	–	+	–
Quercus petraea	+	–	+
Quercus robur	*	–	+
Salix spp	+	*	–
Sorbus aucuparia	+	+	+
Ulmus montana	–	+	–
SHRUBS			
Cornus sanguinea	+	–	–
Corylus avellana	+	–	+
Crataegus monogyna	*	*	+
Cytisus scoparius	*	*	+
Ligustrum vulgare	+	+	+
Lupinus arboreus	–	–	+
Ribes uva-crispa	+	–	–
Rosa canina agg.	*	*	+
Rubus fruticosus agg.	*	*	+
Rubus idaeus	+	–	–
Sambucus nigra	*	+	+
Ulex europaeus	*	*	+
GROUND COVER SHRUBS			
Calluna vulgaris	+	+	–
Clematis vitalba	+	–	+
Hedera helix	*	–	–
Lonicera periclymenum	+	–	–
Solanum dulcamara	+	–	
Vaccinium myrtillus	+	+	

KEY: + observed naturally colonising or direct seeded in trials and large
        scale schemes
     – not observed naturally colonising and not direct sown
     * observed naturally colonising and a species which frequently occurs
SOURCES A-Hall, 1957, B-Richardson et al, 1971, C-Cambridge Direct Tree
Seeding Limited (unpublished data)

The reaction of the spoil and seed sources were important factors in determining the natural colonisation of trees and shrubs. Fewer tree species grew on the highly acid than the alkaline spoils. The colonisation of spoil by trees and shrubs was directly related to the proximity of and type of local seed source. Intensive colonisation was observed where large seed sources were adjacent to tips. However it was also observed that even where good seed sources were present colonisation could be localised or absent indicating that seed and seedling mortality could be high.

Despite the ability of woody species to colonise coal spoil few tips supported a continuous or closed woody cover. Furthermore the timescale required to develop a woody cover ranged from 30 - 80 years. Shrub communities were found on burnt shales at 30 years but generally required a longer period (up to 60 years) to establish on unburnt shales. Vegetation dominated by pioneer trees (e.g. Betula) required 40 or more years following ceasation of tipping. Of the slower growing broadleaves Oak dominant communities were recorded on waste tips of 80 + years. Edaphic conditions were so severe on some tips that progression of the succession was halted at the shrub stage. Richardson et al (ref.2) noted similar timescales for the colonisation of coal shales by woody species in the County of Durham, in North-East England. Typically the early stages of succession with shrub dominated communities occurred between 0-20 years after final tipping, later successional stages with a predominance of pioner trees at 25-50 years and climax vegetation of Oaks at 80+ years. Observations have also been made on colonisation of the pioneer trees Alnus and Betula on coal shales in Scotland (ref. 3) and Yorkshire (ref. 4) present at 7-8 years and 20 years respectively (see Fig. 1).

Nature can be relied on, (with time!) to provide new broadleaved woodland on coal shale. But how can new woodland be created in a timescale that is acceptable to the public, engineers and planners? One technique tested in recent years is the direct seeding of trees and shrubs.

DIFFERENCES BETWEEN THE PROCESSES OF NATURAL COLONISATION AND DIRECT SEEDING

Although the process of natural colonisation and direct seeding both rely on seeds germinating and seedlings surviving and producing a positive growth increment there are marked differences which are summarised in Table 2. Perhaps the most important factors which contribute to the success of direct seeding compared to natural colonisation are the application of controlled rates of seed and the preparation of the seedbed prior to seeding.

TABLE 2
A comparison between direct seeding and natural colonisation

FACTOR	DIRECT SEEDING	NATURAL COLONISATION
SEED INPUT	Seed rain once only. No restriction on species	Yearly seed rain, Species controlled by the availability and proximity of seed source
DATE OF SOWING	Autumn, winter & spring	Seed dispersal period for each species, i.e. late summer through to winter
SEED DORMANCY	Can sow dormant or treated (non-dormant) seed or mixtures	Most species seed is dormant at dispersal
SEEDBED CONDITION	Seedbed prepared to suit species. Specific site problems are remedied	Little or no control
SOWING DEPTH	Controlled	Into cracks or on the surface
SEEDLING ESTABLISHMENT	Rate of growth of seedlings can be encouraged by weeding applying fertiliser, sowing nurse covers etc Even-aged seedling population	Establishment controlled by interaction of yearly weather and seed dispersal patterns and the existing vegetation. Age-structure in seedling population
DEVELOPMENT OF VEGETATION	Depends on the composition of seedling population. All elements of the vegetation initially present but gradually one element (group of species) replaces an earlier one.	Controlled by the recruitment of new seedlings to population. Growth of existing plants can affect establishment of new recruits All elements not present at the same time.
TIMESCALE see text	2-3 years nurse shrub cover 3-8 years pioneer trees predicted 10-20 years to mature tree vegetation	10-30 years shrub and pioneer tree cover 50+ years to mature tree vegetation

DIRECT SEEDING PROJECTS IN BRITAIN, 1980-1986

Research trials and large scale sowings on landscape projects in Britain in the late 1970s and early 1980s (refs 5-7) proved that direct seeding could be successfully employed to establish new woodland. In 1981 the

Scottish Development Agency was interested in exploring new techniques for revegetating mineral waste tips in Scotland. Although information on direct seeding of mineral wastes was available for the United States (e.g. refs 8-10) literature was not available for British species and coal shale wastes. Cambridge Direct Tree Seeding Limited was therefore commissioned by the S.D.A. to design and monitor trials to test direct seeding. Trials were initiated in 1981-3 on a wide range of spoils and observations were concurrently taken from a greenhouse study on the effects of spoil type on germination (ref 11). Early results were very encouraging (ref 12) and by 1984 it was possible to produce a series of preliminary recommendations for seeding on coal and oil shales (ref 13). Following the success of these trials coal shale tips have been direct seeded elsewhere in Britain and a wide variety of plant communities are now present (Figs 2 and 3).

PLANT DENSITY, GROWTH AND SUCCESSION

One noteable feature of vegetation established by broadcast direct seeding is the high density of tree and shrub seedlings (Fig 4). It is necessary to sow high rates of seed to achieve such densities (frequently up to 50,000 to 100,000 plants per hectare) because of the high risk of mortality of the seed and seedlings due to the edaphic conditions and natural agents.

Often rates of growth of seedlings are slow in the first two years after seeding. This appears to be the period of adjustment the seedling's root system requires before shoot extension can be improved. By the third and fourth years there is usually a most dramatic surge in growth (Figs 4 and 5). Rates of growth of the pioneer trees are presented in Table 3. The highest annual increments are for the Alders at 0.40 to 0.70 metres averaged over five years. With the unrestrained form of growth produced from sown stock (strong leading shoots and laterals present all along the main stem) the potential for growth in future years is high.

Comparing the growth of stock established by direct seeding (Table 3) with planted stock (Table 4) would indicate that the average annual increment of sown stock is often higher for broadleaved trees. Growth s for Pinus spp, the only conifers compared, would appear to be similar for both methods of establishment. In the case of one broadleaved tree, Acer pseudoplatanus, the planted stock failed to grow taller than its original height at planting ! When the growth increment of planted stock is adjusted for the original height of the stock the rates achieved by direct sown stock are even better by comparison. It must be noted, however, that

these data have been generated from two different coalfields, each with
diferent coal shales and regional climate. Nevertheless early growth of
direct sown stock would appear to augur well for the future.

TABLE 3
Height growth of pioneer trees four - five years after direct
seeding on coal shale

SPECIES	AGE (YEARS)	AVERAGE HEIGHT GROWTH (m)	AVERAGE ANNUAL GROWTH INCREMENT (metres)
Alnus glutinosa	5	2.0	0.40
Alnus incana	5	3.5	0.70
Betula pendula	5	1.5	0.30
Pinus sylvestris	5	0.7	0.14
Quercus robur	4	1.3	0.33

Note 1. Observations taken from a range of four trial and large scale
sowings on coal shale tips in Central Scotland.

TABLE 4
Height growth of planted stock on coal shale wastes in County Durham
(Data extracted from Richardson et al, 1971)

SPECIES	AGE (years)	AVERAGE HEIGHT GROWTH (metres)	AVERAGE ANNUAL INCREMENT (metres)	ADJUSTED ANNUAL INCREMENT (metres)
Acer pseudoplatanus	10	< 0.9	< 0.09	0
Alnus incana	10	> 2.1	> 0.21	> 0.165
	6½	1.8	0.28	0.185
Betula pendula	10	> 2.1	> 0.21	> 0.15
	8	0.9-1.5	0.11-0.19	0.06-0.13
Pinus nigra calabria	10	1.9	0.19	0.145
Prunus spp	8	0.9-1.5	0.11-0.19	–
Quercus robur	10	1.4	0.14	0.10
Quercus rubra	10	0.9	0.09	0.05

a The adjusted annual increment was calculated by subtracting the height
of the stock at planting from the average height growth then dividing by
the number of years since planting.

b $<$ indicates less than, and $>$ indicates greater than

    The vegetation produced by direct seeding is dynamic i.e. one plant
community is replaced by another as the compostion of the seedling mixture
changes with time. In the early years after seeding a shrub community

develops. The extensive use of nitrogen-fixing shrubs ensures that the
level of spoil fertility is gradually improved and invaluable organic
matter is added through continual leaf-fall. With time the shrub layer is
overtopped by pioneer trees. This process is illustrated using a trial
site on oil shale in Central Scotland (Figs 6-8). The timescale for this
succession is between five to ten times faster that that observed for
natural colonisation.

TABLE 5
A comparison of the vegetation strategies of direct seeding and planting

TIMESCALE (YEARS)	TECHNIQUE	
	DIRECT SEEDING	CONVENTIONAL PLANTING
1	Annual herbaceous nurse cover with emerging tree and shrub seedlings present	Perennial, sometimes annual, grasses and legumes establish
2	Herb-shrub community	Grass-dominated sward develops into which transplants or whips are planted
3	Shrub community begins to overtop the herb layer. Composition of vegetation begins to change	Growth of planted stock is slow, a two layered system develops
4-5	Shrub layer becomes more dense frequently closing gaps in cover, and the pioneer trees begin to emerge above the shrubs. A three layered vegetation of herbs, shrubs and trees is present	Young planted stock overcomes transplant shock and commences active growth, developing pronounced two layered vegetation of grass & trees
6-10	Growth of pioneer trees accelerates to overtop shrub layer. Composition of vegetation shows a reduction in shrub element as young climax trees are nursed by pioneers	Closure of canopy may proceed the composition of the trees remains roughly similar to that at planting

COMPARISON OF DIRECT SEEDING WITH CONVENTIONAL TECHNIQUES

    The different strategies for the establishment of vegetation by direct
seeding and by conventional methods are outlined on Table 5. The type of
vegetation present changes with time, these effects being most marked for

direct seeding when succession from a herb to a pioneer-climax tree
community proceeds. This successional approach which mirrors and
accentuates natural succession using pioneer species was advocated by
J.E.D. Fox (ref 14) in a review of the rehabilitation of mined lands. He
cited further examples where direct seeding was used to speed up
successional change on mined land in Australia.

Biological and ecological considerations

(i) Species. Direct seeding has generally utilised broadleaved trees and
shrubs rather than conifers (Table 1, ref 12). Planting has relied equally
on broadleaved and coniferous trees (ref 2 & 15) and many introduced or non
native species have been planted.

It is easier to include a more diverse range of species in a seeding
mixture than for planting. Up to 30 species have been sown on one site.
This minimises the risk of all species failing on site. Furthermore
greater use can be made of shrubs by seeding simply by increasing the
proportion in the mix. With the higher average prices for nursery grown
stock of shrubs, seeding offers an economic advantage. Species which are
difficult to establish by transplant (e.g. oak, bird cherry) can be readily
established from seed.

(ii) Shoot growth. Growth rates of sown stock appear to be equal to or
greater than those observed for planted stock in the first five years
(refer to Table 3 and 4).

(iii) Root growth. Root growth of direct sown seedlings occurs in all
planes (Fig. 9) and is not confined to one plane as is often found with
notch of slit planted stock (ref 15 & 16). Direct sown and naturally
regenerated stock tend to be more windfirm than planted stock. Wind
socketing and abrasion of the root collar is rarely observed in direct sown
stock.

(iv) Wildlife value. The diversity of species' mixture, the non-uniform
growth of individual species, and stratification of the vegetation (e.g.
fig 7) produce a more desirable habitat for wildlife than the rather
uniform two-layered vegetation often produced by planting. However,
quantitative data are required to compare the wildlife value of vegetation
created by direct seeding and planting.

(v) <u>Plant density</u>. It is much easier to achieve high densities of stock using direct seeding. Planting at densities of greater than 1 to 2 plants per square metre is not usually undertaken because of high costs. By the end of the second year plant densities of 2-10 per square metre are frequently found on direct seeding schemes (C.D.T.S. Ltd, unpublished data).

(vi) <u>Management options</u>. With the high stem density of direct seeding schemes the number of management options and potential for commercial production are increased e.g. thinning pioneer trees to produce small polewood, coppicing of shrubs and pioneer trees for fuel wood, thinning of climax trees for timber production. As a consequence of planting being at regular spacing the management, if implemented, is frequently restricted to thinning or removal of nurse species to allow greater growth of more desirable trees for potential timber production. Given the harsh edaphic conditions of coal shale sites production of any timber may be unrealistic. In this case vegetation could be left to develop naturally with no mnagement input thereby providing a natural wildlife refuge.

(vii) <u>Erosion control</u>. The development of a herbaceous community which is succeeded by a shrub layer ensures a continuous vegetation cover from seeding. Rooting of the shrub layer provides for very effective erosion control on seeded slopes. In contrast the sowing of a grass layer to control erosion is often detrimental to the growth of planted stock.

(viii) <u>Improvement of the soil structure</u>. The contribution of leaf litter from the shrub layer and nitrogen from leguminous shrubs and nitrogen-fixing pioneer trees improves the soil structure under direct sown stands. The absence of such a shrub layer in many planted sites can delay improvements to soil structure.

(ix) <u>Visual amenity</u>. The early development of a green herbaceous layer in the first year followed by an emerging shrub layer provides a strong visual impact prior to the development of the pioneer trees. In the early years the public might not concieve that a woodland is developing, but will realise with time that a young woodland is being created. Clearly planting has an immediate visual impact but this is not sustained if the trees and shrubs do not close canopy and remain moribund. Suitable combinations of direct seeding and planting can be used on sites where there is a need for prudent public relations.

Direct seeding can be implemented immediately after completion of earthworks. Frequently planting is not implemented until 1-3 years after grass seeding. The earlier start which can be taken for direct seeding helps to offset the disadvantages of slower growth in the early years.

TABLE 6 Cost comparison between direct seeding and planting

EXAMPLE SYSTEMS	TYPICAL COSTS LESS MAINTENANCE (£ per hectare)
DIRECT SEEDING	
1. Broadcast seeding + straw mulching for erosion control	2500 - 3500
2. Broadcast seeding tree and shrub mix	1000 - 2000
3. Broadcast shrub mixture + spot sown trees at 2.5 metre centres	1000 - 1500
4. Row seeding at 2.0 metre centres	1000 - 1500
5. Spot seeding at 1.5 metre centres	900 - 1350
PLANTING	
1. Pit planting whips (900-1500 mm size) at 2.5 metre centres	2300 - 2750
2. Pit planting whips (900-1500 cm size) at 1.5 metre centres	5900 - 7150
3. Notch planting transplants (450-900 mm) at 2.5 metre centres	1150 - 1300
4. Notch planting transplants at 1.5 metre centres	2600 - 3140

NOTES: a. Costs are given for 1986 and do not include site preparation (e.g. ripping, cultivations, fencing, application of fertilisers). Costs are compared for a site of size one hectare.

b. Assumes the cost of grass seeding prior to planting is £300 per hectare.

c. Assumes pit dimensions 300 (diameter) x 300 mm (depth) and 1/10th bag of tree planting compost is mixed into the backfill.

(x) Economics.   Seeding systems are very flexible both in terms of seeding pattern and cost (ref 17).   The larger the area to be sown the more economical the treatment because the work can be mechanised using special machinery.   Each organisation will have its own planting costs which have evolved to meet specific site and edaphic problems.   The cost comparison in Table 6 is therefore open to discussion.   However, in the last few years in Britain, C.D.T.S. Ltd have operated a contractual seeding service nationwide, although comparisons vary around Britain it would appear that direct seeding is generally equal to or less than the cost of planting for sites of 1.0 hectares.   Where sites are in excess of 5 hectares the cost of seeding will generally be more favourable than planting.   In calculating the cost of planting is is only fair to include the cost of establishing the grass cover in between transplants.   When direct seeding this latter cost is not included in the cost since all species are sown in one operation.

DIRECT SEEDING IN THE FUTURE

With the recent call for maximising plant production on derelict land (ref. 18) direct seeding would seem an important technique to consider when planing for re-vegetation of wastes from the coal mining industry.   There are a variety of situations and projects where the use of direct seeding can be further explored in Britain:-

Coal recovery sites

Economic factors have led to a recent increase in the scale of recovery of coal from waste tips (see T.K. Macpherson, The Cost-Effectiveness of Rehailitating Colliery Sites through Coal Recovery, in this Symposium). The margin which determines whether these sites can be re-mined is often small therefore budgets for restoration are correspondingly small.   The economies of direct seeding may be great value for rehabilitating these sites.

Abandoned tips following mine closure

Where coal mines have been closed in recent years the active rehabilitation of tips often ceases or is substantially reduced. Inexpensive techniques such as direct seeding can be used to treat large areas using budgets which may already have been fixed.   As the likelyhood of active maintenance and management of these sites is minimal the robust

nature and self-sufficiency of the vegetation produced by direct seeding is
a desirable or perhaps even essential characteristic.

'Soil' improvement of coal shale wastes

Green cropping with nitrogen fixing species has long been practiced in
the reclamation of coal shale wastes in Germany (ref. 19). This practice
could be extended to British conditions using the productive nitrogen
fixing shrubs and trees which can be successfully established by direct
seeding. Following a period during which the organic matter content,
nitrogen status, drainage and soil flora and fauna are improved the direct
sown vegetation can be cleared and more demanding longer term species of
timber value can be introduced. Alternatively agricultural crops could be
established , following the green manuring, to provide an annual or
biennial cash crop. Such an agricultural system would require a cash input
and active management, but an economic return could be possible.

Seed orchard production on coal shales

On many direct seeding schemes in Britain shrubs have flowered and borne
mature seed, sometimes in prolific quantities. That seed has been produced
on these sites is important since the plants will have interacted with the
site conditions to produce seed which is hopefully physiologically, and
perhaps even genetically, more suited to producing plants suited to the
harsh conditions of these sites. The potential for establishing seed
orchards on coal shale to provide improved seed for direct seeding and for
the nurseryman deserves further investigation. If success with such a
project is at all possible an economic return could be generated within a
relatively short timescale from shrub seed production.

Coal shale tips as nature reserves

Many derelict sites, including those generated by the coal mining
industry, often support natural vegetation and act as refuges for plants
which have been excluded by agriculture (ref. 20). These 'natural
reserves' of plant material form an important reservoir of natural genetic
stock. Tips which are hitherto not restored could be sown with tree and
shrub seed collected in the locality thereby increasing the local genetic
reservoir for future generations.

Nurse covers direct sown into planted stock

The ability of nitrogen-fixing shrubs (e.g. Cytisus, Lupinus and Ulex)
to encourage the growth of naturally seeded trees and planted stock is well

documented (refs 20-22). Plantations where the stock is moribund could
well benefit from the introduction of a shrub layer by direct seeding.
Recently planted or new planting schemes could also benefit from the use of
nurse species introduced by direct seeding. Although capital would be
required to implement new works monies would be saved on maintaining
planted stock free of grass competition.

CONCLUSION

It is hoped that the potential for the utilisation of direct tree and
shrub seeding is the treatment of wastes from the coal mining industry has
been demonstrated. It now remains for those engineers, planners, foresters
and landscape architects to explore the possibilities further. Providing
that the correct technical advise is sought direct seeding offers another
option for the revegetation of coal shales and similar spoils.

REFERENCES

1   I.G. Hall, The ecology of disused pit heaps in England, J. Ecol., 45
    (1957) 689-720.
2   J.A. Richardson, B.K. Shenton and R.J. Dicker, Botanical studies of
    natural and planted vegetation on colliery spoil heaps, in: Landscape
    Reclamation, Volume 1, University of Newcastle-upon-Tyne, IPC Science &
    Technology Press, Guildford, 1971, pp 84-99.
3   A.J. Kimber, I.D. Pulford, H.J. Duncan, Chemical variation and
    vegetation distribution on a coal waste tip, J. Applied Ecol., 15
    (1978) 627-633.
4   V.N. Dennington and M.J. Chadwick, The nutrient budget of colliery
    spoil tip sites, J. Applied Ecol., 15 (1978) 303-316.
5   A.G.R. Luke, Sowing to suit. Sowing techniques and nurse crops, GC &
    HTJ, 292 (1982) 13-16.
6   T. LaDell, Techniques No. 44 An introduction to tree and shrub seeding,
    Landscape Design 144 (1983) 27-31.
7   R.N. Humphries and T.F.G. LaDell, The establishment of wooded
    landscapes from seed, Paper presented to the Landscape Institute Golden
    Jubilee Symposium, Reading, July, 1979, 10 pp.
8   W.H. Davidson, Direct seeding for forestation, in: Trees for
    reclamation, Lexington, Kentucky, October 27-29, 1980, Sponsored by
    Interstate Mining Compact Commission and United States Dept. of
    Agriculture, Forest Service. General Technical Report NE-61, pp. 93-97.
9   W.T. Plass, Direct seeding of trees and shrubs on surface-mined lands
    in West Virginia, in: K.A. Utz (Ed.), Proc. of the Conference on
    Forestation of Disturbed Surface Areas, USDA Forest Service, Atlanta,
    Ga, 1976, pp. 32-42.
10 R.F. Wittwer, D.H. Graves and S.B. Carpenter, Establishing Oaks and
    Virginia pine on Appalachian Surface Mine Spoils by Direct Seeding,
    Reclamation Review 12 (1979) pp. 63-66.
11 A.G.R. Luke, Tree and shrub seeding trials on coal shale, oil shale and
    ironstone bings in Central Scotland, pp. 34-48. Unpublished report
    prepared for the Scottish Development Agency, December 1981, pp. 71.

454

12   A.G.R. Luke and T.K. Macpherson, Direct tree seeding: A potential aid
     to land reclamation in central scotland, Arboricultural Journal, 7
     (1983) 287-299.
13   A.G.R. Luke and M.J. Smeeden, Direct tree seeding trials on land
     reclamation schemes in Central Scotland, 1981-1984. Unpublished report
     prepared for the Scottish Development Agency, January 1985, pp. 44.
14   J.E.D. Fox, Rehabilitation of mined lands, Forestry Abstracts 45 (1984)
     565-600.
15   J. Jobling and F.R.W. Stevens, Establishment of trees on regraded
     colliery spoil heaps. Forestry Commission Occasional Paper No. 7,
     Edinburgh, Forestry Commission, pp 48.
16   S. Little and H.A. Somes, Root systems of direct-seeded and variously
     planted lobdolly, Shortleaf and Pitch pines, U.S. Forest Service
     Research Paper NE-26, 1964, pp. 13.
17   A.G.R. Luke, Trees: Naturally! Direct tree seeding, Mineral Planning
     (1983) 40-42.
18   Anon, Tree council calls for clean-up of waste land, Horticulture Week,
     10 October 1986, pp 7.
19   J.M. Bradley, Surface mining and land reclamation in Germany (A
     revision and updating of the report of the same title by E.A. Nephear,
     published in 1972), Cologne, Rheinische, Braunkohlenwerke AG, 1981, pp.
     42.
20   W.S. Dancer, J.F. Handley and A.D. Bradshaw, Nitrogen accumulation in
     kaolin mining wastes in Cornwall I Natural Communities, Plant and Soil
     48 (1977) 153-167.
21   Forest Research Institute, Rotorua, New Zealand, Fertilisers or legumes
     in our Forests?, What's New in Forest Research 46 (1977) 4.
22   S.G. Haines and D.S. DeBell, Use of nitrogen-fixing plants to improve
     and maintain productivity of forest soils, in: Proc. ' Impact of
     Intensive Harvesting on Forest Nutrient Cycling', State University of
     New York, August 13-16, 1979, pp. 279-303.

Fig. 1. Birch woodland which has naturally colonised coal shale.

Fig. 2. Birch and pine five years after seeding onto coal shale, pH 3.5.

Fig. 3. Four year old Pedunculate oak at 2 metres high protected by nitrogen-fixing nurse shrubs.

Fig. 4. High density of Birch seedlings two years after seeding,
Riddochhill, Bathgate.

Fig. 5. By the fourth year Birch saplings provide a dense cover,
Riddochhill, Bathgate.

Fig. 6. Contentibus bing, Mid-Calder - Year 1.

Fig. 7. Contentibus bing, Mid-Calder - Year 4.

458

Fig. 8. Contentibus bing, Mid-Calder - Year 6.

Fig. 9. Seedlings of Common ash and Pedunculate oak excavated from coal shale showing well-balanced root systems.

Reclamation, Treatment and Utilization of Coal Mining Wastes, edited by A.K.M. Rainbow   459
Elsevier Science Publishers B.V., Amsterdam, 1987 — Printed in The Netherlands

MANAGEMENT OF OPENCAST RESTORED LAND FOR CEREAL PRODUCTION

E.J. EVANS, B.Sc., Ph.D., M.I. Biol.
Department of Agriculture, The University, Newcastle upon Tyne, NE1 7RU.

ABSTRACT
     The aim of land restoration following coal extraction is to restore the
land to its former level of agricultural productivity. This is strongly
influenced by the way in which the soil is removed prior to coal extraction, its
subsequent handling and replacement. Use of heavy earth-moving equipment for
soil reinstatement leads to the deterioration of soil structure, characterised
by horizontal layering, the absence of structural fissures within the soil
profile, and increased bulk density of the subsoil leading to reduced air filled
pore space. The problems that arise from these soil conditions on cereal growth
derive from excess water in the top soil resulting in delayed estabishment,
modification of root and shoot growth and lower grain yields.

     An investigation of the effects of installing permanent underdrains prior
to revegetation on soil conditions and crop growth, has been initiated on the
progressively restored Butterwell opencast site and an adjacent unmined control
site in Northumberland. The installed drainage system provide preferential
surface soil conditions for the growth of winter wheat in early spring. This
allowed earlier root penetration and more efficient nutrient uptake which
resulted in enhanced tillering and dry matter production. Site differences
established during this period persisted to give significant yield differences
at final harvest. Wheat yields were consistently higher on the unworked control
site compared to the two restored sites. Early drainage of the Butterwell site
gave substantial yield improvements with the exception of the very dry season
of 1983-84.

     Differences in spring barley yields were less pronounced over the three
sites, although yield levels in general were low. Soil conditions in early
spring on the undrained Butterwell site delayed seed bed preparations and often
resulted in cloddy tilths and low plant establishment.

INTRODUCTION

     It is generally accepted that the quality of opencast restoration has

improved significantly since it was first undertaken in 1941 (Tomlinson, 1980).

Problems associated with the earlier restoration attempts have been identified

(Hunter and Currie, 1956) and implimented in the 1951 Restoration Code and the

Opencast Coal Act of 1958. Considerable emphasis is now placed on detailed soil

surveys and land capability assessments before coaling operations commence. In

many instances considerable improvements to the landscape have been achieved

through the inclusion of colliery waste in   the overall reclamation schemes

(Brent Jones, 1984). In this way maximum use is made of the useful soil making

material available.

     The majority of land for opencast mining is taken out of agricultural

production and is subsequently restored to a standard that allows normal farming

operations to continue. Frequently sites were associated with land of low
agricultural production supporting permanet pastures and rough grazing. Some
opencast sites however, have been situated on the more productive Grade 2 and
3 land supporting mixed farming with cereals rather than grass/livestock as the
dominant farm enterprise.

In England and Wales MAFF act as agents for British Coal during site
restoration and manage the land for a further period of five years post restoration. During this time the land is drained, trees and hedgerows replanted and
other features necessary for efficient farming operations established.
Traditionally a short term glass/clover mixture has been used to restablish
agricultural produciton prior to the installation of a permanent drainage system.
Thereafter, the land is ploughed and a long term grass/clover seed mixture sown.
Due care is taken to avoid damage to soil structure from poaching particularly
during early spring and in the autumn.

The restorative characteristics of grass are well established (Brook and
Bates, 1960), but the performance of arable crops on restored land is less well
documented. Tasker (1957) has argued that arable rotations could be more
beneficial to the development of soil structure than long term leys.

As part of a long term opencast restoration programme within the Faculty of
Agriculture, Newcastle University the usefulness of different grass/cereal
rotations is being evaluated. In this paper a summary is presented of a series
of trials over a period of four years where the growth of winter wheat and
spring barley were compared on restored and unworked land. Further work is in
progress to assess the long term benefits of different grass/cereal rotations.

DESCRIPTION OF SITES

Field trials were initiated on three sites in Northumberland during the
autumn of 1981. Two of these were located on the progressively restored Butterwell Opencast Mine, one of which being drained immediately following restoration
(BD) and the other, on an adjacent area of land, left undrained (BUD). The
restored topsoil and subsoil were moved directly onto the experimental site from
an unmined area left fallow during the previous season. This was done using a
combination of scraper boxes, bulldozers and mechanical graders during the
summer of 1980. An area of 2.8 ha of restored land was drained in the spring of
1981 using 75mm tile drains installed at a depth of 75cm and 10m spacing with
porous black fill to within 35cm of the surface. A third area was established
as an unmined control site on a similar soil type (Dunkeswick Series) at the
University of Newcastle's Cockle Park Experimental Station (CP). The three
sites were sub-soiled during the summer of 1981 and maintained weed free until
sowing.

Detailed analysis of the soil physical characteristics of the three sites
was undertaken one year after restoration. A feature of the restored profiles
was a general increase in dry bulk density and associated reduction in pore
space at depth (Table 1). Significant differences were also observed in the
pore space characteristics. On the restored sites there was an absence of
macropore continuity throughout the profile and an increase in the proportion of
fissures and voids resulting from soil reinstatement.

TABLE 1

Soil Physical Characteristics

	Horizon			
	Ap	Bg	BCg1	BCg2
Cockle Park (CP)				
Depth (m)	0-0.24	0.24-0.43	0.43-0.72	0.72-1.00
Particle size fraction (g g^{-1})				
Sand (63μm-2mm)	0.438	0.462	0.414	0.386
Silt (2 - 62μm)	0.180	0.251	0.231	0.271
Clay (<2μm)	0.243	0.313	0.349	0.330
Dry bulk density (g cm^{-3})	1.36	1.50	1.48	1.53
Pore volume fraction	0.455	0.440	0.426	0.440
Stone content mass fraction	0.026	0.021	0.28	0.042
Butterwell drained (BD)				
Depth (m)	0-0.20	0.20-0.52	0.52-0.82	0.82-1.00
Particle size fraction (g g^{-1})				
Sand (63μm - 2 mm)	0.428	0.388	0.380	0.397
Silt (2 - 62μm)	0.229	0.304	0.335	0.350
Clay (< 2μm)	0.285	0.284	0.268	0.273
Dry bulk density (g cm^{-3})	1.33	1.71	1.72	1.70
Pore volume fraction	0.480	0.345	0.350	0.370
Stone content mass fraction	0.023	0.031	0.050	0.039
Butterwell undrained (BUD)				
Depth (m)	0-0.22	0.22-0.50	0.50-0.82	0.82-1.00
Particle size fraction (g g^{-1})				
Sand (63μm - 2 mm)	0.426	0.401	0.385	0.368
Silt (2 - 62μm)	0.250	0.233	0.243	0.267
Clay (<2μm)	0.306	0.318	0.290	0.300
Dry bulk density (g cm^{-3})	1.44	1.58	1.64	1.66
Pore volume fraction	0.442	0.339	0.400	0.390
Stone content mass fraction	0.031	0.044	0.043	0.056

GROWTH OF WINTER WHEAT

Since the main objective was to study the effects of differing soil
conditions on crop growth, particular care was taken to ensure that the three
sites were managed in an identical manner within any one season. The aim
throughout being to ensure that agronomic inputs did not limit crop performance.
Details of crop monitoring and sampling procedures have been described by Evans
and Brook (1987).

In each of the four seasons 1981-2 to 1984-5 two series of 0.1 ha plots were established on each site. In one series winter wheat was grown continuously while on the other, the crop was sown after a ryegrass ley of between one and three years duration. Individual plots were subdivided for final yield assessment and for sequential sampling throughout the growing season. On each sampling occasion detailed measurements of shoot and root growth was made to assess the effects of site on dry matter production and individual components of growth and yield. At final harvest three areas, each measuring 26m x 2.1m were harvested from each block using a Claas Compact plot combine. Grain samples were retained for moisture and nitrogen determinations. Yield component analysis was based on six samples, each of $0.25m^2$ selected at the same time from the destructive sampling areas.

Grain yield and yield components

Grain yields are prsented in Table 2. In the first year the results, expressed as a mean for the rotational and continuous treatments, showed a significant site variation. Comparison of the two Butterwell sites demonstrates a clear benefit from early drainage. A similar trend was again recorded in subsequent seasons with the exception of the very dry season of 1984 when rainfall during the period April to August was some 100mm below the long term average.

TABLE 2

Effect of site and season on the yield of winter wheat (t ha^{-1} at 85% d.m.)

Site	Season			
	1981-82*	1982-83	1983-84	1984-85
Undisturbed (CP)				
Continuous		8.8	9.5	5.5
Rotational	12.5	10.7	11.5	8.1
Restored drained (BD)				
Continuous		9.2	6.5	3.5
Rotational	10.2	9.2	7.3	9.5
Restored undrained (BUD)				
Continuous		6.9	5.2	3.3
Rotational	6.6	6.6	8.0	4.2
S.E.D.	0.38	0.36	0.44	0.52

*Mean of continuous and rotational treatments

Yields from the continuously grown wheat declined progressively on all three sites and was largely accounted for by the build up take-all (Gammannomyces graminis) and eyespot (Pseudocarposporella herpotrichoides). In 1984-85 an extremely high incidence of take all was recorded on the restored drained site. Satisfactory yields were obtained each year from the wheat established after grass on both the control and drained Butterwell sites, although yields were significantly lower on the undrained Butterwell site, with the exception of the very dry season of 1983-84.

These differences in grain yields were examined in detail for the wheat grown after a grass ley in each of the four seasons. The weight of grain per unit area of ground at final harvest is the product of three primary components, number of ears m^{-2}, number of grains per ear and grain dry weight. In this study differences recorded in grain yield occurred as a result of variation in each of the yield components measured (Table 3). This is consistent with the views of McLaren (1981) who maintains that grain yield is not a direct function of any single yield component. However, the compounded yield component, grain number m^{-2} (the product of ear number m^{-2} and grain number per ear) was highly correlated (r = 0.936) with yield in all seasons (Figure 1).

TABLE 3

Effect of site and season on yield components in winter wheat grown after a grass ley.

Yield component	Seasons			
	1981-2	1982-3	1983-4	1984-5
Ears m^{-2}				
Undisturbed (CP)	475	483	493	457
Restored drained (BD)	417	498	324	481
Restored undrained (BUD)	435	364	309	216
		SED 23.1		
Grains per ear				
Undistrubed (CP)	42.1	48.5	34.9	34.3
Restored drained (BD)	36.9	43.2	45.6	40.2
Restored undrained (BUD)	20.4	34.8	45.5	43.9
		SED 1.01		
Grains m^{-2} (x 10^2)				
Undisturbed (CP)	199.9	234.3	172.1	156.8
Restored drained (BD)	153.9	215.1	147.7	193.4
Restored undrained (BUD)	87.7	126.7	140.6	94.8
		SED 9.03		
Grain dry weight (mg)				
Undisturbed (CP)	45.8	37.5	47.4	45.8
Restored drained (BD)	47.3	40.9	41.8	46.4
Restored undrained (BUD)	44.5	38.6	45.8	37.1
		SED 0.81		

Among the primary yield components, grain yield over the three sites in three of the four seasons was most closely correlated with ear number m^{-2}. In 1981-2 grain number per ear was the dominant yield component. Clearly both ear number m^{-2} and grain number per ear are strongly influenced by site and season.

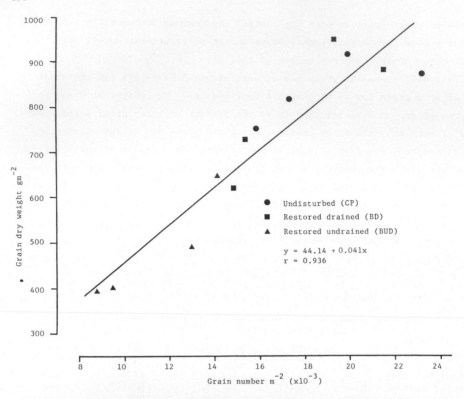

Fig.1. Relationship between grain yield and number of grains m^{-2}.

Development of the ear population

Ear population at harvest is the product of established plant population, tillering capacity and stem survival. All of these parameters were strongly influenced by both site and season (Table 4). Seed rate was adjusted on the basis of seed weight to a target of 400 seeds m^{-2} sown for all treatments. Field emergence varied from 86% down to 20% largely on account of seed bed conditions and the prevailing climate during the autumn. Establishment was particularly poor on the undrained Butterwell site in all seasons, except in the first year after restoration, due to the problems encountered in preparing a suitable tilth.

The maximum number of stems (tillers and mainstems) was generally achieved in early spring, declining thereafter towards final harvest. The ratio of ears surviving to maximum stem number is given as percentage stem survival. Differences in maximum stem numbers between the undisturbed and drained Butterwell sites were generally small, despite wide differences in plant populations. This would indicate that soil conditions at these two sites allowed plants to tiller freely and achieve a satisfactory stem population in early spring. On the undrained Butterwell site maximum stem numbers were significantly lower in 1981-2 after a cold winter with high precipitation and in 1984-5 after a very wet autumn reduced plant establishment. Clearly winter wheat grown on the undrained restored site was more susceptible to adverse growing conditions than either of the other two sites. Only in 1982-3 were crops able to achieve the level of tillering associated with high yielding wheat crops.

TABLE 4

Effect of site and season on plant establishment, tiller production and survival in winter wheat grown after a grass ley.

Growth parameter	Season			
	1981-2	1982-3	1983-4	1984-5
Established population in November				
Undisturbed (CP)	331	171	345	289
Restored drained (BD)	334	161	238	263
Restored undrained (BUD)	239	185	124	79
SED season- 23.6; Site - 20.4				
Maximum stem number				
Undisturbed (CP)	1352	1013	1064	1063
Restored drained (BD)	1273	930	852	1044
Restored undrained (BUD)	623	905	755	321
SED season- 64.0; Site - 55.4				
% stem survival				
Undisturbed (CP)	35.1	47.7	46.3	43.0
Restored drained (BD)	32.8	53.5	38.1	46.1
Restored undrained (BUD)	69.5	40.2	40.9	93.5
SED season- 11.74; Site 10.20				

Further analysis of crop growth suggests that soil conditions during early spring is a major determinant of grain yield on restored opencast land. Early drainage provides an improved soil environment for root growth, particularly in the surface layers where root growth is most active (Evans et al., 1986). This enables the crop to make better use of the available soil nitrogen and thereby encourage shoot growth. A large leaf canopy, in turn, improves the crop's ability to intercept available radiation early in the season, promoting both shoot biomass and grain yield (King and Evans, 1987).

GROWTH OF SPRING BARLEY

Parallel studies were undertaken to compare the growth of spring barley on the three sites described previously. These again examined the performance of continuously grown spring barley with barley following winter wheat in a grass-wheat-barley rotation. Crop management was similar for all crops within any one season. Sampling procedures were similar to the ones already described for winter wheat.

TABLE 5

Effect of site and season on the yield of spring barley (t ha^{-1} at 85% d.m.)

Site	Season			
	1982*	1983	1984	1985
Undisturbed (CP)				
Continuous		5.5	5.8	4.6
Rotational	8.0	5.2	4.9	4.6
Restored drained (BD)				
Continuous		5.1	4.7	4.1
Rotational	7.8	4.6	5.3	5.4
Restored undrained (BUD)				
Continuous		2.9	3.5	1.5
Rotational	5.6	3.1	3.3	2.1
SED	0.47	0.54	0.29	0.31

* Mean of continuous and rotational

Grain yield and yield components

Satisfactory yields of grain were obtained from the undisturbed Cockle Park and the drained Butterwell sites in the first year after restoration (Table 5). The benefits of early drainage was again evident on the restored site with yields from the undrained site being significantly lower than those achieved with early drainage. In subsequent years yields decreased across all sites due to a combination of unfavourable weather and soil conditions influencing crop growth at critical stages of development. In 1983 a period of wet weather delayed sowing until 13 May, well beyond the optimum sowing date for spring barley. In the other two years crops were drilled in April, but growth was adversely influenced by the dry summer of 1984 and the cool wet season of 1985. Extremely low yields were recorded on the undrained Butterwell site throughout this period which again suggests that the adverse soil conditions created by the absence of drainage after restoration was a major limitation on the growth of spring barley.

Differences between continuous and rotationally grown spring barley were less evident than those reported for winter wheat. Take-all and eyespot levels were extremely low, confirming previous observations that barley is less susceptible to these diseases than wheat (Shipton, 1972).

The inlfuence of site on grain yields can be assessed on the basis of ear population and grain number per ear. Treatment differences for the two contrasting seasons 1983 and 1984 are presented in Table 6. The low yields obtained from the undrained restored site arose largely as a result of low ear numbers, especially in 1983. Individual grain weight varied little from site to site. Low plant establishment was a major factor contributing to the low final ear population on the undrained Butterwell site in both seasons.

TABLE 6

Effect of site on ear population and grain number in spring barley in 1983 and 1984.

Yield component	Seasons	
	1983	1984
Ears m^{-2}		
Undisturbed (CP)		
Continuous	670	847
Rotational	610	742
Restored drained (BD)		
Continuous	586	591
Rotational	590	835
Restored undrained (BUD)		
Continuous	393	443
Rotational	394	554
SED	43.1	86.5
Grains per ear		
Undisturbed (CP)		
Continuous	21.2	19.5
Rotational	21.0	20.0
Restored drained (BD)		
Continuous	21.2	21.9
Rotational	20.5	20.1
Restored undrained (BUD)		
Continuous	18.3	20.3
Rotational	19.5	20.5
SED	0.40	0.83

MANAGEMENT CONSIDERATIONS

The work reported here has shown that satisfactory yields of cereals can be achieve from restored opencast land provided an effective drainage system is installed at an early stage of the restoration process. It is of interest to note that drainflow has not been affected by land settlement over the period of study. Detailed examination of soil water movement on the two Butterwell sites by King and Evans (1987) have shown that the overall volume of water retained within the soil profile was not influenced by drainage. However, the soil environment for cereal root growth was greatly enhanced following drainage as a result of improved water movement through the surface layers. This in turn stimulated root growth and increased nitrogen availability and crop uptake.

Such enhanced soil conditions were particularly beneficial during the early spring period for winter wheat. Shoot growth, particularly the leaf canopy was enhanced, tillering increased and the additional assimilate supply enabled more tillers to survive to final harvest. Grain number and grain weight also benefited from the increased assimilate supply.

The performance of spring barley suggests that soil conditions in early spring in the absence of drainage is a major limitation to crop growth. Wet soils delayed drilling and reduced plant establishment while root growth was less vigorous and crops became more susceptibility to drought in early summer. It would therefore appear that restored land is less suitable for spring cereal production.

Satisfactory yield of winter cereals can be achieved on restored opencast land provided sufficient attention is given to the following husbandry inputs.

(a) Seedbed preparation

-subsoiling to reduce compaction below plough depth,

-plough and cultivate the land early to avoid having to work wet soils in the autumn,

-ensure that the number of operations are kept to a minimum to avoid surface compaction.

(b) Time of sowing

-crops should be sown by the middle of September to achieve satisfactory establishment and adequate shoot growth before the onset of winter.

(c) Seed rate

-particular attention should be given to seed rate to enable a target population of 300 plants m^{-2} to be established,

-adjust the weight of seed sown on the basis of its thousand seed weight.

(d) Seedbed fertiliser

-phosphate and potash levels should be based on the results of soil analysis.

-seedbed nitrogen is beneficial for early sown crops on restored land.

(e) Spring nitrogen

-an early application of 50 kg ha^{-1} of nitrogen in March encourages tillering

-the main top dressing should be applied according to crop development,

-a late application of 40 to 50 kg ha^{-1} nitrogen at the time of flag leaf emergence delays leaf senescence.

(f) Crop protection

-wet soil conditions may delay spring herbicide application, therefore weed control programmes should be based on autumn herbicides,

-effective pest and disease programmes are essential if high grain yields are to be attained.

ACKNOWLEDGEMENTS

The financial assistance provided throughout the period of study by the Opencast Executive of British Coal is gratefully acknowledged.

REFERENCES

1   Brent-Jones, E. Land Reclamation in the 80's - The National Coal Board's Techniques. Symposium on the Reclamation, Treatment and Utilisation of Coal Mining Wastes, Durham. (1984). Paper 27.
2   Brook, D.S. and Bates, F. Grassland in the restoration of opencast coal sites in Yorkshire. Journal of the British Grassland Society, (1960) 15: pp116-123.
3   Evans, E.J., Leitch, M.H., Fairley, R.I. and King, J.A. Comparative studies on the growth of winter wheat on restored opencast and undisturbed land. Reclamation and Revegetation. (1986) 4: pp223-243.
4   Evans, E.J. and Brook, R.M. Growth of grass and cereals on restored opencast land. (1987) (in press).
5   Hunter, F. and Currie, J.A. Structural changes during bulk storage. Journal of Soil Science. (1956) 7: pp75-80.
6   King, J.A. and Evans, E.J. The response of winter wheat to nitrogen on drained and undrained areas of restored opencast land. (1987) (In press).
7   McLaren, J.S. Field studies on the growth and development of winter wheat Journal of Agricultural Science. (1981) 97: pp685-697.
8   Shipton, P.J. Take-all in spring-sown cereals under continuous cultivation: disease progress and decline in relation to crop succession and nitrogen. Annals of Applied Biology. (1972) 71: pp33-46.
9   Tasker, T. Restoration of opencast coal land. Agriculture (1957). 64: pp329-332.
10 Tomlinson, P. The agricultural impact of opencast coal mining in England and Wales. Minerals and the Enviornment. (1980) 2: pp78-100.

Reclamation, Treatment and Utilization of Coal Mining Wastes, edited by A.K.M. Rainbow
Elsevier Science Publishers B.V., Amsterdam, 1987 — Printed in The Netherlands

PRE-MINE BASELINE DATA : AN ESSENTIAL TOOL IN RECLAMATION OF

COLLIERY WASTES.

S.O. ADEPOJU   and   G. FLEMING

Department of Civil Engineering,University of Strathclyde,
John Anderson Building, 107 Rottenrow,Glasgow G4 ONG ,
Scotland ,U.K.

SUMMARY
    Successful reclamation and utilization of colliery wastes will
result from EFFICIENT design and execution of a mining plan having
some BASELINE DATA.This must consider the mining and processing
methods,topography,geology,local climate and the physical,chemical
and biological properties of the area to be mined. In addition,the
potential of the mining wastes for possible after_use should be
explored and considered during design,bearing in mind the problems
inherent in restoration schemes. Reclamation process,therefore,
should be regarded as an integral part of the mining process if
cost and time are to be optimized.
    Baseline data outlines the premining conditions,helps determine
reclamation goals and serves as a basis against which reclamation
success can be measured. Recognizing that complexities of soils,
variations with climate,land use and management systems make it
highly difficult to have a simple formula for colliery
reclamation, this paper reviews and outlines some strategies and
approaches that should facilitate colliery waste reclamation and
amelioration programmes. Moreover,it outlines research proposals
initiated to examine the basic improvement in characteristics of
certain colliery wastes when mixed with other materials such as
dredged silt,PFA,sewage sludge and domestic refuse.
    The research objective is to establish guidelines on selection,
control and monitoring of settlement characteristics and other
properties relevant to the proposed after-use.

INTRODUCTION
        Reclamation and treatment of colliery wastes appear to be
extremely difficult projects because there can be as many problems
as there are sites. The engineering and biological properties of
colliery wastes are subject to wide variations since a single
spoil may consist of a complex mixture of shales,clay minerals,
sandstone, quartz dominated mudstones,etc. Although these
problems are varied and often difficult,an indispensable tool in
formulating a solution is experience, combined with access to
BASELINE DATA from the site. Reclamation tasks always remain

interesting and often emotive. It is essential at the onset that land reclamation (treatment and utilization inclusive) be regarded as an integral part of the mining process for ultimate optimisation of cost and time.

There are hundreds of thousands of hectares of wastes across the nation created by extractive industries which,at present,may involve "prohibitive"cost to restore but which would have cost a fraction of the present cost had restoration programme been considered during the design period! The only way mineral extraction can be made environmentally acceptable is to integrate the restoration programme when designing for mining or quarrying It is through this procedure that various expensive problems, such as encountered in acquisition during postmining reclamation,can be alleviated,or even eliminated.

A few case studies have shown how cost and time effective it could be to incorporate restoration and utilization programmes during the design for mining.One such case,cited by Barr (ref. 1), involves a successful integration of reclamation with mining plan in the early seventies.The county planners at Colwick in Nottinghamshire agreed with Hoveringham Gravels on a fourteen year extraction programme.Throughout the extraction period,the recreational after_use governed the plan of excavation and in the end (around 1980) there were two "purpose built" lakes occupying an area of about forty five hectares of what would probably have added to existing derelict land.It could be argued,however,that it is unrealistic for planners to impose arbitrary conditions on the workings,whether in terms of reclamation scheduling with mining thereby leading to time delay,or extra cost during the operation,but the development of the total resource should be planned with minimum disturbance to the environment.

A well planned waste and overburden removal and the proper execution of it would undoubtedly improve the chances of increased reserve recovery and this in turn would allow for utilization of the wastes as part of an integrated development plan. Of course,it is necessary to know well in advance the quantities and production rate of colliery waste materials for design purpose,and in order that sufficient land areas may be created for its disposal,and provision made for handling.

The success of all the foregoing depends much on the amount of baseline data available to the designer and the operator,as well as the thoroughness with which the jobs are carried out. To this

effect,this paper reviews some aspects and outlines some
strategies and approaches that should facilitate reclamation and
amelioration programmes for colliery wastes. Outlines are also
given of some research proposals that look into the improvement of
some basic characteristics of certain colliery wastes when mixed
with other materials such as dredged silt (refs.2-3),PFA and
domestic refuse (ref.4).

FLOWCHART FOR INTEGRATED MINING AND RECLAMATION SCHEMES.

Fig.1.shows the suggested flowchart of various pre and postmine
activities as related to each other for successful operation of
mining and reclamation schemes.

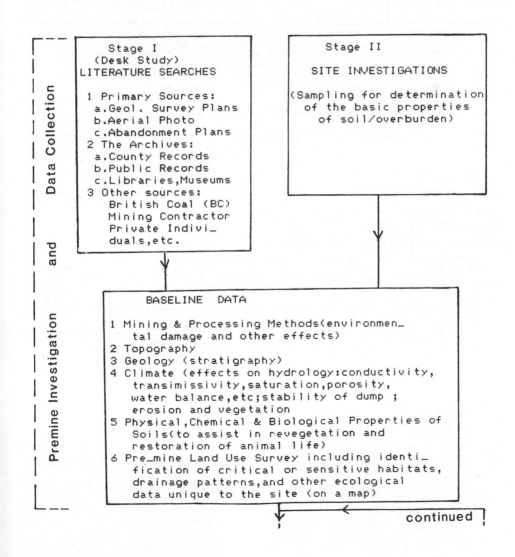

continued

474

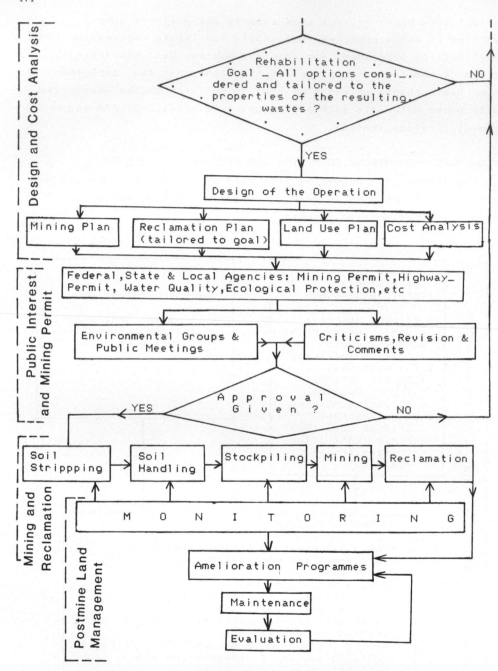

Fig.1. Flowchart for Integrated Mining and Reclamation Schemes.

The flowchart has been developed from concepts contained in a number of references (refs. 6,17 & 22). Fig.1 is a summary of more detailed processes. The Soil Stripping,Handling and Stockpiling stages have been shown in more detail in a flowchart by Schafer (ref. 5) and Holmberg (ref. 6).

SOURCES OF WASTE IN COAL MINES

In 1983,British Coal was reported to be generating approximately 200 million tonnes per year of colliery wastes from opencast operations, and about 50 million tonnes per year from underground mining (ref. 7).The waste production in coaling operation results from three parent sources and are related to the geological characteristics of the locality. These sources are ; the overburden, interburden and the floor materials. Some substandard sections of the deposit which do not measure up to the required specifications also result as wastes. The waste materials may consist of contaminated rocks associated with faults, replacement cavities,kettlebottoms,intercalated bonds of clay,shale or other weak materials,and some are even inherent in the seam.These waste productions occur in the excavation stage and is referred to as the run_of_mine waste. Subsequent to the excavation process,the largest volume of solid waste is produced at the preparation plant where coal is crushed,sized,washed and separated from rock and other impurities.The second stage waste produced is called the washery waste or coal washery discard (CWD).The CWD is separated into two classes (ref.8):coarse (greater than 1mm) and fine (less than 1mm). Coarse refuse is typically conveyed to disposal area by belts,tramways,or trucks while fine waste is transported hydraulically to slurry ponds. It is important to have adequate foreknowledge of the geological type and hence the type of material in the coaling area since it enables a better design for waste handling and treatment.This Knowledge could only be acquired from the first phase of the pre_mine data (see Fig.1.). The properties of the samples taken during formation of data base would give an indication of the would be behaviour of the waste materials and the designer can make provision for handling of any impurities expected to create a nuisance during reclamation process. Adequate sampling should be conducted during drilling operations to have a representative data of the real situation to enable the engineer to be familiar with

the materials,their behaviour,and subsequently the construction
operations necessary from premining to postmining phases to obtain
the desired results.

THE OBJECTIVES OF RECLAMATION

   While some experts believe that prior land use should only
serve as a guide in the development of reclamation and not
strictly adhered to as the ultimate target most mining regulations
set the pre-mine land use as the main target of restoration.
Whiffen and Walker (ref. 9) argue that plans for after use need
not be the catalyst to begin reclamation.In any case,reclamation
is generally undertaken either for economic reasons,or for
aesthetics (topography inclusive). The former vary widely and they
range from the return to economic use by agriculture and forestry
to planning of new industries. Examples include: housing
development,road building,increased amenity of the area,
development of facilities for outdoor recreation and protection
against wind and water erosion. The latter reason might be solely
for preservation of nature and the improvement of visual impact.
The objective of reclamation (or restoration goal, as denoted in
the chart in Fig.1.) needs careful scrutiny as it depends much on
the potential of the mined area and the resulting wastes. As
indicated in Fig.1, if mining permit is not granted for one reason
or another after public inquiries from various interested groups
(viz: environ-mental, ecological,water,private individuals,etc),a
critical point is reached where the plans and goal of restoration
should be reviewed and cross checked with the potential of the
area to be decoaled, and the postmine problems likely to arise or
interfere with other interested groups. Here, a delay of several
months or years may be inevitable. A change in the ultimate goal
of reclamation usually affects mining and land use plans and hence
the cost analyses.

   Should the approval for mining permit be given the next main
task is the mining operations. The execution of this phase (see
Fig.1) should be guided by the objectives of reclamation
predetermined from the pre-mine data collection and investigation
phase. Selective handling and stockpiling of materials are more
often necessary for cases where farming is the ultimate aim for
reclamation. The types of machine employed in excavation, and
material placement method should be carefully chosen to effect the
necessary conditions for plant growth. Too much compaction during

placement could seriously impede root penetration which is essential for provision of water to plants,especially during the dry season. Holmberg (ref. 6) outlined some factors requiring analyses for topsoil storage: the nature of the area permanent or temporary;the quantity and characteristics of material for storage;potential toxicity and stability problems once the material is in place;suitability and feasibility of removing topsoil from the site for later use in reclamation;and the material placement method. Other factors include the recommended slope and aspect of the fill;current vegetative productivity of potential storage areas and predictions on the success of reclaiming the storage area;and access from mine operation to the storage site and removal access if storage is temporary. Other external factors affecting the goal of reclamation include the topography and its effect on future stability; wind and water on storage site; feasibility of erosion control practices against leaching; and drainage. Research has also established (ref. 6) that,in the process of topsoil/overburden respreading,chemically undesirable zones do not need special handling if less than 15% of the total area because they could be diluted enough to be harmless to plant growth and ground water. Aside with the agricultural objectives,if the aim of reclamation is for development (that is,erection of buildings for residential or industrial purposes) then the product of soil reconstruction must satisfy the following requirements:

  i) it must be strong enough to support its own weight and the structural load or wheel load on it

  ii) it must not settle or deform to the extent of causing damage to the structure on it

  iii) it must not undergo excessive swelling or shrinkage

  iv) its strength must be retained permanently,and

  v) its physical and chemical characteristics must be environmentally acceptable.

The importance of baseline data is emphasised if the aim of reclamation is to be successful. This is evident because the aims of reclamation should have been tailored to the predetermined basic and engineering properties of the resulting colliery wastes.

UTILIZATION OF PITS AND WASTES

Finding economic uses for colliery waste materials improves the profitability of mining operations in two ways (ref. 10):

    i) by increasing revenue, and

    ii) by reducing the cost of disposal.

Colliery wastes find uses in four major areas:in reclamation of
voids; as engineering fill; as building materials; and for
extraction of valuable constituents. If no other uses can be found
for colliery waste it must be returned to the parent void which
can only take a limited amount.This is due to the Swelling Index
of colliery overburden being greater than 1.0,and the volume ratio
of overburden to coal (stripping ratio),in cubic metres, generally
varies between 5:1 and 25:1,the average being around 15:1 '
depending on the depth of working. The greater the value of the
two preceding indices the more the volume of wastes to be disposed
of.Other uses are found in engineering fill where low lying areas
can be raised to economic use by utilizing colliery wastes.
Building bricks, concrete and tiles are also manufactured from the
wastes. A new area that may utilize a lot of colliery wastes is in
the construction of double stoppings (ref. 11) for the improvement
of underground mine ventilation, and creation of additional roof
support to prevent ground subsidence.

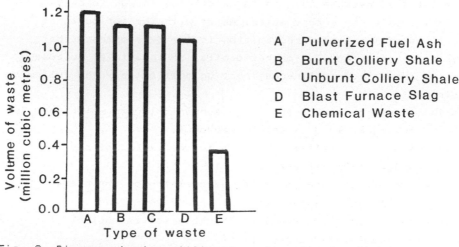

A   Pulverized Fuel Ash
B   Burnt Colliery Shale
C   Unburnt Colliery Shale
D   Blast Furnace Slag
E   Chemical Waste

Fig. 2. Diagram showing utilization of wastes for highway
     construction in Lancashire between 1955 and 1970.

    The Road Research Laboartory has successfully utilized burnt
colliery shale and spent oil shale in road fills and sub_bases
with addition of a small proportion of cement to make it suitable
under freezing conditions. Fig.2 shows the utilization of colliery
wastes for highway construction in Lancashire (ref. 12) between
1955 and 1970. In addition,shale has been found useful in the

manufacture of compost fertiliser,and minestones are also used as site capping materials. Moreover,valuable constituents such as alumina are extracted from the ash product;iron from pyrite; sulphur;and trace metals such as uranium,vanadium and berylium can be separated from the ash after combustion. Other authors (ref. 12) have described in detail some of the extraction processes.

Pulverised Fuel Ash (PFA),another important colliery 'waste', has been successfully utilized in some areas and is still being considered in others (ref.13).It has been found useful in making acoustic plaster(to replace pumice); as parting sand in magnesium and aluminium foundries;for mechanic soap;for metal polish;for water filter;as replacement for Calcium Carbonate in primer paint;and may be used as roofing material to displace limestone. Other possible uses are in sand blasting;in thermal insulation (to replace calcine clay); and for roofing tiles. In addition, PFA has been found useful in projects allied to land reclamation. Its high absorbent capacity may be useful as a drying agent in wet areas,and its property as a pozzolana may be helpful in alleviating the toxicity of chemicals in rock wastes and in rehabilitation of other contaminated lands. These are being considered in a Scottish Development Agency funded project at the University of Strathclyde (ref. 2). The ash has also been successfully utilized in various grouting jobs. Some reclamations without planned after use have been positively counter productive (ref. 14). The pits created by removal of coal can provide a good geological reserve subject to absence of vegetation,and the presence of suitable geological exposures.It can also be used for nature reserves if revegetation and wall stabilisation are carried out. Other uses include industrial development,subject to proximity to urban areas and access to transport network;water storage and landscaping for recreational use if sub water table pit;tipping of domestic or industrial waste if no pollution potential to ground water is likely. It should also be pointed out that the situation of a pit, whether wet or dry,affects the tipped material. Tipping into water produces much more rapid consolidation than tipping into a dry pit. As has been noted,the water tends to break down clay lumps and other similar materials thereby filling the voids within the fill with the finer particles. Further possible uses of various types of workings, based on their physical characteristics,are shown in Table 1 (ref. 15). Again,it is worth reiterating that having enough information

480

from the baseline data would make possible the determination of
the best after use(s) for the site being considered.

TABLE 1
Possible after uses associated with various types of mineral
workings,based on their physical characteristics (after Coppin
(ref. 11)).

POSSIBLE USES	Excavations				Tips	
	Deep		Shallow		Steep	Graded
	Wet	Dry	Wet	Dry		
Original Uses				m		m
Agriculture				M		M
Forestry				M	M	m
Fish Farming	m		m			
Intensive recre- ation & Sport	m	m	M	M		m
Extensive recre- ation & parks	m	M	M	m		m
Nature conserva- tion & wildlife	m	m	m	m	m	m
Water storage & supply	m		m			
Housing &Industry		m		m		
Landfill & Waste disposal*		m		m		

Possibilities: M major
m minor
* temporary use only.

ENVIRONMENTAL PROBLEMS

The severity of the environmental problems created by opencast
and colliery waste depends on the characteristics of the waste
materials and the situation of the site.The problems vary from
physical visual intrusion to slope instability, pollution of
surface run off,land sterilisation,dust blow,spontaneous
combustion,erosion on steep slopes,improper drainage,sand
blasting,etc. Formulation of solutions to most of these problems
would depend on proper analyses of the data from the initial
stage,the foresight and the skill of the designer and the
operators.Generally, the problem of visual intrusion depends on

the choice of dump site,stockpiling location,site screening
technique,material placement method and proper landscaping. Slope
instability caused by erosion can be corrected by erosion control
matting or cementation;the use of some form of dust suppressant
for dust blow; compaction (ejection of air_ within the waste) to
alleviate spontaneous combustion;and control of sand_blasting by
the use of windbreaks,revegetation or other form of stabilising
agents. The problem of acidity can be combatted by the application
of the correct amount of lime,and toxicity may be alleviated by
deeply burying the toxic materials or diluting by mixing. Various
solutions have been suggested in a research report by the
Department of Energy and Transport (ref. 16).

SOIL,OVERBURDEN AND SPOIL SAMPLING
    Compilation of a soil map is essential from the investigation
stage to aid in the sampling process.The soil map contains
information on the colour, texture,structure and drainage status
of different soil horizons.The map will help in identification of
different soils and rocks,and regardless of the sampling method
employed the sampling areas should be thoroughly inspected,and all
types of soil and rocks must be separately sampled for laboratory
testing and analyses.The results of soil and overburden analyses
form the basis on which important decisions and recommendations
would be made about the mining and reclamation techniques and the
ultimate reclamation goals.Experience and a great deal of
ingenuity are required for high precisions. Some errors of
judgement at this stage could prove costly at a later stage of the
operations! The properties of soil,overburden and spoil that
affect the final use are: moiture content,particle size
distribution,texture,stoniness,pH value,the amount of trace
elements,nutrient conditions,densities and the bearing capacity.In
addition,there are other properties that are helpful and on which
routine testing are recommended for spoil whose reclamation plan
was not incorporated in the mining plan.These include the
Atterberg limits,specific gravity and sulphate contents. Fleming
et al (ref. 3) stress the need for a review of guidelines and
standards on these properties to achieve better quality of
recycled soil wastes for use as natural soils. The ultimate
reclamation goal is selected from a list of possible after_uses
but the goal should be flexible enough to accommodate any new
development during mining operations. The properties of soil,

overburden and spoils that affect the final use of a site are
summarised in Table 2 (from Schafer, ref. 17).

TABLE 2
The properties of soil, overburden and spoils that affect the
final use of a site (ref. 17)

INDICATOR PARAMETER	Cover soil quality			Potentially toxic
	Good	Fair	Poor	
Texture Class*	vfsl,fsl,sl, ls,sil	lfs,ls,cl, scl,sicl	s,c,sc, sic	
pH value	5.5-7.0	4.5-5.5,	<5.4,>7.5	4.0,8.5
Moisture content (%)	>10	5-10	<5	
Trace elements,ppm				
Boron	--	<2	2-8	>8
Cadmium	--	<1	1-3	>3
Copper	--	<40	40-100	>100
Lead	--	<20	20-100	>100
Zinc	--	<40	40-200	>200
Pyrite (%)	--	<0.2	0.2-3.0	>3.0

* c=clay, f=fine, l=loam, s=sand, si=silt, vf=very fine

PLANNING AND DESIGN

Jones,(ref. 18),explains the essence of planning when he
wrote,`Restoration of a site has to be planned before work
starts........ Unless arrangements are made in advance for the
materials needed for a good reclamation to be set aside from the
first day of the work,it will not be possible later to obtain the
best restoration possible.` Exhaustive planning and comprehensive
collection of data are essential to make restoration an
economically viable venture.Some experts have even opined that a
satisfactory restoration can be achieved if an opencast operation
is planned and operated from the beginning as if the restoration
were the prime aim and the coal a valuable by_product.This
opinion,however,remains provocative! Based on the data previously
collected,the following decisions and designs have to be
established for reclamation (ref. 19),and incorporated in the
mining plan in the design phase in Fig.1:
    i) the thickness of topsoil/subsoil to be removed and given
        separate storage and protection until resoiling.The least
        amount of material must be moved the least distance for

      cost optimisation
  ii) location and the amount of dumps (inclusive of filling
      jobs),final slopes,vegetation and drainage pattern
 iii) areas of selective mining and the quantity of toxic
      materials to be handled separately
  iv) noise abatement and dust control methods
   v) final level of refilled pit and erosion control
  vi) restoration of drainage system,location of settling pond
      and purification of mine water before discharging
 vii) soil/overburden respreading technique
viii) technique for monitoring progress during various
      operations and after their completion in compliance with
      statutory standards.
  ix) appropriate revegetation for the resulting landform and
      amelioration programmes necessary for the restored
      areas;and
   x) the method of evaluation.
It is often recommended that the mining company maintains a strip
of undisturbed land in the lease as a standard and as evidence for
restoration when comparing with premining conditions (ref. 19).

QUANTIFICATION OF RECLAMATION COSTS AND BENEFITS

   Many of the previous attempts to quantify the costs and
benefits of land reclamation have not been very successful because
it is difficult to attach values to many items involved in the
operation for two main reasons:
    i) some phases of reclamation operation overlap with mining
       proper, hence it can not be completely isolated from
       mining
   ii) reclamation operation is but one part of a continuing
       process of environmental renewal,therefore an attempt to
       isolate it will make it appear very expensive and this
       may be misleading
  Simply,the cost of reclamation,C_R, can be derived from:

$$C_R = C_{MR} - C_{MNR}$$

where C_R is the cost of reclamation;C_{MR} is the total cost of
mining operation requiring reclamation; and C_{MNR} is the total
cost of mining operation not requiring reclamation.
Items of C_R include both the capital (Cc) and operating (Co)

484

costs. Cc includes the cost (C_E) of additional machines
required for topsoil removal and respreading, and the costs
(C_L) required for the rehandling and levelling of the
resulting waste dumps. The cost (C_B) of collection of baseline
data necessary for the determination of the best after use of
the mined area is also part of the capital cost. The major items
of cost for baseline data collection include: soil testing;
climatic studies; hydrology; biologic investigations;
administration; socio_economic studies; and archaeological
investigations (ref. 19).

Co includes the cost of various monitoring programmes (C_M),
laboratory and field testing of materials and their
preparations.
The costing list for reclamation operation itself consists of
rehandling of waste;topsoiling/landscaping; revegetation (C_V);
drainage installation (C_D); and fence installation (C_F). So,
the cost of reclamation can be estimated by

$$C_R = Cc + Co$$

$$= C_B + C_E + C_L + C_M + C_V + C_D + C_F$$

where all the parameters are as defined above. The cost is
affected by a number of factors: the mining method,the depth of
coal seam, the type of waste material,and the ultimate after-use
of the reclaimed site. The result of an investigation (ref.20)
shows that,given the same site and conditions but different
after_uses, restoration for housing schemes appears to be the most
expensive while restoration for forestry appears to have the least
cost.Fig. 3 shows different end_uses in the cost hierarchy.The
most expensive is at the top while the least expensive is at the
bottom.

Quantification of the value of reclaimed land,and the cost
effectiveness or benefits of land reclamation is not easy if an
enhanced commercial value is not involved.Some potential benefits
of a reclaimed site can be expressed in terms of:
    i) elimination of potential hazards
   ii) improvement of local environment
  iii) avoidance of risk of expensive retrofits in latter years
   iv) improved engineering
    v) credibility in legislative testimony and regulatory

hearing,and
vi) avoidance of delay in securing mining permit in latter
projects.

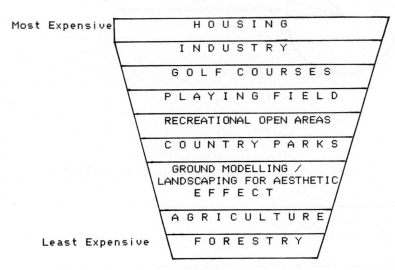

Fig. 3. Different end_uses of mined_land in the cost hierarchy.

The value of reclaimed land (V_R) can,therefore,be computed as:

$$V_R = V_D + C_R(1+i)^n$$

where V_R is the value of reclaimed land; V_D is the value of
derelicted land; C_R as defined above; i is the rate of
return,commensurate with the first four benefits listed above
in addition to the `natural` market interest rate. `i` should also
relate to the market demand for the ultimate after-use; and n is
the post reclamation period in years. An attempt to quote specific
figures in terms of cost for reclamation may be misleading since
the cost varies with different site conditions and the mining
method.It may suffice to indicate,however,that site reclamation
cost may vary between 10% and 26% of the total cost of the mining
contract or even more, with infilling of the final void taking the
largest share. The breakdown of (C_R) for 1973 in the U.K.
(ref. 19) is shown in Fig. 4. The cost of ground stabilisation
by drilling and pressure grouting has also been reported (ref.21)
to range from £25 (1985 pounds sterling) per sq.metre of building
area for a thick seam at greater depth,to £10 per sq. metre for a
thin seam at very shallow depth. Table 3 shows some unit costs for

specific rehabilitation works on non_colliery areas.

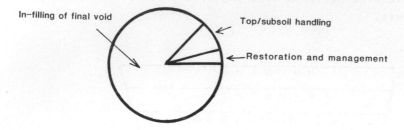

Fig. 4. Distribution of reclamation costs.

TABLE 3

The unit costs for some rehabilitation and ground stabilisation works.

Period	Nature of Job	Unit Cost*
1979	Exavation and drying of dock silt (cu.m.)	£12.00
1980	Site development work on loose or soft natural soils (sq.m.): (a) potential factory floor	£24.71
	(b) developable plot area	£ 9.22
1982	Redevelopment of former industrial (crucible_ manufacturing) sites (cu.m.):	
	(a) excavation and disposal off site	£ 5.93
	(b) filling (grade I)	£ 9.52
	(c) filling (grade II)	£ 8.11
1984	Rehabilitation of gaswork site:	
	(a) site fencing (per metre)	£14.04
	(b) excavation and filling (cu.m.)	£12.40
1985	Dynamic Compaction (sq.m.):	
	(a) employment of < 8_tonne tamper with 50_tonne crane (excluding on/off)	£2-£2.50
	(b) 8-15_tonne tamper with 100_tonne crane	£2-£3.50
	(c) > 15_tonne tamper with 100-150_tonne crawler or tripod (ref. 24)	negotiated

*Cost depends on the site conditions.

CONCLUSIONS

The most obvious conclusion to be drawn from various problems inherent in the reclamation,utilisation and treatment of colliery wastes is that many of these difficulties can be alleviated if enough pre_mine data are obtained, and adequate sampling and proper data analyses are carried out to provide the designer and the operators with the physical,biological and engineering behaviours of the resulting colliery wastes.The integration of the reclamation plan with the mining would also lessen many of the problems. Mining and restoration techniques should be selected to suit the ultimate goal of reclamation which in turn should be tailored to the predetermined characteristics of the resulting wastes. The mining technique should allow for concurrent restoration of the area being decoaled and should give room to minimum environmental disturbance possible.

Colliery wastes find uses in many areas,the choice is made based on the characteristics of the resulting wastes.The cost of reclamation is not easily quantifiable since it overlaps with some phases of mining operation. Therefore,all possible options should be examined in term of market demand for ultimate after-use to get optimum cost.The application of Computer Aided Design (CAD) and Computer Aided Management (CAM) is advocated.Interactive Surface Modelling (ISM) would alleviate problems faced in general volume calculations,contouring,waste disposal scheme monitoring and the mineral extraction operations.Further development should be explored in the use of ISM to calculate and represent graphically volumes and voids in various stages of open pit mining,and the final form of the landscape. Research currently underway on reclamation at Strathclyde University aims to survey the derelicted (mineral working related) lands and restoration pattern in Scotland; assessment of restoration techniques and the improvement of colliery waste by addition of other materials such as PFA,dredged silt and domestic refuse; and the laboratory and field studies of settlement characteristics of different colliery overburden materials. The research aims to explore the potentials of surface mined lands and quarry voids,and the possible after_uses of demineralized lands. Basic information are sought on the behaviour of strip mining wastes such as colliery shales, mudstones and peat, and their interactions with other materials listed above,for the purpose of obtaining guidelines on selection, control,and monitoring of settlement characteristics and other properties relevant to the proposed after-use.

The results of the study and the techniques developed should be transferrable to other countries including the developing countries.

REFERENCES

1   J. Barr , Derelict Britain, Penguin Books,1969.

2   G. Fleming, J. Riddell and P. Smith, Feasibility Study on the
    Use of Dredged Material from the Clyde Estuary for Land
    Renewal,Report to the Scottish Development Agency by Civil_
    Engineering Dept.,Strathclyde University,June 1985.
3   G. Fleming,J. Riddell and P. Smith, Land Renewal with Topsoil
    from Clyde Port Dredgings, ASCE Proceedings,Water Forum `86,
    World Water Issues in Evolution,Vol.I,Long Beach,California,
    August 1986
4   G. Fleming, P. Smith, M. Chafer and A. Dickson, Landfill
    Technology Research,First Interim Report,Dept. of Civil
    Engineering,University of Strathclyde,Glasgow,Oct. 1985.
5   W.M. Schafer, Cover_soil management in Western surface mining
    reclamation, Symposium on surface mining hydrology,
    sedimentology,and reclamation. Lexington: University of
    Kentucky,1979.
6   G.V. Holmberg, Land use,soils and revegetation, Surface Mining
    Environmental Handbook,Elsevier Inc.,New York,1983.
7   Mining Technology, September 1983, 337-342
8   Mining Enforcement and Safety Administration (MESA)
    Information Report,1975, IR 1023. Federal Interest in Coal
    Mine Waste Disposal.
9   P. Whiffen and D. Walker,Rehabilitation at Stonyfell Quarry,
    Quarry Management,May 1984.
10 S.C. Brealey, Colliery tips and tailings disposal, Mining
    Technology, September 1983.
11 S.O. Adepoju, Double_Stopping System For Mine Ventilation,
    M.Sc.Thesis,Graduate School,West Virginia University,
    Morgantown,U.S.A.,August, 1981.
12 D. Tattersal, The reclamation of Lancashire ,December 1970,
    Manchester Statistical Society.
13 J.J. Bakker, Utilization of flyash ,Reasearch Topic,University
    of Strathclyde,1969.
14 C.D.W. Savage and H.G. Keefe, Reclaimed Land ,Sir Alexander
    Gibb & Partners.
15 N.J. Coppin, After_uses for pits and quarries, Quarry
    Management and Products, September 1981.
16 Departments of Environment and Transport, The environmental
    impact of large stone quarries and open_pit non_ferrous metal
    mines in Britain,Research Report 21, 1976.
17 W.M. Schafer, Guides for estimating cover_soil quality and
    mine_soil capability for use in coal strip mine reclamation in
    the Western U.S., Reclamation Review 1979,vol.2, 67-74.
18 B.E. Jones, Methods and Costs of Land Restoration , 16th
    Course in Quarry Management,The Quarry Managers` Journal:
    Institute of Quarrying Transactions,October 1971.
19 W.L.G. Muir, Reclamation of surface_mined land , Miller
    Freeman Publications Increase, 1979.
20 J. Casson, Dereliction _ Its nature and potential,
    Proceeedings of the Derelict Land Symposium;the University
    of Leeds,1969.
21 A.J. Smith, Land Reclamation After Mining, Colliery Guardian,
    October 1985.
22 M.V. Symons, Sources of information for preliminary site
    investigation in old coal mining areas, Proceedings,Conference
    on Large Ground Movements and Structures,UWIST,Cardiff,1977.
23 F.J. Brenner, Land Reclamation After Strip Mining in the
    U.S.A., Mining Magazine;September 1985.
24 A.D. Crossley and G.H. Thomson, Land redevelopment involving
    ground treatment by dynamic compaction,Technical Notes (TN5)
    ICE Conference on Building on Marginal and Derelict Land
    May 1986.

Reclamation, Treatment and Utilization of Coal Mining Wastes, edited by A.K.M. Rainbow
Elsevier Science Publishers B.V., Amsterdam, 1987 — Printed in The Netherlands

RECLAMATION OF MANNERS AND PEWIT COLLIERIES, ILKESTON, DERBYSHIRE

P E WRIGHT [1] and K SHIPMAN [2]

1 Reclamation Group Leader, Planning Department, Derbyshire County Council, County Offices, Matlock, Derbyshire, DE4 3AG (Great Britain)

2 Senior Assistant, Reclamation Group, Planning Department, Derbyshire County Council, County Offices, Matlock, Derbyshire, DE4 3AG (Great Britain)

SUMMARY

The 54 hectare reclamation scheme included the site of the Pewit Colliery which closed in 1875 and had subsequently been used for household refuse disposal; the Manners Colliery; several old railway embankments; a disused canal; a large number of abandoned allotments built on top of old ironstone spoil and a reputed total of 27 ironstone and coal shafts.

The site has been reclaimed for a mixture of industry and recreation and includes a 16 ha industrial site; a five hole extension to an adjacent golf course; an enlarged and deepened subsidence flash to improve the fishing facilities and footpaths alongside and across the cleaned up canal. In addition a 160,000 cubic metre capacity balancing reservoir has been created to serve the industrial estate.

The scheme has been carried out with Derelict Land Grant in conjunction with the Department of the Environment. The paper describes the problems of acquisition, phasing, design details, construction and overall costs.

INTRODUCTION

The site of the old Manners and Pewit collieries is situated on the west side of Ilkeston in Derbyshire. The area has a long history of mining both for coal and ironstone and the remains of the infrastructure that used to serve it including the old canal and railways, were still evident ten years ago when the County Council first began preparations for a reclamation scheme. The Pewit colliery had been closed for over a hundred years and its remains had been covered by refuse and random tipping and then by water as the site settled following other mining activities. The Manners Colliery had closed more recently in the 1940s but its tips and foundations were still very evident. In between the two sites were some old ironstone spoil tips which had been levelled and turned into allotments, long since abandoned, with high overgrown privet hedges and abandoned cold frames. Interlaced between them all were a maze of old railway embankments and cuttings.

During the 1970s the coal content of the refuse tip on the Pewit site was explored and some coal recovery was carried out. This operation was slow and by re-opening the refuse tip made the site even more of a mess and this, together with the takeover of some allotment plots for scrap metal recovery gave the area a very rundown appearance. As it is situated close to large areas of housing it was hoped that a reclamation scheme could make the area more attractive and provide some additional recreational facilities.

Acquisition

By 1980 most of the land required for the scheme was already in public ownership, either by Erewash Borough Council or Derbyshire County Council. The allotment area was owned by a multiplicity of people, some known, some unknown. To get over this difficulty a Compulsory Purchase Order was sought under Section 112 of the Town and Country Planning Act 1971. As it was anticipated that this procedure would take some time a decision was taken to start the Pewit half of the scheme first with the Manners Colliery and allotments to be completed later.

Department of the Environment Approval

During this time the Department of the Environment were kept informed of the proposals. They gave approval in principle that an acceptable scheme for the area, which had been officially accepted as "derelict", would attract Derelict Land Grant.

CONSULTATIONS

East Midlands Electricity Board

In the centre of the site was a building occupied, under a long term lease from the Council, by the East Midlands Electricity Board, see Plan A. This building was the old Ilkeston Power Station and had been built on part of the Manners Colliery in the early part of the century. As the EMEB had no intention, then, of moving from the site the reclamation scheme would have to be designed around them. At the same time access through the site would have to be maintained during the consultation period; also four underground power cables leading from a transformer yard attached to the building crossing the site would need to be maintained and eventually diverted.

Coal Seams

It was known that two coal seams outcropped in the Manners part of the site and it was originally intended to remove this coal by opencast methods. However doubts as to the economics of removing sufficient coal to prevent a high wall being left through the middle of the site plus the cost of having to compact all the overburden meant that the decision was made to leave the coal in the ground. Boreholes had led us to believe that the outcrop would be covered by a fill area but this was proved wrong and a portion of one seam had

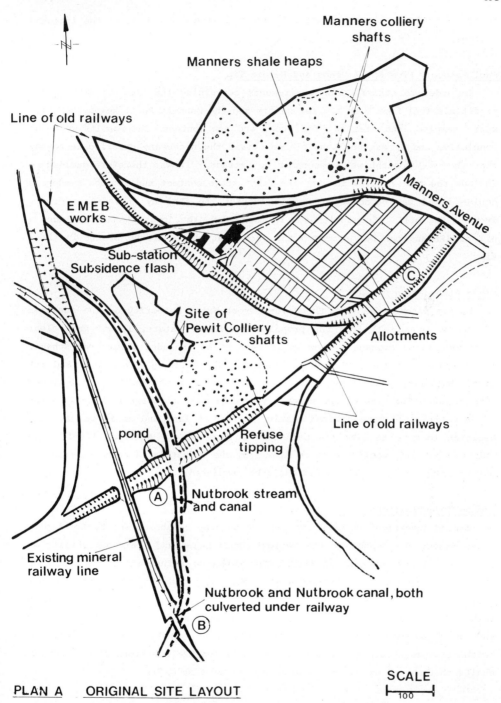

Manners colliery shafts

Manners shale heaps

Line of old railways

Manners Avenue

E M E B works

Sub-station
Subsidence flash

Site of
Pewit Colliery
shafts

Ⓒ

Allotments

pond

Refuse
tipping

Line of old railways

Ⓐ Nutbrook stream
and canal

Existing mineral
railway line

Nutbrook and Nutbrook canal, both
culverted under railway

Ⓑ

PLAN A   ORIGINAL SITE LAYOUT

SCALE
├──┤
100

to be removed so that there would be no coal within 2 metres of the finished surface.

Fuel Recovery From Spoil Heaps and Refuse Tip

In order to satisfy DoE requirements concerning the need to extract any profitable fuel from both the refuse tip and the Manners spoil heaps, the tips were sampled and tested for coal and ash content. The results were not conclusive and it was not possible to obtain the necessary certificate to say that the coal content was uneconomic to extract. It was therefore decided to include the test results in with the first stage contact and let the tenderers decide if it was economic or not to extract the fuel by including a credit item in the bills of quantities. In the event, the lowest tender included a nil credit and the DoE accepted the fact that there was no profit in these tips.

Flood Attenuation

During discussions with the Severn Trent Water Authority on the question of the storm outfall from the proposed industrial estate, it was made clear that some flood storage system would need to be included within the scheme. The catchment of the Nutbrook had been rapidly built up since 1945 and the Water Authority considered that further development would require flood attenuation. The size of the necessary storage area to cater for the 40 acres of industry would have covered nearly an acre of land and so it was an easy decision to make to site the balancing feature on the Nutbrook itself. In this way not only would the storm flow from the industrial area be catered for but the flow in the brook itself could be regulated.

Ground Investigations

Ground investigation were carried out both by boreholes and by trial pit. These proved the depths of the various spoil heaps and their suitability as covering material or general fill. The amount of refuse and its location on the Pewit site was also investigated. The content and depth of the railway embankment at 'A' on Plan A was also tested as it was hoped that this could be reshaped and extended to form a balancing dam across the line of the Nutbrook and the Nutbrook Canal. A search was also made of old maps and records to establish more about the history of the site and the positions of the several shafts that showed up in the records held by British Coal.

British Rail

One mineral line was still active through the site and both the canal and brook were culverted through the railway embankment at 'B' on Plan A. The canal had originally been bridged when the railway was built but after the canal closed a culvert was constructed on the line of the canal but for some reason the level of this culvert was approximately 1 metre too high and no water ever passed through it. Instead the canal water joined the brook and used the brook culvert, flooding the adjacent fields after heavy rain. To alleviate the flooding a new culvert was intended, set at a lower level. When the original canal culvert had been installed the remaining space beneath the bridge was left as a footpath and it was intended to put a new culvert beneath the footpath. British Rail gave extremely stringent conditions for the construction of the proposed 1.8m dia. culvert as inspection had revealed that the abutments were old and crumbling.

DESIGN

The design of the scheme was divided into three distinct areas of work which were all interconnected; the Pewit area and the rearrangement of the refuse; the Manners area and the design of the area for industrial development; and the balancing dam and associated drainage works.

Pewit Site

The household refuse had been tipped in various areas on the Pewit site and to attain the contours that would suit the proposed golf course the material had to be excavated from various parts of the site and placed in the available areas for fill which was mainly in the railway cutting at 'C' on Plan A. Covering material was obtained from the various railway embankments plus a certain amount of material from the Manners tips. It was important not to take too much material from the Manners area as this would affect the earthworks balance from the industrial plateau. It is important at this stage to make clear that in the majority of reclamation schemes an on-site balance of earthworks is a prerequisite of any design. Although material could be imported onto the site, if available at the right time and of an acceptable quality, no material would be exported from the site.

Manners Site

The existing overall gradient across the area designated for the industrial plateau was 1:20 and the County Council's Economic Development Working Party, the client of this case, required a maximum gradient of 1:50. This meant that, in order to save too much excavation into the natural ground

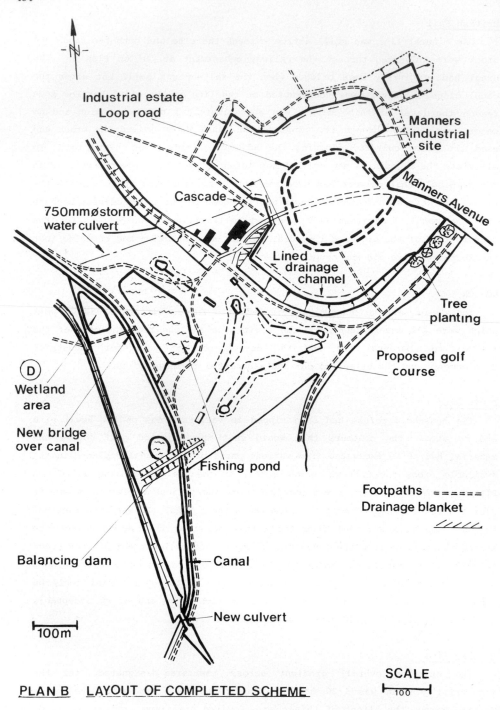

Industrial estate
Loop road

Manners
industrial
site

Manners Avenue

Cascade

750mmⵁstorm
water culvert

Lined
drainage
channel

Tree
planting

Proposed golf
course

D

Wetland
area

New bridge
over canal

Fishing pond

Footpaths ======
Drainage blanket
//////

Balancing dam

Canal

New culvert

100m

SCALE
100

PLAN B   LAYOUT OF COMPLETED SCHEME

to the north and east, sufficient spoil would have to stay on the Manners site
to achieve this gradient and a cut-fill balance. The design was made more
difficult by the variable nature of the materials being used; burnt and
unburnt colliery spoil; natural ground and tips of blue bind from the
ironstone workings, so that an accurate estimate of the expected loss of
volume due to compaction could not be made. An average figure of 10% was
eventually chosen but in the event this proved too low and additional fill had
eventually to be won from all corners of the site.

Balancing Dam and Associated Drainge

The requirements of the Severn Trent Water Authority was that all storms
in excess of a 1 in 2 years storm had to be stored. This meant that storage
for approximately 160,000 cubic metres of water had to be designed. Because
this was in excess of 25,000 cubic metres a panel engineer had to be engaged
to advise and approve the designs. The borehole data suggested that the
material in the old railway embankment would be suitable as fill for the new
embankment although it would need additional compaction. The embankment wsa a
maximum 5.8m high incorporated a drainage blanket into the downstream toe and
included a 1.3m dia. throttle culvert plus a spillway in case the culvert
became blocked. Floods in excess of 1:100 a year are designed to fill the
balancing area and overtop the dam and so a special grass seed mix that can
withstand water flows for a short period (Ref: 1) was used to ensure that the
topsoiled embankment will carry this flood without damage. The area on the
upstream embankment which is most affected by the rise and fall of water level
was protected by Enkalon, a proprietary thin nylon mattress material through
which grass can grow, to give added protection.

The outfall from the throttle culvert was led into a redug channel on the
line of the old canal. However to assuage the desire of the adjacent riparian
landowner the line of the old stream had to be maintained as well and a feeder
into this stream was provided. After a trial excavation beneath the old
railway bridge it was agreed with British Rail that as long as the excavation
for the new culvert proceeded carefully, excavating only enough trench to
place one length of pipe at a time, no further precautions need be taken.

CONSTRUCTION

Manners/Pewit Stage 1 Civil Contract

The contracts for the civil works were based on the ICE Conditions of
Contract and the Civil Engineering Standard Method of Measurement. The Stage 1
contract was let to A F Budge Ltd and work started in August 1982.

(i) <u>Old Shafts</u>. The old plans of the site and British Coal's advice suggested that there were 13 shafts on the Stage 1 site. Extensive excavations in the purported area of the shafts revealed two only. In addition some late investigation had revealed the existence of two more shafts close to the downstream toe of the balancing dam and it was considered advisable to investigate these more fully. Unfortunately the shafts were outside the limits of the county Council ownership and the owner was proving rather intractable at that time. Therefore powers were obtained under the Land Drainage Act and notice was given and entry made. The shafts were found to be shallow bell pits and were infilled and capped.

(ii) <u>Buried Bridge Abutments</u>. Excavation on the line of the railway embankment preparatory to the construction of the new dam revealed some extensive foundations for the bridge that had once crossed the canal. These were circular wrought iron cylinders approximately 0.6m dia, heavily braced, in three rows which proved to be very difficult to remove.

(iii) <u>Balancing Dam Material</u>. The suitable material in the old railway embankment was found not to be as extensive as had been hoped and extra material for the balancing dam had to be found from borrow pits on the Stage 2 part of the site.

(iv) <u>Culvert Under Railway Bridge</u>. When construction of this culvert started, British Rail engineers rapidly called a halt as the foundations of the bridge were not the same as had been found in the trial pit excavation. This had, it seemed, found the only real foundations, the remainder were non existant. In order that the work could proceed the whole length of the culvert excavation had to be sheet-piled and the piles left in with the whole trench back-filled with concrete. Because of the low clearance under the bridge a special gantry had to be designed to place and drive the piles and to manoeuvre the 1.8m dia. pipes, each one weighing 6 tonnes.

(v) <u>Fishing Pond</u>. The subsidence flash was already in use as a fishing pond but was heavily infiltrated with reeds, and to improve its quality and size a nominated sub-contract was let to E G Harris Ltd to remove the reeds and de-silt the pond by means of a suction dredger. The silt was pumped over to the other side of the canal to a low lying area at 'D' on Plan B. Originally it was meant to leave the silt to dry out and then spread it but before this happened the whole of the settling lagoon became covered in reeds and it was decided that it would prove an attractive additional wet-land habitat for the area. Minimum work to tidy the banks was carried out and the area left to naturalise.

(vi) <u>Topsoil</u>. The final works under this contract included the tidying up of the bank of the old canal with the construction of paths alongside it, together with other footpaths linking the site with other footpath systems. The whole area was given a covering of topsoil brought in from Carsington where at that time the dam construction was under way. The original planning permission for the Carsington Reservoir project included an agreement that the topsoils which would otherwise have been buried under the proposed new lake should be made available at no cost to the planning authorities. This meant that for the cost of haulage, good quality topsoil was made available. This is an example of good long term planning as topsoil is a valuable commodity which should be preserved at all costs.

Manners/Pewit Stage 1 Landscape Contracts 1 & 2

Immediately the dam was completed a small landscape contract was let to grass the whole embankment. This was carried out in the Autumn of 1983. The following year a larger landscape contract was let to establish grass and plant trees on the remainder of the Stage 1 area.

During the course of the Stage 1 civil contract the Compulsory Purchase Order for the allotment area was determined and it was possible to clear the whole area of the dense privet hedges. This enabled some of the allotment topsoil to be recovered for use on the Pewit site as well as opening up this part of the site for some additional ground investigation.

Manners/Pewit Stage 2 Civil Contract

(i) Earthworks. The contract for the second stage of the civil works was again won by A F Budge and work started in February 1985. To achieve a satisfactory surface for the industrial plateau the contract included for all spoil in the colliery spoil heaps and the allotment area to be excavated either to a depth of 2 metres below finished level on to original ground and the whole of the fill to be compacted in 200mm layers with 4 passes of a 72T vibrating roller or equivalent.

(ii) Groundwater. The allotment area was composed of uncompacted blue bind shale from former ironstone workings. On the interface with original ground this material was found to be extremely wet and the contractor had to leave areas to drain before he was able to complete its excavation. In one area which had to be built up to form an embankment with a maximum height of 9m the issue of water was so persistent that an extensive drainage blanket was placed under the toe. This was formed with a layer of clean limestone 150mm thick sandwiched between two layers of Terram 1000 drainage membrane.

(iii) Shafts. Excavations to find the 14 shafts suspected in the site were more successful than on the Stage 1 contract. A total of 9 were found and these were capped at original ground level with a reinforced concrete cap 2½ times the diameter of the shaft. One shaft found on the site had been backfilled by the NCB in the 1960s. Data on the backfilling was sparse except that it was known that the shaft lining had collapsed during the operation and a 'tulip' failure had occurred. the NCB had then completed the backfilling of the shaft with readymixed concrete but it was not known how much concrete was in fact used in the backfilling and hence nobody was sure how secure was the treatment. Eventually it was decided to attempt to check the depth of concrete in the shaft as if it was a concrete pile. A Transient Dynamic Response test was carried out and no harmonics were found (Ref 2). This suggested that the concrete had a depth in excess of 15 times the diameter of the shaft which was approximately 4 metres. This gave a minimum depth of concrete in the shaft as 60 linear metres and therefore it was decided that the shaft was adequately secure.

(iv) Roadworks. During this contract the access to the EMEB works had to remain open at all times until the alternative route was completed. This alternative route was the base course of the proposed estate road which would be completed under a separate contract engineered by the County Surveyor. Included in the reclamation contract was all the road drainage except the final gully connections. On completion the base course was given a temporary tar spray and chip wearing course. Once the temporary new road had been constructed diversion of two of the four electricity cables was carried out together with diversions of the water and telephone services.

(v) Drainage. the drainage of the industrial area was designed to take the storm flows for the plateau in the interim period between completion of the reclamation works and the full development of the site. The drainage pattern is therefore designed to take a 95% runoff via unlined and lined open ditches with a final reinforced concrete cascade outfall into the permanent storm water sewer. From experience it was considered better to have a good failsafe drainage system from the start which would need little maintenance in the period before the site is developed when there are few people around either to notice maintenance problems or carry out repairs.

Manners Pewit Stage 2 Landscape Contract

The second stage civil work was completed with the topsoiling of embankments prior to the letting of the final landscaping contract. The site was handed over to the County Council's Economic Development Working Party in September 1986 for the remainder of the roadworks to be completed and the estate sold off for development.

Costs

The whole of the reclamation works described in this paper have been grant aided by the Department of the Environment from their Derelict Land Grant Budget

The costs were as follows:

		£
Land acquisition		242,000
Site investigation		12,000
Civil Works		1,792,000
Landscape Works		91,000
Diversions - Water	46,000	
Electricity	156,000	
Telephones	4,000	206,000
Provision of foul sewer outfall		15,000
Miscellaneous		10,000
Grant aided portion of Design and Administrative Costs		212,000
		2,589,000

		£
Value of site on completion of reclamation:		
Industrial Land: 40 acres @ £28,000/acre		1,120,000
Land for golf course and recreation area @ nil		-

The civil works costs do not include for the whole of the roadworks, that is above sub-base level on the southern loop road and the whole of the northern loop road. This work was separately financed by the County Council. Because of this the value of the industrial land for Department of Envrionment purposes is not the "fully serviced" value.

This breakdown of costs show clearly that to create acceptable land for industrial development from derelict land in this area of England requires an input of Derelict Land Grant to offset the loss making element of the scheme. The overall cost of £48,000 per hectare is not abnormally high for this type of work and gives an indication of the scale of expenditure that needs to be sustained to reclaim the remaining derelict land in the country.

ACKNOWLEDGEMENTS

The opinions expressed in the paper are the Authors' and do not necessarily reflect the views of Derbyshire County Council. The Authors wish to acknowledge the assistance given to him by the County Planning Officer and other members of the Planning Department in the preparation of this paper.

500

REFERENCES

1. CIRIA Technical Note 71 "A guide to the use of grass in hydraulic engineering practice" 1976 Whitehead, E and Nichssons of Rothwell.

2. Civil Engineering April 1982 "Integrity testing" R T Stain (Test consult/ CEBTB Ltd)

Reclamation, Treatment and Utilization of Coal Mining Wastes, edited by A.K.M. Rainbow   501
Elsevier Science Publishers B.V., Amsterdam, 1987 — Printed in The Netherlands

AN INVESTIGATION INTO THE RECLAMATION OF OPENCAST BACKFILLED SITES DESTINED FOR ROAD CONSTRUCTION IN THE UNITED KINGDOM

R. N. Singh[1], F. I. Condon[1] and S. M. Reed[2]
1 Department of Mining Engineering, University of Nottingham, University Park, Nottingham, NG7 2RD, United Kingdom.
2 British Coal, Opencast Executive, United Kingdom.

SUMMARY
    The paper presents a current research programme into the stability and development of backfilled opencast mine sites in the U.K. Post-mining usage of restored surface mine sites are detailed, with particular reference to structural development. Standard backfill compaction specifications are presented along with factors which affect backfill settlement. Two case studies regarding backfill settlement at sites destined for road development are detailed.

SURFACE COAL MINING OPERATIONS IN THE UNITED KINGDOM

    The exploitation of near surface coal deposits in the United Kingdom accounts for approximately 14% of the nations total mined output. The opencast coal industry is a very profitable and valuable sector of British Coal's operations, providing much needed supplies of coking coal and anthracite. Currently, 14-15 million tonnes of coal are won each year by opencast methods. Site areas are generally geologically or physically constrained to between 10 and 800 hectares, the average being 140 hectares. In an average year at least 1400 hectares of land may be converted to areas of opencast backfill, (Singh, Condon and Denby 1986). In 1985/6, 7,800 hectares of land in England and Wales were subjected to opencast operation and 9,000 undergoing rehabilitation, (National Coal Board, 1986). Over 400 hectares of derelict land left over from previous industrial activities were also under restoration. Overburden to coal ratios may be as high as 25:1, which can result in the formation of considerable depths of unconsolidated backfill materials, (Singh, Denby and Reed, 1985).

SETTLEMENT OF OPENCAST MINE BACKFILLS

    The total settlement of a backfill mass is dependent upon the size of the individual rock masses within the fill. The larger the rock fragments then the greater the bulkage and consequently the greater the degree of settlement possible. Total settlement is also determined in part by the fill depth. Factors affecting the rate of settlement include groundwater recovery, fill properties, (rock types etc) and time, (Reed and Singh 1986). Compaction is a method of reducing void ratios within the fill, in turn this will lead to reduced magnitudes of settlement. By reducing backfill permeabilities, a slower groundwater percolation will result leading to reduced effects of collapse settlement.

DEVELOPMENT OF RESTORED OPENCAST MINE SITES

The post-mining development of opencast mine sites may consist of one or a combination of the following;

o   Agriculture, (the predominant after-use).

o   Recreation.

o   Development, (buildings, roads or other civil engineering structures).

o   In Scotland, forestry is a common after-use.

By far the majority of sites are returned to agricultural use by a well established procedure, (Brent-Jones 1984). Opencast reclamation where possible has been advantageous to the local community as a method for restoring derelict areas or as a means of providing recreational facilities.

COMPACTION OF BACKFILLED OPENCAST MINE SITES

Pressures on land use and the proximity of many opencast deposits to urban environments have resulted in increasing numbers of mine sites being ultimately required for some form of structural development. Mine operators are under great pressure to restore a backfilled site with the confidence that the surface will be adequately stable to enable the construction of buildings, roads, railways or service installations such as water or gas mains, electricity pylons etc. On some sites where an intention has been declared to use the backfill surface for building purposes, then fill compaction has been introduced as part of the standard site operations, to the Department of Transport's Specification for Road and Bridge Works 1976, (Series 600).

Need for controlled compaction:-

The requirements of a planned development on the operations of opencast sites stem from the desire of one of the interested parties to reduce the resulting backfill settlement. Controlled placement of backfill involves the restoration of the site in layers, with fill placement by scrapers or trucks and the use of roller based compacting plant. Controlled placement results in a slower restoration of the site and the need for additional machinery. To avoid delaying the excavation operation stockpiling maybe necessary. Stockpiling requires space on site and involves extra cost in the re-handling of excavated material.

Compaction specifications:-

No specific level of compaction has as yet been defined, so that the most common specification for compaction of opencast backfill is that of the Department of Transport's Specification for Road and Bridge Works, and it's revisions; this being a widely recognised compaction standard and as such is generally acceptable to the prospective developer. The Specification defines suitable material for the formation of road embankments and the minimum requirements for compaction. Procedures for compaction are outlined by machine type and classification, and define the minimum number of machine passes required, together with the maximum thickness of each compacted layer.

Two distinct types of compaction project have been formulated. Road schemes involve the monitoring of the compaction process by consulting engineers acting on behalf of the road authorities, while compaction of industrial and domestic building land is controlled by engineers acting on behalf of both the mine operator and the subsequent land owner. Compaction is carried out by layer placement of excavated material by either scrapers or trucks, followed by the controlled running of compaction plant e.g. smooth drummed vibrating roller, tamping foot roller etc. either as towed or self-propelled units.

Table 1. shows the number of sites which have contained compaction schemes as part of their operations. The majority of sites in this instance have required road diversions as a means of avoiding sterilisation of coal reserves. Compaction on road construction routes is generally confined to the line of the road, together with formation of a suitable embankment. Sites restored for building construction may be partly or wholly compacted.

Site	Dates	Compaction Details.
Ketley Grange	1970-73	Route of M54 - road corridor
Dora	1974-78	Road scheme - corridor
Old Park	1974-78	Telford road scheme - corridors.
Clares Lane	1975-77	Telford road scheme - corridors.
Mitre East	1977-79	Building land - specific compaction
Waverley	1977-80	Building land - full extent
Westerton	1977-81	Road scheme - corridor
Shilo South Ext.	1978-81	Route of A6096 - corridor
Blindwells	1978-	Route of A1 (By-pass) - corridor
Crosstree Farm	1979-81	Route of A55 (By-pass) - corridor
Holly Bank	1979-84	Building land - specific areas
Cadgerhall	1980-	Reinstallation of C125 - corridor
Natsfield	1982-85	Building land - complete site
Flagstaff	1985-	A42 interchange scheme - corridors
Barnabas	1986-	Building land - complete site
Lounge	1986-	Road Compaction. A42 corridor.
Dixon	1987-	Road Compaction.

Table 1 Summary of some opencast backfill compaction schemes.

Series 600 has been drawn up as a way of surmounting the problem of specifying suitable end result specifications for the various fill materials which may be employed in the formation of roads and embankments in the U.K., and the problems involved in enforcing them. A number of situations are considered, but for the compaction of opencast fills it is usually the definition of the material, the method of construction of road embankments and the methods of compaction under Series 600 which are applied.

Fill materials:-

Fill material is deemed to be unsuitable if it is comprised of:

(a) material from swamps, marshes and bogs.

(b) peat, logs, stumps and perishable material.

(c) material susceptible to spontaneous combustion.

(d) material in a frozen condition.

(e) clay with a liquid limit exceeding 90% and/or a plasticity index exceeding 65%.

(f) material having a greater moisture content than that permitted in the contract, (based on DTp specifications obtained from standard compaction tests, (Condon 1986).

Opencast backfills in England and Wales are not recorded as being susceptible to spontaneous combustion, and thus are only unsuitable for use in road embankments and other compaction projects when the broken rock is in a frozen condition. This prevents the placement of opencast backfill for compaction during periods of freezing in the winter months. A case study detailed later in this paper concerns a site where a road is to be constructed upon fill susceptible to spontaneous combustion of up to 70 m. depth. (This site is however the only recorded site in the U.K. where the fill is susceptible to spontaneous combustion).

The procedure by which a fill may be compacted under the Series 600 specification depend upon the fill material. Opencast fills are generally well graded due to the crushing of rock from site handling.

FILL COMPACTION AND RESTORATION RESEARCH.

Two projects have been sponsored by the British Coal Opencast Executive in the field of opencast backfill restoration and stability. The first examines the effects of the re-establishment of the groundwater regime on restored opencast mine sites, Reed (1986). In particular the occurrence of collapse settlement in backfill materials. The second project makes an assessment of compactive methods relevant to the treatment of backfill destined for structural development, Condon (1986). The following case studies are extracted from these research projects.

SITE A, Minor Road Development on a Deep Restored Backfill.

An investigation is being conducted into the stability of a variable thickness opencast mine backfill destined for surface development. A minor road is to be constructed over the fill. An instrumentation scheme was devised consisting of magnetic extensometers/standpipe piezometers to monitor vertical settlement and groundwater recovery within the fill. The depth of the fill varied from 30 to 70 metres in the area of instrumentation. Prior to working the site, piezometer instruments in the area indicated a groundwater level, 10 metres below the surface. The site consisted of three synclinal basins, two of which drained indirectly into a large mining excavation to the North, which was undergoing restoration. The excavation in this region of the mine had approached 250 metres. It was expected that the final levels of water over the site could in certain areas recover to within metres of the surface.

The Diversion of a Public Road

In order to work the site the removal of a public road was necessary, and a condition of the planning consent was that this road should be replaced on the termination of mining. This investigation was concerned with a temporary replacement road constructed over the backfill, which will later be replaced by a permanent road at a later date.

The Construction of the New Road

The new road is to be constructed over normal backfill of up to 70 metres depth. The standard specification regarding for the design of this type of road is that the upper 16 metres of the fill is treated as follows. The layers are levelled by a bulldozer not less than 14 tonnes in weight and must undergo 8 passes of a single axle vibrating roller of not less than 8 tonnes in weight, 2 metres in width and with an operating frequency of not less than 2000 cycles/minute. Scrapers and earthmoving plant are not considered acceptable as compactive machinery.

Monitoring Scheme

The proposed scheme involves installing four magnetic extensometers/ piezometers, (figure 1), in the positions shown in figure 2. The details of magnet horizons are presented in table 2. These types of extensometer instrumentation are frequently used in settlement monitoring schemes, (Charles et al 1984, Reed 1986). The instrument monitors the settlement of different horizons within the fill together with groundwater position. The value of this particular scheme is that results can be used to prove whether any damage to the road's structure is due to the settlement of the backfill.

506

Table 2 Initial Magnet Positions and Water Levels, Site C.

Magnet No.	E1	E2	E3	E4
	(Depth below top of Standpipe, metres)			
Datum	37.569*	70.829	32.478	59.792
SP1	31.670	62.197	24.275	48.694
SP2	16.389	51.718	18.572	38.407
SP3	11.346	42.204	13.462	28.250
SP4	6.593	31.860	7.136	17.968
SP5	1.338	20.599	0.782	6.662
SP6	-	15.031	-	2.479
SP7	-	8.633	-	-
SP8	-	1.242	-	-

* Access Tube broken at this level

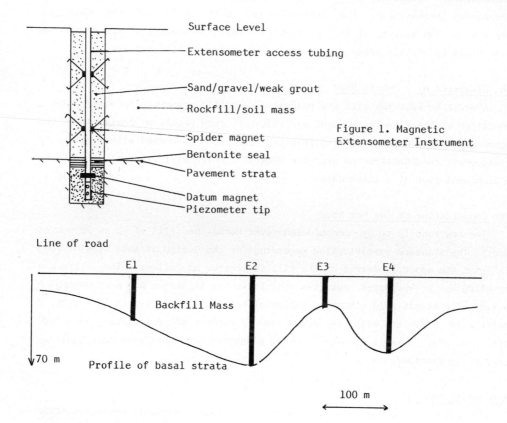

Surface Level

Extensometer access tubing

Sand/gravel/weak grout

Rockfill/soil mass

Spider magnet

Bentonite seal

Pavement strata

Datum magnet

Piezometer tip

Figure 1. Magnetic Extensometer Instrument

Line of road

E1   E2   E3   E4

Backfill Mass

70 m   Profile of basal strata

100 m

Figure 2. Instrumentation plan, Site A, minor road development

The settlements recorded on uppermost magnet of each instrument, 1 year after the commencement of monitoring are; E1, (40 metres depth), 40 mm. E2, (70 metres depth), 55 mm, and E4, (60 metres depth), 5 mm. There has been no correlation between settlement and the position of the water table. On instrument E4 it is interesting to note that initially movements of up to 72 mm were recorded principally in the lower parts of the fill. These movements have however been countered by, presumably, heave movements at a later date. It may be possible on the second set of readings showing magnitudes of 51-72 mm settlement that the datum magnet was misread.

The following conclusions can be drawn from this work;

a). The completion of groundwater recovery within the fill occurred within 18 months of placement. The present groundwater levels are approximately 5 metres below the original pre-mining levels.

b). The rapid recovery may have initiated collapse settlements in the fill prior to instrumentation, however since monitoring commenced no collapse settlement has been observed associated with a further 5 metres of recovery.

c). There has been no severe differential settlement resulting from the variance in fill depths.

d). The stability of the fill provides a sharp contrast to previous observations, in particular the absence of both vertical and shear movements in deeper fill materials, (Reed and Singh 1986).

SITE B, Major Road Development.

An investigation is currently being conducted on an opencast mine being compacted for structural development. The investigation is comparatively recent with only the installation of instrumentation having being carried out, to the date of this paper, (August 1986).

The fill which is found to be unsatisfactory under the Series 600 requirements, and all other fill not compacted is to be placed in areas in 1m or 5m layers by layer tipping from dump trucks or scrapers. Thus the site, though small, encompasses three compaction regimes.

In addition to the requirements of Series 600 further conditions for the placement of the backfill have been written into the contract for the site. The purpose of these is to ensure stability of the road embankment, by concurrent placement of surrounding fill and by the establishment of a stable base across the whole extent of the site. The proper integration of previously compacted fill with newly placed material is also required. A further requirement refers to the increased differential settlement encountered across the fill/solid interface on other sites, and requires the benching of high and final walls.

Analysis of results

The groundwater table had already recovered to 20 metres below the restored fill surface by the time that the instrumentation had been installed. Consequently a great degree of collapse settlement might have occurred prior to monitoring owing to rapid recovery. Alternatively to this but likewise not proved was that the method of backfilling in 6-8 metre layers was a very effective system and that normal dump truck traffic provided excellent compaction, (Ferguson 1986). The fill had actually been in position for 6 months prior to instrumentation. Over the period of one year since the commencement of monitoring the water level rose by an average of 5 metres, with most of this recovery occurring in the first six months. The water recovery appears to have ceased within one year from the commencement of the monitoring programme.

Settlement Results.

Over the period of one year of monitoring some suprising results have been obtained. Initially it was thought that Extensometers E2 and E4 would have larger settlements owing to the deeper fill. This has however been shown to be untrue. Above all, the most suprising observation, (although from the point of view of road construction ideal), is the lack of vertical settlement recorded on all the instruments, particularly those in deeper fill. Unfortunately a blockage appeared shortly after installation in Extensometer E3, and consequently the results on this instrument have been lost. The settlement results for each of the other extensometers are presented in tables 3, 4 and 5.

Magnet	SP1	SP2	SP3	SP4	SP5
	Total Vertical Settlement (mm)				
Month No.					
1	0	+6	-9	-12	+2
4	0	-18	-30	-35	-39
11	0	-14	-25	-34	-40
12	0	-15	-23	-26	-40

Table 3 Settlement Results, Ext. E1.

Magnet.	SP1	SP2	SP3	SP4	SP5	SP6	SP7	SP8
	Total Vertical Settlement (mm)							
Month No.								
1	+6	0	-27	-33	-35	-30	-30	-29
4	+2	-3	-26	-11	-7	-18	-29	-32
11	+4	-4	-29	-14	-13	-29	-45	-49
12	0	-7	-30	-17	-17	-34	-50	-55

Table 4 Settlement Results, Ext. E2.

Magnet	SP1	SP2	SP3	SP4	SP5	SP6
	Total Vertical Settlement (mm)					
Month						
1	-23	-20	-24	-19	-15	-16
4	-55	-61	-72	-49	-54	-51
11	0	-2	-2	-1	-4	0
12	0	+2	-5	-3	-1	-5

Table 5 Settlement Rates, Ext. E4.

+ denotes heave movement.
- denotes settlement.

Benching is to be concurrent with the placement and compaction of the road embankment, with no bench being advanced more than one step above the raising fill.

The compaction specification defines moisture contents such that the the the maximum permitted moisture shall be:

Cohesive material;   1.0 x Plastic limit
Well/uniformly graded granular material;   Optimum Moisture Content + 2%

No material is to be compacted if the moisture content is less than the Optimum - 2%. The optimum moisture content is to be determined by the contractor, using Test 12 of BSS 1377, (British Standards 1967).

Instrumentation of Backfill

Instrumentation is planned for the monitoring of the backfill, both in the interests of the Site Consulting Engineers, who wish to gain experience of the problems associated with opencast backfill, the Department of Transport and the Opencast Executive. An instrumentation scheme has been designed which will monitor settlement through levelling and the use of magnet extensometers.

Levelling

50 settlement stations are planned altogether, 30 along the line of the major road, and a further 20 in the industrial zone to the North East of the site. Stations are planned to be set in concrete at overburden level and extended through the subsequently laid soil layers.

Extensometers

Extensometer instruments have been installed to monitor three methods of fill placement, with magnets installed every three metres.

Piezometers

Up to 6 standpipe piezometers are planned for placement in the solid areas a surrounding the excavation.

All instrumentation and site layout are presented in figure 3.

Current Site Status

The site consists of three coaling areas of which the first two to be worked are the largest. The state of the site, (August 1986), is that the first coaling area has been fully backfilled with controlled compaction to form the road embankment. The minimum depth of the area is 4 m, and the deepest part of

510

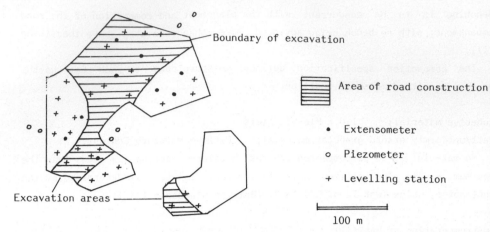

Figure 3. Instrumentation plan, Site B, major road development

the fill is 29 m. The second area of the site is presently been worked and a full half of the area has been backfilled with controlled compaction. The site should be completed by Spring 1987 at the latest. The first area has been fully instrumentated.

Areas of interest on the site lie in the monitoring of the compaction scheme, the values of void ratio obtained from the compaction and any correlation between this and the ultimate settlement of the fill, and in the scale of settlements involved.

CONCLUSIONS

The following conclusions can be drawn from the study to date.

i). Specifications for the compaction of opencast mine backfills require greater evaluation in particular with reference to the effectiveness of compaction methods.

ii). The stability of a restored fill is dependent on many site specific characteristics, which must be taken into account irrespective of which compactive measures have been utilised. Such factors include fill depth, fill area, presence of groundwater, rock types (ease of degradation) etc.

iii). A costing analysis is required to evaluate the feasibility of compactive measures previously applicable to road embankments, dams and other civil engineering structures which may or may not be applicable to the depths of opencast backfills which may be encountered.

ACKNOWLEDGEMENTS.

The authors would like to thank British Coal for their co-operation and continual support for this project. Gratitude is extended to Prof. T. Atkinson, Head of Department of Mining Engineering, Nottingham University for his support. This work could not have been carried out without the kind and enthusiastic support from many members of British Coal. Specific thanks are due to Headquarters Staff, Geotechnical Engineers, Geologists and Site Staff. Opinions expressed are those of the Authors and do not necessarily reflect the opinions of British Coal.

REFERENCES

1. Brent-Jones, E. (1984). Land Reclamation in the 80's. The National Coal Boards Approach. Paper 27, Symposium on the Reclamation, Treatment and Utilisation of Coal Mining Wastes, Durham, England, Sept 1984.
2. British Standards Institute. (1976). BS 1377. Methods for testing soils for civil engineering purposes. 1976.
3. Condon F. I. (1986). Restoration of Opencast Mine Sites for Future Development. University of Nottingham M.Phil Thesis, 1986.
4. Department of Transport. (1986). Specification for Road and Bridge Works HMSO Fifth Edition. 1976
5. Ferguson, D. (1986) Personal Communication.
6. National Coal Board. (1986) Reports and Accounts 1985/6 Hobart House London.
7. Reed S. M. (1986). Groundwater Recovery Problems associated with Opencast Mine Backfills. University of Nottingham, Ph.D Thesis. 1986.
8. Reed S. M. and Singh R. N. (1986). Groundwater Recovery problems associated with opencast mine backfills. International Journal of Mine Water. Vol 5. No. 3. September 1986. p46-70
9. Singh R. N. Condon F. I. Denby B. (1986). Investigations into techniques of compaction of opencast mine backfill destined for development. Geotechnical Stability in Surface Mining Syposium. Nov 6,7, 1986. Calgary , Canada.

Reclamation, Treatment and Utilization of Coal Mining Wastes, edited by A.K.M. Rainbow
Elsevier Science Publishers B.V., Amsterdam, 1987 — Printed in The Netherlands

The Cost-Effectiveness of Rehabilitating Colliery Sites through Coal Recovery

Tom Macpherson BA, MICE, MRTPI

ABSTRACT

In Scotland, in recent years, colliery rehabilitation schemes have been arranged so to maximise the revenue produced from the sale of extracted coal. The experience of the Scottish Development Agency is drawn upon to review the treatment and cost of colliery rehabilitation schemes carried out since 1975 by traditional methods. Five recent schemes based on coal recovery/rehabilitation methods are then examined and compared with traditional methods in relation to cost-effectiveness and site problems.

The wider costs and employment implications for both methods are briefly examined and the contract arrangements adopted for coal recovery/rehabilitation schemes reviewed.

Finally, low-cost vegetation techniques which can be used in conjunction with colliery rehabilitation are discussed in relation to their development in Scotland and their applicability at coal recovery sites.

1. COLLIERY REHABILITATION IN SCOTLAND

In 1975 the Government made the Scottish Development Agency responsible for clearing all derelict land in Scotland done through public funding[1]. Since then, colliery rehabilitation has accounted for just under 20% of the Agency's total spend on land reclamation. As the Agency directly funds rehabilitation contracts as the client no derelict land grant applies in Scotland - as occurs in England.

Whilst a few colliery rehabilitation schemes are designed in-house, most are prepared by private consultants or local authority rehabilitation teams. This is done on the basis of obtaining competitive tenders which allows the Agency, as Employer, to accept the best offer.

The consultant or local authority appointed acts as Engineer for the Works so that the task of Agency technical staff is to initiate schemes by issuing design briefs, setting acceptable costs and ensuring that contract implementation is achieved timeously and within budget. During the past 11 years good progress in treating colliery dereliction has been achieved to the point whereby 60% of all significant colliery dereliction has been

514

removed. Figure 1 shows this progress and, by projection, predicts how
all significant colliery dereliction could be removed by 1991, if past
rates of rehabilitation are maintained. Figure 1 includes, in the area of
dereliction, some less obtrusive colliery sites under 1 hectare in size
which are either naturally revegetated and too remote to (arguably)
justify treatment. These figures assume that currently active collieries
remain so up to 1991. Nevertheless, in 1986 some 150 million tonnes of
surface colliery spoil remained to be treated.

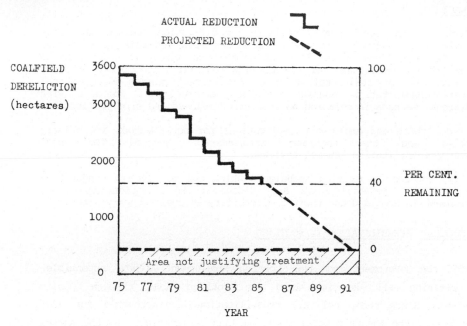

Figure 1   Reduction in Coalfield Dereliction in Scotland since 1976

Here, it should be mentioned that the largest Scottish tips are not of
colliery spoil. These tips occur to the west of Edinburgh where some 90
million tonnes of spent oil shale lies in eight major tip complexes. The
comparatively slower rate of rehabilitation of spent oil shale, when
compared to coal spoil, is due in part to tip owners "hope value" of the
materials potential for construction infill. The rehabilitation of spent
oil shale sites is not referred to further, other than to note that this
material has been used in greater quantities than coal spoil in Scotland
in civil engineering works over the past decade. Figure 2 shows the main
coal and oil shale fields within which the Scottish tips are located.

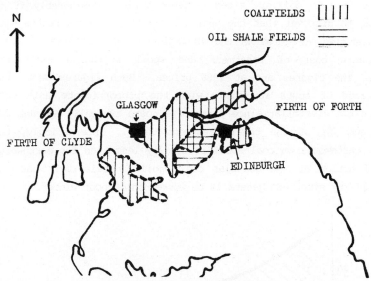

Figure 2   Location of main coal and oil shale fields in Scotland.

2. TRADITIONAL COLLIERY REHABILITATION SCHEMES

Reviewing 26 Agency funded colliery rehabilitation schemes implemented by traditional methods has made it possible to identify their main cost characteristics. Scheme selection has concentrated on rehabilitated sites where the final landform was agriculture/forestry, industry or recreation. Infrastructure costs included in the contracts reviewed have been excluded. Operations which have been included are colliery building demolition and clearance, spoil recontouring, earthworks, shaft capping, fencing and cultivation work. Sites with particular problems such as burning spoil or serious ground instability have been excluded. Figure 3 illustrates the type and extent of each site afteruse for all colliery sites rehabilitated since 1975, based on 40 completed schemes.

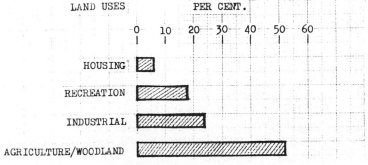

Figure 3   Colliery Rehabilitation Site Afteruses

As the size of the 26 colliery sites reviewed varies considerably it is not surprising to discover that the unit rehabilitation costs reduce with increasing site size. Figures 4A and 4B indicate the economies of scale related to unit cost of treatment and spoil regrading earthworks respectively. The figures are at 1986 prices. Both figures show that whilst the range of costs is fairly wide the proportionate unit cost reductions are not dissimilar since earthworks costs typically account for between 40% and 70% of the total treatment cost[2]. Interestingly, the unit cost for earthworks on coal spoil is usually less than for earthworks in road construction as the granular coal spoil is relatively easy to handle and colliery sites can generally be worked throughout winter.

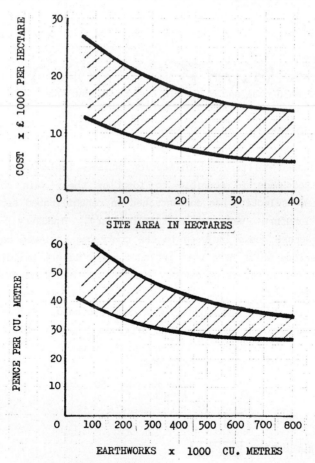

Figure 4A (above) Shows how unit rehab. cost reduces as site size increases
Figure 4B (below) Shows how unit earthworks cost reduces with quantity

Efforts to move away from rehabilitation for agriculture in Scotland commenced in the early 1980's. This was done partly in response to agricultural advice that such rehabilitation was of only marginal farming benefit and partly due to evidence that in several cases it was more cost-effective to invest more heavily in rehabilitation treatment which would support subsequent built development. Such development would furthermore reduce the ongoing maintenance costs associated with agricultural rehabilitation.

Although rehabilitated sites with a topsoil covering may reach 80% of the productivity of natural grazing land, directly seeded spoil and subsoil covered sites may only reach between 20% and 50% comparative productivity[3]. Also, where development is inappropriate the difference in cost between rehabilitation for agriculture and for amenity is so significant that such additional investment for agriculture is sometimes difficult to justify given that the rents obtained for the rehabilitated farming land may be as low as £80 per hectare per annum – often less than 1% of the rehabilitation cost and sometimes less than the cost of measures necessary to prevent re-acidification.

Moving towards preparing more sites for development was also favoured by the Agency as it provided a platform for stimulating further economic activity within the land treated (using its other powers for business development). Whilst some rehabilitated sites offer land for immediate development, others have been made to cater for future development by creating flat landforms which are easily developed and screened by tree planted peripheral mounds. Where isolated industry is intended, landscaping to the outer mound faces is done to a good standard since these areas are continually seen by passing public. When industry eventually establishes within the basin created, the bunds are usually well covered with maturing vegetation. Such sites are particularly suitable for locating "bad neighbour" industries or businesses which require a high degree of security.

This rehabilitation strategy continues with the recent addition of coal recovery/rehabilitation which, when viable, allows the fine coal particles within the spoil to be recovered and sold to reduce rehabilitation cost. Tips contain a certain amount of coal which has either been dumped with excavated spoil or has not been separated in the coal preparation process. Coal may be present in significant proportions in some of the older tips[4]. As coal preparation processes improved over the years considerably less coal is left in the spoil but tips with a recoverable coal content of as low as 7% may, in certain circumstances, still be viable.

Figure 5   Traditional rehabilitation at Whitrigg Colliery, near Bathgate, for industry. Concealed industrial land lies behind the mound between the arrows when viewed from adjoining local road and M8 motorway.

3. Coal Recovery/Rehabilitation

The oil crisis of the early 1970's stimulated interest in using more indigenous coal and this, in turn, led to an increase in recovering coal from tips. Subsequently, increasing costs in the coal industry meant that by 1983 UK coal prices averaged £38 per tonne. By then, 70% of coal was used for electricity generation, thus reinforcing the demand for coal of small particle size which could readily be supplied from colliery tips. Both of these factors contributed towards making coal recovery/rehabilitation projects viable. Also, about this time the technical feasibility of recovering coal improved with the increasing availability of mechanical/chemical equipment for treating the effluent from coal washing plants. Previously, effluent could only be clarified in large space consuming lagoons which could not always be accommodated within smaller colliery sites. An added impetus in Scotland to recover coal in this way stems from the property that the sulphur content of Scottish coal varies from between 0.6% and 0.8% compared with an average of 1.6% for the UK.

Before 1983 few coal recovery schemes included final land rehabilitation and those commercial ones which did achieved fairly poor standards - thus adding a certain notoriety to such operations. Perversely, many government funded rehabilitation schemes completed during this period, were on "coal rich tips" where the primary function was environmental improvement, leaving the coal unrecovered. Although no tip reworking proposals have yet been made, the time may be approaching when recovering coal from previously rehabilitated sites will be viable. Since 1983, each new tip rehabilitation project is examined to assess its coal recovery potential and, to date, the Agency has promoted five such schemes. Whilst none has been completed yet (at October 1986) sufficient cost information has emerged to allow comparison with traditional rehabilitation.

Revenue from coal sales is a complex matter and can depend on several site specific factors such as the total spoil available, its coal content, calorific value and ease of recovery. Equally, proximity of site to the coal user and market price are critical. A further important factor may be the level of royalty payment due to British Coal as such payment normally becomes a condition of the transfer of the land - if minerals are likely to be removed from the site for sale. In all five cases included in Table 1 this royalty amount ranged from either zero (no royalty due to British Coal) or less than £5 per tonne: the final figure depending on negotiations with British Coal.

Comparisons between the two column totals in Table 1 indicates the major savings (£1,081,000) which result from rehabilitation by coal recovery as compared to traditional methods.

TABLE 1

Site (Location)	Area (ha)	Coal Recovery/Rehab. Phase 1 Cost (£1000)	Phase 2 (Cost £1000)	Total CR/R Ph 1 + Ph 2 Cost (£1000)	Traditional Rehabilitation Cost (£1000)	Saving per site £1000
Canderigg (Hamilton)	12.0	69 (Cr)	72	3	180	177
Minto (Fife)	19.7	80	70	150	256	106
Meikle Earnock (Hamilton)	13.5	10 (Cr)	76	66	189	123
Blairhall (Fife)	28.3	150 (Cr)	170	20	340	320
Easton (Bathgate)	22.0	*200 (Cr)	120	- 80	275	355
				159	1,240	

All figures at 1986 prices   (Cr) = Credit

Traditional Rehabilitation Cost = estimate based on Fig 4A

*Provisional Figure

4. ADVANTAGES AND DISADVANTAGES OF RECOVERY/REHABILITATION

The significantly lower treatment costs achieved through coal recovery/rehabilitation are not solely explained by the revenue generated from coal sales. Other advantages which stem from rehabilitation through coal recovery are:

 (i) greater reduction in tip volume

(ii) additional employment

 (i)   With traditional rehabilitation, spoil compaction from earthworks can account for a 15% to 20% volume reduction in the size of the regraded tip. The final figure depends on the initial "looseness" of the material and the total volume regraded to obtain satisfactory contours as dry spoil generally remains at its angle of repose when tipped[5]. At coal recovery sites the total tip content is normally processed so that spoil compaction may reach as high as 40%. This major reduction results from compaction of all of the tip material and the removal of coal, which alone, can account for a 10% reduction in tip volume.

Such a volume reduction can either minimise or remove the need to spread spoil on adjoining land. Alternatively, where the site boundary is fixed, a relatively lower final landform profile may be achieved[6]. Figure 6 illustrates the benefit of this at a typical Scottish tip where there is scope for increasing the basal area around the tip for spoil spreading.

Figure 6   A 40% volume reduction at this coal tip at Gilmerton, near
Edinburgh, would reduce the tip height to a quarter of its
original height and increase the basal area by a factor of 1.6.

(ii) Of the projects already undertaken the average employment per site is 40 man week years which is more than four times that for traditional rehabilitation. The labour cost of this can be valued at about £0.6 m per site so that coal sales, in effect, both support an additional labour input of about £450,000 per site and reduce the overall cost of the physical works by the savings indicated at Table 1.

Disadvantages

In general, coal recovery/rehabilitation contracts take longer to place than traditional ones. Where local authorities officials and elected members have come to expect traditional reclamation difficulty may be encountered satisfying some of them that the longer coal recovery/rehabilitation route is the more cost-effective one. Objections are made that the extended contract period and environmental disruption make it electorally unattractive and the additional jobs provided are short lived. The relatively higher number of lorry movements through trucking coal from the site is frequently singled out as a major disadvantage.

Additional time for scheme design and tender document preparation has also to be allowed since site investigation is important in providing vital information to tenderers on the quantities and heat value of coal in the spoil. This is essential to obtaining competitive tenders.

Planning constraints on operating times and specifying that mechanical/chemical methods are used for clarifying recycled washing water (rather than lagoons) are examples of conditions which may have to be incorporated in contract documents. Additional time may also have to be built into the tendering period to allow contractors to investigate local coal markets to gauge demand and price. Consequently, the minimum periods to allow for contract document preparation and tendering have to be about 16 weeks and 6 weeks respectively.

Figure 7   Lagoons like this may not always be feasible or permitted by
            the planning authority.

5.   ARRANGING CONTRACTS

To date, tender documents have been modelled on conventional reclamation contracts based on the Institution of Civil Engineers Conditions of Contract modified to cover specific conditions relating to coal recovery. An important objective in framing the contract is to ensure that the full credit value from the coal sale is realised. Although some contractors favour tendering on a variable quantity/rate basis for the coal sold, a fixed sum offer is preferable from the Employer's viewpoint. This makes both selection of the best offer straightforward and guarantees that the work will be completed for a fixed amount. Where the best offer is one where the fixed sum is paid to the contractor, budgeting for this amount is essential at the contract acceptance point.

As most companies specialising in coal recovery prefer to leave the final landscaping work to others, contracts are conveniently arranged in two phases.

Phase 1 - A coal recovery/regrading phase to achieve a predetermined landform within a 2/3 year period, including any ancillary works such as shaft capping and drainage.

Phase 2 - A final landscaping phase to achieve final treatment over a 4 to 6 month period, including cultivations, grassing, planting and fencing.

Phase 1

Under this phase the coal recovery contractor is also allowed to extract and sell any other minerals from the site in addition to recovered coal. This often comprises coal slurry dug directly from lagoons and/or burnt coal spoil which can be used for construction elsewhere. In recent contracts typical royalties payable to British Coal which the contractor allowed for in his tender were

coal recovered from site	-	£5.00 per tonne
coal slurry dug from site	-	£2.50 per tonne
burnt spoil extracted from site	-	£0.25 per tonne

524

The Phase 1 tender documents comprise drawings, specification and a bill of quantities. The fixed sum offer is obtained on the basis that the final landform and ancillary works are completed in accordance with the drawings and specification.

Items for ancillary works are priced in the bill of quantities so that competitive rates are available which can be applied if the works content is subsequently varied. As the sites included in Table 1 are moderately sized in coal recovery terms, no attempt was made to exclude reputable small sized firms from tendering. However, the precaution was taken to obtain a performance bond to remove the risk of the successful contractor simply removing the coal and abandoning the site before satisfactorily completing the ancillary works. The bond amount is sufficient to carry out any uncompleted earthworks required to achieve the final contours.

Figure 8   Completion of phase 1 coal recovery/regrading at Meikle Earnock tip, Hamilton. Phase 2 will include the cultivation of the main site area to the right of the ditch which, in turn, will be piped and gravel backfilled.

Whilst contractors are free to select their own washing plant the use of settling lagoons may be disallowed so that suitable mechanical/chemical equipment is required. The choice of the most suitable type of washing plant depends on several factors. Three of the most important are the total volume of spoil to be processed, the percentage coal content and the coal particle size characteristics.

With large tips the installation of a high cost "dense medium" plant may be justified as its higher operating efficiency allows it to recover about 2% more clean coal than most other types of plant. Tips in excess of 500,000 tonnes may be suitable for such a plant and consequently this type of operation is of interest to the larger specialist contractors. With lesser sized "coal rich tips" a cheaper but simpler form of plant (e.g. a barrel washer) may be viable, albeit less coal is recovered. Where the coal is predominantly fine (less than 2 mm) more sophisticated plant may be necessary. At the Minto site, in Fife, the coal was finer than could be efficiently recovered by a barrel washer alone so that a hydrosizer plant was introduced in conjunction with the barrel/cyclone circuit. At this plant the hydrosizer recovered 70% of the total recovered coal[7].

Securing any mineshafts (or adits) is best done during the the Phase 1 earthworks operation, when the shaft is most accessible. This is usually the time when the excavation level over the shaft is at its lowest.

Figure 9   The washing plant at Minto Bing, Fife showing the coal spoil
           supply gantry on the right the barrel washer in the background
           and the two hydrosizer units in the foreground.

Whilst the most suitable method of securement (e.g. infill, grouting, capping) depends on the size and type of shaft, the surrounding ground conditions and the final land use, securement costs can vary considerably. Again, this is a complex matter which cannot be dealt with here in detail other than to comment that the cost per shaft can vary from between £5,000 to £20,000.

Phase 2

This phase includes cultivation, landscaping and finishing works and the contract is placed after receiving competitive tenders, in the usual way, on the basis of remeasurable works. It typically includes for surface ripping, subsoiling, fertilising and grassing, where this is appropriate. Where a soil covering is not applied the incorporation of lime into the surface spoil becomes important and is always included where there is a need to raise the spoil's pH level. The quantity of lime required varies from site to site and this aspect of spoil treatment is critical at sites where the spoil is particularly acidic. For example, at Baads Bing in the Lothians an initial dosage of 50 tonnes per hectare was applied and further yearly incremental additions needed to maintain a neutral condition to avoid acid regeneration[8]. However, whilst it can be tempting to add large amounts of lime at this stage it is likely that overliming will be wasteful in that the leaching out of lime will occur faster than the rate of oxidation of the pyrites with the result that yearly applications of lime will still be required.

At larger sites, it may be possible to phase final restoration so that spoil within parts of the site can be completely processed and returned to the final landform for early landscaping. Where this is done it may be necessary to engage a nominated sub-contractor who can work under the main contractor appointed under Phase 1.

6. CULTIVATION AND VEGETATION ECONOMIES

Attention, so far, has concentrated on cost savings arising from landforming, without considering the final cultivation/vegetation costs. At some sites the potential for savings here is limited, due to the inherent potential spoil acidity. Like many other UK spoils, some Scottish spoils suffer from vegetation being destroyed by acid run-off resulting from the oxidation of pyrites. Whilst most spoils in Fife are mildly acid some spoils in the Lothians can be highly acid with pH values as low as 2.8. The high cost of treating and maintaining vegetation on such spoils has occupied scientists and horticulturists for decades and the literature on this topic is extensive.

In a Scottish context, this paper concludes by explaining various trial work currently under way aimed at investigating low cost vegetation techniques. Most of this work is in the development stage and no single technique has been applied over a full site. This work carries the underlying presumption that covering spoil with topsoil (either stripped from adjoining 'reception' land or by importation) is impractical or uneconomic.

(i) Adding Sewage Sludge

The benefit of using sewage sludge as a soil conditioner/nutrient additive has been recognised for many years as it can add nitrogen phosphorous and organic matter to nutrient deficient spoils. In 1983 the Water Research Centre was commissioned by the Agency to report on available sources of sewage sludge in central Scotland[9]. This work was subsequently extended in 1984 to field trials[10]. Although monitoring of this work continues it has become clear that sewage sludge can be effectively used as a spoil conditioner/nutrient additive where:

  (i)   The sludge is in cake form as this is more efficient in applying organic matter than the liquid form.

 (ii)   The sludge source is close to the site to minimise transport cost.

(iii)   The sludge cake is stored in a "dedicated heap" so that it is free from 'foreign' objects which might foul the traditional muckspreaders used on site.

(iv) Its organic nitrogen level is high enough, and its heavy metals content low enough, to achieve a balance which keeps the latter within permitted levels and the former to increase the organic nitrogen level to about 1,000 kilogrammes per hectares.

Early results suggest that sites can be successfully treated with sewage sludge for under £90 per hectare (1986 prices) which is about 10% of the cost of using imported natural topsoil.

(ii) Tree and Shrub Seeding

In 1981 the need to minimise the amount of rehabilitated grassland led to investigating alternative cheaper forms of vegetation. At that time a series of trials with direct tree and shrub seeding were established at various coal spoil and oil shale sites. Here, the idea was to develop a method for covering large rehabilitated areas with trees and shrubs which would naturally develop into areas of self sustaining woodland. These trials were designed by A G R Luke of Cambridge Direct Tree Seeding Ltd on the basis that the areas could develop into mature deciduous woodland in the long term[11].

Briefly, this work concentrated on "pre-treating" the seed of various species before seeding, to break dormancy, so that a major proportion of these germinated relatively quickly after broadcasting the seed. Seed mixtures were designed to contain a variety of species which would encourage a phased woodland development to occur in quicker succession than nature allows. That is to say, woody shrubs develop in the initial phase while pioneer trees develop subsequently to be followed by more mature species such as oak.

As the success of these trials will take some years to evaluate, the current period from 1986 to 1988 is being used for assessment. Early indications suggest that whilst the technique may not have any particular advantage over other techniques on moderately acid coal spoils it can be successful when applied to low acid coal spoils and all spent oil shale.

In addition to being considerably cheaper than planting, the technique has the added advantage that trees and shrubs which grow from seed in problem spoils develop a tolerance which cannot be as readily achieved by transplants. In other countries, notably North America and West Germany, direct seeding is frequently used to establish woodland on rehabilitated sites[12]. This technique, therefore, is very suitable when used in conjunction with coal recovery/rehabilitation.

(iii) Topsoil Substitutes

The high cost of providing topsoil has led to investigations into 'manufacturing' topsoil substitutes. Glasgow University's Department of Agricultural Chemistry conducted and reported on trials using wood waste (sawdust and bark) added to coal spoil. Although wood waste made no direct nutrient contribution, its addition considerably enhanced the water and nutrient holding properties of the spoil. The most satisfactory mixture devised comprised coal spoil, wood waste and sewage sludge. Whilst the proportions of each will vary depending on the nature of each constituent, it was concluded that larger sized wood waste particles were better than smaller ones. The most successful ratios of wood waste to coal spoil range from 1:1 to 1:2[13].

However, subsequent costing applied to full scale site applications showed that the transport and mixing costs were such that cheaper alternatives could be found so that this technique was not pursued further.

In a similar context, trial work is currently under way using dredged silt from the River Clyde by landing dredged river bottom deposits for use as a topsoil substitute. This work, carried out by Strathclyde University, is progressing well and has shown that the dredged material contains no significant toxins but does require some nutrient addition. As the particle sizes of the dredged material vary it is possible to allow the materials to dry out and be mixed so to achieve a satisfactory overall particle size distribution. The cost for uplifting this material from a riverside landing area, in a condition suitable for site use, is approximately £6 per tonne (1986 prices).

6. <u>CONCLUSION</u>

Rehabilitation of colliery sites often allows several options for treatment. Where the spoil contains a reasonable proportion of coal, say over 7%, it is worth investigating the viability of coal recovery/rehabilitation as this can often defray the whole or part of the earthworks costs. Although such a route takes longer than traditional rehabilitation it can be more economic and provide about three times more employment.

The final use of sites can also be a critical cost factor in rehabilitation: posing the question whether to spend relatively large sums on vegetation and maintenance or increasing investment to achieve long term development potential for a site? Where grassland treatment is necessary then topsoil substitutes can be considered, particularly where suitable waste materials are readily available at low cost and close to the site. Where recreation is required, it is likely that a good quality topsoil will be necessary to withstand the intensive wear on the grass.

Coal recovery/rehabilitation schemes which are properly controlled need no longer suffer from the stigma with which they were once associated and the resulting rehabilitation cost savings can make a useful contribution to other forms of derelict land clearance.

REFERENCES

1. Scottish Development Agency Act 1975 (Sections 7 and 8)

2. Macpherson T, (1980) "Land Renewal in Scotland" Chartered Municipal Engineering Volume 107.

3. Doubleday G P, Personal Communication to Scottish Development Agency (1978).

4. Cottrell S D, (1986) "Where There's Muck - Introducing Tip Washing" Land and Minerals Surveying Volume 3.

5. Thompson G M, Rodin S, (1972) "Colliery Spoil Tips - After Aberfan" Institution of Civil Engineers Paper 7522.

6. Arguile R P, (1974) "Clearance of Dereliction in Industrial Areas" Institution of Municipal Engineers: Monograph No 21.

7. Alexander Russell PLC, Personal Communication (1986).

8. Backes C A, Pulford I D, Duncan H J, (1985) "Neutralisation of Acidity in Colliery Spoil Possessing PH-Dependent Charge" Reclamation and Revegetation Research 4.

9. Swift D, (1983) "Sources of Sewage Sludge in Central Scotland for use in Land Reclamation" - Water Research Centre Report to Scottish Development Agency.

10. Bayes C D, (1986) "The Use of Sewage Sludge in Reclamation of Colliery Waste Tips - Summerlee Field Trial, Lanarkshire" - Water Research Centre Report to Scottish Development Agency.

11. Luke A G R, Macpherson T, (1983) "Direct Tree Seeding: a Potential Aid to Land Reclamation in Central Scotland" Arboricultural Journal Volume 7

12. Luke A G R, Humphries R N, Harvie H J, (1982) "The Creation of Woody Landscapes on Roadsides by Seeding - a Comparison of Past Approaches in West Germany and United Kingdom" Reclamation and Revegetation Research 1.

13. Pulford I D, (1981) "Investigation into the Suitability of Mixtures of Waste Materials as Substitutes for Topsoil" (Glasgow University Report prepared for Scottish Development Agency).

Reclamation, Treatment and Utilization of Coal Mining Wastes, edited by A.K.M. Rainbow   533
Elsevier Science Publishers B.V., Amsterdam, 1987 — Printed in The Netherlands

SHORT-TERM DURABILITY OF CEMENT STABILISED MINESTONE

M D A THOMAS, R J KETTLE and J A MORTON
[1]Department of Civil Engineering and [2]Geological Sciences, Aston University, Birmingham, UK.

SUMMARY

The performance of cement stabilised minestone in a standard short-term durability test is compared with various physical and geochemical properties of the raw minestones. The durability has been assessed by measuring both the strength loss and dimensional stability of CSM specimens following 7 day periods of immersion. Correlations are presented between these data and the fines content, plastic limit, slake durability and sulphate content of the raw minestone. A 'best' fit multiple regression equation is presented from the results of this study which enables the performance of CSM to be predicted in this short-term immersion test.

INTRODUCTION

Minestone, or unburnt colliery shale, is principally composed of the shales, mudstones and seatearths of the Coal Measures with subordinate sandstone and limestone resulting from the driving of drifts and shafts. The mineralogy is dominated by quartz and the clay minerals; kaolinite, illite, chlorite and a mixed-layer illite - montmorillonite (ref. 1). The general composition of minestone is largely controlled by the geographical location of a mine or the geological setting of the seam.

Minestone has found many outlets within the construction industry and perhaps the most demanding of these is its use as a cement-stabilised road-base material. Early studies (ref. 2) indicated that some minestones could be stabilsed with economic cement contents (ref 3) to achieve the compression strength requirements of the Department of Transport specifications (ref. 4) for Cement-bound materials. Subsequent field trials (refs. 5, 6) also met with some success and, following improvements in the availability of suitable mix-in-place and continuous mixing plant, the NCB have utilised cement stabilised minestone (CSM) in many construction applications including hardstandings, coal stacking areas and haul-roads (refs. 6-8).

The design of pavements incorporating cement-bound materials is usually carried out in accordance with Road Note 29 (ref. 9) and the construction is controlled by the requirements of the Department of Transport Specification for Road and Bridge Works (ref.4). Three categories of cement bound materials are currently included in these specifications but it is unlikely that minestone would meet the grading requirements for Lean Concrete (ref. 7) and hence is normally considered for use as a Cement-bound Granular Material or a Soil-Cement. Both these materials are required to achieve an average 7-day crushing strength of at least 3.5 MN/m^2 for cubical specimens or 2.8MN/m^2 for cylindrical specimens with an aspect ratio of 2. The strength results must also satisfy uniformity requirements expressed through the coefficient of variation. Further, any material used within 450mm, of the surface must be non-frost susceptable as determined by the TRRL Frost Heave Test (ref.10)

The strength of CSM can be controlled by the cement content, although such relationships are a function of the grading and mineralogy of the raw minestone (ref.11). Unfortunately, the strength requirements of the DOT Specification are not sufficient to ensure a lasting structure (ref. 12) and the need for the inclusion of durability criteria still exists (ref. 11). Future specifications will require CSM specimens to have proven resistance to increases in moisture content as assessed by the Immersion Test of BS1924 (ref 13), a minimum Immersion Ratio of 80% is now specified (ref.14). The extreme variability encounted with raw minestone necessitates the rigorous testing of each sample to be utilised. Consequently there has long been a need to establish the critical parameters of the raw material from which the performance of CSM can be predicted. This study is concerned with isolating those properties of the raw material that affect the short-term durability of CSM exposed immersion in water.

EXPERIMENTAL METHODS

Raw Minestone Analysis

Sixteen minestones were sampled from a number of colliery sites across the U.K. Standard Classification tests such as particle size distribution, Atterberg Limits, specific gravity and pH were carried out following the procedures of BS1377 "Methods of Testing Soils for Civil Engineering Purposes" (ref. 15) as modified for the characteristics of minestones (ref. 16).

The total sulphate and water soluble sulphate contents were determined in accordance with the Road Research Laboratory Report No. LR324 (ref. 17) since these methods were modifed for colliery shale. Relative clay mineral proportions were determined by X-Ray Diffraaction Analysis, while the resistance to slaking was assessed using a modified slake durability test. The details of these methods have been presented elsewhere (ref. 18) and are not included here.

CSM Preparation

The raw minestones were mixed with 10% cement at the optimum moisture content and cylindrical specimens, 50 mm dia x 100 mm, were prepared at 95% of the maximum dry density. The reduced target dry denisty was chose to avoid difficulties in the compaction of the more plastic minestones. With these very high target densities can lead to extreme density gradients through the compacted specimen. Such variability can result in considerable variation in the performance of CSM (ref 11).

All the specimens were sealed and cured at 20°C at the moulding moisture content. After 7 days the specimens were removed from the curing environments and selected for testing compressive strength and immersion resistance at 7 days. Further compressive strength determinations were performed following 14 days curing at constant moisture content.

Briefly , the Immersion Test involves soaking the CSM specimens under water for 7 days.

During this period linear dimension and mass changes were monitored and after 7 days the residual compressive strength was determined. This strength, expressed as a percentage of the strength of specimens cured at constant moisture content for 14 days, gives the Immersion Ratio.

DISCUSSION OF RESULTS

The particle size distributions are given in Table 1 and other physical properties determined are presented in Table 2. These properties show substantial variation between samples. Minestone 61 and 62 (from the North Eastern Coalfield) are on the one extreme, ganular and cohesionless, whereas minestones 31 41 (from Staffordshire) and 51 (from Nottinghamshire) are highly plastic with a considerable proportion of material finer than 63 microns. The plasticity of the minestones appears to be a function of the grading rather than the clay mineralogy, shown in Table 3, in agreement with a previous study (ref. 11).

TABLE 1

Particle Size Distributions of Raw Minestones

		Percentage finer than ;					
		--------- mm ---------			-------------- microns -------------		
Sample	37.5	20	5	600	63	20	2
Specified		45	25	8	0		
31	98	94	78	54	44	38	13
32	92	75	40	15	5	-	-
41	97	92	73	45	33	27	8
51	100	95	77	52	34	26	12
52	91	64	27	10	6	-	-
61	100	84	45	20	10	-	-
71	95	85	53	20	12	6	1
81	89	80	63	40	15	6	3
62	93	80	39	18	10	-	-
72	94	80	50	17	10	-	-
73	95	78	49	20	12	4	1
91	98	87	58	22	8	-	-
92	97	85	70	50	40	36	15
93	90	72	42	16	9	-	-
94	99	95	76	37	25	18	6
95	100	93	67	33	20	12	4

The modified slaking index ranged from 30 to 93% and 3 distinct classes of minestone can be drawn from the data. Minestone 31, having an index of 30, was totally isolated from the other minestones. Those with indices between 50 and 70 were all from the Staffordshire, Yorkshire or Nottinghamshire Coalfields. These minestones are associated with lower rank coals than the minestones from the North Eastern, Kent and South Wales regions, which had indices in excess of 70%. This index has a reliable negative correlation with the plasticity and sub-63 microns content of the minestone. However, no relationship was found between clay mineralogy and slake durability as has been suggested for North America Shales (ref. 19).

The sulphate content and pH are also presented in Table 3 and suggest further disparity

TABLE 2

Physical Properties of Raw Minestones

| Sample | Atterburg Limits % | | | Air Dry m/c % | Specific Gravity | Modified Slaking Characteristics | |
	Plastic Limit	Liquid Limit	Plasticity Index			Index	m/c %
31	30	53	23	10	2.26	30	23
32	20	31	11	2.5	2.39	60	13
41	28	48	20	6	2.34	55	18
51	22	36	14	3.3	2.52	62	15
52	21	37	16	3.8	2.51	61	15
61	17	25	8	2	2.53	72	14
71	18	26	8	2.2	2.55	79	11
81	25	34	9	2.5	2.48	71	13
62	18	26	8	3.9	2.59	79	11
72	16	24	8	1.3	2.63	93	7
73	22	32	10	2.6	2.55	86	9
91	19	33	14	3.1	2.41	90	9
92	27	48	21	9.7	2.32	57	24
93	20	30	10	5.2	2.49	88	12
94	22	36	14	7.3	2.54	60	13
95	22	34	12	4.8	2.48	66	13

TABLE 3

Chemical and Mineralogical Properties of Raw Minestones

| Sample | pH | Sulphate (%SO3) | | * Clay Mineralogy % - age :- | | | |
		Total	Water Sol.	K	C	I	M
31	2.9	2.35	0.7	14	5	53	28
32	7.7	0.08	0.05	16	8	54	22
41	6.9	0.04	0.03	15	6	45	35
51	7.6	0.11	0.08	14	8	48	30
52	7.8	0.05	0.03	12	9	54	26
61	7.5	0.29	0.13	19	10	44	27
71	4.1	0.96	0.39	9	7	63	22
81	6.7	0.8	0.31	13	5	52	30
62	6.9	0.45	0.23	22	8	43	27
72	8.9	0.26	0.06	8	6	56	30
73	5.1	0.4	0.17	8	4	60	28
91	7.8	0.07	0.02	17	9	53	21
92	6.6	1.89	0.19	23	6	50	21
93	7.5	0.14	0.07	47	14	25	14
94	6.3	0.44	0.22	16	8	44	32
95	6.9	0.24	0.12	17	9	43	31

* K = Kaolinite, C = Chlorite, I = Illite, M = Mixed-layer Illite-Montmorillonite

among minestones. The pH of minestone is controlled by sulphuric acid production, the latter being a consequence of pyrite alteration and therefore the pH will depend upon the initial pyritic sulphur content and the environmental conditions within the tip (ref. 20).

The engineering properties of the cement stabilised minestones are shown in Table 4. The required cube strength of 3.5 mPa of the Department of Transports specifications (ref. 4) was converted into a target strength of 2.03 mPa for the cylindrical specimens compacted at 95% of the maximum dry denisty used in this study. This conversion figure is based on strength comparisons between similar cubical and cylindrical specimens and details are given elsewhere (ref. 18).

TABLE 4

Properties of Cement Stabilised Minestones.

Sample	Compaction Char. OMC %	Dry Den Mg/m	Specimen Mass g	Ultimate Crushing Strength 7day MN/m	14 d...	------------ Immersion ------------ UCS MN/m	I.R. %	Expansion %	mass inc%
31	22.7	1.52	348	1.44	1.59	0.00	0	2.46	7
32	8.8	2.11	426	2.27	2.56	1.94	76	0.14	1.4
41	15.5	1.82	391	1.34	1.62	0.29	18	1.88	5.4
51	11.5	1.98	411	1.22	1.94	1.32	68	0.08	3.2
52	10.2	2.06	423	1.63	2.14	1.71	80	0.08	2.2
61	9.5	1.88	384	2.45	2.60	2.10	81	0.06	3.6
71	9.1	2.06	419	2.45	2.86	1.71	60	0.4	3.7
81	10.9	1.89	390	1.53	2.16	1.35	62	0.42	3.6
62	9.8	1.85	383	2.07	2.32	1.86	80	0.16	3.1
72	7.5	2.28	455	5.61	5.87	5.79	99	0.02	1.3
73	8.9	2.18	442	3.79	4.37	3.68	84	0.08	1.4
91	8.6	1.99	401	4.14	4.18	3.94	94	0.02	1.5
92	22.1	1.52	345	0.66	0.85	0.30	36	1.37	5.9
93	13.3	1.8	379	3.16	3.57	3.26	91	0.07	1.9
94	12.2	1.89	395	2.03	2.41	1.69	70	0.11	3.7
95	12	1.93	402	3.96	2.98	2.09	70	0.2	1.6

The results show that there is a tendency for the coarser, less plastic minestones to acheive a higher compressive strengths. Minestoner containing high proportions of fines and active clay fractions require high moisture contents for compaction and therefore achieve correspondingly low dry densities. The reduced dry density alone is sufficent to reduce the strength of stabilised materials but further consequences are; mixing and compaction problems due to clay aggregations - and increased surface area, the formation of colloidal aggregations - producing areas where stabilisation is due only to the migration of lime and thirdly the retarding of primary hydration due to localised lime depletion resulting from any pozzolanic reaction with clay minerals.

Minestones with high modified slake indices tended to produce stronger CSM's than those that broke down rapidly in the slake durability test. Thus it would appear that the strength of CSM depends not only of the proportion and plasticity of the fines but also on the engineering properties of the coarse fraction. In fact, criteria can be chosen from these raw minestone indices to solely preclude the CSM's that failed to achieve a crushing strength of 2.03 MPa at 7 days. These are:-

1. Slaking Index   greater than 55%
2. Plastic Limit   below 22%
3. Plastic Index   below 15%
4. Sub-63 microns content   below 25%

The immersion ratio ranged from zero to 99% showing the extreme variability in the performance of cement stabilised minestone. All the specimens had measured immersed strengths below the 14 day strength of equivalent specimens cured at constant moisture content. All specimens imbibed water and displayed volume expansion. The swelling is undoubtedly the cause of the strength loss and the relationship between the Immersion Ratio and the linear expansion after 7 days immersion is shown in Figure 1. Consequently the properties of the minestone which influence the swelling potential of CSM will similarly affect its degree of strength loss when immersed.

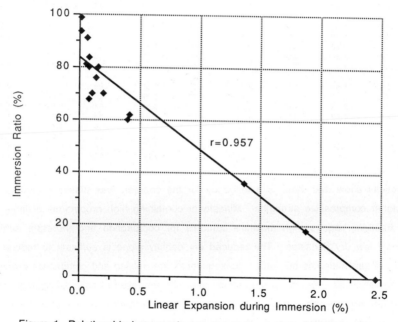

Figure 1   Relationship between Immersion Ratio and Swelling During Immersion.

The causes of swelling in CSM are complex but to date a combination of mechanisms have been proposed (refs. 11, 21, 22). These are:-

1. Swelling of expansive clay minerals
2. Sulphate attack on hydrated cement and/or clay minerals
3. Slaking of mudstones.

No relationship is apparent between the immersion resistance and the relative proportions of expansive illite-montmorillonite. However, by considering the relationships presented in Figures 2, 3 and 4, showing the immersion resistance of CSM against the Sub-63 microns content, plasticity and slaking resistance of its ingredient minestone, it is possible to

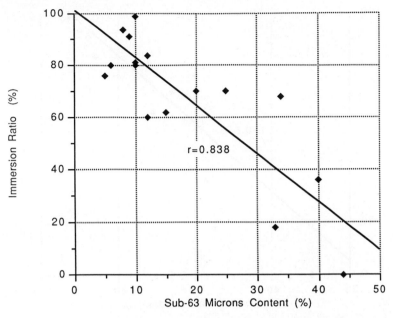

Figure 2 Relationship between Immersion Ratio and Sub 63 Microns Content.

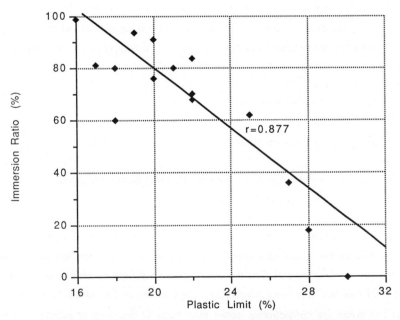

Figure 3 Relationship between Immersion Ratio and Plastic Limit.

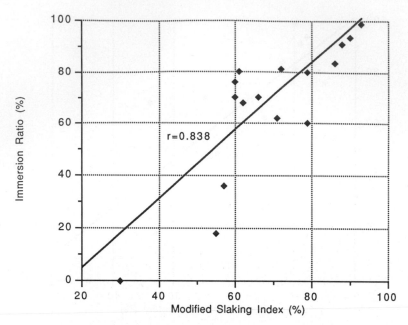

Figure 4 Relationship between Immersion Ratio and Modified Slaking Index.

summarise that the coarser grained, less plastic minestones with high slake durabilities produced CSM's with a higher resistance to the affects of immersion than did the finely graded, plastic, and "soft" minestones. None of these indices alone is capable of selecting minestones with adequate durability when stabilised, but, it is possible to suggest a limiting value for each index to achieve an appropriate level of Immersion Ratio. These are:

1. Sub 63 microns content   below 20%

2. Plastic Limit   below 25%

3. Modified Slaking Index   in excess of 60%

These limits are successful in separating the 'durable' CSM's from those with immersion ratios below 80% with one exception. Minestone 71 appeared to be physically suitable for cement stabilisation but the CSM produced from this material recorded an immersion ratio of only 60%. A closer examination of the minestones passing the above criteria shows that to some extent the swelling of the stabilised minestone during immersion is controlled by the sulphate content of the raw material, in particular the water soluble sulphate content as shown in Figure 5. Minestone 71 has total and water soluble sulphate contents of 0.96 and 0.39% (expressesed as % SO_3) and these are considerably higher than those of the other physically "accpetable" minestones. The rate of sulphate attack will depend on the solubility of the sulphate minerals and hence the expansion during a 7 day immersion period is likely to be largely controlled by the

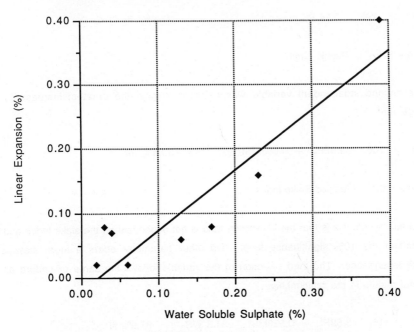

Figure 5 Relationship between Expansion following 7 days Immersion and Water Soluble Sulphate content for Coarse Grained CSM

water soluble sulphate content.

Predicting the Immersion Ratio from Raw Material Indices

Many single soil parameters may be used to predict the Immersion Ratio with limited accuracy. In order of decreasing correlation coefficient these are:-

Index   Corr. Coeff.(r)

Plastic Limit	- 0.877
Liquid Limit	- 0.860
Moisture content after slake test	- 0.850
Sub 63 microns	- 0.838
Modified Slake Index	0.837
Optimum Moisture Content	-0.837

A multiple regression was performed on the data presented in Tables 7.1 to 7.4. The first independant variable was chosen to give the greatest percentage fit ($100r^2$) and gives the following relationship:-

$$I R = 196 - 5.94 PL \qquad 77.0\% \text{ fit} \qquad (1)$$

where P L = Plastic Limit

The second independant variable was selected to give the greatest increase in the percentage fit:-

$$I R = 105 - 3.87PL + 0.663MSI \qquad 83.4\% \text{ fit} \qquad (2)$$

where MSI = Modified Slake Index

The null hypothesis is that the immersion ratio is not dependant on the slake index and can be rejected at the 10% significance level. No other parameter offers a more statistically significant dependance. The third selection of the variable follows the same procedure as the second and results in the relationship:-

$$I R = 115 - 3.62PL + 0.529MSI - 34.3 WS \qquad 87.0\% \text{ fit} \qquad (3)$$

where WS = water soluble sulphate (as % SO_3)

Again the null hypothesis can be rejected at the 10% significance level. No further parameter can be added to the model in order to statistically increase the percentage fit at the 10% significance level.

The plastic limit and modified slake index represent the sensitivity of a CSM to the affects of the physical swelling mechanisms in the fine and coarse fraction respectively whereas the water soluble sulphate content provides an indication of the expansion due to sulphate attack during the 7 day immersion period.

CONCLUSIONS

The short term durability of CSM is controlled largely by the properties of the raw minestone such as; grading, plasticity, slaking resistance and sulphate content. Predictions of both expansion and strength loss due to immersion can be made from these paramenters. However, none of the sixteen CSM's studied had ceased to expand at the termination of the immersion test. It is unlikely that the full extent of the damage resulting from changes in moisture content can be realised by short periods of immersion. In particular, the full consequences of the sulphur bearing mineralogy may not be realised for months rather than days (refs. 12, 18) and an investigation of long-term durability and geochemical stability is required if the full effects of the chemical attack are to be determined.

ACKNOWLEDGEMENTS

The authors are indebted to British Coal for the funding of the research project. The views expressed are those of the authors and not necessarily of British Coal.

REFERENCES

1   R. J. Collins, 1986. "A method of Measuring the Mineralogical Variation of Spoil from British Collieries", Clay Minerals, II, pp 31 - 50.

2   R. J. Kettle and R. I. T. Williams, 1969. "Preliminary Study of Cement Stabilised Unburnt Colliery Shale", Rds. and Rs. Constr., Vol 47, No. 559, pp 200 - 206.

3   J. Kennedy, 1979. "The Economic Viability of Using Cement Stabilised Materials for Pavements", Cem. Conc. Res., Tech. Note DN/1048.

4   Department of Transport, 1977. "Specification for Road and Bridge Works", HMSO, London.

5   R. J. Kettle and R. I. T. 1978. "Colliery shales as a construction material", Proc. Int. Conf. Use of By-Products and Wastes in Civil Engineering, Vol II, Ecole Nationale des Ponts et Chaussess, Paris. pp 475 - 81.

6   D. A. Tanfield, 1978. "The Use of Cement Stabilised Colliery Spoils in Pavement Construction", Proc. Int. Conf. Use of By-Products and Wastes in Civil Engineering, Vol III, Ecole Nationale des Ponts et Chaussees, Paris, pp 481 - 501.

7   A. K. M. Rainbow, 1982. "Colliery Spoil - Its Product, Properties and Use in Cement Stabilised Form", Seminar on Waster Materials in Concrete, Cement and Cpncrete Research (TDH8610).

8   W. Sleeman , 1984. "Practical Application of Cement Bound Minestone within the British Coal Mining Industry", Proc. Int. Sym. Reclamation, Treatment and Utilisation of Coal Mining Wastes, National Coal Board, pp 53.1 - 53.19.

9   Department of the Environment, 1970. "A guide to the structural design of pavements for new roads." Road Note 29, HMSO, London.

10 D. Croney and J. Jacobs, 1970. "The Frost Susceptibility of Soils and Road Materials", Report LR0, RRL.

11 A. T. McNulty, 1985. "The Durability of Cement Bound Minestone", Thesis, PhD, Aston University, Birmingham.

12 J. M. Morton, R, J. Kettle and M. D. A. Thomas, 1984. "Some Experimental Observations on the Potential Expansion of Cement Bound Minestone", Proc. Int. Symp. on the Reclaimation, Treatment and Utilisation of Coal Mining Wastes, National Coal Board, Durham University, pp 56.1 - 56.8.

13 British Standards Institution, 1975. "Methods of Test for Stabilised Soils", BS1924.

14 A. K. M. Rainbow and W. Sleeman, 1984. "The Effect of Immersion in Water on the Strength of Cement Bound Minestone", Proc. Int. Symp. on the Reclaimation, Treatment and Utilisation of Coal Mining Wastes, National Coal Board, Durham University pp 55.1 - 55.15.

15 British Standards Institution, 1975. "Methods of Test for Civil Engineering Purposes". BS1377, BSI, London.

16 National Coal Board, 1971. "Application of British Standard 1377:1967 to the Testing of Colliery Spoil", NCB, London.

17 P T Sherwood and M D Ryley, 1970. "The Effect of Sulphates in Colliery Shale on its Use for Roadmaking", Report LR324, Road Research Laboratory.

18 M D A Thomas, 1986. "The Performance of Cement Stabilised Minestone", Thesis PhD., Aston University, Birmingham, U.K.

19 R C Deen, 1981. "The need for a Schema for the Classification of Transitional (Shale) Materials", Geotech. Testing J., GTJODJ, Vol. 4(1), March, pp 3-10.

20 R M Henderson and P J Norton, 1984. "A Method of Predicting The Pollution Potential of Surface Mining Backfill", Int. Mine Water J., Vol. 3(1).

21 P T Sherwood, 1962a. "The Effect of Sulphates on Cement and Lime Stabilised Soils", Rds. and Rd. Constr., Feb., pp 34-40.

544

22 P T Sherwood, 1968. "The Properties of Cement Stabilised Materials", Report LR205, Road
Research Laboratory.

Reclamation, Treatment and Utilization of Coal Mining Wastes, edited by A.K.M. Rainbow
Elsevier Science Publishers B.V., Amsterdam, 1987 — Printed in The Netherlands

THE WETTING EXPANSION OF CEMENT-STABILISED MINESTONE – AN INVESTIGATION OF THE CAUSES AND WAYS OF REDUCING THE PROBLEM

C.E. CARR and N.J. WITHERS, British Coal Corporation, Coal Research Establishment, Stoke Orchard, Nr. Cheltenham, Gloucestershire, United Kingdom

SUMMARY

Minestones from Snowdown, Parkside, Wardley and Newcraighall collieries were tested to assess their suitability for stabilisation with cement and to establish a range of conditions under which they may be used with minimum risk, particularly with respect to wetting expansion.

All the materials, except the Newcraighall Minestone, proved suitable for stabilisation when compacted to maximum dry density within a moisture content range of optimum for compaction and optimum +2%. It is important not to work below the optimum moisture content. as this can result in significantly lower strength and increased expansion when exposed to free moisture.

Two types of expansion were identified which could affect cement-stabilised Minestone (CSM). One in the short term, due to hydration of the clay mineral within the Minestone, and another which occurs in the longer term due to sulphate attack upon the cement matrix of the CSM.

Expansion due to hydration can be reduced by the addition of sand to the mix, and it may be possible to minimise the risk of long term sulphate attack by the use of sulphate resisting cement or by the addition of pulverised fuel ash (PFA) to the CSM. However it is considered that the most important factor for the successful utilisation of CSM is good site control, ensuring that the material is compacted to its maximum density and within a moisture content range of optimum and optimum +2%.

INTRODUCTION

Maximising the utilisation of coal mining waste or Minestone is clearly beneficial, both in terms of potential savings in spoil disposal costs and as a means of reducing the environmental impact of coal mining. The major outlet for Minestone is in bulk fill applications such as highway embankments and land reclamation schemes (ref. 1).

In general, Minestone is a relatively weak material, mechanically, and therefore cannot be considered as a good quality aggregate for use in the construction industry. However with the addition of cement (5-10%) many Minestones can be strengthened sufficiently, when compacted, to enable them to meet the UK Department of Transport specified 7 day strength requirement (3.5 N/mm^2) for cement stabilised materials (ref. 2). This enables it to be used in more demanding applications such as a structural layer in roads, hardstandings and parking areas (ref. 3).

The use of cement-stabilised Minestone (CSM) was developed over a number of years in various applications, both within the British Coal Corporation (Yorkshire and North East England) and outside, particularly in Kent (ref. 4). However problems of deformation of the pavement at construction joints were encountered at some of these sites and this was particularly evident on the Canterbury by-pass in 1980 (ref. 5). Snowdown Minestone was used at Canterbury and deformation occurred within a few days of laying which was attributed to expansion of the CSM sub-base by water uptake. At other sites the deformation occurred after a much longer period of time. A research programme was therefore initiated to assess the suitability of a number of Minestones for cement stabilisation, to investigate the cause of the expansion and to establish the conditions under which CSM could be used with minimum risk of wetting expansion.

Four Minestones were tested which were chosen on the basis of their potential for utilisation in this application and to give good geographical coverage of the county. In addition, samples of CSM, from sites where expansion problems had occurred, were examined and tested. The programme includes long term monitoring of laboratory samples of CSM and this work is still on-going. This present paper reports the results of the investigations obtained so far and current views on the causes of the expansion and ways of reducing the problem.

MINESTONE CHARACTERISATION

Four Minestones were tested: Snowdown (Kent), Parkside (Western Area), Wardley (North East Area) and Newcraighall (Scottish Area) and these were characterised by a range of standard soil mechanics tests (ref. 6).

The particle size distribution of the minus 50 mm screened fractions from the Snowdown and Wardley Minestones, as used in this application, did not meet the grading requirement for cement bound granular material (ref. 2) as they were short of fines. However this should generally not prove to be a problem as the degree of compaction that the materials are subjected to when being laid on site should increase their fines content sufficiently to enable them to comply with the specification.

Other characterisation data (Table 1) indicated that Newcraighall Minestone differed from the others in that it had higher water absorption and Atterburg limits. The other Minestones possessed similar properties, with Snowdown Minestone having the highest relative density and lowest water absorption.

Results of compaction tests (ref. 7) on the Minestones with and without cement, are given in Table 2. The addition of cement made little difference to the optimum moisture content for compaction or maximum dry density.

TABLE 1

Properties of selected Minestones

Minestone	Coefficient of Uniformity	Relative Density	Water Absorption (%)	Liquid Limit	Plastic Limit	Plasticity Index	Linear Shrinkage (%)
Snowdown	7	2.49	4.6	28	19	9	4
Wardley	11	2.26	6.8	27	19	8	4
Parkside	18	2.43	6.9	30	21	9	6
Newcraighall	19	2.58	20.3	40	27	13	7

TABLE 2

Optimum moisture content and maximum dry density values for selected Minestones

Minestone	Optimum Moisture Content for Compaction (%)		Maximum Dry Density (Mg/m^3)	
	With Binder	Without Binder	With Binder	Without Binder
Snowdown	6.2	5.5	2.25	2.26
Wardley	7.5	8.0	2.08	2.06
Parkside	8.0	8.1	2.07	2.07
Newcraighall	16.0	15.2	1.69	1.71

ASSESSMENT OF THE SUITABILITY OF THE SELECTED MINESTONES FOR STABILISATION

The speed at which deformation of the CSM pavement had occurred at Canterbury indicated that the prime cause was expansion of the CSM due to hydration of the clay material within the Minestone. The main emphasis of the initial test programme was therefore to study the strength characteristics of the various CSM mixes when prepared over a range of initial moisture contents and also to examine their expansion behaviour when exposed to free water.

All the CSM compacts were prepared using ordinary Portland cement as the binder, at a level of addition of 10%.

Unconfined compressive strength

Test cubes (150 mm) for 7 day compressive strength determinations were prepared from CSM mixes produced at the optimum moisture content for compaction (OMC), OMC + 2% and OMC - 2%. The specimens were compacted to refusal using a vibrating hammer (ref. 7).

Each Minestone exceeded the DoT strength specification of 3.5 N/mm^2 (ref. 2) when stabilised with 10% cement and compacted at OMC (Table 3). However when compacted at moisture contents above or below OMC there was a decrease in strength, particularly below OMC. There was a reduction in strength of about 80% when the Snowdown CSM was prepared at OMC - 2%.

TABLE 3

Unconfined compressive strength of cement stabilised Minestone (10% ordinary Portland cement)

Minestone	Moisture Content (%)	7 Day Unconfined Compressive Strength (N/mm^2)
Snowdown	OMC - 2%	1.7
	OMC	8.9
	OMC + 2%	7.3
Wardley	OMC - 2%	2.3
	OMC	5.6
	OMC + 2%	4.8
Parkside	OMC - 2%	2.7
	OMC	5.2
	OMC + 2%	4.6
Newcraighall	OMC - 2%	1.4
	OMC	3.6
	OMC + 2%	2.7

OMC: Optimum moisture content for compaction

Wetting expansion

The wetting expansion behaviour of the range of CSM materials was assessed by monitoring unconfined cylindrical CSM compacts (150 mm diameter, 150 mm length) when totally immersed in water. The compacts were first cured in sealed bags for 7 days prior to mounting them in brass frames (Figure 1) in which the expansion could be monitored by measuring the deflection of a plate in contact with top of the compact by means of a dial gauge. The experimental arrangement was set up within a controlled temperature room (20°C ± 2°C).

Fig. 1. Rig for monitoring expansion of cement stabilised Minestone

Expansion of the compacts generally levelled off at 6-8 weeks. All the cement stabilised Minestones expanded to varying degrees, even when compacted at OMC (Table 4). The Snowdown CSM exhibited the least expansion (0.1% at OMC) and the Newcraighall the highest (>5%). In the latter material, the extent of expansion was such that the compact disintegrated. The influence of the initial moisture content of the CSM mix on its properties is again evident

from the results obtained with the Snowdown CSM; there being a 60% increase
in the level of expansion when the compact was prepared from a mix having a
moisture content of OMC - 2%.

Further testwork on wetting expansion was undertaken to assess the
relative importance of the fines fraction of the Minestone in relation to the
wetting expansion properties of CSM. Clearly, the fines would contain
clay-sized material (<2 µm) which should be potentially the more active
component in connection with hydration and expansion but it should be noted
that Minestone is predominantly based on clay mineral throughout its complete
particle size range. The minus 0.5 mm fraction was screened from Snowdown
Minestone and replaced by the same amount of equivalent sized silica sand
(considered non-expansive). CSM compacts were then prepared with 10% cement
addition for strength tests and expansion monitoring. The replacement of the
minus 0.5 mm fraction of Snowdown Minestone, which represented only 8% of the
total material, with the sand, resulted in a 50% reduction in expansion and an
increase in 7 day strength of 22% (Table 5).

TABLE 4

Short term wetting expansion of cement stabilised Minestone
(10% ordinary Portland cement)

Minestone	Moisture Content (%)	Mean Linear Expansion (%)
Snowdown	OMC - 2%	0.16
	OMC	0.10
	OMC + 2%	0.07
Wardley	OMC	0.21
Parkside	OMC	0.66
Newcraighall	OMC	5.00*

OMC: Optimum moisture content for
compaction

*   Specimen disintegrated before
expansion was completed

TABLE 5

The properties of cement stabilised Snowdown Minestone after
replacement of the minus 0.5 mm fraction with sand

	Optimum Moisture Content (%)	Maximum Dry Density (Mg/m^3)	7 Day Unconfined Compressive Strength (N/mm^2)	Mean Linear Expansion (%)
Snowdown (No replacement)	6.2	2.25	8.9	0.10
Snowdown (-0.5mm replacement with sand*)	5.4	2.32	10.3	0.05

* 8% of total Minestone replaced

Effect of sand addition on the properties of cement stabilised Minestone

The addition of an inert filler should reduce the potential for the
Minestone to expand and sand addition has, in fact, been used commercially in
Kent. Therefore tests were carried out to investigate the effect of sand
addition on the properties of CSM by preparing a range of mixes from Parkside
Minestone with sand (Zone 4) additions of 25 to 75%.

Data on the properties of the various mixes are presented in Table 6. As
might be expected the addition of sand progressively reduced the expansion and
increased the strength of the CSM with increasing level of addition.

EXAMINATION OF SITE SAMPLES OF CEMENT STABILISED MINESTONE

During the course of the test programme, the opportunity arose to examine
samples from CSM sites where problems of expansion had occurred. At two of
these sites, the Escrick to Stillingfleet road in North Yorkshire which had a
CSM sub-base, and the CSM lorry park at Horden colliery in Durham, the
pavements had performed successfully for a number of years before being
deformed by expansion of the CSM. Examination of the samples taken from these
two sites in the laboratory revealed the presence of extensive deposits of a
white crystalline material throughout the matrix of the materials.

TABLE 6

Effect of sand addition on the properties of Parkside Minestone
stabilised with 10% ordinary Portland cement

Sand addition (Zone 4) (%)	Optimum Moisture Content for Compaction (%)	Maximum dry Density (Mg/m^3)	7 Day Unconfined Compressive Strength (N/mm^2)	Mean Linear Expansion* (%)
Zero	8.0	2.07	5.2	0.66
25	7.5	2.12	7.2	0.21
35	8.0	2.13	N.D.	0.10
50	8.2	2.16	8.2	0.03
75	8.8	2.00	10.3	0.005

ND: Not determined
* Specimens made at their optimum moisture contents

Examination of the white deposits using scanning electron microscopy and X-ray diffraction showed that the material consisted of needle crystals of hydrated calcium aluminate tri-sulphate ($3CaO. Al_2O_3. 3CaSO_4. 31H_2O$) better known as the mineral ettringite (Fig. 2). Chemical analysis of the CSM indicated that there had also been a significant increase in the acid soluble sulphate content of the Minestone since the material had been laid, e.g. the Horden CSM contained $0.4 - 1.2\%$ sulphate (as SO_3) whereas the Wardley Minestone used in this CSM had an initial sulphate content of <0.1%.

The CSM from both sites had become very brittle and they easily crumbled. These observations indicated that there had probably been a deterioration of the cement bond due to sulphate attack. The formation of ettringite is a well known form of expansive attack in concrete and occurs when sulphate solutions react with the tri-calcium aluminate component in the cement. The generally accepted mechanism, though it may be an oversimplification, is that the resulting crystals occupy a volume larger than the original aluminate, thus expansive stresses are created which can result in the eventual disintegration of the concrete.

Bar scale: 10 μm

Fig. 2 Ettringite crystals within cement stabilised Minestone matrix

The presence of the extensive deposits of ettringite within the site samples indicated that CSM may suffer from more than one type of expansion; in the short term from hydration of the clay minerals within the Minestone and in the longer term from the formation of ettringite as a result of sulphate attack upon the cement binder.

LONG TERM PROPERTIES OF CEMENT-STABILISED MINESTONE AND EFFECT OF BINDER TYPE

Following the examination of the site samples, further experiments were undertaken to study the strength and expansion properties of CSM over a much longer monitoring period. This was in order to determine the extent to which these properties were affected by any increase in sulphate level and subsequent ettringite formation when under laboratory conditions. In addition experiments were set up to assess the effect on CSM properties, of using different binders for the stabilisation.

The usual way of preventing sulphate attack problems in concrete is to use sulphate resisting cement (SRC); this cement contains less sulphate-susceptible tricalcium aluminate. Therefore CSM mixes were prepared using SRC as binder. Wardley Minestone was used in this testwork and the level of binder addition was reduced to 8%. The incorporation of pulverised fuel ash (PFA) can also improve the sulphate resistance of concrete (refs. 8,9). This has been attributed to the reaction of the PFA with the free lime within the hydrated cement and the resultant reduction in permeability and possibly to the formation of protective calcium silicate hydrate films around the vulnerable aluminate components of the cement (ref. 10). A mix was prepared from Wardley Minestone containing 10% PFA and 8% OPC. A further mix was prepared to assess the effect of lime addition, this contained 10% lime and 8% OPC.

All the CSM test specimens were compacted at their OMC. As in previous testwork, 150 mm cube specimens were used to monitor strength development (cured in sealed polythene bags) and 150 mm diameter cylindrical specimens used to monitor expansion properties (immersed in water). For comparison purposes, the properties of compacts prepared from stabilised sand and gravel were also monitored. Throughout the experiments a check was maintained on the total sulphur, pyritic sulphur and sulphate contents of the CSM compacts.

All the CSM compacts met the 7 day strength requirement of 3.5 N/mm^2 (Fig. 3). There was no significant difference in the strength achieved using OPC and SRC as binder. The addition of 10% lime to the mix promoted a small but consistent increase in strength throughout the monitoring period. However the use of PFA produced a significant increase in the rate of strength development with a strength increase of about 70% being achieved after 1 year's curing compared to the normal CSM.

Expansion monitoring indicated that there was no significant difference in the properties of the CSM prepared from OPC and SRC up to a curing period of 1 year (Fig. 4). However it should be noted that no increase in sulphate (acid or water soluble) content of the CSM or ettringite formation was detected during the monitoring period.

An interesting result was obtained with the mix containing lime. After generally following the trend in expansion for the CSM prepared from OPC only, there was a relatively sudden reduction in the dimensions of the compacts under test after about 30 weeks curing. This resulted in a reduction in the level of expansion, reached previously, of about 30%. Though the actual dimensional changes being monitored in these experiments are very small, i.e. fractions of a mm, the experimental procedure used has generally proved to be sensitive and reliable. This shrinkage of the compact is considered to be a real effect and it has also been encountered in other tests using CSM mixes containing lime. The cause of the effect and why it should so suddenly occur is unclear but it may be related to flocculation of the clay mineral within the Minestone by the lime.

The incorporation of PFA within the CSM mix resulted in significant reductions in wetting expansion (50-60%) throughout the monitoring period in addition to its beneficial effect on strength development.

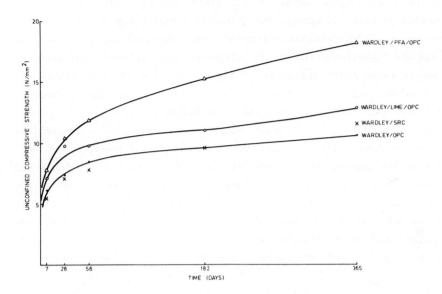

Fig. 3.   Unconfined compressive strength of cement-stabilised Minestone compacts

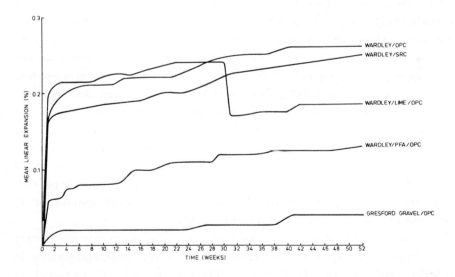

Fig. 4   Expansion characteristics of cement stabilised Minestone compacts

DISCUSSION

Of the Minestones tested in the programme, the Newcraighall material proved unsuitable for stabilisation. Its 7 day strength at OMC only just exceeded the specification value of 3.5 N/mm^2 and it exhibited excessive expansion when immersed in water. The characterisation tests showed that this Minestone had the highest water absorption, OMC and Atterburg limit values. It also had the highest coefficient of uniformity, i.e. highest fines content. These results reflect the influence of the active clay fraction within the material and, when considering Minestone for stabilisation, these parameters would be useful indicators for assessing their suitability.

The Snowdown, Parkside and Wardley Minestones all had similar properties. However the Snowdown Minestone had the lowest water absorption and OMC, together with the highest maximum dry density. It is therefore probably the best suited for stabilisation, particularly with regard to its potential for expansion and this was confirmed in the later expansion tests.

All the CSMs tested exceeded the Department of Transport's specification for 7 day strength when compacted at OMC. However, when made at moisture contents other than at OMC, strengths decreased, particularly when prepared at moisture contents below OMC. At OMC, the Snowdown CSM exhibited the highest strength of the materials tested but this showed an 80% reduction in its strength when compacted at 2% below OMC.

All the CSMs tested showed some propensity for expansion when immersed in water with the Snowdown material exhibiting the best performance, having a short-term expansion value of about 0.1%. Though the CSMs expanded they all remained structurally sound except Newcraighall. It should be noted that it is not possible to equate the levels of expansion which occurred in the laboratory samples to acceptable levels for 'on site' construction purposes. The expansion is also monitored using unconfined compacts and therefore the expansion results should only be used as a means of comparing the relative performance of the Minestones under test.

The influence of moisture content on CSM properties was also highlighted in the expansion results for the Snowdown Minestone. When compacted at 2% above the OMC, the expansion was reduced by about 30% as compared to that which occurred when compacted at OMC. However at a moisture content of 2% below OMC the extent of expansion increased by 60%.

The strength and expansion results illustrate the importance of correct moisture content with regard to CSM performance on site. Poor performance is particularly likely to be encountered if the CSM is laid and compacted at moisture contents below OMC and a recommended working range would be OMC to OMC + 2%. The results obtained with the Snowdown CSM and all the earlier characterisation tests would seem to indicate that the problems encountered at the Canterbury by-pass may have been as a result of poor site control rather than as a result of an inherent weakness within the material itself.

From the results of the sand replacement tests it is apparent that fines fraction of Minestone has a major influence on the expansion of the material. Although Minestone can contain up to 60% clay mineral, most is bound up within the coarser fractions and therefore is not as readily available to free moisture. It is only within the fine particle size range of the Minestone that a discrete clay fraction is found, and it would appear from the results that this fraction of the Minestone can be responsible for causing a disproportionate amount of the short term swelling. However removal and replacement of the fine particle size fraction of a Minestone, whilst being a relatively simple operation on a laboratory scale, would not be a practical or economic operation for large scale site work.

The straight addition of sand to the mix may be a more realistic option. Significant decreases in expansion together with increases in strength were achieved but high levels of sand addition were required. The reason for the decrease in expansion levels appears to be the result of a simple dilution effect by the sand, resulting in the mix containing less Minestone which is capable of expanding. The corresponding increase in strength results from the inherent strength of the sand which acts as a fine aggregate. The disadvantage of sand addition is clearly that less Minestone would be used. The decision to use sand within a CSM mix would therefore have to be judged upon the quality of a proposed Minestone or the intended end-use of the CSM and the availability of cheap, good quality sand.

Laboratory tests and examination of site samples indicated that two types of expansion can occur within CSM. Expansion can occur in the short term as a result of hydration of the clay mineral within the Minestone, particularly if the material is laid below OMC. Expansion can also occur over a longer period as a result of sulphate attack upon the cement binder. This leads to the formation of ettringite which has a large crystal volume and, as found in concrete systems, the long term expansion of the CSM may be attributed directly to the crystallisation and growth of this material within the CSM matrix. Alternatively, it may be that the sulphate attack simply weakens the cement bond between the Minestone particles and hence increases the susceptibility of the CSM to further expansion by hydration. The increase in sulphate content within the CSM is presumably associated with oxidation of residual pyrite within the material.

The alternative use of sulphate resisting cement (SRC) as the binder in the Wardley CSM made little difference to the long term strength development of the material or expansion properties compared with the Wardley/OPC CSM mixes. However it should be noted that, so far within the monitoring period, there has been no significant increase in sulphate content or ettringite formation.

The lack of ettringite formation in the CSM mixes used for the long term strength determinations may have been due either to the time periods for the tests being insufficient for the ettringite to form in any significant amount or that ettringite formation may be dependent upon having access to free moisture. All specimens of CSM for strength determination are cured and stored in airtight plastic bags, with no access to additional moisture. The expansion monitoring for this part of the test programme is still ongoing and further checks will be maintained on sulphate/ettringite formation within the immersed specimens and what effect this has on expansion properties.

The use of PFA within the binder system produced some encouraging results. The addition of 10% PFA to the Wardley/OPC CSM mix promoted an increase in long term strength development of about 70% and a corresponding reduction in the level of expansion of 50-60%. Although it is recognised that the addition of 10% PFA to the Wardley/OPC mix effectively raises the binder content of the mix, due to the pozzolanic nature of the PFA, the beneficial effects that it has upon the properties of the CSM seems disproportionate to the amount of extra cementation that would reasonably be expected from the addition of 10% PFA. This implies that as well as imparting extra strength to the CSM through pozzolanic activity, the PFA improves the CSM in other ways as well. As yet this action is unclear but this will be studied in more detail in further investigations of the use of PFA within CSM.

CONCLUSIONS

(i) Acceptable cement-stabilised Minestones (CSM) can be prepared from the Snowdown, Wardley and Parkside Minestones when compacted within a moisture content range of optimum moisture content for compaction (OMC) to OMC + 2%. However the Newcraighall Minestone is unsuitable for cement stabilisation due to low strength and excessive expansion when exposed to free moisture.

(ii) Two types of expansion were identified within CSM. Firstly a short term expansion due to hydration of the clay minerals within the Minestone and secondly expansion may occur in the longer term from sulphate attack upon the cement matrix of the CSM, resulting in the formation of expansive ettringite crystals.

(iii) It is possible to minimise the amount of expansion that occurs in the short term by the addition of sand to the CSM but high levels of addition (>30%) are likely to be required. It may also be possible to reduce the risk of expansion due to sulphate attack by the use of sulphate resisting cement as an alternative binder or by the addition of pulverised fuel ash to the mix.

(iv) The most important factor for ensuring the successful use of CSM is probably good site control, especially with regard to ensuring maximum compaction and maintaining the CSM mix within a moisture content range of OMC to OMC + 2%.

(v) Because of the 'unforgiving' nature of Minestone and the fact that good site control is not always the case, CSM is perhaps best suited for lower grade applications where small amounts of expansion may not be a problem, or on sites with a limited design life.

ACKNOWLEDGEMENT

This paper is published by permission of the British Coal Corporation but the views expressed are those of the authors and not necessarily those of the British Coal Corporation.

The authors would like to thank Dr. A.K.M. Rainbow, Minestone Services for his advice and support during the project and Mr. D.D. Skeen, Coal Research Establishment for his practical assistance.

REFERENCES

1   D. Turnbull, The Role of the Minestone Executive in British Mining and Civil Engineering, Proc. Symposium on the Reclamation, Treatment and Utilisation of Coal Mining Wastes, Durham, England, Sept. 10-14, 1984, British Coal Corporation, London, pp 1.1 - 1.12.

2   Department of Transport, Specification for Road and Bridge Works, 1976, Her Majesty's Stationery Office, London, Clause 806.

3   Minestone Services, Cement Bound Minestone Users Guide for Pavement Construction, August 1983, British Coal Corporation, London.

4   W. Sleeman, Practical Application of Cement Bound Minestone within the British Coal Mining Industry, Proc. Symposium on the Reclamation, Treatment and Utilisation of Coal Mining Wastes, Durham, England, Sept. 10-14, 1984, British Coal Corporation, London, pp 53.1 - 53.19.

5   T.V. Byrd, Sad Canterbury Tale, New Civil Engineer, Nov. 1980.

6   British Standards Institution, Methods of Test for Soils for Civil Engineering Purposes, BS 1377, 1975, London.

7   British Standards Institution, Methods of Test for Stabilised Soils, BS 1924, 1975, London.

8   E.G. Barber et al, PFA Utilisation, Central Electricity Generating Board, 1972, p. 20.

9   C. Plowman, The Chemistry of PFA in Concrete - A Review of Current Knowledge, Proc. AshTech '84, Second International Conference on Ash Technology and Marketing, Barbican Centre, London, England, Sept. 16-21, 1984, Central Electricity Generating Board, pp 437-443.

10 F.M. Lea, The Chemistry of Cement and Concrete, Third Edition, 1970, Edward Arnold, p. 442.

Reclamation, Treatment and Utilization of Coal Mining Wastes, edited by A.K.M. Rainbow   561
Elsevier Science Publishers B.V., Amsterdam, 1987 — Printed in The Netherlands

ASSESSING THE DURABILITY OF REINFORCEMENT MATERIALS IN MINESTONE

D.C. READ and C.E. CARR

British Coal Corporation, Coal Research Establishment, Stoke Orchard, Cheltenham, Gloucestershire, United Kingdom.

SUMMARY
     A field trial is being carried out to assess the durability of a range of reinforcement materials buried in a Minestone for various exposure periods. This study is also aimed at characterising the Minestone environment so that possible causes of any degradation can be identified.

     The field trial is being supplemented by a series of laboratory studies in which the influence of various fill conditions on the durability of specimens of metal reinforcement buried in them is being assessed. The conditions being investigated include the influence of fill type, burial time, compaction, moisture content, temperature, standing water, flowing water, water table fluctuations and the influence of a drainage layer. The types of fill used in this study include a number of Minestones, burnt shales, pulverised fuel ash, gravel, chalk and limestone.

     The effect that any corrosion may have on the mechanical properties of the metal and non-metal reinforcement specimens is being assessed by strength loss measurements. In addition the degree of moisture uptake within the non-metal specimens is being determined. The rates and type of attack, that the metal specimens suffer, are being assessed by weight and thickness loss measurements and by examination by microscopy and X-ray diffraction.

     Within the laboratory studies, the rates and type of attack are also being continually monitored by computerised electrochemical techniques. These include electrochemical noise, and zero resistance ammetry. So far, the corrosion data produced by the electrochemical techniques correlate well with that produced by the other methods.

INTRODUCTION
     The disposal of Minestone separated from the associated coal in a coal preparation plant usually represents a significant add-on cost to the coal industry. A potential method for reducing these costs is to established commercial outlets for the Minestone. The major outlet for Minestone is in bulk fill applications but other uses include cement-stabilised Minestone, brickmaking and lightweight aggregate manufacture (Ref. 1). Research programmes are now in progress aimed at developing other uses for Minestone.

One of these programmes is aimed at demonstrating the technical suitability of Minestone as fill in reinforced earth type structure such as bridge abutments and retaining walls for embankments (Refs. 2, 3). Part of this programme involved the construction of five operational reinforced Minestone retaining walls at various British Coal Corporation sites. The sites included Bedwas Colliery, Oxcroft Coal Disposal Plant, Newmarket Silkstone Colliery tip, Prince of Wales Colliery and Donisthorpe Colliery. The structure at Donisthorpe used a reinforcement of galvanised steel. The others used the non-metal reinforcements, 'Tensar', 'Paraweb' and 'Fibretain'. These are described below.

Despite these successful applications of Minestone, the UK Department of Transport's current specification for reinforced earth retaining walls for bridge abutments and embankments (Ref. 4) still excludes the use of Minestone as the fill for such structures.

These structures are generally expected to have service lives in excess of 120 years and a particular concern is that Minestones may cause unacceptable degradation of the reinforcement before this time. Metal reinforcement is potentially the more vulnerable as metals often suffer some corrosion when buried underground and, in some environments, it could become severe.

To determine the extent to which Minestone environments pose a potential corrosion problem, a programme of research was initiated to assess the long term durability of a selected range of reinforcement materials when buried in Minestone. The programme is in two phases. The initial phase entails a field trial which involves the long term burial of specimens of candidate reinforcement materials in one particular Minestone. The specimens are being retrieved after different exposure periods for examination and testing. The programme has been extended in a second phase to include a laboratory burial exercise. This will enable a more detailed investigation of the corrosion behaviour of steel reinforcement in a range of Minestones and other fills to be carried out with attention also being given to the influence of variation in burial conditions.

This present paper outlines the programme and some of the techniques which are being used to assess the durability of the candidate reinforcement materials.

FIELD TRIAL

The objective of the field trial is to monitor and characterise any degradation of a range of reinforcement materials when buried in a Minestone under typical site conditions. During the trial, specimens of the reinforcement were to be retrieved at various time intervals so that the onset and progression of any degradation could be assessed. The nominated exposure times were 1, 2, 5, 10 and up to 20 years.

Reinforcement test specimens

The candidate materials under test are shown in Fig. 1 and include both metal and non-metal specimens.

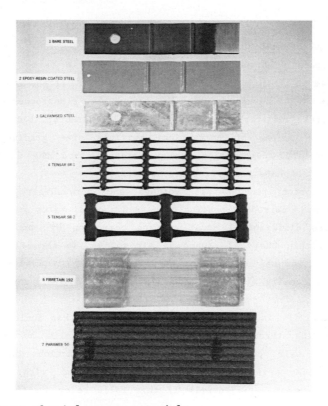

Fig. 1   Specimens of reinforcement materials

The metal specimens are 200 mm sections of 40 mm wide, 5 mm thick reinforcing strapping made from killed hot rolled carbon steel, type 34/20. The black steel specimens still retained the mill scale caused by the hot rolling process whereas this had been removed by acid pickling for the galvanised and epoxy resin coated specimens. The galvanising was carried out by hot zinc dipping. The non-metal specimens include two types of 'Tensar Geogrid', 'Fibretain' and 'Paraweb'. The two 'Tensars' are both uni-axially orientated polyethylene mesh of different mesh size. 'Fibretain' is a continuous strap of undirectional glass fibres bonded by resin whereas 'Paraweb' is a continuous strip of parallel bundles of uni-directional Terylene yarn encased in polyethylene.

Prior to burial the specimens of reinforcement material were weighed and their dimensions measured. Some specimens were retained as controls and some of these were subjected to tensile strength testing according to BS18 (Ref. 5) using an Avery Denison 500 kN capacity machine for the metal specimens and an Instron 100 kN capacity machine for the non-metal specimens.

Characterisation of the Minestone fill

The Minestone fill material was characterised in terms of those parameters believed to influence underground corrosion. It has been shown (Ref. 6) that such corrosion may be caused by electrochemical or biological attack and that the relevant fill characteristics to be measured include degree of compaction, particle size grading, fill resistivity, chemical composition, redox potential and the degree of contamination by sulphate reducing bacteria (SRBs).

Good compaction reduces air voidage and permeability of the environment towards corroding species such as dissolved salts and oxygen. It also reduces differential aeration type corrosion by reducing the inhomogeneity within the fill caused by air voids. In these tests the degree of compaction was assessed by comparing the in-situ density with the maximum dry density measured in the laboratory according to BS 1377 (Ref. 7).

Fill resistivity is a measure of the difficulty that ionic species have in diffusing through the fill. It usually falls as moisture content, ionic salt content and degree of compaction increases. Generally a lower resistivity is often associated with a more aggressive environment, particularly if this falls below about 2000 Ωcm (Ref. 6). In these trials the resistivity was determined by measuring the potential gradient through the fill when a known current was passed through it using the Wenner four electrode method according to DOT BE3/78 (Ref. 8).

The chemical components that have the most influence on corrosion include chloride, water soluble sulphate, pH, moisture, acid soluble sulphate, soluble iron and organic carbon. The first four being particularly relevant to electrochemical attack whilst the last five are particularly relevant to attack by sulphate reducing bacteria (SRBs). SRBs tend to contaminate most earth type materials but only become active under certain environmental conditions and when certain nutrients are available, i.e. those listed above. They can corrode iron and leave a film of biological detritus rich in iron sulphide. The degree of contamination by SRBs was assessed using the 28 day culture growth method according to ASTM D993-58(1970) (Ref. 9).

The suitability of an environment for metal attack by SRBs is normally assessed from its redox potential, i.e. its reducing or oxidising ability. This parameter is quantified in terms of the potential difference between an inert electrode and a reference electrode when buried in the fill environment. It usually needs to be above 430 mV to prevent attack by SRBs (Ref. 6). In these trials the redox potential was measured according to DOT BE 3/78 (Ref. 8) using two inert platinum electrodes and a copper/copper sulphate reference electrode.

The above characterisation of the fill was carried out at the time of burial and after five years during the retrieval of the third batch of specimens.

Burial of reinforcement specimens

The burial was carried out in 1981 in an embankment of Minestone at Donisthorpe colliery. A 13 metre by 4.3 metre, 2 metre deep trench was excavated in this embankment and five groups of specimens were laid on the flat bottom. Each of these five groups contained five specimens of black steel, five of galvanised steel, five of epoxy coated steel and five of 'Fibretain'. In addition each group contained single 1 metre lengths of 'Paraweb' and the two types of 'Tensar'. Two successive layers of Minestone were then compacted onto these specimens using a 10 tonne road roller, to a depth of 0.5 m. Further specimens were laid so that there were five separate sets of replicate specimens buried at depths of 0.5 m, 1.0 m, 1.5 m and 2.0 m. (Fig. 2)

So far specimens have been retrieved after exposure periods of 1, 2 and 5 years by excavating pits into the embankment. After retrieval, loosely adhering material was removed by wiping and, in the case of metal specimens by water washing. Corrosion products were removed from the metal specimens by chemical cleaning methods. This involved soaking in Clarkes solution (inhibited hydrochloric acid) in the case of black steel and saturated ammonium acetate solution in the case of galvanised steel.

Fig. 2   Specimens being buried in the field trial

After cleaning, the metal specimens were subjected to weight loss and tensile strength loss measurements. Additionally the corrosion products adhering to some of the uncleaned specimens were analysed by scanning electron microscopy and X-ray diffraction. Corrosion penetration measurements were also carried out on uncleaned specimens by optical microscopy of polished sections.

An assessment of the thickness of the coating on the epoxy coated specimens was also made using a magnetic paint thickness meter.

Samples of the non-metal specimens were subjected to moisture uptake measurements. Other samples were inspected for surface damage, using an optical microscope, and were then subjected to tensile strength testing.

LABORATORY TRIAL

The laboratory studies are aimed at providing a more convenient and controlled environment in which to study the corrosion behaviour of metal reinforcement within a wide range of Minestones and to compare this behaviour with that in currently acceptable fills such as gravel and limestone. The Minestone used at Donisthorpe has also been included so that a comparison can be made between field and laboratory conditions.

The metal specimens include the black and galvanised type of specimens used at Donisthorpe and in addition, included bright steel in which the mill scale had been removed from the black steel reinforcement specimens by acid pickling. Bright steel is included in order to assess how galvanised steel would behave after the zinc coating had been corroded away.

Specimen burial

Typically twelve specimens have been buried in each fill environment, three were black, three were galvanised and six were bright. One example is shown in Fig. 3. The various fills are contained in plastic tanks set up in the laboratory. The arrangement of tanks is shown in Fig. 4. The fills included Minestone from five of the major coal producing areas and a colliery spoil from Belgium where this material has been extensively used as a fill for reinforced earth structures. Pulverised fuel ash (PFA), three burnt shales, chalk, gravel and limestone are also used.

Fig. 3 An arrangement of metal specimens in a fill

Fig. 4 The arrangement of tanks used in the laboratory trial

Prior to burial these fills and the metal specimens were characterised in a similar manner to the fill and metal reinforcement used at Donisthorpe. The fills were laid in the tanks at the optimum moisture content (OMC) for compaction and were compacted to refusal with a vibratory hammer.

Redox and resistivity probes were also laid in the fills so that any changes in the environment can be monitored.

One of the Minestones was also laid at different levels of compaction and moisture content so that the influence these parameters have on corrosion could be assessed. Other parameters being investigated with this Minestone include those of particle size grading and temperature. Metal specimens have also been laid across an interface between this Minestone and gravel (Fig. 3) so that the possible influence of the drainage layer which would generally be incorporated into reinforced Minestone structures, can be assessed.

Environments associated with standing water, a fluctuating water table and flowing water were also simulated in the laboratory experiments. The extent to which such water regimes would occur within reinforced earth structures is uncertain but such environments may influence corrosion behaviour and thus a study of their possible effects was included in this programme.

Each of the fill conditions was triplicated so that retrievals could take place from each of the fill/specimens systems at three different time intervals. Retrieved specimens were examined in a similar manner to the specimens from Donisthorpe, i.e. weight and strength loss, corrosion penetration and corrosion morphology. In this way the extent of any corrosion can be related to exposure time.

Electrochemical corrosion monitoring

The influence of exposure time on the extent of corrosion is also being assessed by continually monitoring this with an on-line corrosion monitoring system, based on electrochemical techniques, wired to the specimens. Such techniques are based on corrosion being an electrochemical process, in that, corrosion occurs when metal atoms discharge electrons and go into solution as positively charged metal ions. The corrosion rate is the rate at which this occurs and is controlled by the resistance the metal surface has to the process. This resistance is known as the charge transfer resistance and can be determined by various well established techniques. These techniques usually involve measuring the ratio of an applied electrical pressure across a corroding system, to the resultant current flow through it and include galvano-static polarisation (Ref. 10), linear polarisation (Ref. 11) and A.C. impedance methods (Ref. 12). Such techniques have been used previously to measure metal corrosion rates within various candidate fill materials (Ref. 13) and within an embankment of pulverised fuel ash (Ref. 14).

The equipment used in this current study was developed by the Corrosion and Protection Centre Industrial Services (CAPCIS) at the University of Manchester Institute for Science and Technology (UMIST). It is a multi functional unit that had been developed for monitoring the type and rate of corrosion in a variety of different systems, e.g. chemical process plant, re-bars in concrete and dewpoint corrosion associated with flue gas (Ref. 15).

The CAPCIS unit incorporates a number of other electrochemical techniques for measuring corrosion rates besides the more common methods based on charge transfer resistance measurements. They include electrochemical noise and zero resistance ammetry techniques.

Electrochemical noise is a generic term describing low level random fluctuations of the free electrochemical corrosion potential or current. The potential noise technique provides data which can be correlated with the mode of corrosion attack such as pitting corrosion and crevice attack. The ratio of the potential noise to current noise is the electrochemical impedance. Usually this can be equated to the charge transfer resistance, which is inversely proportional to the corrosion rate.

A Zero-Resistance Ammeter (ZRA) measures the coupling current between two identical electrodes, e.g. metal specimens, while imposing a pseudo zero-resistance to its flow. In theory, two such electrodes should maintain the same free corrosion potential and hence no current should flow. However, in practice, the two free corrosion potentials will fluctuate slightly and independently such that, at any instant of time, there will be a potential difference between the two electrodes. When the two electrodes are coupled together through a ZRA, they are forced to a common potential and the difference and fluctuations are manifested as a current flow. The magnitude of this current can usually be directly related to the corrosion rate. Short term fluctuations appear to reflect localised corrosion events on the electrodes in a manner similar to the potential noise output. The current noise/coupling current ratio equates to the localised attack/overall attack ratio and hence represents a pitting index (Ref. 15).

These electrochemical techniques are being used in this study and are triplicated so that simultaneous corrosion data could be obtained from all three types of metal specimens.

A solid state multiplexor is used to sequentially connect each fill/metal system to the monitoring equipment for 24 hours at a time and restart the sequence every month. The equipment is being controlled by a computerised data management system which operates the multiplexor and stores the electrochemical data onto disc file. The system also carries out statistical analysis of the data and concatenates consecutive data from each fill/metal system from which it computes trends in corrosion behaviour.

So far, the electrochemical noise data has been found to correlate well with the level of localised attack observed on retrieved specimens. Also the ranking order, in terms of fill aggressiveness towards metal, as deduced by weight loss measurements, appears to be similar to that predicted by the ZRA technique. However it was unable to produce meaningful data from those environments that were either dry or uncompacted. The reason for this was believed to be poor fill electrical conductivity.

The AC impedance and linear polarisation techniques used to measure charge transfer resistance are generally found to be unsuitable for all the environments studied in this programme. The main reason for this appears to be that corrosion rates are too low to be accurately resolved by these techniques.

CONCLUSIONS

The field trial has enabled a substantial data base to be built up on the durability of a range of reinforcement materials when buried in one particular Minestone. The laboratory trials enabled this data base to be extended to other Minestones, other fills and various fill conditions.

The laboratory trials are also demonstrating that electrochemical noise and ZRA techniques are useful methods for assessing the type and rate of metal corrosion.

ACKNOWLEDGEMENTS

This paper is presented by permission of the British Coal Corporation but the views expressed are those of the authors and are not necessarily those of the British Coal Corporation.

The reinforcement specimens were procured by Minestone Services and the co-operation of the Reinforced Earth Company in supplying specimens of metal reinforcement and of Messrs. Netlon, Pilkington Bros. and ICI in supplying the 'Tensar', 'Fibretain' and 'Paraweb' materials respectively, is gratefully acknowledged.

The authors would like to thank Dr. A.K.M. Rainbow, Minestone Services for his advice and support during the project and Mr. D.D. Skeen and Mr. N.J. Withers for their practical assistance.

REFERENCES
1. C.E. Carr, M.J. Cooke, The preparation of mineral products by the heat treatment of coal mining wastes, Symposium on the reclamation treatment and utilisation of coal mining wastes, Durham, September 1984, Paper 39.
2. A.K.M. Rainbow, An investigation of some factors influencing the suitability of Minestone as the fill in reinforced earth structures. NCB, London, 1983
3. A.K.M. Rainbow, NCB uses colliery waste for reinforced earth, Ground engineering, April, 1982
4. Department of Transport (1986), Specification for highway work, HMSO, August, 1986, pt. 2, Sect. 600, 'Earthwork'.

572

5. British Standards Institution, Methods for Tensile Testing, BS18: Part 2: 1971.
6. C.J.F.P. Jones, Earth reinforcement and soil structures, Butterworths 1985.
7. British Standards Institution, Methods of test for soil for civil engineering purposes, BS 1377:1975.
8. Department of Transport (1978), Reinforced earth retaining walls and bridge abutments for embankments, Tech. memo. BE3/78.
9. American Society for Testing and Materials. Standard test method for sulphate reducing bacteria in water and water formed deposits, ASTM D993-58(1970)
10. K. Hladky, L.M. Callow, and J.L. Dawson, Corrosion rates from impedance measurements, Br. Corrosion Journal, Vol 15, No. 1, 1980.
11. D.A. Jones, N.D. Greene, Electrochemical measurements of low corrosion rates, Corrosion, Vol. 22, (1966), pp 198
12. C. Gabrielli, Identification of electrochemical processes by frequency response analysis, Solartron Instruments, 1980. Tech. Report No. 004/83
13. G. Friant, The study of the corrosiveness of black and red shales in relation to the metal straps from the point of view of their use as infill in reinforced earth. Symposium on the reclamation treatment and utilisation of coal mining wastes, Durham, September, 1984, Paper 10.
14. P.N. Braunton, W.R. Middleton, Assessment of the corrosion of mild steel in reinforced earth structures back filled with pulverised fuel ash, Ash Tech '84, Second international conference on ash technology and marketing, London, September, 1984.
15. J.L. Dawson, W.M. Cox, D.A. Eden, K. Hladky, D.G. John, Corrosion monitoring in process plant using advanced electrochemical techniques, Il Giornale delle Prove non Distruttive - Vol. 2, 1986.

Reclamation, Treatment and Utilization of Coal Mining Wastes, edited by A.K.M. Rainbow   573
Elsevier Science Publishers B.V., Amsterdam, 1987 — Printed in The Netherlands

POLYMERIC MESH ELEMENT REINFORCEMENT OF REINFORCED MINESTONE

R.K. TAYLOR[1], T.W. FINLAY[2] and D.A. FERNANDO[2]
[1]Department of Engineering Science, University of Durham (England)
[2]Department of Civil Engineering, The University, Glasgow, (Scotland)

SUMMARY
    An investigation into the behaviour of compacted colliery discard
(minestone) reinforced with polymeric mesh element reinforcement is reported.
Two types of mesh element were used, having different vertical profiles, and
the tests performed included 100 mm diameter triaxial tests, 300 mm square
shear box tests, and pull-out tests.
    Drained triaxial tests were conducted on unreinforced samples, and on
samples containing 0.1%, 0.12%, and 0.14% of elements by weight.   Shear box
tests were done on unreinforced samples and on samples with 0.12% mesh elements,
this value having been found from the triaxial tests to give the optimum
strength.   Pull-out tests were carried out with isotropic and anistropic mesh
element strip material using load control with different strip widths and fill
moisture contents.
    Results are presented which show substantial improvements in the shear
strength parameters of the reinforced samples, indicate differences arising from
the chosen test methods, and lead to the conclusion that the interaction mechanism is by the formation of aggregations between the fill and the reinforcement.

INTRODUCTION

    In recent years polymers have been used to enhance the properties of soil,
primarily because of economical considerations and versatility of the polymers
in the more general areas of soil stabilisation.   At present testing methods
and design procedures have not been formulated.   A comprehensive treatment of
the likely tests for polymers to be used in civil engineering construction has
been summarised by Hoare. (ref.1.).

    This work is primarily concerned with the assessment of shear strength
parameters, ductility and pull-out resistance of polymer-reinforced minestone,
and is based on an investigation by Fernando. (ref.2.).

    The minestone used in this programme was from the Abertillery colliery,
South Wales.   Particle sizes of 20 mm and less were investigated.   The
polymers used were supplied by Netlon Ltd.   Isotropic and anisotropic meshes
were used, the isotropy being a function of whether the mesh had been formed by
equal or unequal stretching during manufacture.   The elements used for the
shear strength investigations comprised 40 mm square isotropic mesh elements of
two prototype sizes, viz.. 7 and 12, with type 7 having apertures 6 mm by 4.7 mm
and type 12 having apertures 12 mm square, type 7 being thinner than type 12.

574

The strips used for the pull-out tests were of type 7 mesh both isotropic and anisotropic, and type 12, anisotropic, in lengths up to 400 mm.   Fig.1 shows the isotropic mesh elements while figs. 2 and 3 show the type 7 and type 12 strips used in the pull-out tests.

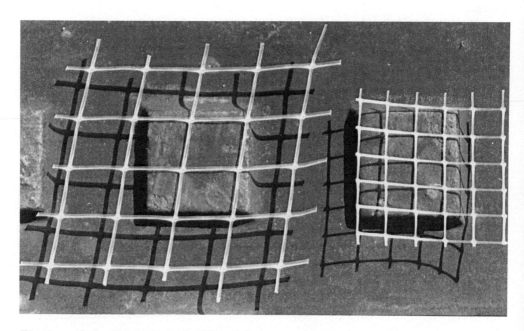

Fig.1. - Mesh elements, types 12 and 7

Fig.2. - Mesh strip, type 7

Fig.3. - Mesh strip, type 12

SHEAR STRENGTH

The tests carried out to investigate the effect of introducing polymer mesh elements to minestone during compaction included standard compaction tests, drained triaxial tests, large drained shearbox tests, and unconfined compression tests.

Compaction tests

Standard compaction tests to BS1377:1975 were carried out on unreinforced minestone, and on minestone containing varying percentages of polymer mesh elements. The unreinforced minestone yielded a maximum dry unit weight of 16.15 kNm^{-3} at an optimum moisture content of 11.30%. The minestone at a moisture content of 11% reinforced with mesh elements gave the results in table 1, from which the optimum percentage of reinforcement was seen to be in the range 0.12% to 0.14% of total sample weight.

TABLE 1
Effect of reinforcement content on maximum dry unit weight

% Reinforcement	γ_d kN/m^3
0.056	15.94
0.100	16.05
0.110	15.95
0.126	16.08
0.159	16.07
0.216	15.67

Triaxial, shearbox and unconfined compression tests

The triaxial tests were standard drained tests on 100 mm diameter samples, the mesh elements being placed randomly within the reinforced samples to avoid the creation of preferred shear planes. The samples were prepared by static compaction in four layers at the density and moisture contents found from the standard compaction tests. Types 7 and 12 were used at 0.10%, 0.12% and 0.14% mesh element concentration at cell pressures of 50,100, 150, 200 and 250 kNm^{-2}.

Large 300 mm square shearbox tests were carried out on reinforced samples containing 0.12% concentrations of types 7 and 12 elements using normal pressures of 50, 100, 150, 200 and 250 kNm^{-2}.

Unconfined compression tests were carried out to assess the increase in ductility of reinforced samples which had been suggested by the triaxial test results.

The results of the above tests are summarised in Table 2.

TABLE 2

Summary of shear strength results

Reinforcement		Triaxial		Unconfined Compression		Shear Box	
Type	%	c'(kNm^{-2})	φ'degrees	UCS (kNm^{-2})	ε%	c'	φ'
–	0	12.50	34.80	28.5	12	0	37.6
7	0.10	5.3	37.70	23.7	8		
7	0.12	6.8	36.80	266.8	7	5.4	37.7
7	0.14	28.9	32.80	102.3	34		
12	0.10	21.2	35.30	16.1	43		
12	0.12	61.6	34.60	238.2	26	19.8	37.4
12	0.14	24.9	34.70	24.8	31		

PULL-OUT

Tests were carried out to find the optimum size of the mesh elements to resist pull-out, this being a function of the overburden pressure (in the low range, 0-15 kNm^{-2}), soil properties and polymer types. In the tests, embedded lengths were varied along with polymer types and tests were carried out at moisture contents of 7, 10 and 11% and at different overburden pressures. The results are shown in table 3.

DISCUSSION

Shear strength

Table 2 is a summary of the shear strength test data, as obtained from the statistical best fit straight line as proposed by Bland (ref.3.). From the drained triaxial tests it can be seen that the major increases in the c'values, compared with the unreinforced samples, are observed with the type 12, 0.12% reinforced sample and in the case of type 7, in the 0.14% reinforced sample. It is also noted that the changes in the Ø' angle are small with respect to the changes in the c' values.

The shearbox tests carried out with 0.12% of mesh elements indicate virtually no change in Ø' values between unreinforced and reinforced samples, but show an increase in c' for both the reinforced samples, the greatest increase being with the type 12 mesh elements.

The unconfined compression test results confirm the increase in ductility observed with drained triaxial test values (fig.4.). This graph shows that for the case considered, the deviator stress increases with strain for the reinforced samples, but for the unreinforced sample, the stress drops after reaching the peak at a relatively low strain.

Pull-out

    Pull-out test data are summarised in Table 3 in terms of both the total pull-out force, T_p, and the pull-out force/10 mm width, T_{PN}.

    Figure 5 shows that for the isotropic strips, a decrease in moisture content and an increase in normal stress leads to increases in the pull-out load/unit width.

    Figures 6 to 9 show the relationship between pull-out force/10 mm width and embedment length.

    The general trend in all of the pull-out tests was a linear increase in pull-out force/unit width as the embedment length increased.

    For <u>equal lengths of embedment,</u> the type 7 strips offered a pull-out resistance which was, on average, 4 times higher than the type 12 strips

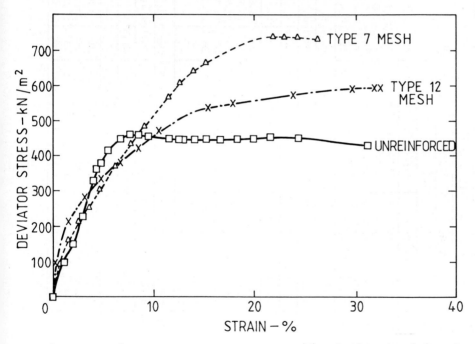

Fig.4.   Stress/strain of unreinforced and 0.14% mesh element reinforced samples at σ_3 = 250 kNm^{-2}.

when installed in the 6 kN/m pulled direction, and 4.9 times higher when installed in the 5.3 kN/m pulled direction.

    For an <u>equal number of meshes embedded</u> in the direction of pull, the pull-out resistance/unit width of the type 7 strips was 1.6 times greater than for the type 12 strips, when installed in the 6 kN/m pulled direction, and 2.4

TABLE 3

Pull-out test results

Notation

- N_L = No. of embedded mesh elements – length
- L = embedded strip length (mm)
- N_w = No. of embedded mesh elements – width
- W = embedded strip width (mm)
- σ_n = overburden pressure (kN/m²)
- m.c. = fill moisture content (%)
- T_p = pull-out force at failure (N)
- T_{PN} = T_p/10 mm strip width (N)

Isotropic strips – type 7 – o'all length 400 mm

N_L	L	σ_n=10.3, m.c.=10.0		σ_n=12.3, m.c.=10.0		σ_n=10.3, m.c.=6.9	
		T_p	T_{PN}	T_p	T_{PN}	T_p	T_{PN}
4	18.8	11.4	3.80	10.3	3.43	10.3	3.43
6	28.2	12.3	4.10	12.3	4.10	11.4	3.80
8	37.6	13.3	4.43	14.3	4.77	13.3	4.43
10	47.0	15.3	5.10	15.3	5.10	16.3	5.43
12	56.4	16.2	5.40	19.2	6.40	18.2	4.07

Anisotropic strips – type 7 – overall length 250 mm – σ_n = 5, m.c. = 11

Pulled direction = 6 kN/m ($N_w;W$)

N_L	L	2;12		3;18		4;24	
		T_p	T_{PN}	T_p	T_{PN}	T_p	T_{PN}
2	9.4	3.1	2.58	12.6	7.00	16.1	6.71
3	14.1	5.0	4.17	18.5	10.28	17.0	7.08
4	18.8	5.9	4.92	19.5	10.83	25.9	10.79
5	23.5	8.9	7.42	24.0	13.33	26.9	11.21
6	28.2	11.0	9.17	26.0	14.44		
8	37.6	15.0	12.50				
10	47.0	19.0	15.83				

Pulled direction = 5.3 kN/m ($N_w;W$)

L	2;9.4		3;14.1		4;18.8	
	T_p	T_{PN}	T_p	T_{PN}	T_p	T_{PN}
12	8.8	9.36	8.8	6.24	15.6	8.30
18	10.9	11.60	16.1	11.42	18.5	9.84
24	12.7	13.51	21.9	15.53	21.0	11.17
30	14.1	15.00	29.0	20.57	26.9	14.31
36	17.0	18.09				

Anisotropic strips – type 12 – overall length 250 mm – σ_n = 5, m.c. = 11

Pulled direction = 6 kN/m ($N_w;W$)

N_L	L	2;24		3;36		4;48	
		T_p	T_{PN}	T_p	T_{PN}	T_p	T_{PN}
2	24	8.3	3.46	15.1	4.19	13.2	2.75
3	36	12.2	5.08	17.1	4.75	20.1	4.19
4	48	14.5	6.04	22.5	6.25	24.7	5.15
5	60	16.1	6.71	25.9	7.19	28.4	5.92

Pulled direction = 5.3 kN/m ($N_w;W$)

N_L	L	2;24		3;36		4;48	
		T_p	T_{PN}	T_p	T_{PN}	T_p	T_{PN}
2	24	6.8	2.83	14.2	3.94	20.1	4.19
3	36	11.2	4.67	16.1	4.47	24.0	5.00
4	48	13.2	5.50	17.5	4.86	24.0	5.00
5	60	16.0	6.67	18.6	5.17	22.0	4.58

times higher when installed in the 5.3 kN/m pulled direction.   It seems, therefore, that both the number of meshes embedded lengthwise, and the aniso- tropy of the strips affect the pull-out, with the type 7 strips offering the greatest pull-out resistance when installed in the 5.3 kN/m pulled direction.

The other factor which affects the pull-out is the number of mesh widths used in the strips, and it appears that, in general, a strip which is 3 mesh widths wide offers the greatest pull-out resistance per unit width.

For the strips tested in pull-out it can be concluded that a type 7 strip, installed in the 5.3 kN/m pulled direction, with a width equivalent to 3 mesh widths will offer a greater pull-out resistance/unit width than the type 12 strip.

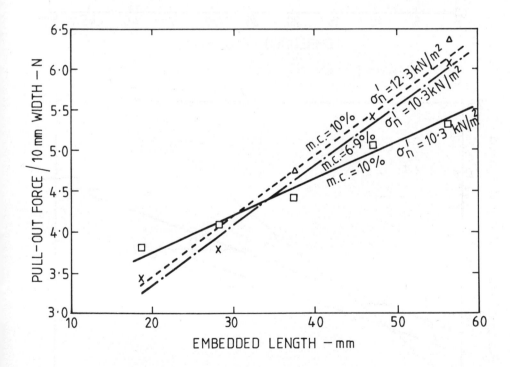

Fig.5. - Pull-out force v length for isotropic type 7 strip

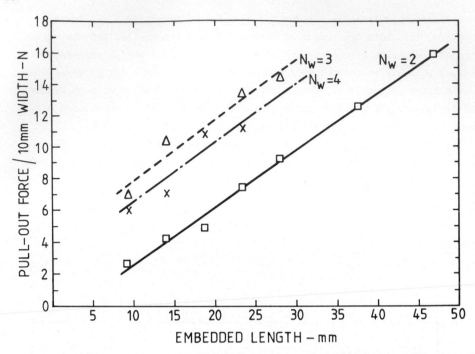

Fig.6. - Pull-out force v length for anisotropic type 7 strip 6 kNm^{-1} direction.

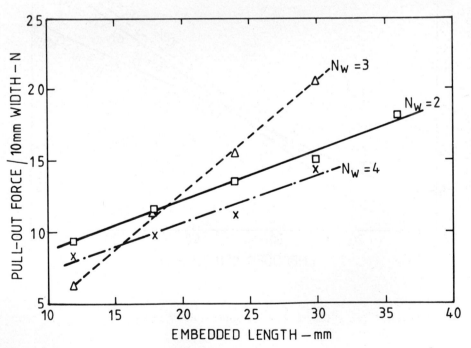

Fig.7. - Pull-out force v length for anisotropic type 7 strip 5.3 kNm^{-1} direction.

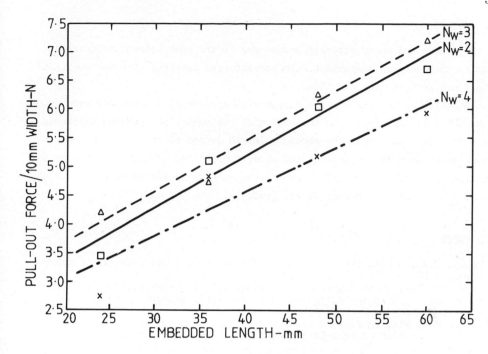

Fig.8. - Pull-out force v length for anisotropic type 12 strip 6 kN/m pulled direction.

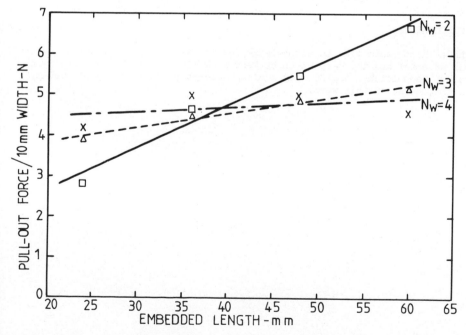

Fig.9. - Pull-out force v length for anisotropic type 12 strip 5.3 kN/m pulled direction.

CONCLUSIONS

Ductility and shear strength parameters in the mesh element reinforced samples increased in comparison with unreinforced samples, with the greatest effect occurring in the c' values.

In pull-out, the anisotropic strips with a width of 3 apertures were found to be the most efficient, and the fact that the number of embedded apertures was more significant than the embedded length seemed to indicate that the mechanism of interaction between the minestone and the elements was due to the formation of aggregations and interlock with the apertures, a mechanism which had also been noted by Mercer et al. (ref.4.).

REFERENCES

1 Hoare, D.J., Geotextiles - Compatability and use, 1986, Civil Engineering April, 1986.
2 Fernando, D.A., Mesh element reinforcement of colliery spoil, M.Sc. Project by Advanced Course in Engineering Geology, 1984, University of Durham, England.
3 Bland, J.A. Fitting failure envelopes by the method of least squares, Quarterly Journal of Engineering Geology, 1983, Vol.16, pp.143-147, London, England.
4 Mercer, F.B., Andrawes, K.Z., McGown, A., Hytiris, N. A new method of soil stabilisation. Symposium on polymer grid reinforcement in Civil Engineering, 1984. Paper 8.1 London.

ACKNOWLEDGEMENTS

The authors would like to acknowledge Messrs. Netlon Ltd. for supplying the polymer material and the help given by Messrs. McEleavey, Richardson and Swann, laboratory technicians, engineering geology laboratory, University of Durham, during the course of this work.

Reclamation, Treatment and Utilization of Coal Mining Wastes, edited by A.K.M. Rainbow
Elsevier Science Publishers B.V., Amsterdam, 1987 — Printed in The Netherlands

REINFORCED MINESTONE USING SPECIAL DESIGNED REINFORCEMENT

R.B. SINGH[1], T.W. FINLAY[1] and A.K.M. RAINBOW[2] .

[1]Department of Civil Engineering, University of Glasgow, Glasgow, G12 8QQ,
(U.K.)
[2]British Coal Minestone Services, Philadelphia, Tyne & Wear, DH4 4TG, (U.K.)

SUMMARY
    This paper presents the results of a preliminary study of a triangular-
shaped polypropylene earth anchor for use in compacted minestone and other fill
materials. Problems associated with the anchor design are presented together
with load/displacement curves. A comparison is made with an equivalent steel
anchor, and steel strip material as used in reinforced earth, and comparisons
are made with theoretical predictions.

INTRODUCTION

    A report in 1981 by Murray and Irwin (ref.1) describes the use of a novel

form of earth retention system in the form of round steel bars, bent at one end

to form triangular or Z shaped anchors, which can be easily installed in

compacted fill. Compared with steel strips commonly used in earth reinforcement

systems, the Anchored Earth system was reported as offering low fabrication

cost, high pull-out resistance, and the avoidance of joints where

corrosion might occur.

    Despite the apparent advantages of the system, the problem of corrosion

still exists, although it can be partly overcome by providing a corrosion

allowance on the round bar. British Coal Minestone Services have had a

preliminary investigation carried out on an anchor element manufactured from

polypropylene, and have tested the element in compacted minestone. This paper

reports the progress of the investigation to date.

ANCHOR ELEMENT

    Polypropylene was chosen, after consultation with I.C.I., as being a

suitable material to resist corrosion while at the same time offering a high

strength. The triangular anchor, shown in fig.1, was developed as a result of

work reported by Singh et al (ref.2) on steel semi-Z shaped anchors in clay.

FILL MATERIAL

    Although the completed investigation will cover a range of minestones, the

tests reported were done only on minestone from Cardowan in the Scottish area.

584

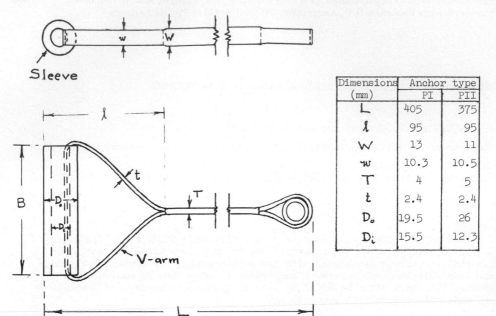

Fig.1. Details of polypropylene anchors PI and PII

Dimensions (mm)	Anchor type	
	PI	PII
L	405	375
ℓ	95	95
W	13	11
w	10.3	10.5
T	4	5
t	2.4	2.4
D_o	19.5	26
D_i	15.5	12.3

The engineering properties were determined in accordance with the N.C.B. Technical Memorandum (ref.3). The natural moisture content was 11.8%, and the 2.5 kg. rammer compaction test gave a maximum dry density of 1.6 Mg/m^3 at an optimum moisture content of 11.2% with effective shear strength parameters of $c' = 7$ kN/m^2, $\emptyset' = 44°$.

PULL-OUT TEST RIG

A steel box with internal dimensions 550 mm wide by 420 mm long by 254 mm high was used to contain the compacted fill with the anchor installed at mid-height and protruding through a slot in the front face of the box. The anchor was subjected to displacement-controlled loading, the pull-out force and anchor displacement being measured by means of a load cell and a linear transducer respectively.

THE TESTS

Although the original intention had been to compare the pull-out resistance of the polypropylene anchor embedded in each of the three selected minestones, it became obvious after a few tests that the anchor would have to be modified to overcome a weakness in its design. This weakness caused splitting of the anchor at the Y-junction, and certain modifications were carried out to rectify this problem.

All the tests were carried out in Cardowan minestone at a moisture content of about 12%, compacted to 95% of the maximum dry density obtained from the 2.5 kg rammer compaction test (1.6 Mg/m^3). A surcharge loading of 10 kN/m^2 was applied during each test.

The original anchor, P.I. was tested to a displacement of 60 mm (test P.I./1) before a modification to the pull-out rig was carried out to allow for a greater extension. It was then re-used in test P.I./2. The load/displacement behaviour indicated an anomaly at about 60 mm displacement and the maximum load was reached at 100 mm displacement. Investigation of the anchor on recovery from the fill showed that the anchor had split at the Y-junction and the sleeve had also deformed badly - hence the anomaly.

A modified anchor design P.II. was then manufactured with a thicker sleeve and a thicker section as shown in fig.1.

Testing of this anchor (test P.II./1) appeared to give a normal load/ displacement curve, but once again it was observed that a split had developed at the Y-junction. To prevent a recurrence of this problem, a strengthening sleeve was placed around the anchor at the junction, and test PII/2 was carried out. Although this test was satisfactorily completed, signs of opening up at the junction at the pulling end were noticed, and a further sleeve was placed at this point before completing a final test, PII/3.

To provide a comparison with an anchor made from another material, a steel anchor was produced having the same basic shape as the polypropylene one, and a similar test was carried out (test S/1).

The load/displacement curves, for each of the five tests described, are shown in fig.2.

THEORY

A theoretical analysis proposed for the TRRL anchor (ref.4) yielded a pull-out resistance of 531 N compared with the experimental value of 1950 N.

Previous work by Singh et al (Ref.2), on semi-Z shaped anchors in clay led to a theoretical approach which predicted the pull-out capacity with a reasonable degree of success. A modification of this theory, based on observed failure surfaces in the present test series overestimated the pull-out capacity and suggests that further work is required on this aspect of the investigation.

DISCUSSION

The effects of the various modifications to the anchor are clearly shown in fig.2. The final strengthened anchor test P.II./3 reached an ultimate load of 1900 N at a displacement of 78 mm compared with 1960 N at a displacement of 96 m for the steel anchor test, 5/1. Although the steel anchor is initially more rigid than the polypropylene it appears that both anchors eventually achieve approximately the same elasticity after a displacement of about 60 mm.

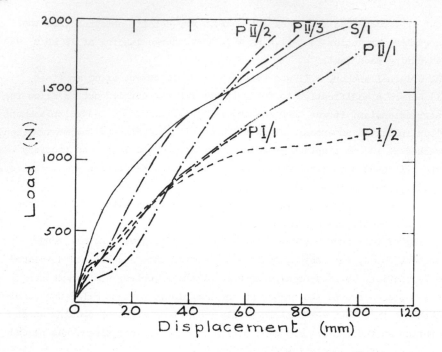

Fig.2.   Load/displacement curves from anchor tests

CONCLUSION

Preliminary pull-out tests of a triangular polypropylene anchor in Cardowan
colliery spoil highlighted design deficiencies. When these were overcome, the
behaviour of the polypropylene anchor compared favourably with an equivalent
steel anchor. Triangular anchors, however, undergo greater displacements than
the steel strip material associated with normal reinforced earth construction.
It is suggested that further work needs to be done if a proper comparison is to
be made.

REFERENCES

1. Murray, R.T. and Irwin, M.J., A preliminary study of TRRL anchored earth.
    TRRL Supplementary Report SR 674; 1981.
2. Singh, R.B., Finlay, T.W., Sutherland, H.B., Fabric studies in model
    investigations of anchored earth. Proc. VIIth Int. Working Meeting on Soil
    Micromorphology, Paris, July, 1985.
3. NCB 1971 Application of B.S.1377-1967 to the testing of colliery spoil.
    National Coal Board Technical Memorandum 1971.
4. Jones, J.F.P., Earth Reinforcement and Soil Structures. Butterworth & Co.
    1985.

Reclamation, Treatment and Utilization of Coal Mining Wastes, edited by A.K.M. Rainbow   587
Elsevier Science Publishers B.V., Amsterdam, 1987 — Printed in The Netherlands

THE SLIDING RESISTANCE BETWEEN GRID REINFORCEMENT AND
WEATHERED COLLIERY WASTE

R. W. SARSBY
Reader, Department of Civil Engineering and Building, Bolton Institute
of Higher Education, Bolton, England.

SUMMARY
    Drained direct shear box tests have been conducted to investigate the
frictional resistance developed when weathered minestone slides over grid
reinforcement.   All samples were soaked and were left for various periods of
time under an applied normal stress. The resultant data indicated that the
fill undergoes significant degradation during this process and the friction
angle for the fill/reinforcement interface decreases significantly.

INTRODUCTION

    As part of an ongoing research programme into the use of waste materials
as bulk fill in reinforced soil construction direct shear box tests have been
undertaken on weathered coal mining spoil.   The use of such a waste material
could lead to significant cost savings and would definitely produce
environmental and community benefits.   The reinforced soil technique relies
on the transfer of shear stress between the tensile reinforcement and the
surrounding fill and previous work has indicated that, for a wide variety of
fills, grids are a particularly efficient form of reinforcement because of the
interlock that develops with granular materials.   The efficiency of a
reinforcement is characterised by the coefficient of interaction, α, which is
the ratio of the friction coefficient for fill on reinforcement to the
coefficient for the fill alone.   Earlier work on 'fresh' minestone, i.e.
material sampled directly after screening and washing, concluded that at low
stress levels the frictional interaction between minestone and grids was as
good as that for conventional, good-quality fills - (ref. 1).   However at
higher stress levels there was significant crushing of the colliery spoil and
the coefficient of interaction fell well below that obtained with strong
aggregates.   Since mining waste is prone to crushing and weathering this
study has been concerned with the effect of soaking and prolonged loading of
this type of fill.

TEST PROGRAMME

    It was decided to investigate the shearing behaviour of weathered colliery
spoil for two reasons - the existence of very large quantities of this material

as opposed to freshly-screened waste and the degradation of 'fresh' minestone which was observed in previous work.   In addition to investigating the immediate shearing behaviour of this weathered spoil it was decided to observe how rapidly this material undergoes further degradation and how its shear strength behaviour (in relation to soil reinforcement) would change. Consequently three series of tests were conducted using the direct shear box apparatus :-

      i)   Series A - "immediate" drained tests on weathered minestone.

      ii) Series B - tests on material which had been soaked and then left under a normal stress for two days.

      iii) Series C - soaked samples which were subjected to a normal stress for two weeks before being sheared.

Previous work on the interaction between conventional fill materials and grid reinforcement had shown that the relative size of the grid apertures to the mean particle size influenced the coefficient of interaction - (ref. 2). Consequently in this investigation two different sizes of grid were used, i.e. a fine grid and a coarse one.

All specimens were sheared at a rate (which was determined from consolidation data) which would maintain drained conditions throughout, and the normal stress range employed was 25 to 200 kN/m^2.

FILL

The minestone was obtained from Bickershaw Colliery at Leigh, Lancashire. The material was unburnt and was taken from a spoil heap which had been in existence for a period of years so that the spoil had been subjected to weathering.   Sufficient waste was taken from the heap so that a fresh batch could be used for each test specimen in the programme.   In the laboratory, particles larger than 40 mm were removed (this was a negligible amount) and all of the minestone was thoroughly mixed before being divided into batches and stored in sealed bags.

Grading tests on the minestone indicated that it was a sandy gravel with a 50% particle size of about 7 mm. and approximately 3% fines.   The natural moisture content was 10.4%.   The maximum bulk unit weight (at natural moisture content) was determined as 17.8 kN/m^3 using the vibrating hammer method.

REINFORCEMENT

The reinforcements employed were polypropylene geogrids manufactured by Netlon Ltd..   For the fine grid (SR1) the apertures were approximately 8.5 mm broad by 48 mm long whereas with the coarse grid (SR2) the corresponding dimensions were 17 mm and 88 mm respectively.   These reinforcements were chosen because of their chemical inertness.

APPARATUS

All of the tests were conducted in a direct shear box with plan dimensions
of 300 mm by 300 mm.   The test arrangement has been previously described by
Sarsby and Marshall (ref. 1).   For the tests with reinforcement a clamp is
incorporated into the lower half of the box so that the grid is held in the
plane of sliding, i.e. at the mid-height of the box, with compacted fill on
either side of the grid.

TEST PROCEDURE

To form a specimen the minestone was placed and compacted (at natural
moisture content) in four equal layers.   The compaction was undertaken using
a vibrating hammer whereby a specified weight of fill was compressed to the
requisite layer thickness so that all samples were produced with a bulk unit
weight equal to 95% of the maximum achievable.   The second layer of spoil was
compacted so that it was flush with the top of the bottom half of the box.
For reinforced specimens the grid was placed on top of this layer so that
shearing failure would occur by the fill sliding along the upper surface of
the reinforcement.   When the fill was placed in the upper half of the box it
interlocked with both the mesh and the underlying minestone.

After sample preparation the carriage containing the shear box was flooded
with water and the specimen was left to soak and become saturated.   It was
then consolidated under the requisite normal stress and the consolidation data
was used to check the rate of shearing to be used.   Each specimen was left
for a selected period of time under normal stress before being sheared to
failure.

After the shear stage the apparatus was dismantled and soil was removed
from the shear zone and re-sieved to observe for degradation.

TEST DATA
Stress-strain behaviour

Typical data are presented in figures 1 and 2.   The behaviour of samples
tested "immediately" after preparation is typical of that associated with the
shearing of dense granular media, i.e. a well-defined maximum shearing
resistance with a subsequent decrease to a residual state, and material
dilation at failure.   This is true for both plain and reinforced specimens at
all stress levels although the deformation required to mobilise the maximum
resistance is greater for the reinforced fill.   Similar trends were observed
with the 'fresh' discards tested previously although for that material the

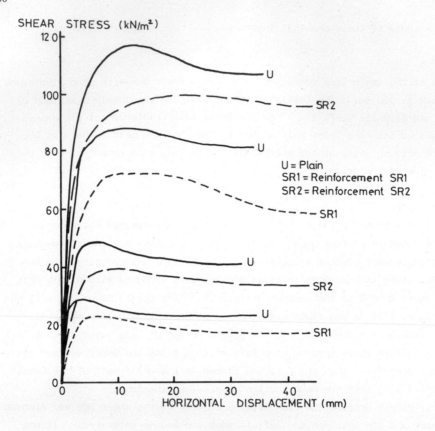

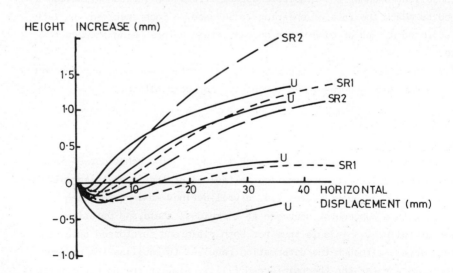

Fig.1   Shearing   Behaviour   of   Series   A   samples ('immediate' tests)

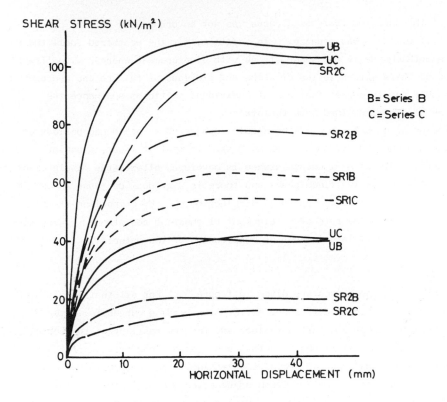

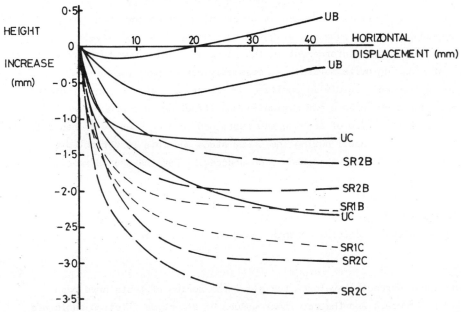

Fig. 2   Shearing Behaviour of Series B and C samples (prolonged soaking)

peak of the shearing resistance curve was not so pronounced but the overall
amount of dilation was greater.   In addition, for the weathered spoil there
is a smooth change in the shear stress with displacement whereas with 'fresh'
minestone there were a series of steps and plateaus as failure was approached
- probably due to abrupt fracture of individual particles and hence the
absence of a well-defined peak resistance.

The shearing behaviour of the samples subjected to prolonged periods of
soaking and applied stress is very much akin to that of loose, granular
aggregate, i.e. the maximum resistance is developed after large displacement
(there being no well-defined peak) and there is a gradual decrease in sample
volume to a constant value at large displacements.   By comparison to the
'immediate' tests the rate of development of shearing resistance is very slow
and there is a very large amount of compression of the material.

Crushing

Typical grading curves are given in figure 3.   For all test specimens
there was a significant difference between the grading curves for material
taken from the shear zone after failure and for the spoil prior to compaction.

From these data it is observed that considerable degradation of the
weathered spoil has occurred during the testing.   In the preliminary
compaction tests the amount of crushing observed was negligible.   The curves
for the immediate tests (series A), which are shown in figure 3a, show that
plain samples suffered a large amount of crushing during the shear stage -
this degradation being enhanced slightly by the normal stress level.   The
effect of including reinforcement on the plane of sliding was to increase the
amount of crushing quite markedly - presumably due to particles in the shear
zone being 'locked' rigidly in position by the grid.   The coarseness of the
grid does not seem to have had a significant effect on this process.

Even a short period of soaking and sustained loading (series B) has a large
effect on the plain spoil during the shear stage and this is reflected in the
volume decrease of these specimens when sheared.   For the soaked specimens
(series B and C) the change in the grading curve caused by the inclusion of
reinforcement is small with only a slight reduction in the proportion of gravel
as compared to samples tested 'immediately'.   However there is a large
difference between the volume change behaviour of these tests.

Friction

The shear strength envelopes for all three series of tests have been
plotted in figure 4 and the resultant values of effective friction angle are
contained in Table 1.

For the plain minestone the failure envelopes are curved with the friction

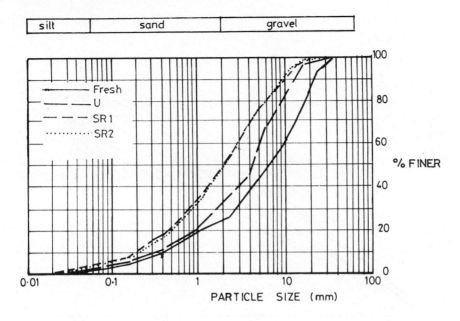

a) Series A

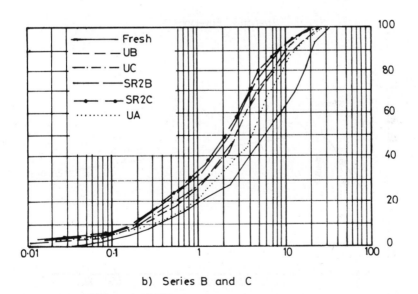

b) Series B and C

Fig. 3   Grading   Curves   of Material from the   Shear   Zone

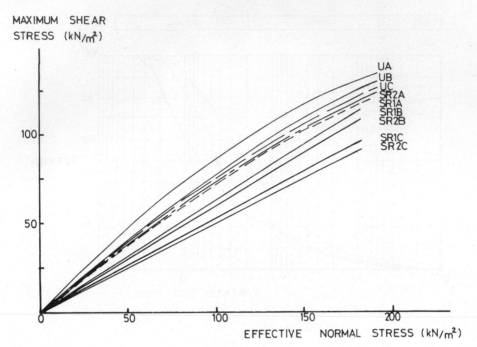

MAXIMUM SHEAR STRESS (kN/m²)

UA
UB
UC
SR2A
SR1A
SR1B
SR2B

SR1C
SR2C

100

50

0

50   100   150   200

EFFECTIVE NORMAL STRESS (kN/m²)

Fig. 4   Shear   Strength   Envelopes

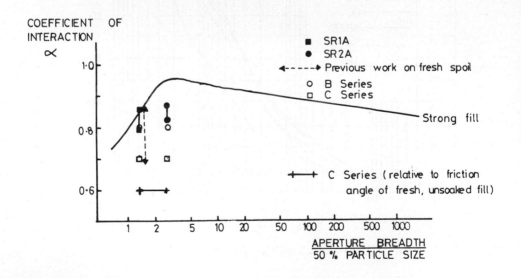

COEFFICIENT OF INTERACTION
∝

1·0

0·8

0·6

■ SR1A
● SR2A
◄-----► Previous work on fresh spoil
○ B Series
□ C Series

Strong fill

+----+ C Series (relative to friction
        angle of fresh, unsoaked fill)

1   2   5   10   20   50   100 200   500 1000

APERTURE BREADTH
50 % PARTICLE SIZE

Fig. 5   Effect of Grain Size on the   Interaction   Coefficient

TABLE 1

Effective shear strength parameters

Series	Type of test		$\emptyset'$ $(c' = 0)$ Effective Normal Stress (kN/m^2)		
			50	100	150
A1	'Immediate'	– plain	44	41	38
A2	"	– reinforcement SR2	37	36	34
A3	"	– reinforcement SR1	36	35	34
B1	Short soak	– plain	39	38	36
B2	" "	– reinforcement SR2	31	31	31
B3	" "	– reinforcement SR1	32	32	32
C1	Long soak	– plain	39	37	35
C2	" "	– reinforcement SR2	27	27	27
C3	" "	– reinforcement SR1	28	28	28

angle decreasing with stress level – particularly for the 'immediate' tests.
For the soaked, reinforced samples the friction angle is constant over the
applied stress range and there is negligible difference between the volume
obtained for the two types of grid.

As stated previously the coefficient of interaction, α , is a measure of
the efficiency of soil reinforcement and in figure 5 the empirical relationship
between α and mesh/particle size is plotted.   For the 'immediate' tests the
mean α value is approximately 0.85 and for the grids employed this compares
favourably with the performance of conventional, strong aggregates.   However
the soaked samples (series B and C) give lower values of α , i.e. in the range
0.80 (series B) to 0.7 (series C) when compared to the corresponding soaked,
unreinforced specimens.   If the friction angle for soaked, reinforced samples
is compared to the data for 'immediate' tests on plain spoil then drastically
reduced α values are obtained, i.e. in the range 0.70 to 0.60.   Direct shear
tests on the interaction between these grids and clay soils have generated
values in the region of 0.5 and for sand sliding along solid sheets of the
polymeric material from which the grids are formed the α value has been
determined as 0.63.   Viewed in this light the weathered spoil interacts very
poorly with grids by comparison to conventional fills if the minestone
undergoes soaking and prolonged stress and possible progressive deformation
due to the loss of dilation.

CONCLUSIONS

1. Densely-compacted, weathered minestone interacts well with grid
    reinforcement when subjected to shearing shortly after placement.   The
    material crushes during shearing and the presence of reinforcement

596

enhances this degradation.

2. If densely-compacted weathered spoil is soaked and left under stress for a
    period of time then the material is highly prone to crushing with a
    resultant loss of dilation and very poor interaction with grid
    reinforcement.

ACKNOWLEDGEMENTS

The Author would like to express his gratitude to the staff at Bickershaw
Colliery and to Mr. B. Porter for all his hard work in conducting the shear box
tests.

REFERENCES
1   R.W. Sarsby and C.B. Marshall, The Interaction between Unburnt Colliery
    Waste and a Polymer Geogrid, Symposium on the Reclamation, Treatment and
    Utilisation of Coal Mining Wastes, Durham, England. September 1984.
2   R.W. Sarsby, The Influence of Aperture Size/Particle Size on the Efficiency
    of Grid Reinforcement, Second Canadian Symposium on Geotextiles and
    Geomembranes, Edmonton, Canada, September 1985.

Reclamation, Treatment and Utilization of Coal Mining Wastes, edited by A.K.M. Rainbow   597
Elsevier Science Publishers B.V., Amsterdam, 1987 — Printed in The Netherlands

J P Hollingberry, Director, Wimpey Laboratories Ltd

INFILLING OLD MINE WORKINGS AND SHAFTS

INTRODUCTION

The back filling of worked out areas in some mines has always been part of the mining process to make these areas safe, to dispose of mine wastes and improve ventilation. These methods have been modified in Civil Engineering and building in order that stabilisation can be achieved to avoid damaging settlement as a result of the collapse of shallow mine workings.

1. The Enquiry

This would normally come from a client who has retained a mining consultant or engineering geologist to carry out the initial site investigation and feasibility study. Details required by the specialist contractor to prepare a tender include the following:-

1.1 Access

The nature and location of the site. As quite heavy plant will be used to complete the work and a large number of lorry loads of bulk materials will be delivered to the site, adequate road access and running surfaces will be required.

1.2 Structures

Outline details of the proposed new construction and any existing buildings on the site.

1.3 Services

A large number of holes will be drilled at close spacings across the site. Therefore it is essential that the location of all services is known and accurately plotted on all contract drawings. Note should also be made of any overhead power and telephone lines. A large quantity of water will be required for mixing of grout and flushing for drilling. Adequate sources must be identified and if not available in the immediate area, provision has to be made for importation in tankers.

1.4 Geology

For many sites where coal has been extracted, no mining records are available and preliminary site investigation details are essential to determine the likely cost of the infilling within reasonable bounds. This should include:

- Type and depth of superficial deposits
- Description of the coal measures and extraction detail
- Depth, dip and thickness of coal seams
- Standing water level.

Although much information is available on the general geology in most mining areas in the UK, verification and checking is needed at each individual site.

Superficial deposits can range from natural sands, gravels and boulder clay to made ground comprising waste products from industry. Occasionally they may include significant obstructions for drilling such as steel in slag.

1.5 Working Methods

For the purposes of this paper, a typical greenfield site is considered where a 1m thick coal seam was worked at a depth of 30m and a mixture of pillars, partially collapsed workings and voids have been identified during the site investigation.

1.5.1 Drilling

Assuming that a heavy structure is to be constructed on the site, a 6m grid of 50mm primary holes would be specified. These to be drilled down to the pavement at the level of the coal seam and 40mm grout tubes fed down these holes to the workings. While it is normally possible to open hole drill in the coal measures, casing would be required to advance holes through unstable superficial deposits. This would normally be conventional casing for rotary drilling or be of heavy duty quality for rotary percussive drilling. If grout acceptance is high, secondary intermediate holes will be drilled in those areas affected to achieve a final pattern of holes at 3m centres. Where necessary, boreholes can be inclined to avoid surface obstructions so that the point of interception at the workings is at the plan location required.

If significant voids are expected and/or the seam dips steeply, 100mm dia. boreholes will be drilled at 1.5m centres around the perimeter of the site especially down dip. Viscous grouts with gravel addition will be placed down these holes to form rough barriers in the workings to minimise flow of the infill grout away from the area of the site.

Where consolidation of two seams is required then the top seam would be consolidated as a first phase. After the primary grout has set, holes would be redrilled and deepened to intersect the lower seam.

Occasionally, methods would have to be modified to cope with artesian water pressure.

1.5.2 Grouting

Generally grout strengths of 1 N/mm^2 are specified for most sites.

For perimeter holes typical grout proportions would be 1:5:6 cement/pulverised fly ash/washed building sand. Where large open voids are encountered, gravel would be introduced simultaneously with the grout in a proportion of about 3 to 1.

For infill grout typical proportions would be 1:12 in dry workings, 1:9 in flooded, cement/pulverised flyash.

Water content would be adjusted to the minimum to provide a suitable pumpable mix.

Grouting would preferably commence at the down dip end of the site and proceed systematically up dip, driving water out ahead of the grout. Infill grouting would follow behind the formation of a perimeter barrier.

On completion of the grouting to refusal in the workings, grout tubes would be withdrawn and a connection made to any casing so that the specified pressure can be applied to check the grouting.

1.5.3 Shaft Drilling and Grouting

Similar drilling and grouting methods may be suitable for shafts to a depth of 30m, but if deeper shafts are to be consolidated, heavier rotary drilling plant will be required to get a suitable grout delivery pipe to full depth. To ensure that the drillrig is stable and safe, a working platform would be erected over the shaft. On completion a reinforced concrete capping may be required, ideally founded on competent rock.

1.5.4 Test Grouting

On completion of all grouting, test grouting would be carried out on holes at random positions across the site to test the efficiency of the works and to determine if further treatment is necessary.

1.6 Bill of Quantities

The Bill should include items to cover the following operations:-

a)	Transportation of plant equipment and crew to and from site	Sum
b)	Transportation of rotary rig for shaft drilling to and from site	Sum
c)	Provision of access to site including suitable running surfaces	Sum
d)	Establishment of necessary offices and services	Sum
e)	Insurances and Bonds	Sum
f)	Provision of daily reports and borehole logs	Sum
g)	Moves of drilling plant and set up at each borehole	Number
h)	Drilling 100mm perimeter holes including allowance for casing	per m.
i)	Drilling 50mm infill holes including allowance for casing	per m.
j)	Set up rotary drill rig at shaft including platform	Sum
k)	Drilling 107mm holes in shaft including casing	per m.
l)	Making connection for grouting	Number
m)	Batching, mixing and placing perimeter grout	per tonne

n)	Batching, mixing and placing infill grout	per tonne
o)	Batching, mixing and placing grout in shaft	per tonne
p)	Placing gravel	per tonne
q)	Provision of grouting materials – cement	per tonne
	– p.f.a	per tonne
	– sand	per tonne
	– pea gravel	per tonne
r)	Check grout holes	per hole
s)	Standing Time:- Infill Drill Rig	per hour
	– Rotary Drill Rig	per hour
	– Grouting Plant	per hour

2. Plant

2.1 Drilling

A variety of drills are used for this work but most quarry blast hole rigs suitably modified to handle casing have the capacity to drill to the required depth at high economic penetration rates. If crawler mounted they are able to move across most sites. For shaft work higher powered rotary rigs are employed either crawler, tractor or lorry mounted. Air or water flush may be required.

2.2 Grouting

Again various batching mixing and pumping units are used, but each should have the capacity to place up to 20 tonnes of dry materials per hour. High shear mixers are linked to storage pan mixers delivering by rotary or ram pumps to the injection points. Gravel is normally added at the top of each borehole, mixing with the grout in a suitable delivery hopper.

3. Supervision

To be an economic operation high production is required in all areas and to maintain high drilling and grouting output, experienced competent on-site supervision is essential. On site regular maintenance of plant minimises time lost due to breakdown. Back up plant must be available. Experienced supervision is also of value when unexpected ground conditions are encountered during the course of the work or there may be the need to modify working methods to produce an effective consolidation treatment to cover unexpected openings such as adits, bell pits and wells.

Reclamation, Treatment and Utilization of Coal Mining Wastes, edited by A.K.M. Rainbow
Elsevier Science Publishers B.V., Amsterdam, 1987 — Printed in The Netherlands

"SOURCES STUDY" - A STRATEGY FOR THE IDENTIFICATION AND SELECTION OF COLLIERY SPOIL FOR USE IN THE INFILLING OF ABANDONED LIMESTONE MINES IN THE BLACK COUNTRY.

D.W. STEVENS B.Sc(Eng.) and K.L. SEAGO B.Sc(Hons.) MICE, MIHT, MIWPC, AIArb.

Ove Arup & Partners, Birmingham.

SUMMARY
Because colliery spoil is used as an ingredient in "rock paste", a study was undertaken to prepare a strategy for the selection of possible sources located in the Midlands area. The work was done in sufficient detail to assist the selection of the most appropriate sources for the infilling of a particular mine or mines. For each of the preferred sources an attempt was made to accumulate enough knowledge to support a planning application (if necessary) to work that source. Discussions with the relevant local authorities and landowners was central to the work, since viability of sources relates as much to consent to working as to the physical suitability of materials. This paper describes the different aspects of the study and includes a description of the environmental, engineering and planning considerations that had to be investigated for each of the sources. The 'strategy' for the selection of sources has been put into operation, as colliery spoil is presently being used as an ingredient in the "rock paste" at the Littleton Street Mine, in Walsall.

INTRODUCTION
In 1978 subsidence and structural damage at the Old Park Industrial Estate located at Wednesbury in the Black Country Borough of Sandwell prompted the commissioning of a major research project to identify the scale and attendant risks of the legacy of abandoned limestone workings in the area. The cost of the study was shared by the Department of the Environment (DoE) and the local authorities with an interest; i.e. the Metropolitan Boroughs of Dudley, Sandwell and Walsall, together with the now disbanded West Midlands County Council.

The findings of the Study were published in 1983 (see Ref.1) and appropriate remedies were suggested to accommodate the different forms of subsidence which could possibly arise from the collapse of the mines. One of the remedies proposed infilling the voids with a pumpable material known as "rock paste". This is a mixture of colliery spoil and water, with supplementary strengthening additives. (The development of rock paste is covered by authors Figg and Cole in another paper to be presented at this

602

symposium). Rock Paste is an economic infill material which has an additional attraction in that not only does it tackle the dereliction associated with the abandoned limestone mines, but also the problems associated with coal mining dereliction in the form of spoil heaps.

Research work carried out in 1984 and 1985 into the pumpability and flow characteristics of the rock paste gave confidence of its viability as an infilling material. Consequently Ove Arup & Partners (Arup) were commissioned in 1985 to prepare a strategy for the selection of sources of colliery spoil to be used in the making of rock paste. Arup employed Land Use Consultants (LUC) to advise on the restoration aspects of the work, particularly at the Sources sites following the removal of the spoil heaps. The Study was funded by Derelict Land Grant and Walsall Metropolitan Borough Council agreed to co-ordinate the Works on behalf of the remaining affected Black Country boroughs.

The work was to be done in sufficient detail to assist the selection of the most appropriate sources for the infilling of a particular mine or mines. For each preferred source enough knowledge was to be accumulated to support a planning application (if necessary) to work that source. Discussion with the relevant local authorities and landowners was to be central to the work, since viability of sources relates as much to consent to working as to the physical suitability of materials. To ensure that the strategy remains up to date, it would be produced in such a way as to accommodate periodic review.

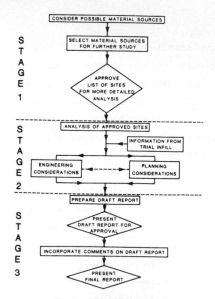

Fig. 1 – Stages of Work

The Study, itself, was split into three finite stages. Fig.1 illustrates a flowchart of these stages, but in brief they were:

Stage 1 - an initial sift of a large field of identified sources.

Stage 2 - a detailed assessment of a short list of potential sources.

Stage 3 - the formulation of a strategy to bring forward fill material from preferred sites.

PROCEDURES FOR THE STUDY

Stage 1

Objectives

The objectives for Stage 1 were to consider and select possible material sources for study and to carry out initial environmental and planning appraisals in order to take forward a list of sites for more detailed engineering and environmental analysis. The purpose of this approach was to minimise the extent of engineering and site investigation at this early stage with a large number of potential sites under consideration.

Identification of Sites

The Study used, as a basis, the sites already identified in the Tarmac Study (Ref.2). Tarmac identified a list of 58 individual sites, mainly of colliery spoil but with several containing sands, slag or pulverised fuel ash (pfa). Sources of sands, slags and pfa were not considered in Arup's study as these materials do not have the characteristics of colliery spoil to make rock paste. They can be used as infill material but with the use of different techniques outside the scope of this paper.

Arup concluded at an early stage of the Study that the general impression of a large number of prominent sites in urgent need of early reclamation was no longer the case in the Staffordshire and North Warwickshire coalfields. There are certainly a number of sites which would still fit this description, but much has been done over the last 15 years (and even in the time since the Tarmac study was undertaken) in dealing with derelict land and in reclaiming colliery spoil tips in particular. Furthermore, although the stark profiles of prominent tips are usually regarded as eyesores, total removal is rarely a desirable objective. New land uses on reclaimed land can frequently benefit from the landscaping opportunities which reclamation offers. It was clear that local authorities in the source areas would need to be satisfied that excavation of material for limestone mine infilling would have such beneficial results for individual source sites.

Reconnaissance of Sites/Meetings with Interested Parties

Preliminary site reconnaissance was then undertaken to establish the extent and broad environmental characteristics of each site. Fig.2 indicates

604

the location of certain sites investigated in Stage 1. Meetings were held with County and District Council Planning Departments and with British Coal – Minestone Services, Deep Mines (Western and South Midlands Area), and the British Coal Opencast Executive, Central West. On the planning side these discussions clarified the status and, where appropriate, policy framework for individual sites. Site appraisal allowed a preliminary view of the engineering characteristics, possible working constraints, and environmental impact and potential benefits of excavation.

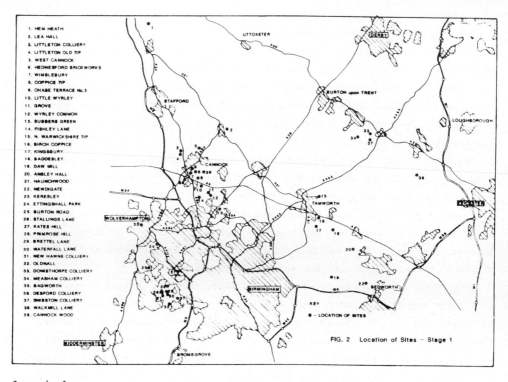

1. HEM HEATH
2. LEA HALL
3. LITTLETON COLLIERY
4. LITTLETON OLD TIP
5. WEST CANNOCK
6. HEDNESFORD BRICKWORKS
7. WIMBLEBURY
8. COPPICE TIP
9. CHASE TERRACE No.3
10. LITTLE WYRLEY
11. GROVE
12. WYRLEY COMMON
13. SUBBERS GREEN
14. FISHLEY LANE
15. N. WARWICKSHIRE TIP
16. BIRCH COPPICE
17. KINGSBURY
18. BADDESLEY
19. DAW MILL
20. ANSLEY HALL
21. HAUNCHWOOD
22. NEWDIGATE
23. KERESLEY
24. ETTINGSHALL PARK
25. BURTON ROAD
26. STALLINGS LANE
27. KATES HILL
28. PRIMROSE HILL
29. BRETTEL LANE
30. WATERFALL LANE
31. NEW HAWNE COLLIERY
32. OLDNALL
33. DONISTHORPE COLLIERY
34. MEASHAM COLLIERY
35. BAGWORTH
36. DESFORD COLLIERY
37. SNIBSTON COLLIERY
38. WALKMILL LANE
39. CANNOCK WOOD

KEY
● – LOCATION OF SITES

FIG. 2   Location of Sites – Stage 1

Appraisal

The appraisal was recorded in order to produce a preliminary ranking. In doing so, the appraisal and discussions identified particular sites where there was no merit in proceeding with further investigation as the authorities would oppose extraction. For example, at Ansley Hall and Haunchwood in Warwickshire, where the sites had just been reclaimed; at Waterfall Lane (West Midlands) where work was in progress; and at West Cannock (Staffordshire) where design work for reclamation was well advanced and where shale was required for use in the reclamation project. This process also identified the need to distinguish between existing colliery shale tips and run-of-mine dirt which still is being produced at several of the collieries listed. This was particularly important as several sites (e.g. Littleton, Birch Coppice, Baddesley) had potential source material in both forms. The number of sites appraised in Stage 1 are summarised in Table 1.

TABLE 1 -

Summary of sites appraised in Stage 1.

	Run-of-Mine Dirt	Large Sites (>50,000m³)	Small Sites (<50,000m³)	Already Reclaimed
16 sites in Staffordshire	3	12	1	-
11 sites in Warwickshire	4	5	-	2
10 sites in West Midlands	-	-	10	-
5 sites in Leicestershire	5	-	-	-
TOTAL:	12	17	11	2

Conclusions of Stage 1

Stage 1 of the Study concluded by reducing the number of sites from 42 to 16 with 21 sites shelved and the Leicester sites identified as possible long term reserves. However, two promising additional sites (Cannock Wood and Walkmill Lane) were identified and included in the list for further study along with four run-of-mine dirt sources.

The full list taken forward for further study was:

Derelict Sites	Operational Tips
Little Wyrley	Littleton New Tip
Littleton Old Tip	Baddesley Tip
Walkmill Lane	
Wyrley Common	
Wimblebury	
Chase Terrace 3	
Burton Road	
Stallings Lane	
Kingsbury	
Primrose Hill	
Grove	
Cannock Wood	

Run-of-Mine Dirt

Littleton
Hem Heath
Lea Hall
Baddesley

Stage 2

Objectives

The objectives for Stage 2 were to study in detail the planning and environmental aspects of each site brought forward from Stage 1. In so doing the detailed appraisal identified further sites where there was no merit in proceeding with further investigations. Once this situation was reached for any particular site it was shelved in a similar manner to that in Stage 1.

Development Planning Appraisal

For each short listed site, development plans, both Structure Plan and, if relevant, Local Plans, were consulted to establish development policy for sites and surrounding areas. In some cases only strategic policy applied; in most cases, detailed development control policies in local plans overlapped as a Subject (County Council) Land Renewal Plan covers land also detailed in District Plans.

Having established the planning context, the advice of the local and mineral planning authorities was taken at a series of meetings with District and County Officers. Parallel to these discussions, contact was made separately with the National Coal Board, where it appeared that sites within the ownership of the Board might be progressed without a planning application, but within the procedures of the General Development Order. (Note: Subsequent to the Study, amendments to the planning law have come into force resulting in planning applications having to be submitted for operational tips). These meetings provided guidance on the preferred route to be followed with (if necessary) a planning application, and whether a scheme would be mineral workings or land reclamation.

The final set of planning information required, which overlaps considerably with engineering assessments, was to establish who was going to take a lead in order to activate a fill recovery scheme when a demand was established. Through meetings with the local authorities, NCB and other landowners, it was possible to establish a preferred route to be taken. The variations in those procedural routes are considerable: some are complex actions requiring the consent of several parties; in other cases one landowner could co-ordinate all the necessary inputs to commence a scheme. It would be wrong to assume that the recommended actions are wholly cut-and-dried; the intention was to identify sites that are procedurally straightforward, and where not so, to suggest issues that will need early resolution.

Engineering Appraisal

The short listed sites were assessed to determine whether the material contained within the tip was suitable for use as rock paste. Other aspects of engineering interest were also covered, for example accessibility, contractual arrangements and costs.

To assess the suitability of fill a visual inspection of each source was made. Records were made of such factors as the presence of burnt shale, overburden depth and characteristics, surface debris and rubble content. Samples of material were taken for laboratory testing to determine engineering properties. A comparison was made of the test results with criteria defining suitability for rock paste. A broad assessment was formed of the relative qualities of fill materials at the source sites.

The approximate available quantity of fill at each site was determined. Topographical surveys were carried out where necessary to supplement existing information on levels. The volumes of material were then assessed taking account of overburden and burnt shale, if present, and discard of screenings and rubble.

Site access was considered together with routes away from site towards the Black Country. It is evident that access onto certain sites is straightforward, whereas at others it will be necessary to negotiate entry across private land. General routes away from sites towards the nearest Motorway and the Black Country were assessed. Note was taken of local authority comments where road improvement requirements were indicated.

Discussions with the local authorities determined their probable requirements for working of sources, and where limitations would need to be incorporated in contract arrangements. Account has been taken of wheel washing, working hours and plant disposition and appropriate conditions to be included in any future contract documents for the winning of the material identified.

From the several aspects assessed a generalised cost of recovering and transporting fill has been calculated. In addition, the cost of site restoration has been worked out based on the indicative level of restoration finance agreed with the DoE.

Environmental Appraisal

Fieldwork was undertaken to inspect each of the short listed sites in turn and to record and assess the landscape context, site conditions, and the potential impacts of both shale recovery and restoration operations. This process started with a broad appraisal of the landscape character of each site in its existing condition and the relationship - visual, structural and functional - to the surrounding landscape. More detailed inspection of each site was then made to assess site conditions and resources - the nature of dereliction, landform shape and stability, drainage patterns, substrata conditions, existing or establishing vegetation patterns and potential conservation areas. Other site constraints such as the presence of mine shafts and statutory services were collated and recorded at this stage. In most cases this process identified that the site area as defined by reclamation opportunity was inevitably greater than that of immediate engineering interest for shale recovery. However, at the same time it was recognised that shale recovery operations would require room for working the site and disposing of discard. Indicative working plans were

prepared to assess the extent of working impacts, taking account of
essential flexibility in the handling and disposal of discard, the
protection of conservation areas, and access to and from the site.

Finally, the working constraints and opportunities were related back to
the planning context to establish restoration principles for each site. This
includes indicative landform shapes related to extraction/discard volumes
and to restored land use, the broad pattern of landscape structure keyed
into surrounding context and any on-site conservation areas, and the need
for any specific substratum treatments to suit specified after-use.

Conclusion of Stage 2

At the conclusion of Stage 2 three groupings of sources sites had been
identified within the short list of possibilities: seven preferred sites;
four reserve sites; and a number of sites as strategic back-up. Fig.3
indicates the category and location of the preferred sites.

Key

1. Little Wyrley
2. Grove
3. Baddesley
4. Kingsbury
5. Cannock Wood
6. Walkmill Lane
7. Primrose Hill

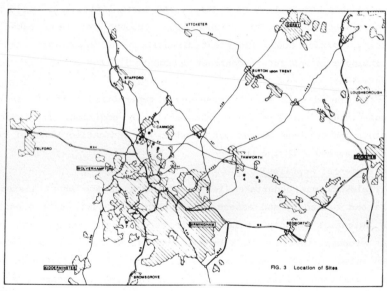

FIG. 3 Location of Sites

POTENTIAL SOURCES

Preferred Sites

The seven preferred sites identified by the Study included:

Little Wyrley	)	
Walkmill Lane	)	
Grove	)	- Staffordshire
Cannock Wood	)	
Primrose Hill		- West Midlands
Kingsbury	)	- Warwickshire
Baddesley	)	

The quantities of material which should be available for the making of
rock paste are summarised in Fig.4 and the costs associated with the winning
and transporting of the colliery spoil are identified in Table 2.

TABLE 2 -

Summary of the costs associated with the winning and transporting of colliery spoil.

SITES ITEM	PRIMROSE HILL	KINGSBURY	BADDESLEY	WALKMILL LANE	LITTLE WYRLEY NO.3	GROVE	CANNOCK WOOD
Cost/m³	5 £/m³	£/m³	£/m³	£/m³	£/m³	£/m³	£/m³
Dudley	4.75	5.50	5.30	5.10	5.23	5.40	4.70
Sandwell	-	4.50	4.30	4.60	4.47	4.65	4.00
Walsall	-	4.70	4.50	4.10	3.73	3.90	3.20

(NOTE: Prices base date - December 1985. Costs for making and pumping rock paste are not included).

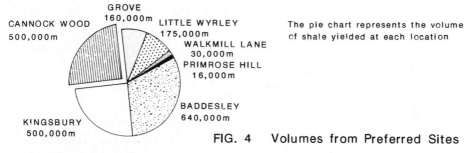

GROVE
160,000m
CANNOCK WOOD
500,000m
LITTLE WYRLEY
175,000m
WALKMILL LANE
30,000m
PRIMROSE HILL
16,000m
BADDESLEY
640,000m
KINGSBURY
500,000m

The pie chart represents the volume of shale yielded at each location

FIG. 4   Volumes from Preferred Sites

Reserve Sites

A further list of four sites has been identified as reserves to the seven preferred sites. The list comprises one derelict site, Wimblebury where 400,000m³ of colliery spoil is available, together with three run-of-mine dirt sources.

Both the Western Area and the South Midlands Area of the British Coal Board have indicated that run-of-mine dirt will be available from certain of their operational collieries. In the case of the Western Area, the two collieries of interest are Littleton and Lea Hall.

The material available from Littleton is the "fines" discard from the coal preparation plant. In the case of Lea Hall it is the "coarse" discard. Laboratory testing of these two types of discard has been carried out, and at this stage in time it would seem possible to mix the two materials to produce a blend suitable for rock paste. Further investigation is necessary to ascertain how these materials can be transported to and handled at the infilling site.

The main source of run-of-mine dirt available in South Midlands Area is Baddesley Colliery. Here the discard produced is a mixture of both fine and coarse material.

It is evident that all three sources of run-of-mine dirt could play a major part in providing a "back up" to the preferred spoil tip sites. They

could also be invaluable where "small" quantities of material are necessary
to complete an infilling contract. In this case, preference could be given
to run-of-mine dirt as opposed to "opening up" another tip where it is known
that the total available quantity will not be needed.

Strategic Back-Up

Hem Heath

Stoke City Council confirm that in the immediate future (next three
years) all material from Hem Heath is required to form peripheral screening
banks to recently agreed tipping areas. In the longer term all interested
parties (NCB, Stoke City Council and Staffordshire County Council) agree
that use of run-of-mine dirt from Hem Heath could be considered (if still
required for infilling) as Hem Heath's tipping life is limited to about 10
to 15 years maximum. The City Council would not be prepared to interrupt the
current bank building works because of the need for early screening, but
longer term extraction, subject to acceptable haulage routes, would
certainly be given serious consideration.

The Leicestershire Sites

At the outset of the Study, five operational colliery sites in
Leicestershire were considered. From discussions with NCB it was established
that none of these sites was viable, mainly due to commitments on coal
recovery from tips. The distance factor was working against Leicestershire
and addition of on-site complications effectively ruled sites out.

However, during the course of the Study, more was learned of the pattern
of production decline in the Leicestershire coalfield and several additional
sites were identified that may become of relevance once operations cease.
The opening of the M42 could also change transportation costs significantly.
If a large scale mine infilling programme develops, there will be merit in
reviewing Leicestershire sources to see if derelict sites can be reclaimed
ahead of taking run-of-mine dirt. For this reason, potential Leicestershire
sources should be considered as a strategic back-up.

STRATEGY RECOMMENDATION

Appraisal of potential source sites yielded a preference for seven
locations, plus four definite reserves (3 run-of-mine; 1 derelict).
Additionally, a strategic reserve for the 1990's has been identified. Fig.5
lists these sites in order of availability. The order also reflects the
policy preference of tackling dereliction before moving on to take
run-of-mine dirt.

Fig. 5 - Strategy for Source Site Working.

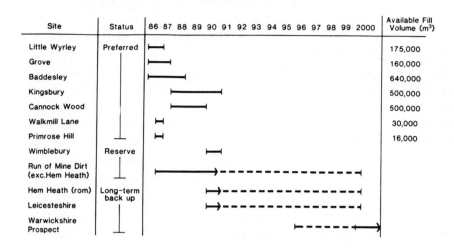

Site	Status	86 87 88 89 90 91 92 93 94 95 96 97 98 99 2000	Available Fill Volume (m³)
Little Wyrley	Preferred		175,000
Grove			160,000
Baddesley			640,000
Kingsbury			500,000
Cannock Wood			500,000
Walkmill Lane			30,000
Primrose Hill			16,000
Wimblebury	Reserve		
Run of Mine Dirt (exc.Hem Heath)			
Hem Heath (rom)	Long-term back up		
Leicesteshire			
Warwickshire Prospect			

The aggregate volume of fill available from the seven preferred derelict sites - approximately 2.0 million cubic metres - may well be inadequate if infilling of mines on a wide scale is the appropriate remedy and resources are available for the work. This adds significance to run-of-mine dirt sources; and strengthens the need for regular review of the sources strategy as preferred remedies develop.

The strategy assumes commencement of demand in 1986 for the Littleton Street infilling scheme, and thereafter a continuing demand at a similar rate of infilling progress. Overlap of source availability provides for choice and flexibility in the programme.

It must not be assumed that over-abundance of colliery spoil will be conveniently available for filling limestone mines. Arup research indicates that there will probably not be a shortage of material given the likely rate of demand, but:

i)   if the pace of infilling quickens significantly, or;

ii)   if fill is required for other major projects,

a shortfall of appropriate and cheap material in the Midlands is possible. To avoid such a situation, careful monitoring of supply and demand would be prudent.

If infilling needs take up the sources in the chronological order indicated, the early benefit will be to badly derelict sites, most of which are in the urban fringes north of the West Midlands conurbation. The logic of solving this dereliction problem is clear, given the Government objectives of improving coalfield environments and tackling the limestone mines problem.

The additional cost of reclaiming sources sites to a reasonable standard is modest and adds very little - approximately 4%- to the all-up cost of mine infilling.

A further meeting with the DoE officials was held when a list of preferred sites had been made. It was agreed that where a source site is derelict, and recovery of fill provides an opportunity to reclaim the site, the standard of reclamation attached should be analogous to a normal DLG-funded scheme providing an open-space or agricultural land-use. By adopting this policy, the DoE is ensuring that the twin objectives of land reclamation and fill provision can be comprehensively met; and that the resources implications are even-handed for the source and infilling areas.

The policy of addressing dereliction rather than using material from operational collieries should be cost effective up to 1990. Based upon the infilling programme as presently assessed, by 1990 the large, accessible derelict sites would have been tackled. Beyond 1991/2 available fill sources would mainly be operational collieries. This will, in due course, raise interesting questions about alternative disposal points and the economics of spoil disposal and fill requirements. An alternative to run-of-mine dirt in the 1990's may come from derelict sites further afield than earlier sources, for example, in Leicestershire. Development of the motorway network will significantly influence transport and environmental costs of bringing fill to the Black Country.

To some extent, the preferred order of source site use is based on an assessment of the ease of setting schemes in hand. It should not be assumed that this will be a simple or speedy process. All the source sites have complications to some degree - technical, procedural, or both - and lead times for producing fill at source should be carefully assessed. It would also be beneficial to have a choice of sources for a particular mine infilling. This would give flexibility during the lengthy infilling contracts; and the existence of choices may influence the negotiating stance of source material owners.

TAKING FORWARD A SOURCES STRATEGY

Work on a sources strategy has proven that at several locations in the Midlands appropriate reserves of colliery spoil are available over the next 10-15 years. To activate those sources will be, in some instances, a complex procedure. The willingness of parties involved in participation at the source end of operations may depend on the commitment they see to schemes in the Black Country.

During the sieving process to derive a short list of preferred sources several prominent, derelict sites have been excluded - mainly due to deficiencies of material or access. Many land reclamation issues relevant to some of the rejected sites will be capable of resolution through other initiatives. At several locations the gravity of the land dereliction problem is such that early progress would be highly desirable in a context other than a fill requirement.

This study work has established a strategy to be followed for activating sources, contingent upon identified demand in the Black Country. For the preferred sites, a sequence of actions to be followed has been identified. Two particular aspects of those actions are critical. Firstly, several of the major sources will take time to set up and run to the point of producing material. Careful attention will be needed to the lead times at sources to ensure that material can be made available at an adequate rate when an infilling demand develops. Secondly, we would advocate flexible designs for source site working and reinstatement. The requirement of consistent fill material will mean that final landforms will not be known until well into works programmes. Nonetheless, broad principles of reinstatement and standards have been developed in the Study, and it is considered that interested parties will know what is to be achieved by sources working. While it cannot be presumed that any particular source site will automatically gain immediate planning approval (where appropriate), considerable care has been taken during the course of the Study to discuss individual sites with respective planning authorities, and to incorporate officers' views in these feasibility studies, covering planning and outline design of both excavation and reclamation works.

To avoid any dismatch of the respective lead times at source site and mine, it would be useful to have advance design work completed on two or three sources by mid 1986 as the demand for fill builds up. The funding implications of this design work are not major, and routeing of funds through the source local authorities would draw them into the limestone programme at the time when their assistance could be of maximum benefit.

This Study identifies a sequence of sites to be worked when demand arises. There is merit in adhering to this sequence as far as possible. A commitment to the strategy is the policy overlap where the limestone programme meets surface reclamation at colliery sites. It is important that the DoE continue to make an input, ensuring that the overlap of derelict land grant expenditure continues to address two policy objectives to mutual benefit.

REFERENCES

1 Ove Arup & Partners (1983): A Study of Limnestone Workings in the West Midlands.
2 Tarmac Construction Limited (1984): A Study of Materials for Infilling the Black County Limestone Mines.
3 Ove Arup & Partners, Land Use Consultants (1985): Infilling of Limestone Mines Using Colliery Spoil : Identification and Selection of Sources of Material.

ACKNOWLEDGEMENTS

The authors thank the Department of the Environment, who funded both the initial study and the various follow-up works through Derelict Land Grant aid, and the Metropolitan Boroughs of Dudley, Sandwell and Walsall.

Reclamation, Treatment and Utilization of Coal Mining Wastes, edited by A.K.M. Rainbow   615
Elsevier Science Publishers B.V., Amsterdam, 1987 — Printed in The Netherlands

THE INFILLING OF LIMESTONE MINES WITH ROCK PASTE

P.A. BRAITHWAITE AND T. SKLUCKI

Ove Arup & Partners, 2, Duchess Place, Edgbaston, Birmingham B16 8NH.

SUMMARY

   Rock paste, a mixture of colliery spoil and water, has been developed as a
proposed infill material for limestone mines.   During 1985, a trial was
organised by Ove Arup & Partners to use the material to infill a section of
the Castle Fields limestone mine in Dudley.   Rock Paste was mixed under
production conditions and pumped underground where its performance was
monitored and evaluated with remote instrumentation.   This included video
cameras amongst other techniques.   The mine was successfuly filled and
proposals are now in hand to commence the first infilling "proper" of the
Littleton Street mine in Walsall.   By the date of the Conference, it is
anticipated these works will be well advanced and it will be possible to
update Delegates to the Conference at that time.

INTRODUCTION

   Infilling of limestone mines is part of the project to investigate and

treat abandoned limestone workings in the West Midlands (ref. 1).   The main

objective of this project is to remove many hectares of land from the "blight"

of development restrictions imposed because of the risk of surface instability

caused by collapses within the workings.

   The project is funded by the Department of the Environment through Derelict

Land Grant aid administered via the Metropolitan Boroughs of Dudley, Sandwell,

Walsall and Wolverhampton.

   Stabilisation of workings by infilling is one of the treatment options

available.   The particular option chosen for a mine depends on many factors

including the condition of the mine, depth of mine and the land use of the

overlying ground.

   This paper describes the development of one particular method, using rock

paste, which is appropriate for bulk infilling of mines with large

interconnected voids which are either air or water filled.

   Rock paste which is basically a mixture of colliery spoil and water, is

designed to be a cheap infilling material which flows as a plastic mass and

can be made and placed using conventional civil engineering plant.   The main

advantages of using rock paste are;

   i)   colliery spoil can be obtained very cheaply

   ii) few injection points are required, compared with say infilling using
        sand-flushing, as rock paste will flow plastically to fill a cavity

iii) there is the added environmental benefit of removing dereliction at
the site of the colliery shale source while returning the land above
the mine to beneficial use .

Early work on investigating the production of rock paste was carried out by
the Building Research Establishment at Watford and Cardington (refs 2 and
3).   The development of rock paste as a construction material is described by
Cole and Figg (ref 4)

During 1985 a trial infilling contract was undertaken to fill a section of
mine to prove that it was possible for a general contractor to mix and pump
rock paste and that the flow characteristics of rock paste in the mine were as
theoretically predicted.   This paper outlines the main knowledge gained, and
describes the setting up of the first contract to infill a complete mine.

Limestone mining in the West Midlands was mainly in the Upper Wenlock and
Lower Wenlock Limestone which are seams in the Wenlock Series of Silurian
Age.   These seams are 4m to 8m thick and 9m to 13m thick respectively.   They
are separated by 30m to 35m of Nodular Beds, a calcareous mudstone with
limestone nodules.   Above the Upper Wenlock Limestone is approximately 3m of
"passage beds" which are a transition to the Ludlow Shales.   The character of
the rocks is described further in ref 1.

TRIAL INFILLING AT CASTLE FIELDS MINE
Purpose of the Trial
The trial infilling project was to take the technique of bulk infilling
with rock paste from the experimental approach of BRE (ref 2) towards its
implementation as a mine treatment under the usual civil engineering contract
system.   For sake of economy it was desired to put infilling works out to
competitive tender by general civil engineering contractors.   The type of
contract chosen for the trial infilling was an admeasurement contract under
general conditions based on the Conditions of Contract (Fifth Edition)
approved by the ICE jointly with the ACE and FCEC (ref 5).   The project was a
trial of this type of contract as well as of the method of infilling.

Dudley Metropolitan Borough Council undertook the trial infilling project
in the common interest of the Boroughs of the West Midlands affected by the
abandoned limestone workings.

Site Description
The location chosen for the trial infilling was the Sports Centre site at
Dudley (National Grid Reference: SO 952 907), as shown on fig 1.   Castle
Fields Mine in the Upper Wenlock Limestone is beneath the site.

While most of the abandoned limestone workings in the West Midlands are now
completely flooded, part of Castle Fields Mine is above the level of the

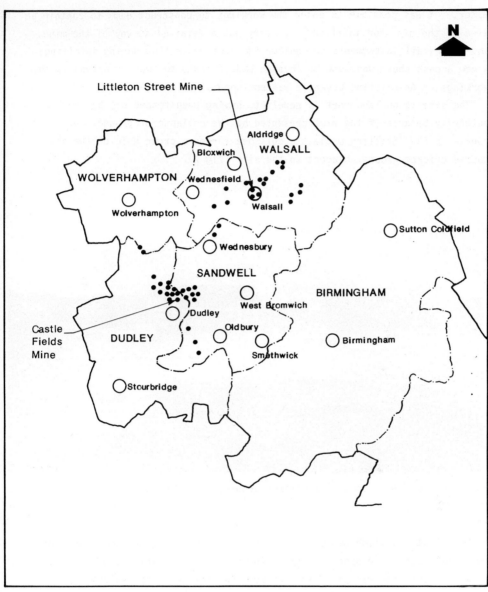

Littleton Street Mine

Aldridge

Bloxwich WALSALL

WOLVERHAMPTON Wednesfield

Wolverhampton Walsall

Sutton Coldfield

Wednesbury

SANDWELL

BIRMINGHAM

West Bromwich

Castle Fields Mine

Dudley

DUDLEY Oldbury Birmingham

Smethwick

Stourbridge

N

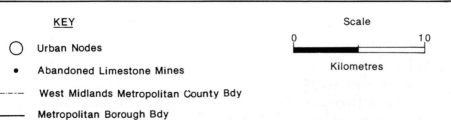

KEY

Scale

0　　　　　　　　　　10

○ Urban Nodes

● Abandoned Limestone Mines

Kilometres

–·–·– West Midlands Metropolitan County Bdy

——— Metropolitan Borough Bdy

Fig.1 Location of abandoned limestone mines in the West Midlands

water. It was possible to enter the workings to construct dams to contain an area of the mine for infilling, to carry out a detailed survey of the mine, and to install instruments to monitor the rock paste flow during infilling. A new access shaft was sunk by Thyssen Ltd. in 1983 to improve access to the workings.   An existing borehole was equipped as an emergency exit.

The site is on land that is derelict, having been fenced off by the local authority because of the risk presented by the collapse of ground. During the course of the infilling project a collapse formed a crown hole on the old County cricket ground adjacent to the site (fig 2).

Fig. 2 - Crown Hole in Dudley cricket ground, 25th May 1985.

Work by the British Geological Survey (ref 6) showed that there were many roof falls occurring sporadically in the mine, some of which exceeded a tonne in mass. The mine was considered to be in a state of progressive deterioration.

Objectives of Trial Infilling

The objectives were to show that

(i)   Using plant similar to that used by BRE (ref 2) a contractor could take colliery spoil from a spoil tip, mix it with water to produce a rock paste of controlled quality with a shear strength in the range 1 to 3 kN/m^2 and pump it through a steel pipeline at least 100 m long to give at the delivery end paste suitable for infilling a mine.

(ii)   Having demonstrated (i), a contractor could inject about $30,000m^3$ of rock paste into a mine via boreholes, maintaining as an objective a minimum throughput of $60m^3$ per hour for 10 hours per day. The mine infilling would be recorded by instruments and by closed circuit television.

(iii)   Having achieved (i) and (ii), a contractor could make other rock pastes of suitable quality for filling a mine, using colliery spoil materials from several different sources and mixtures of these with selected granular materials.

(iv)   Rock pastes pumped into a $70 \ m^3$ capacity water tank through a tremie pipe would not disperse in the water but spread as coherent bodies.

Contract Organisation Project objectives were to be met by three phases of a trial infilling contract:

Phase 1   :   Site Establishment and Proving the Plant (i)

Phase 2   :   Infilling the Mine (ii)

Phase 3   :   Material Test Pumping (iii) and (iv)

Phases 1 and 2 were priced on a bill of quantities whereas Phase 3, being more experimental in nature, was carried out on daywork priced on a schedule of rates.

The contract was let, after competetive tendering, to R M Douglas Construction Ltd. of Birmingham. The trial infilling contract value was around a million pounds.

As work was carried out in the mine, the site was deemed an active mine under the Mines and Quarries Acts and the contract documents provided for the statutory requirements.

Trial Infilling Area of the Mine

The part of Castle Fields Mine used for trial infilling is shown on fig 3. It is approximately the shape of an equilateral triangle with a side length of 100m. The area dammed off for the infilling is approximately 6600 m^2.

The mine was worked by the room and pillar technique with a fairly regular arrangement of pillars in rows aligned south west - north east. The trial infilling area was mined in about 1888.

The pillars are rectangular in shape and typically 5m to 6m wide and 4m to 8m long. Rooms are 6m to 8m wide (Fig. 4). The rate of extraction, the proportion of the mine plan area that was excavated, was 80% and the average height of the workings was approximately 5m.

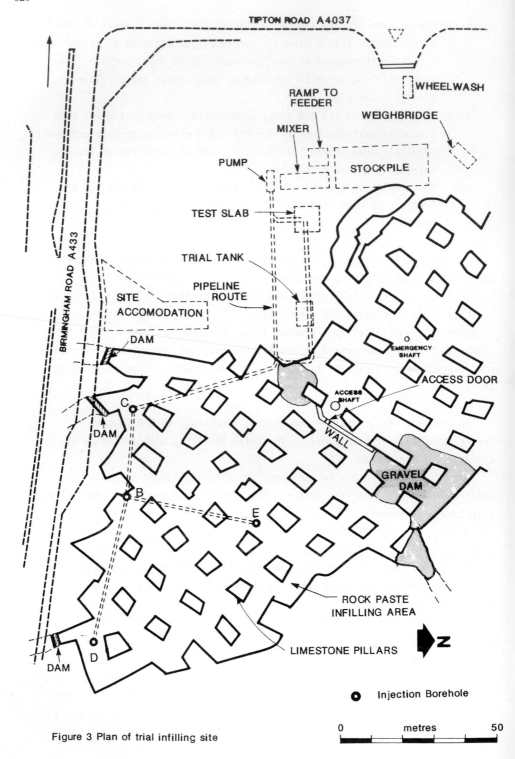

Figure 3 Plan of trial infilling site

Fig. 4 - Castle Fields Mine during infilling, 8th July 1985.

On the northwest side of the trial infilling area a barrier was constructed
to contain the infilling material.   This was a reinforced concrete wall
flanked by grouted gravel dams.   In the wall, opposite the mine access shaft,
was a doorway.   The doorway was used for access to the trial infilling area
for observation of the rock paste flow until the paste had reached the
concrete wall.   Grouted gravel dams, constructed entirely from the ground
surface, were used to form the barrier in areas of the mine where the roof was
in dangerous condition.

On the south side of the mine is a solid rib left by the mine operator to
"protect" the Birmingham Road.   This rib is pierced by three gallieries
leading to workings further south.   These workings were filled with sand in
1964/1965 and the galleries had been blocked by concrete faced gravel dams.

On the northeast side is a solid limestone rib, dividing the infilling from
another part of the mine.   This rib is pierced by a single gallery at the
northwest end, which was sealed by a subsidiary grouted gravel dam.

The mine area chosen for the trial infilling has a cover (depth of the mine
roof from ground surface) of between 18m and 24m.

The mine floor over a large part of the centre of the infilling area was covered in sand flushed from a neighbouring 1973/74 infilling contract to a depth of up to a metre. This part of the mine floor was nearly horizontal at about 150m O.D.   The height of the mine cavity was generally about 5m.

On the southern edge of the mine the floor rose steeply to around 155m O.D. Here the floor was of bare rock or roof fall debris.

Scattered through the mine were heaps of rock debris from roof falls, some older and some younger than the sand fill. Over the falls were rises in the roof, called chimneys, some of which were 5m higher than the general roof level.

Four boreholes for rock paste injection were drilled into the section of mine dammed off for the trial. These are boreholes B, C, D and E shown on fig 3. The boreholes were lined with 200mm plastic well casing for 10m above the mine roof and then to the ground surface with steel casing. The upper end of the steel casing had a shoulder suitable for seating the rock paste pipeline couplings. The casing was cement grouted.

Ten observation boreholes were drilled into the mine infilling area. The boreholes were lined with cement grouted 102mm plastic well casings.

Plant and Site Establishment

Colliery spoil was obtained from tips approximately 35 km from the infilling site.   Wheeled tractor loading shovels were used for excavation of 3m high benches.   Spoil was fed through mobile screening plant which was fitted with a vibrating screen with a 50mm mesh.

The contractor provided a covered stockpile at the infilling site for the colliery spoil, with a capacity equivalent to a day's consumption. Water was obtained from a borehole drilled into a flooded part of the mine.

The Barber Greene continuous mixing plant, fig 5, was that used by BRE for their experiments, modified by fitting a two bin feeder with variable speed belts. The bins were charged with a wheeled tractor loading shovel.

The continuous mixer incorporated a mixing trough 2.1m long, 1.1m wide and 0.6m deep with two longitudinal contra-rotating shafts. The shafts, each carrying 14 pairs of paddles, were driven by a 30kW motor.

The feeder was capable of supplying colliery spoil at a rate of up to 160 tonnes per hour.   The mixer water system had a pump with a capacity of 45 m^3 per hour feeding a spray bar in the mixing trough.

The mixer outlet was fitted with a hydraulically operated diverter to direct flow to either the pump hopper or to a skip for sampling or disposal. The device comprised two inclined plates, mounted back to back so that the ridge between the plates could be made to slide from one side of the falling stream of paste to the other.

Fig. 5 - Rock paste plant for trial infilling during heap test pumping

The pipeline for distribution of rock paste to injection boreholes was built of 200mm nominal diameter seamless steel pipes with two bolt, two piece pipe couplings of forged steel with rubber gaskets. Bends were hot formed to full circular diameter. Pipe wall thickness was 6.3 mm.

The pipeline was supported on each side of couplings by timber sleepers. Steel strap pipeline restraints were screwed to the sleepers and hooped over the pipeline.

Colliery spoil mixed with water to produce the rock paste for infilling the prepared mine void was obtained from two sources in Staffordshire, a tip at the disused Coppice Colliery site and one at Littleton Colliery. The Coppice site was developed for Phase 1. During Phase 2 infilling the first 20% of void was filled with paste made from Coppice spoil. A second source was developed at Littleton to provide the necessary production capacity. Both sites were operated until 76% of the void had been filled, when the available spoil at Coppice was exhausted and the plant transferred to Littleton for the remainder of Phase 2.

Observations of Rock Paste Placement

A closed circuit television system was installed to observe the flow of paste in the mine. Forty two monochrome television cameras each with a 500 watt flood lamp were fixed to scaffold props in the mine, fig 6.

Fig. 6 - Installing closed circuit television system in Castle Fields Mine

Fixed in view of the cameras were 82 props, fig 4, with elevation scales graduated in metres and decimetres. The scales were printed black on white styrene sheeting, and stapled to softwood boards wired to the props. The scales were all fixed so that the zero graduation corresponded to 150m O.D.

Scale figures 0.1m high were legible on the television monitor at distances of up to 15m, the decimetre graduations were discernible at distances of up to 30m and the metre graduations were discernible at a range of about 50m.

In the site accomodation were two video cassette recorders and monitors. A control unit allowed switching either recorder to any of the cameras in the mine and a sequence of switching could be programmed. Superimposed on the video images were the camera number, date and time.

The cameras and lights were kept powered throughout infilling to avoid condensation within the cameras. Eventually, all cameras and lights in the mine were inundated and earth leakage breakers in each camera circuit tripped.

The advance of rock paste over the mine floor is shown on fig 7. The edge of the rock paste was plotted by direct inspection and by reference to the television views of the mine. Paste from injection borehole D had travelled approximately 80m across the mine floor by 19 July 1985 after 16 days pumping. Individual lobes of the paste were recorded as travelling at speeds up to 0.3m per minute. Injection was then moved to borehole B. On 26 July 1985 the paste had reached the vicinity of the concrete barrier wall and the door through the barrier was closed. Around 30% of the mine void had been filled. It did not prove necessary to utilise injection boreholes C and E to fill the mine.

The boreholes that had been drilled into the mine enabled the paste level to be determined in the later stages of infilling and after infilling had been completed. In addition to the ten observation boreholes, use was also made of the injection boreholes when not being used for paste injection. On fig 8 are profiles of rock paste infill on the section line shown in fig 7.

The sonde used to dip the boreholes to determine paste elevation is a modified version of a magnet settlement sonde. The modification , designed to a specification by Ove Arup & Partners, converts the reed switch sensor into a waterproof contact switch. The sensor is carried on a tape graduated in centimetres. Accuracy of measurement was found to be 0.03m.

To investigate the development of earth pressure and pore water pressure in the infill medium, pressure cells were installed in the mine.

Fourteen total pressure cells and four piezometers were installed. The total pressure cells were oil filled flat jack cells. The piezometers, pore water pressure cells, had high air entry porous ceramic elements. Both types of cell had pneumatic transducers. Pneumatic tubing was laid from the cells to a duct in the concrete barrier well and thence up the shaft to a terminal panel. Across the mine floor the tubing was buried or covered in sand.

The total pressure cells were fixed to steel frames clamped to scaffold props set vertically in the mine. The cells were set with the plane of the cell horizontal. The piezometers were in open topped 50 litre steel cylindrical containers full of saturated sand.

Instrumentation and Measurements of Plant

In addition to the usual contract records such as daily returns of plant and labour, the Contractor was required to submit a daily log of pumping time with reasons for any stoppages, results of checks and measurement of the plant performance, estimates of rock paste production, and records of material sampling and testing.

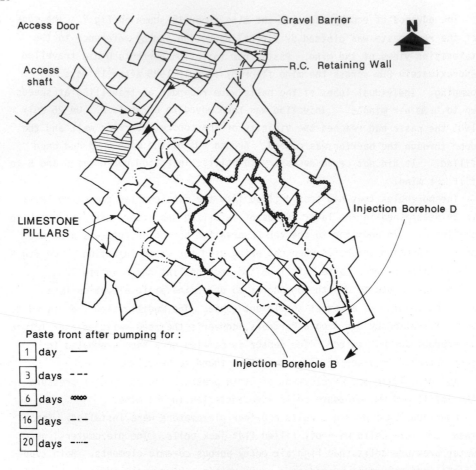

Access Door

Access shaft

Gravel Barrier

R.C. Retaining Wall

LIMESTONE PILLARS

Injection Borehole D

Paste front after pumping for :

1 day —

3 days - - -

6 days ∞∞∞

16 days —·—

20 days —····—

Injection Borehole B

Fig 7   Progress of rock paste over mine floor

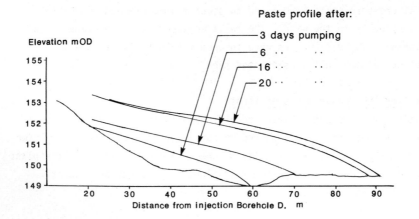

Paste profile after:

3 days pumping

6 ··   ··

16 ··   ··

20 ··   ··

Elevation mOD

155
154
153
152
151
150
149

20   30   40   50   60   70   80   90

Distance from injection Borehole D. m

Fig. 8   Profiles of rock paste (on section marked on Fig. 7)

The rock paste production plant was instrumented to indicate material feed rates, electrical power and energy consumption, water consumption, pump running time and pump hydraulic oil pressure. The rock paste pipeline was instrumented with pressure guages. Indicated values were generally recorded hourly during pumping.

Chart recorders were connected to transducers for mixer power, pump hydraulic oil pressure, and pipeline pressures. Chart recordings were made for a few minutes each hour and for more extended periods in experimental parts of the work.

The pipeline wear was recorded by weighing pipes before and after use and by monitoring pipe thickness with an ultrasonic gauge.

During the mine infilling, at approximately daily intervals, samples were taken of colliery spoil and rock paste for soil mechanics classification and chemical testing (with methods modified as recommended by the NCB, ref 7, where appropriate).

Rock paste strength was monitored during infilling by using a plate penetrometer, the modified Proctor Penetrometer Apparatus used by BRE, ref 2. This was applied to the paste either in the mixer outlet hopper, or in a half oil drum when paste was sampled at the injection borehole.

Results of the Trial Infilling

The trial was successful as all objectives were realised. The first phase proving of the plant culminated in a one day performance test during which the contractor pumped 600 cubic metres of suitable rock paste through 159 m of pipeline in nine hours continuous operation.

The infilling of the mine took three months. In seventy four days of pumping, $29100m^3$ of rock paste were injected. The rock paste behaved broadly as predicted, flowing as a plastic material around the mine pillars. The paste rose into chimneys in the mine roof; cameras and observation boreholes had been positioned in certain of these.

Only two boreholes were used for the infilling of the mine. Rock paste was pumped 176 m to injection borehole D and 159 m to injection borehole B. Pumping consumed energy at the rate of 1.7 kWh/m^3 when pumping to borehole D and 1.4 kWh/m^3 when pumping to borehole B.

The mean pumping rate was 53 cubic metres per hour. The time spent pumping was 75% of the time available for pumping. The remaining 25% may be divided into 15% plant downtime (both paste production and spoil handling plant) and 10% clearing paste blockages.

The wear rate of the pipeline was found to be in the range 0.2% to 0.5% pipe weight loss per thousand cubic metres of rock paste pumped. Wall thickness reductions were consistent with the weight loss measurements.

Provision was made for injection of relatively small quantities of water, up to 1 m^3 per hour in total, to be injected at up to three locations on the pipeline via manifolds with six ports to the pipeline spaced about the circumference. Water injection was found to reduce the pressure in the pipeline, the reduction taking place within seconds of water injection being started. The reduction occurred for a quantity of water injected which if added at the mixer would not measurably decrease paste strength and pumping resistance in the pipeline. Continuous use of water injection was not necessary for the pumping of rock paste of the required strength over the distances required for the trial infilling at practical flow rates. It was very useful however to be able to quickly apply water injection on those occasions when the paste flow rate dropped after some particularly strong paste entered the pipeline. This portion of paste could then pass through the pipeline, a process taking a few minutes, without adjustment of the quantity of water added at the mixer. Such an adjustment to the mix could result in several m^3 of paste weaker than desired being pumped.

The paste blockages were predominantly in the pump hopper. If a relatively weak paste was introduced into the hopper, then on the pump piston suction stroke the paste segregated and coarse particles or aggregations were left in the hopper. This material was crushed by the movement of the pump valve and formed a hard pack which restricted flow of plaste into the pump cylinders. A high pressure water jet was found to be useful in clearing the pump hopper and attention to control of paste consistency reduced the frequency of occurrence of blockages.

The pump downtime, though initially around 12% of available pumping time, fell to about 2% as teething problems were resolved. Downtime of other plant was dominated by a breakdown of the water abstraction pump and a breakdown of the mixer motor.

A number of heap tests were also carried out, fig 5 and ref 3. In those a large quantity, approximately 20 m^3, of rock paste was pumped onto an 8 m square concrete slab. The dimensions of the heap were analysed to derive the strength of the paste. The chief use of heap testing was to investigate different paste mixtures and changes in spoil quality.

During Phase 3 of the contract, rock pastes where successfully made and pumped with colliery spoils from Cannock Wood Tip and Grove North Tip in the Cannock Chase District of Staffordshire and from Baddesley Colliery Tip in Warwickshire. An attempt to handle recently washed spoil from Birch Coppice Colliery was unsuccessful as the spoil would not pass freely through the plant. Pastes were successfully made and pumped with a mixture of Grove spoil and a quarried sand, and with mixtures of Baddesley spoil and pulverised fuel ash.

Three types of paste were pumped to an underwater discharge in the trial tank, and did not disperse in the water.   Paste profiles determined by sounding the tank were consistent with theory (ref 4).   One of these trials was observed by representatives of the Severn Trent Water Authority.

Rock paste consolidation and strength gain in the mine are being monitored and analysed, preparatory to a further treatment of the trial infilling section of the mine.   This may be carried out in conjunction with infilling of some other part of the mine.

Colliery spoils coarser than those pumped by BRE, ref 2, were among those used for the trial infilling.   The resulting rock pastes were generally denser and with lower moisture content than the paste pumped by BRE.   All spoils that had been screened with mobile plant at the source tip were found suitable for feeding and mixing at the infilling site.

In a full scale infilling operation, it was estimated that infilling costs would be around ten pounds per cubic metre.

On completion of Phase 2, the 4ha source site at Coppice was landscaped and developed as a football field and that at Littleton was returned to use for spoil disposal.   Sites at Yield Fields Hall Farm on the border of Cannock Chase and Walsall were evaluated for use as a source early in the project and while not developed as a source they were landscaped and 10ha brought into agricultural use.   The infilling site in Dudley was cleared and part has been improved to provide a car park, pending further treatment of the mine to deal with residual voids.

INFILLING LITTLETON STREET MINE

After successful completion of the trial infilling of part of Castle Fields Mine, agreement was reached between the Department of the Environment and Walsall Metropolitan Borough Council for Ove Arup & Partners to proceed with the design for infilling the Littleton Street Mine, Walsall (fig 1), which is a completely water filled mine.

While it was accepted that there would be some trial elements in the first use of rock paste to infill a water filled mine, the main objective was to treat the mine and bring back the surface to full beneficial use, so removing the constraints on redevelopement over some 21 hectares of land.

Site Description

Littleton Street Mine is a room and pillar working, about 70m below the surface, in the Lower Wenlock Limestone.   The mine is on the northern edge of Walsall town centre, and covers a plan area of about 13 hectares.

This mine was chosen to be the first for infilling as there was a detailed knowledge of the mine and also because of the proximity of the mine to the town centre.

Mining History Originally there were two separate mines, Littleton Street Lime Works which opened in 1804 in the south of the area, see fig 9, and Portland Street Lime Works to the north, which started some time after 1846. The two mines were merged in 1886 to form the Hatherton and Portland Limestone Works. These works were finally abandoned in 1903.

A large collapse in the Portland Street Lime Works occurred in 1861, as a result of pillars being "robbed", which was conmmon practice by miners at this time. The collapse caused a surface depression which was about 65m in diameter and 25m to 30m deep at its lowest point. A number of houses were destroyed by this event.

Ground Conditions The section in fig 10 shows the sequence of strata above the mine. Below the surface deposits of fill and drift the Ludlow Shales outcrop, in the north west of the site to a maximum thickness of 10m. Up to 7m thickness of Upper Wenlock Limestone outcrops below the drift deposits over the central area of the site, while the Nodular Beds directly underlie the drift deposits in the south east corner. The Nodular Beds are about 30m thick and overlie 6m to 9m of Lower Wenlock Limestone, below which lie the Wenlock Shales.

Extent and Condition of the Mine A site investigations of the mine, which included 16 holes drilled using rotary non-coring techniques, was carried out during 1985 to determine the condition and extent of the mine. The investigation revealed that the extent of the Lower Wenlock Limestone mine closely corresponds to that shown on the 1903 mine abandonment plan.

The water filled mine was remotely surveyed using an ultrasonic scanning tool and close circuit television cameras (ref 8). An area of collapsed workings was identified in the north western part of the mine, the site of the collapse recorded in 1861. Roof falls were also identified at three other locations in the mine. The remainder of the mine was found to be in much the same condition as it must have been when abandoned.

The volume of the mine has been calculated to be about 550,000 m^3. The mine generally slopes at 4° to the north west. In areas where roof falls have taken place, the roof is now on average 6 m above its original level.

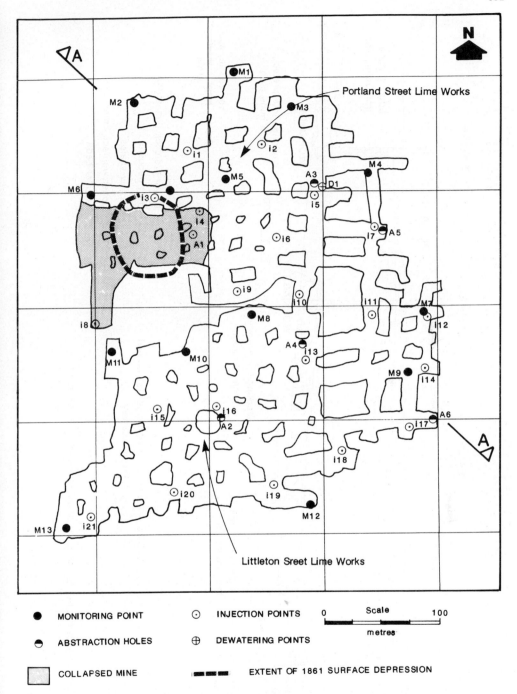

MONITORING POINT

INJECTION POINTS

ABSTRACTION HOLES

DEWATERING POINTS

COLLAPSED MINE

EXTENT OF 1861 SURFACE DEPRESSION

Scale 0 — 100 metres

Fig 9. Showing Littleton Street mine plan and location of infilling holes

632

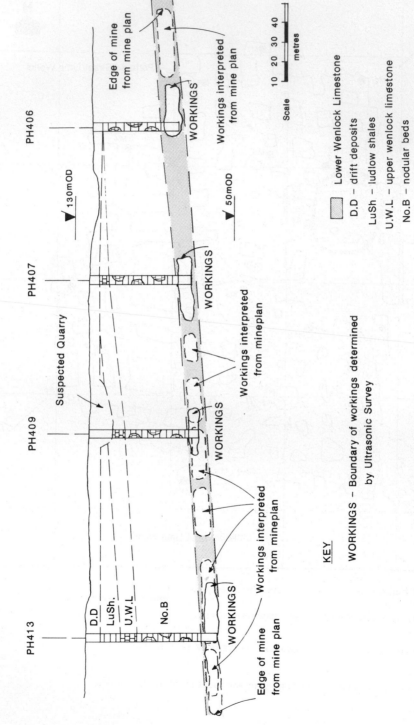

Fig.10. Section showing the sequence of strata above Littleton Street mine

KEY

WORKINGS – Boundary of workings determined by Ultrasonic Survey

Lower Wenlock Limestone

D.D – drift deposits
LuSh – ludlow shales
U.W.L – upper wenlock limestone
No.B – nodular beds

Scale

10 20 30 40
metres

Rock Paste as Bulk Infill

The decision to use rock paste as bulk infill was taken on the basis that this was the cheapest of the infilling materials and also on the success of the trial at Castle Fields Mine.

Walsall M.B.C. wish to redevelop the land above the mine as early as is practical after infilling is completed. To meet this requirement, rock paste consisting of colliery shale, water and small percentages of pulverised fuel ash (pfa) and lime will be used. The addition of pfa and lime (ref 4) will ensure a gradual increase in strength with time and also the amount of consolidation will be reduced as compared with the rock paste used in the Castle Fields Mines trial.

Design of the Infilling Works

As the whole of the Littleton Street Mine is to be infilled, there is no requirement for constructing dams.

Rock paste will be injected into the higher part of the mine and will flow down-dip to the lower extremities. The paste will then rise and fill the higher parts of the mine, "backing-up" towards the injection point.

The injection point outler, in the mine, will be arranged so that the rock paste does not drop through water a distance greater than 1.5m to avoid dispersion of the finer material of the paste. The outlet pipe will be periodically raised during infilling to keep it clear of hardened paste.

The layout of injection points and observation boreholes is shown on Fig 9. Infilling will commence from injection points I10 and I22.

To ensure that the rock paste has the desired flow characteristics, a shear strength of between 1.5kN/m² and 3 kN/m² is required at the injection point.

As the mine is water filled and the ground water is at about 119.5 m OD, about 4m below surface, precautions have been taken to protect nearby basements from being flooded in the event that infilling should raise the groundwater.

If abstraction of water from the mine for use in making rock paste does not maintain the groundwater level at or below 120m OD, then a dewatering system will be brought into use. Water pumped out of a borehole will be discharged into an existing stormwater sewer via a treatment system which will reduce the suspended solids to less than 100 milligrams/litre.

Sources of Colliery Spoil

The sources of colliery spoil were identified as a result of a study of source sites in the West Midlands (ref 9).

In order to ensure enough suitable material will be available for infilling Littleton Street Mine, three sources will be used. These sources are:

Source	Estimated Yield
Little Wyrley No. 3 Colliery	175000 m^3
Grove Colliery	160000 m^3
Baddesley Colliery	640000 m^3

Once extraction of the shale has been completed at Little Wyrley No. 3 Colliery and Grove Colliery, the sites will be restored to agricultural use at Little Wyrley and to allow redevelopment at Grove Colliery. Only part of the spoil heap at Baddesley Colliery will be used, the remainder will be used for another infilling contract and restoration will take place at a later date.

Infilling Plant Requirements   The plant required for making and placing rock paste is similar to that used at the trial, with the exception that additional equipment is needed to introduce lime and p.f.a. to the mix. A schematic arrangement of plant for feeding and mixing the rock paste constituents is shown in fig 11.

The design of the pipeline is left to the contractor as part of his temporary works design. The pipeline must safely contain the rock paste at the maximum pressure which the method of placement could impose, as the pipeline will be laid along roads, pavements and other public areas.

Pipeline pressures will be continuously monitored at 100m intervals and also within 10 m of each pump. If the pressure in any part of the pipeline should exceed the maximum safe working pressure, or have a sudden loss in pressure, then the pump and all feeds into the pipeline will be automatically cut off.

As the pipelines are exposed to the public, additional protection around the pipeline and instrumentation will be provided to both ensure safety and prevent vandalism.

Water injection, of up to 10% of the water added at the mixer, is allowed along the length of the pipelines.

Quality Control of Rock Paste   During Phase 1 of the Contract, colliery spoil from each of the three sources will be mixed with water, lime and p.f.a. and pumped through 200m of pipeline and into $5m^3$ containers and to heap tests. Some of the containers will be periodically rocked from site to side, to enable monitoring of the flow characteristics.

635

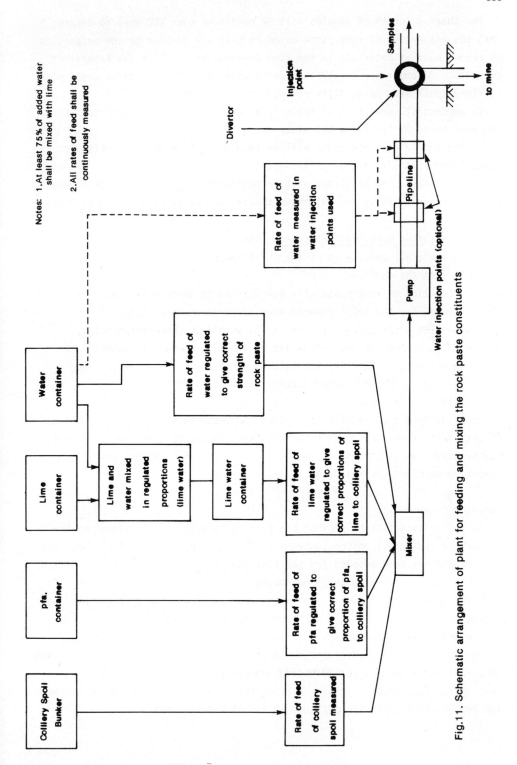

Notes: 1. At least 75% of added water shall be mixed with lime

2. All rates of feed shall be continuously measured

Fig.11. Schematic arrangement of plant for feeding and mixing the rock paste constituents

The shear strength of samples will be monitored over 100 days to ensure that the properties of rock paste mixed in bulk are similar to the properties of pastes made of materials in the same proportions mixed in the laboratory (ref 4). A comparison of strength gains with time of rock pastes mixed with colliery spoil from three different sources will also be made.

On successful completion of these "placement tests", the contractor will commence Phase 2, infilling the mine.

During infilling, rock paste will be tested for bulk density, moisture content and particle size distribution.

The rock paste for infilling will be monitored by plate penetrometer shear strength tests taken at the injection points, and by carrying out heap tests.

Monitoring Rock Paste Placement   As there is no access into the mine for personnel, all monitoring of the infilling process will be carried out remotely from the surface.

The elevation of rock paste will be recorded by using sondes similar to the one developed for the trial lowered down observation boreholes.

Close circuit television cameras may be used down observation holes. However, the quality of the recordings will not be as good as those of the trial as there is no facility for setting up powerful flood lamps. The recordings can only provide qualitative information as elevation scales cannot be installed in the mine.

It is planned periodically to monitor the progress of infilling by the use of ultrasonic surveying. This technique has proved to be extremely useful, during site investigations of limestone mines, for enabling the extent and condition of mine workings to be assessed (ref 8).

Contract Organisation

As the trial infilling project had been successfully administered by the use of the ICE Conditions of Contract (Fifth Edition) (ref 5), this same form of contract has been adopted for the Littleton Street Mine infilling. All of the work is priced as a bill of quantities.

Eight general contractors were invited to tender and the contract was awarded, in October 1986, to Fairclough Civil Engineering.

The works are in two phases.

Phase 1 covers the establishment of both the infilling and source sites and construction of mine injection points, observation boreholes and water abstraction facilities. Trials to prove that the plant is capable of mixing and pumping rock paste are included in the 24 weeks of Phase 1.

Phase 2 allows for infilling the mine using rock paste of controlled quality and restoring the land at the source site. The contract allows 80 weeks for this phase.

The mine cavity will be deemed to be filled at any injection point when a specified water pressure can be maintained at the head of the injection point. Once the final injection point is filled, the contractor will visit each injection point to "top-up" with rock paste before he leaves the site.

CONCLUSIONS

The trial infilling at Castlefields Mine, Dudley, was successful. Rock pastes, made from colliery spoil and water, were mixed and pumped using modified and normal civil engineering plant by a general contractor. The rock paste flowed within the mine, as predicted, as a plastic material.

Experience gained during the trial has been used to plan the first major infilling contract, at Littleton Street Mine, Walsall. A two year contract has been let to infill Littleton Street Mine with $550000m^3$ of rockpaste, consisting of colliery spoil, water, lime and pfa. Apart from some preliminary trials to prove that the plant is capable of mixing and pumping the modified version of rock paste, the contract is fundamentally a normal "ICE 5th" civil engineering contract.

Infilling the mine will allow redevelopment of the surface to commence without the developer having to take special precautions because of the presence of abandoned limestone mine workings.

ACKNOWLEDGEMENTS

The authors thank the Department of the Environment, who direct the funding of the infilling works through Derelict Land Grant aid, and the Metropolitan Boroughs of Dudley and Walsall who promoted the works, for permission to publish this paper.

We would also like to thank the contractors, R.M.Douglas Construction for the trial infilling and Fairclough Civil Engineering for the Littleton Street infilling works, for demonstrating the adaptability of the industry.

REFERENCES

1.   K.W. Cole, P.A. Braithwaite, P.C. Dauncey and K.L. Seago, Removal of actual and apprehended dereliction caused by abandoned limestone mines, in Proc. Conf. on Building on Marginal and Derelict Land, Glasgow, 7-9 May 1986, Instn. of Civil Engineers, Westminster, 1986, pp. 99-124.

638

2.   D.L. Hills and W.H. Ward, The Use of Colliery Spoil to Backfill
      Abandoned Limestone Mines, Part 1: Large Scale Pipeline Pumping Trials,
      Building Research Establishment Report N189/84,
      Department of the Environment, Westminster, February 1985.
3.   W.H. Ward and D.L. Hills, The Use of Colliery Spoil to Backfill
      Abandoned Limestone Mines, Part 2: Heap Experiments and their
      Application to Mine Filling, Building Research Establishment Report
      N189/84, Department of the Environment, Westminster, February 1985.
4.   K.W. Cole and J. Figg, "Improved rock paste": A slow hardening bulk fill
      based on lime, pulverised fuel ash and colliery shale, in Proc. 2nd.
      Int. Symp. on the Reclamation, Treatment and Utilisation of Coal Mining
      Wastes, Nottingham University, 7-11 September 1987, Elsevier, Amsterdam,
      in press.
5.   Permanent Joint Committee of the Institution of Civil Engineers, the
      Association of Consulting Engineers, and the Federation of Civil
      Engineering Contractors, Conditions of Contract and Forms of Tender,
      Agreement and Bond for use in connection with Works of Civil Engineering
      Construction, 5th. edn., June 1973 (Revised January 1979), Institution
      of Civil Engineers, Westminster, 1979.
6.   A. Miller, J.A. Richards, D.M. McCann and K.T. Muirhead, Microseismic
      Investigations at Castle Fields Abandoned Limestone Mine, Global
      Seismology Report No. 263, British Geological Survey, NERC, Edinburgh,
      November 1985.
7.   National Coal Board Joint Working Party, Application of British Standard
      1377:1967 to the Testing of Colliery Spoil, Technical Memorandum,
      National Coal Board, London, May 1971.
8.   P.A. Braithwaite and K.W. Cole, Subsurface investigations of abandoned
      limestone mines in the West Midlands of England by use of remote
      sensors, in Proc. Conf. on Rock Engineering and Excavation in an Urban
      Environment, Hong Kong, 24-27 February 1986, Institution of Mining and
      Metallurgy, London, 1986, pp. 27-39, also in Trans.Instn. Min. Metall.
      (Sect.A:Min.industry) V95, October 1986, pp A165-214.
9.   D.W. Stevens and K.L. Seago, "Sources Study": The identification and
      selection of sources of colliery spoil for use in the infilling of
      abandoned limestone mines in the Black Country, in Proc. 2nd. Int. Symp.
      on the Reclamation, Treatment and Utilisation of Coal Mining Waste,
      Nottingham University, 7-11 September 1987, Elsevier, Amsterdam, in
      press.

Reclamation, Treatment and Utilization of Coal Mining Wastes, edited by A.K.M. Rainbow
Elsevier Science Publishers B.V., Amsterdam, 1987 — Printed in The Netherlands

METHODS OF DEVELOPMENT ABOVE ANCIENT SHALLOW PILLAR-AND-STALL COAL WORKINGS

I E HIGGINBOTTOM

Wimpey Laboratories Limited, Beaconsfield Road, Hayes, Middlesex, UB4 OLS (England)

SUMMARY
   A review is given of the main options available to designers and planners when selecting foundations and layouts appropriate to differing conditions on sites underlain by ancient shallow pillar-and-stall coal workings. Some of the principal difficulties and limitations are discussed and examples are described.

INTRODUCTORY COMMENTS

   This paper reviews some thirty years' experience in interpreting the results of documentary and borehole investigations at sites affected by ancient shallow coal workings in Britain, and in assisting developers to design structures and layouts suited to the underground conditions. Geotechnical, financial, social and environmental factors may each influence the final solution, but seldom recur in the same combinations and every site is in some sense unique. All decisions must therefore be taken with the fullest possible knowledge of the mining situation.

   The term 'shallow' is here used to imply conditions under which the behaviour of the ground is dominated by the risk of roof collapse in old pillar workings, rather than by other closure processes such as the crushing of pillars or heave of the floor. There is general agreement that, except in a few special situations, roof failure is the most important process affecting ground stability above ancient pillar workings, and since it usually follows roughly predictable patterns is is more readily allowed for in design than are some other mining-related factors.

   Although shortage of land has forced the construction industry to make greater use of sites above old shallow workings, experience of their long-term behaviour, especially under surcharge loads, is relatively limited and mostly confined to the post-war period. In spite of a wider awareness of the problem, its existence may still be identified only after acquisition of the site, especially where the developer is not resident in a mining area and lacks the relevant experience. He will often order a geotechnical investigation designed to examine other aspects of the site, such as the bearing capacity of the surface deposits, and the physical techniques used may be unsuitable for extensive exploration in the Coal Measures. Evidence of old workings may be found by accident, or insight on the part of the site investigation specialist, who may nevertheless be powerless to define the problem fully because the brief from his client is narrow and based on a pre-conceived view of the questions to be addressed.

In Britain, warning of a possible problem should normally appear at the stage of planning application, when the planning authority refers the application to British Coal for comment. However, old workings are notoriously ill-documented and it is rare for shallow underground plans to survive (if they were ever made in the first place) in a suffiently detailed and reliable form to be useful for design purposes. Usually British Coal are forced to rely on the large-scale (1/10000) Geological Survey maps of the coalfields. These may themselves be untrustworthy, especially in areas which were built-up at the time the survey was made, but in open country the field geologist was sometimes able to infer the positions of coal outcrops from the surface evidence of shallow workings, which might then have been up to a century less old and relatively more visible. Very often however the maps are unreliable in respect of coal outcrop positions, although these will have a critical influence on the assessment of site stability and layout. Many of the large-scale geological maps are old and they do not necessarily represent the present state of knowledge. The British Geological Survey should be approached for more recent information at the planning stage although the author believes that this ideal is not routinely observed. The removal of the British Geological Survey from old mining centres such as Leeds, the general curtailment of the mapping programme, and the early retirement of many highly experienced field staff, has not improved access to this information.

It is not surprising therefore that the initial opinions expressed by British Coal should often, and quite rightly, be loaded on the side of caution. They might deter a would-be developer, in which case there is much to be said for the engagement of an independent mining consultant, familiar with the historical development of the coalfield, to give a more detached appraisal of the real likelihood of risk. Very often a site will prove better than at first predicted, after all appropriate studies have been completed.

Apart from the cost of detailed investigations, which should be taken into account when acquiring the site, another financial constraint placed on developers in recent years has been the declining popularity of multistorey structures, especially for housing. Special foundations to solve the problems of shallow workings are more economical where valuable superstructures are placed on relatively small sites. With low-height development over comparatively extensive sites it has previously been rarely possible to solve the problems at seam level, but rising property values combined with falling real costs of infilling, are changing this situation and the consolidation of shallow workings under two-storey housing is now more widely adopted. Analysis of a few recent examples suggests that it may typically add no more than about 3 to 5 per cent to the price of a house, depending partly on local property values.

Five basic procedures are available to minimise the risk from ancient shallow workings (Fig. 1). They may be used singly or in combination, according to the special requirements of individual structures, or in response to changing geological conditions across a large site. These procedures are listed below, roughly in order of increasing complexity:

1. avoid the problem by locating structures on shallow foundations over safe areas of the site;

2. remove the workings by bulk excavation;

3. form a stiff foundation near ground level to span a possible surface collapse;

4. transfer the structural loads below the workings by piles or piers;

5. infill the workings and found at a higher level.

642

Much of the author's experience has been gained in collaboration with his colleagues J P Hollingbery (ref.1) and P F Winfield (ref.2), whose papers (also in this volume) are complementary to his and may be referred to for additional details and case histories, especially in the fields of special foundations for high buildings and the consolidation of old workings.

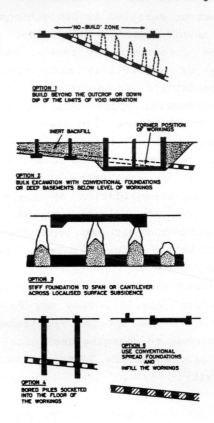

Fig. 1. Foundation design options over old shallow pillar workings

FOUNDING ON SAFE AREAS

This assumes that the only real hazard is that of span failure and at its simplest it entails accurately locating the outcrop of the worked seam and defining a hazard zone, down dip of the outcrop, in which roof collapse might propagate to the surface (Fig. 1 Option 1). Structures may then be placed either outside the coal outcrop or down dip of the hazard zone, and the latter may be relegated to open space. The problem is to estimate the

ultimate height of the void propagation, expressed as a multiple of
the original working height, which is assumed to equal the seam
thickness. Values given in the literature (e.g. refs 3 - 6) range
from 2.5 to over 10, depending on assumptions as to the final shape
of the collapse and the appropriate "bulking factor", i.e. the
percentage by which the volume of the fallen roof debris exceeds
the volume of the same material before collapse. Because it is
rare for the geometry of the ancient workings and the mechanical
properties of the roof to be known with anything like enough
confidence to allow a more sophisticated analysis, the concept
that, through bulking, the void becomes stowed with its own debris,
so that its upward propagation is a self-limiting process, is the
main one of practical value in the hazard assessment.

As the dip steepens, so the width of the hazard zone reduces,
and with it the area lost to development. The relationship between
the dip and the width of the hazard zone for various ratios of rock
cover to seam thickness is shown in Fig. 2. In the British
coalfields, dips are only locally steeper than 10 degrees and are
often less than 5 degrees. The hazard zone rapidly widens as the
dip flattens and becomes wasteful of space if unused, especially
when conservative depths of safe rock cover (D_C) are applied, say
10 times the seam thickness. It is common practice therefore tc
adopt a lower value for D_C, such as 6 times the seam thickness, tc
determine a 'no-build' zone, outside which a transition zone, with
stiff raft foundations or infilling, may extend to a D_C value of at
least 10 times the seam thickness. This assumes that the surcharge
load is effectively dissipated at seam level, as would normally be

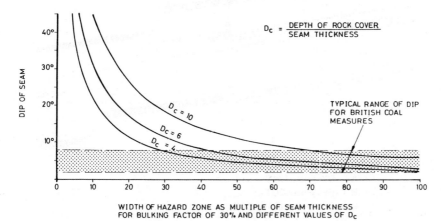

Fig. 2. Dip and hazard width relationships

644

true for low-height developments. It is a matter of observation that void propagation to the surface is rare where D_C equals 4 or more.

Since this is not an exact science, predicting the widths of the hazard and transition zones can entail somewhat qualitative and subjective judgements, particularly as to whether small residual risks are acceptable when balanced against the costs and relative repairabilities of different superstructures, or the expense of consolidating the workings.

Problems arise with closely-spaced seams whose respective hazard zones may overlap over a wide area, and where void migration from a lower set of set of workings may interact on a higher set. In the latter situation the author has sometimes contemplated a worst case, in which the seam thicknesses are added together and considered to act at the higher (shallower) level. This must be conservative, but the author doubts whether more realistic solutions can always be proved with enough confidence for additional precautions such as structural stiffening or infilling to be dispensed with.

In defining the hazard zone it is rarely possible to fix the limits of the workings confidently, without a very costly and detailed drilling programme, unless a natural feature such as a fault, a washout, or the water table (Fig. 3) has determined them.

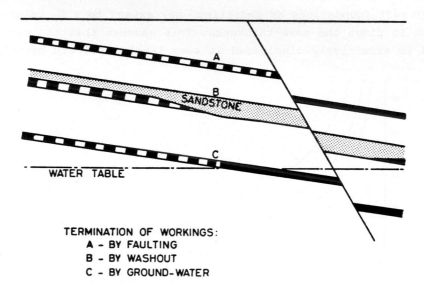

TERMINATION OF WORKINGS:
A - BY FAULTING
B - BY WASHOUT
C - BY GROUND-WATER

Fig. 3. Working limited by natural factors

Except in such cases, it is common practice to disregard the boundaries of the workings when 'zoning' the site for development. The cost of the additional investigation required to prove them with confidence is usually not thought justifiable in relation to the possibility of reducing the area of the hazard zone. Very rarely however, it may be possible to prove by drilling that a mine boundary shown on an abandonment plan is accurate enough to be used in the hazard assessment.

The presence of superficial deposits can modify the stability assessment. It would be unrealistic for example to ignore the beneficial effects of a thick cover of strong glacial till, which can have mechanical properties not greatly different from those of weathered Coal Measures. On a site at Airdrie (ref. 7), the till cover was judged to have two-thirds the strength of the underlying rock, and since the till thickened considerably in one direction it enabled important economies to be made in developing the site. On the other hand, saturated granular deposits may suddenly collapse into a void at rockhead, greatly extending the area of surface subsidence, which in the absence of such material, would normally be smaller in width than the underlying working.

Figure 4 shows an example of hazard zoning on a site near Newcastle upon Tyne. Although the dip was virtually flat, the site rose steeply from south to north to create the thicknesses of cover shown. Only seam 1, the lowest and 0.6m thick, was extensively worked, in 16 out of 24 holes. Seam 2 (0.45m thick) was solid in each of 26 holes and was judged unworked. Seam 3 (0.3m thick) was apparently worked in only 2 out of 23 holes, possibly on the line

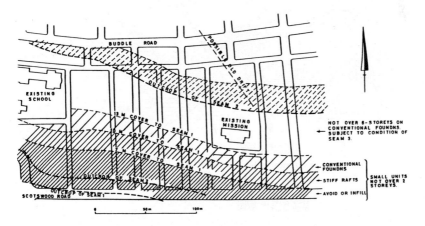

Fig. 4. Zoned foundation layout, Newcastle upon Tyne

of an old adit known hereabouts. The site was zoned for development as follows: open space or consolidation of workings where less than 7m of cover was present over Seam 1; small units (e.g. semi-detached 2-storey houses) on stiff rafts where there was from 7m to 10m of cover; similar small units on conventional foundations where between 10m and 13m of cover was present above seam 1. These limits make allowance for an average of about 2.5m of glacial till over the site. Outside the 13m line no constraints were imposed except for a height limit of 6-storeys, with the condition of Seam 3 to be further investigated.

Figure 5 shows how conditions on a housing site near Chesterfield were interpreted for planning purposes. Although 41 shafts and an adit were known, only 8 boreholes out of 31 to the Deep Hard (1.8m thick) found evidence of workings, apparently fully closed. One borehole out of 11 to the Piper (2.4m thick) found similar evidence of workings. It was concluded that the workings were very localised around a large number of individual shafts, that they were very ancient and that no voids of any size remained open, the roof of weathered mudstone having collapsed. Therefore the whole site was utilised, with buildings on stiffened foundations being located virtually up to the seam outcrops. There was no drift cover to the site, but areas of backfilled opencast workings (shown on Fig. 5) were present. The following constraints were observed:

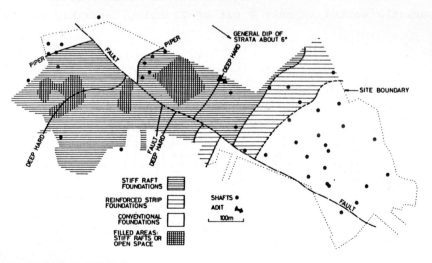

Fig. 5. Zoned foundation layout near Chesterfield

1. from outcrop to depth of cover equal to six times the seam thicknesses (also on opencast areas) : 2-storey semi-detached houses on stiff rafts

2. where depth of cover was from six to nine times the seam thicknesses: 2-storey semi-detached houses on reinforced strip foundations

3. with a depth of cover exceeding nine times the seam thicknesses: any type up to 3-storey on conventional foundations.

EXCAVATING THE WORKINGS

Bulk excavation (Fig. 1 Option 2) may be economically attractive to depths well over 6m where the site is restricted in plan area, as in the case of individual high-rise buildings. Conventional earthmoving equipment will usually operate successfully in the weathered mudstones and shales which often form the roof strata above old shallow workings. The latter may therefore be dug out over a large part of the hazard area, or even the whole of it provided the seam is relatively thin, say not more than 1m. The excavated material may not all be suitable for replacement and that from seam level at least will usually have to be carried to tip, which may be impracticable in a congested urban setting.

This is a self-evidently attractive solution where a deep basement or underground car park can be incorporated in the scheme. Failing that, bases may be constructed in the floor of the seam, and columns and walls carried up in stages to correspond with a phased backfilling of the site. For low-height structures, a stiff raft or beam foundation, to guard against differential settlement, may be constructed at ground level after controlled compaction of the backfilling.

Bulk excavation may be advantageously combined with a site regrading scheme. An example near Bristol is shown in Figure 6, where it was proposed to remove shallow workings and place the overlying material in an adjacent valley to improve the previously adverse site gradients. The seam on the downthrown side of the fault had been shown by drilling to be unworked. This method has also been used to form a foundation below the large thicknesses of old wastes and disturbed ground, rather then actual voids, associated with the working of some very thick seams, such as the

Staffordshire Thick Coal, where bulk excavation to a depth of 9m has been carried out by the author's firm. The liability of the Thick Coal wastes to spontaneous combustion (where they are not already burnt out) was an important factor in this case, necessitating the use of an imported inert backfill.

It is possible to offset the expense of bulk excavation by selling the residual coal, but the financial analysis must take account of the royalty (about £16/tonne in March 1987) payable to British Coal, the uncertainties as to how much residual coal is actually present, and the need for careful selective digging of the coal pillars. Other potential adverse factors are possible difficulties with planning consent (especially under urban conditions), the need for stockpiling areas on site, and the delay to construction whilst the various formalities are in progress.

Fig. 6. Bulk excavation of workings to regrade site

STIFF FOUNDATIONS NEAR GROUND LEVEL

It is possible to design stiff foundations which will span a typical crown-hole or cantilever over it without overturning, provided the hole is not anomalously large (Fig. 1 Option 3). Design criteria adopted by the author's firm for house foundations, in conjunction with the National House-Building Council, assume an unsupported span of 3m or a cantilever under corners of 1.5m in two directions. Reliance on this method is not of course recommended where there is a known possibility of void migration to the surface, because of the risk to human safety and to services, paved areas, etc. It is quite commonly used however where there is a residual uncertainty as to whether old workings are completely absent from the site, or where they are thought to be largely closed and the problem is essentially one of inequalities in bearing capacity due to the undisturbed and weathered character of the roof strata. It has also been employed where the consolidation, under a surcharge load, of ancient shallow total-extraction workings seemed possible. A stiff grillage of beams may be used, but for house foundations it is more usual to employ a raft stiffened with downstand beams around the perimeter and under party walls. However, flat-soffitted rafts on a polythene slip layer have been used in Bathgate (West Lothian) (ref 8), as a precaution against the possibility of horizontal strains caused by the failure of pillars over large areas. These might also be appropriate where there is a danger of modern deep longwall mining triggering or renewing collapse of old shallow workings, and it is desired to resist both types of movement. This would be unusual in the case of one or two-storey housing, but is common practice for taller structures.

In some coalfields (e.g. Midlothian) the seams may have nearly vertical dips and have been extensively worked almost to outcrop. The pillar of coal left at rockhead, can, if it fails, drop for a large distance in the plane of the seam. Where structures have to be carried across the coal outcrops, as in the case of road works, beams or stiffened slabs bearing on the adjacent bedrock are generally the only practicable precaution.

PILES OR PIERS TAKEN BELOW WORKINGS

This approach (Fig. 1 Option 4) may be combined with that of bulk excavation, as already described. Where this is not possible however, piling to the base of the seam might be an economic alternative, though normally only for high, or otherwise costly, superstructures. If old workings are also present in a second deeper seam which is shallow enough to threaten the piles, piling may have to be combined with infilling of the deeper workings before a secure foundation can be achieved. A piling method that can be depended upon to reach the floor of the workings will be required. Moreover, piles will have to be able to resist the vertical and horizontal forces that may be imposed by movement of the workings and it is therefore usual to make them of large diameter, so that a cast-in-situ bored pile is a common choice. Piles may need to be sleeved to withstand dragdown forces and to prevent loss of concrete into cavities. Piling is usually economic only through superficial deposits and Coal Measures clay-rocks (shales and mudstones), which are generally susceptible to

augering. It has been widely used to transfer structural loads to
the base of ground badly disturbed by mining, such as that over the
Thick Coal wastes in the West Midlands, in which case, the
condition of the Heathen Coal, some 6m deeper, becomes important
and may call for infilling.

Where substantial thicknesses of sandstone overlie the workings,
piles are not usually economic and infilling becomes the cheaper
alternative. Also, no matter what the rock type, as the depth
increases, so the cost of infilling reduces against that for piling
until it becomes the more economical solution. If a suitable rock
socket is formed at the pile base, most Coal Measures rocks will
carry a load in end bearing that will allow the permissible working
stress in the concrete shaft to be developed. A large-diameter
pile shaft will permit in situ visual inspection or testing of the
rock, subject to the necessary safety precautions.

A combination of piling with infilling of the workings was used
on a major part of the St. Helens Central Redevelopment project
(Fig. 7), where conditions varied rapidly across the site due to
the presence of six coal seams dipping at about 16 degrees to the
east. Below the eastern half of the site, three seams were
extensively worked and left standing in pillars. These were fully
consolidated down dip from their subcrops below the superficial
deposits, to a point at which 12.2m of rock cover was present above
the highest seam. Pile foundations were then used to transfer the
structural loads through the weak superficial deposits to rockhead
above the consolidated workings. The presence of a massive
sandstone (Pemberton Rock) above the uppermost seam would in any
case have prevented economic piling to the base of the workings in

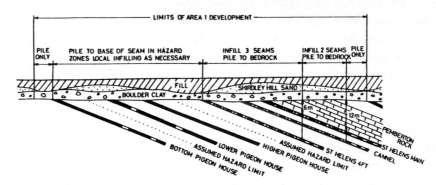

Fig. 7. Piling combined with infilling to suit varying geological
conditions, St. Helens Central Redevelopment

this area. In the western half of the site the condition of the remaining three seams was obscure. Very sporadic evidence of working was found in the lower seams but the highest in this group appeared unworked. Hazard zones above these seams were defined, and since they had mudstone roofs, piles located within these zones were taken to the base of the seams, with provision (which in the event was not required) for localised infilling around the pile bases should workings be encountered.

Where it is possible to resite the structure, and the strata have a significant dip, piling costs (and those of other special foundations) can be reduced by moving the site towards the outcrop. Fig. 8 illustrates this diagrammatically with reference to pile lengths and is based on an actual case at Brislington, where the effect was increased by the natural slope of the ground.

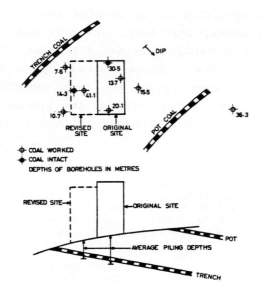

Fig. 8. Moving structure towards outcrop to reduce foundation costs

INFILLING THE WORKINGS, WITH CONVENTIONAL FOUNDATIONS AT A HIGHER LEVEL

This eliminates the problem by treating it at its source. The contractual questions to be addressed and the techniques commonly used in Britain are described by J P Hollingbery (ref. 1). The following additional points are made here because they are related to the behavioural characteristics of shallow pillar workings.

The object of infilling is to prevent further movement in the old workings, so that they can continue to support the surface, the loads being carried by the intact Coal Measures. Strong grouts are therefore unnecessary. However, the grout strength may need increasing where concentrated loads have to be carried close to the roof of the workings, for example below pile toes if it is not practicable to extend the piles into the floor of the seam.

Perimeter holes are commonly drilled at a distance outside the structure of one-third to one-half the depth to the seam, by analogy with the lateral 'draw' which characterises subsidence over total extraction workings. However, the analogy is not accurate because of the different failure mode shown by pillar-and-stall workings, whereby surface instability normally affects a plan area smaller than that of the underlying void. The practice of infilling an area around the structure might therefore seem extravagant. It continues to be adopted however, and is probably justified where costly structures are involved, because complete infilling affords protection against all forms of movement in the workings and not merely those caused by roof collapse. The additional cost becomes relatively less important as the site increases in size, since the ratio of the treated area to that of the structure itself then falls rapidly and approaches unity for large sites.

REFERENCES

1. J.P. Hollingbery, Infilling old mine workings and shafts, 2nd Intnl. Symp. on the Reclamation, treatment and utilisation of coal mining wastes, Univ. Nottingham, England, September 7-11-1987, Elsevier, Amsterdam, 1987, pp.
2. P.F. Winfield, Foundations for sites over natural voids and old mine workings, ibid, pp.
3. P.R. Healy and J.M. Head, Construction over abandoned mine workings, Construction Industry Research and Information Association, Spec. Pub. 32 (1984).
4. P.J. Piggott, K. Wardell and P. Eynon , Ground movements arising from the presence of shallow abandoned mine workings, Proc. Conf. on Large ground movements and structures at University of Wales Institute of Science and Technology, Pentech press, London, July 1977, pp. 749 -780.
5. D.G. Price, J.L. Knill and A.B. Malkin, Foundations of multi-storey blocks on the Coal Measures, Q. Jl. Engng. Geol. 1 (4) (1969) 271 - 322
6. R.K. Taylor, Characteristics of shallow coal-mine workings and their implications in urban redevelopment areas, in: F G Bell (Ed), Site Investigations in Areas of Mining Subsidence, Newnes-Butterworths, London, 1975, pp. 125 -148.
7. D.G. Price, Engineering geology in the urban environment, Q.Jl. Engng. Geol. 4 (3) (1971) 191 - 208.
8. anon. Rafting over Bathgate's old mines, New Civil Engnr, 25 February 1982, p. 26.

Reclamation, Treatment and Utilization of Coal Mining Wastes, edited by A.K.M. Rainbow   653
Elsevier Science Publishers B.V., Amsterdam, 1987 — Printed in The Netherlands

FOUNDATIONS FOR SITES OVER NATURAL VOIDS AND OLD MINE WORKINGS

PETER F WINFIELD FIStructE
Grove Structural Consultants

SUMMARY
    Three methods are considered for constructing buildings on sites where
natural voids or old mine workings are present. Five case histories are
presented of tall buildings constructed in the United Kingdom with the loads
from the foundations taken below the voids or workings, and one where the
buildings were constructed above old, worked and flooded coal seams.

INTRODUCTION

    When natural voids or old mine workings are present under a site,
conventional structures may still be built, but consideration has to be given
to three possible methods for constructing the foundations. These methods are
as follows:-

    (a) Form the foundations below the cavities.

    (b) Consolidate the voids and provide spread foundations near ground level.

    (c) Form foundations near ground level to bridge over the voids.

    The choice for a particular site will depend on the relative costs of
alternative methods, on the magnitude of the loads to be supported and the
importance of the structure.

    The money available for the usual type of site investigation is often
insufficient for the exact locations and the sizes of the cavities to be
determined, before the foundation design is prepared. When this is the case
the foundation must be of a type that can be modified as further information
about the cavities is discovered as work proceeds. This will avoid costly
interruptions to the work that would result if fundamental changes to the
foundation design had to be made.

    The extent of cavities under a site is often greater than the initial
evidence would lead one to expect.

    It may be possible to erect structures on conventional foundations near
ground level without any special precautions, if the voids or old mine workings
are at considerable depth, and if it can be shown that subsidence of materials
above the cavities is unlikely to take place. This may be the case if a strong
bridge of competent rock is present over cavities of relatively small plan
area. On the other hand it is not unusual for a development project to be
abandoned because the cost of the foundations make the scheme uneconomic. This

should not be regarded as the end of the site for building purposes, as land values change rapidly in the cities. As land becomes even more scarce a high return on a development may well justify the expense of special foundations.

Land which is commonly known to be difficult to develop because of possible subsidence may be purchased cheaply, and this would help to off-set the high cost of the foundations.

The types of foundations shown as (a), and (b), above are most suitable for heavy structures with small plan areas. Consolidation work or the work necessary to transfer the loads to below the cavities is concentrated on a small area of the site. High bearing pressures are frequently possible and the foundation may be loaded to its full capacity with resulting economy. Light buildings covering large areas of a site may require virtually the same foundation, but it will be greater in plan area, and is unlikely to be loaded to anywhere near its full capacity.

Housing projects planned to be of the low rise type have been changed to high rise after the discovery of old mine workings. This has reduced the cost per dwelling for foundations, and the same number of dwellings has been provided. There remain many sites in our old cities which are only suitable for development using high rise construction.

FORM THE FOUNDATIONS BELOW THE CAVITIES

The loads from the superstructure may be supported below the cavities, either by using piles or piers, or by bulk excavation to remove the cavities.

Bulk excavation for spread foundations

A bulk excavation nine metres deep was made to allow spread foundations to be constructed below the pavement of old pillar and stall workings in coal at Leeds. The excavation through boulder clay and coal measure shales, sandstone and fireclay, was made entirely by machine without the use of explosives. A ramp was constructed into the excavation as work proceeded to facilitate the removal of spoil by lorries. The site was large enough for the sides of the excavation to be battered through the boulder clay, although nearly vertical sides were possible in the coal measure rocks. No temporary timbering was necessary, and water entering the excavation was removed by pumping without difficulty.

The boulder clay was removed from the site, the sound rock was retained as backfill to be supplemented by imported burnt shale wastes to make up the quantity required.

The bases were founded below the worked seam which was completely removed, and column and wall stems were used for the full height of some nine metres to ground level without intermediate tie beams. This was done for simplicity of

Construction, the stems have a cross section of sufficient area for the full working stress in the concrete to be developed without reduction for slenderness.

The stems were constructed in three lifts of approximately three metres each, the backfill was replaced in three stages so that the work was never higher than three metres above a working platform.

A special floor was constructed below the ground floor to support the services fully in the event of the deep backfill subsiding under the service ducts. Foundations for the tower crane were also taken below the worked seam, the scaffolding was founded on the backfill and examined regularly for signs of settlement.

This type of construction is worth considering when the cavities are less than about ten metres below ground level and when the dig can be made without recourse to the use of explosives. Care should be exercised to ensure that the most suitable plant is provided for the work as each new material is encountered.

Clearly if old workings are flooded, and dewatering is not easy, this is unlikely to be an économic type of foundation.

Large diameter piled foundations

At Dudley we constructed five blocks of high flats over the Thick, Heathen and Stinking seams of coal. Large diameter bored piles have been used to transfer the column loads to below old workings.

Subsidence caused by collapse into cavities may impose large horizontal loads onto piles, and for this reason it will not always be possible to use them. However if subsidence is nearly complete and further large movements are not expected to take place piles may provide the best foundation.

At Dudley and for similar contracts, we have used large diameter bored piles of minimum shaft diameter of 1050 mm. This size ensures that the shaft has a high resistance to horizontal shear, it can be heavily reinforced if necessary, and may be capable of supporting a large vertical load so reducing the total number of piles required for a structure. The toe of the pile will be founded below the cavities and this usually means boring through bedrock, with a 1050 mm diameter pile a man may enter to inspect the work or to use a jack hammer or explosives to loosen the rock to facilitate mechanical boring.

The pile shaft has to be sleeved through cavities so that a permanent shutter is provided for the concrete, a light steel lining is used inside the main casing. Even when a cavity is not encountered in the pile bore it is advisable to use a full length permanent lining, as the lateral pressure of the wet concrete may be sufficient to break through the side of the bore into a cavity. This may result in damage to the shaft of the pile as the concrete above the level of the cavity may arch across the pile, and not fall down the shaft to replace concrete lost to the cavity. Such a defect is unlikely to be detected by the site personnel and could lead to structural failure. A continuous core has been extracted from some piles by diamond drilling as a method of examining the completed pile shaft. This is an expensive operation, whereas light steel linings can be installed quite cheaply, particularly when the quantity required is known well before the piling commences.

It is usually cheaper to provide the pile with a socket into rock below the old workings than to attempt to found at a higher level by under-reaming a large base to the pile. Both methods were tried at Dudley, where it was shown by plate bearing tests that an end bearing pressure of some 4280 kN/sq m could be adopted in the socket when the flight auger reached virtual refusal. This pressure allows the full strength of the pile shaft to be developed with resulting economy. It is sometimes advisable to probe below the intended toe level of a pile to ensure that no soft material or voids exist below apparently sound rock. This may be done by jack hammer used both vertically and on the rake.

Before sending a man down a pile shaft into old workings, a check should be made on the condition of the air using a gas meter to detect if explosive methane gas is present. A supply of fresh air will be required at the bottom of the shaft. During piling operations for the airport at Glasgow natural gas entered a shell pile and ignited at ground level before the core of concrete was placed. An additional hazard in the work may be the boreholes of the site investigation, if the position of a pile happens to coincide with that of a borehole. Gas or water under artesian pressure may enter the pile hole from levels below the toe of the pile.

Some volatile coals are susceptible to spontaneous combustion when in contact with free air. When excavating or piling through these coals it will be necessary to install thermometers in the coal to observe any rise in temperature. The work should be completed as quickly as possible and the coal sealed from the air. If ignition does take place it is usual to inject carbon dioxide into the coal in an attempt to replace the free air, and contain the fire. Backfill placed into excavations made into the volatile coals must be incombustible and we have used waste moulding sand for this purpose.

Although piling may provide a satisfactory foundation there are certain disadvantages to be considered. Not the least of these is that an estimator is unable to predict with accuracy the final cost of the piling work as so much will depend on the time taken to overcome obstructions, form rock sockets, to install the permanent linings, and on the prediction of the lengths of the piles.

Disturbance to the ground caused by the piling rig may induce collapse of old workings and endanger the rig. We have injected a cheap fill into some cavities to strengthen the working area for piling.

The pile caps and ground beams are usually conventional in design with some additional tensile reinforcement in the beams to ensure that the pile heads are well restrained and maintain their relative positions in the event of the piles being subjected to lateral loading.

At Dudley the Thick Coal wastes were removed from the site of one block and piling was done from the bottom of an excavation eight metres deep. Column and wall stems some eight metres long were therefore required to support the ground floor. The wall stems were designed to act as pile caps in addition to their usual purpose, saving the cost of conventional caps. No caps were required under the column stems, the columns being cast onto starters left in the piles.

Small diameter pile foundations

At a site in Kent work had commenced on the foundations of a block of 15 storey flats before evidence of workings in chalk were discovered. The site had some ten metres of dense gravel overlying chalk. As part of the site investigation the area occupied by the building had been probed on a eight metre square grid. The probes penetrated three metres into chalk and no evidence of solution cavities or mine workings was found. It was decided to support the block on driven piles, terminating in dense gravel. After approximately one third of the piles had been installed and after a period of particularly heavy rain a large subsidence occurred on the working area, taking one apparently satisfactory pile with it. A few weeks later a similar subsidence took place to one side of the site of the block.

A television survey carried out from the bottom of a number of boreholes showed that mining of the chalk had taken place. To avoid working below the standing water level the miners had kept the mine high in the chalk, leaving a roof of chalk only 600 mm thick. It was therefore not surprising that the roof collapsed and the gravel poured into the mine when disturbed by the heavy vibration from the driven piling rig.

To avoid changing the basic design of the foundations it was decided to change from driven piles founded in the gravel to permanently lined bored piles of 450 mm diameter founded three metres into the chalk. The piles carried the same working load so no change to the pile caps and ground beams was necessary, and they in turn provided an additional investigation of the chalk.

To protect the bored piles from damage in the event of further collapse of the gravel into the chalk the loose gravel and cavities were consolidatd with a cement and flyash grout. A further intensive probing of the site in the area of the driven piles was carried out to ensure that they were not similarly undermined. An area around the block extending some six metres away from it was intensively probed also to ensure that no cavities existed there. In the event of a collapse of the chalk here the gravel could move both vertically down and also laterally from under and around the piles.

It would appear that if piles are to be used in similar circumstances, it is advisable to found them in chalk as they then form their own probe, and this we have done at Grays Thurrock. Also at Grays we have built plate rafts high up on gravel, but only after probing the site on a five metre grid closed to two and a half metres square at the corners of the building and under important structural elements.

Pier foundations

We produced a scheme for constructing a shopping centre at Newcastle over old pillar and stall workings at about ten metres below ground. In spite of there being some six metres of sandstone above the workings, crownholes had appeared on the site and conventional foundations would have been unsafe. The scheme involved the construction of only eight massive piers founded below the workings and isolated from horizontal forces by compressible material around them. A tower was to be erected on each pier and the entire structure hung from the towers. Given a sympathetic architect such structures can be made to look attractive. However after several years of delay the developer decided to instruct Wimpey Laboratories Ltd to fill the workings, and a conventional shopping centre was erected on the site.

CONSOLIDATE THE VOIDS AND PROVIDE A SPREAD FOUNDATION NEAR GROUND LEVEL

Probably the most obvious way of dealing with cavities under a site is to fill them by the injection of a material from ground level and to provide a conventional foundation. The consolidation of old mine workings and natural voids is the subject of a separate paper by my colleagues at Wimpey Laboratories Ltd.

If the cavities have been consolidated it is usual to keep the foundations as high as possible to take advantage of the dispersion of the load that will take place before the critical depth is reached. It is considered bad practice to use piles over consolidated cavities unless a thick layer of rock exists between the toes of the piles and the cavities. This is because piles concentrate the loads at depth and may break through into any small cavities the grout has not been able to reach. The natural raft of rock is particularly necessary over old mine workings as loose stowage and soft seat earth may not be improved by the consolidation work.

The foundation engineer should be aware of the likelihood of cavities existing on any site where the ground conditions make excavation and tunnelling easy. This is where good progress may be made with hand tools, and where the ground is dry and able to span over the workings without danger to people using them.

When working in parts of Nottingham where soft sandstone is near the surface we probe below each base with a jack hammer for a depth of three metres. Several old cellars have been found which belonged to large houses formally occupying the site. The cover of soft sandstone is removed by machine and the cellar filled with mass concrete and the bases constructed at the original formation level.

FORM A FOUNDATION NEAR GROUND LEVEL TO BRIDGE OVER THE VOIDS

When a site investigation has revealed the exact extent and location of cavities it may be possible to leave the cavities intact and to bridge over at ground level.

This may be done when the cavities exist as a result of a man-made structure being buried under the site, e.g. a tank, tunnel, well, basement or similar structure.

The foundation may comprise beams spanning over the cavities to abutments on each side, or a cellular box structure spanning two ways across the cavities to stable ground on each side.

The jackable cellular raft foundation may be suitable for a site undermined by old pillar and stall workings where subsidence is thought to be complete but precautions against further differential subsidence are required.

Old mine shafts are a hazard on sites and should be located & consolidated before work starts on the foundations.

Reclamation, Treatment and Utilization of Coal Mining Wastes, edited by A.K.M. Rainbow
Elsevier Science Publishers B.V., Amsterdam, 1987 — Printed in The Netherlands

BUILDING MATERIALS FROM INDUSTRIAL WASTE OF COAL BASED POWER PLANT

Dr Desai Mahesh D.,[1] and Mr Raijiwala D.B.,[2]

(1) Professor, Department of Applied Mechanics, S.V. Regional College of
     Engineering & Technology, Surat 395 007. State Gujarat, India

(2) Senior Lecturer, Department of Applied Mechanics, S.V. Regional College
     of Engineering & Technology, Surat 395 007, State Gujarat, India.

1.   Need for Housing

Providing housing is one of the basic preventive measures against disease
death, social unrest and social upheaval because the house and its neighbour-
hood mould the character of the people. From the reports of National
Buildings Organisation, it is estimated that the housing shortage in 1990 will
be 22.3 million in rural areas, and 6.9 million in urban areas in India.
The average number of persons per dwelling was 4.67 for one room, 2.84 for two
rooms. Percentage distribution of census houses by predominant materials of
wall and roof in rural areas is as under (Survey by NBO, India)- 11.6% houses
have grass, leaves, bamboos, 58.3% houses have mud, unburnt bricks etc, 30.1%
have burnt bricks, stone walls, 50.4% have a roof of grass, leaves, thatch,
49.6% have roof tiles, shingles, A.C. sheets, R.C.C. etc.

In India 40% of the population live in a one room house, 28.6% live in a
two room house, 14.1% in a three room house, 7.8% in a four room house and
9.5% in a five or more room house, in rural areas.

This reflects very evidently, that there is an accute shortage of
housing in developing countries and research towards cheaper building
components should be accelerated rapidly.

2.   Building Materials

While reviewing housing projects, the major component (34%) of building
materials is in the forms of building bricks/blocks/cellular blocks etc. The
burnt clay bricks and sun dried bricks/blocks laid in mud/lime mortar will be
the generalised construction practice all over the country. For such types of
construction and manufacture of burnt clay bricks mainly depends on the
availability of good soil. The soils of India are broadly classified [5] as
alluvial, latteritic marine, desert deposits, black cotton soil and red murrum.
The expansive black cotton soil covers 3000000sq km area of Indian peninsula.
Alluvial silt and fine sand is non cohesive/dilatant, unsuitable for moulding
standard bricks/blocks. These bricks have distortion in shape, low compressive

strength, high apparent porousity heavy efflorescence, large breakage while in transit (12 to 15%) (Table-1) [11].

Table:1

Test Results of Bricks of South Gujarat Region
(Bricks from different manufacturers were marked A to H)

Mark on Sample	No of Samples Tested	Average Compressive Strength MN/m.sq.	Water Absorption %	Efflor- escence	Dimensional tolerances within limits
A	18	3.10	26.74	Very High	No
B	20	3.38	23.26	High	No
C	20	2.84	24.17	Very High	No
D	18	5.81	14.99	Moderate	No
E	20	3.52	22.97	High	No
F	20	4.46	20.49	High	No
G	20	6.12	19.76	Moderate	No
H	20	4.21	21.31	Very High	No

The results eventually reflected that most of the bricks are not within the specified limits of Indian Standard Code of Practice (1077-1976), min. compressive strength is 3.5 MN/m.sq. and water absorption is less than 20%. The range of market cost varies from 400 to 500 Rs (25 to 35 pounds)/1000 bricks. The price inflation is because the suitable soil is to be excavated∠ * soil strata. There will be a huge loss of agricultural land which cannot be regenerated. Hence a phase has come for the search of alternative raw materials which can replace a long timely raw material namely soil mass. ∠*after removing the top layer of expansive....

3.   Newer Raw Materials

Alternative raw materials that have been used in experimental studies are lime, flyash, gympsum, sulphur, kiln ash, cinder, crushed stone, surkhi, etc., During the last few decades, a prominent cheap raw material as industrial effluent is - flyash.

Flyash is a finely divided residue resulting from the combustion of grounded or powdered bituminous coal or sub-bituminous coal (lignite). It is extracted by cyclone separation or electrostatic precipitation. At present 40 thermal power plants in India are yielding 9.2 million tonnes of flyash a year. A thermal power plant of 250 MW, 3320 Kcal/kg having an ash content of 39%, the quantity of coal required will be 240 tonnes/hour with maximum rating. The mechanical collector having 80% efficiency will collect 68 tonnes/

hour flyash in hoppers, whereas electrostatic precipitator having 95% efficiency will collect 80 tonnes/hour of flyash. Thus for a 250 MW thermal power plant with an annual load factor of 57% (5000 hours/year), the flyash output will be 400000 tonnes/annum. It has been reported[1] that Rs 15 (1 pound) has been spent for dumping/disposing of one tonne of flyash, but it acquires a large quantum of agricultural land and the problems of pollution are unimaginary.

4.   Indian Flyashes

While comparing Indian Flyashes with the flyashes of different countries it has been revealed that Indian flyashes are richer in silica, alumina contents but possesses lower amounts of ferric, sulphuric anhydrides. (Table-2).[13]

Table 2:

Variation in chemical composition of Indian Flyashes

Chemical Constituent	Content % Maximum	Content% Minimum	Average %	Standard deviation
Loss on Ignition	16.60	0.30	5.96	4.68
SiO_2	66.74	37.15	55.34	6.76
Al_2O_3	28.87	18.31	23.72	3.01
Fe_2O_3	21.94	3.23	9.39	5.65
CaO	10.80	1.30	3.22	2.55
MgO	5.25	0.80	2.13	1.46
SO_3	2.91	Traces	0.94	0.75

Mineral composition[13] of flyash varies with fineness, mineral matter, combustion state of coal. The major mineral forms in the Indian coals are hydrated silicate (kaolinite), carbonate (calcite), sulphate (pyrites), quartz, feldspar etc. The specific surface of Indian flyashes using Blaine's air permeability apparatus ranges between 2194 to 6842 cm.sq/gm. It is greatly affected by unburnt carbon content, particle shape, particle size and sp. gravity of flyash. The pozzolanic activity of flyash can be estimated by lime reactivity tests and for Indian flyash, the strength is always greater than 4 N/mm.sq.

An experimental study[14] was conducted by Department of Applied Mechanics, S.V. Regional College of Engineering & Technology, Surat-395007 for

664

the manufacture of lime-gypsum-flyash bricks/blocks for Ukai Thermal Power
Plant. (Table -3)

Table 3:

Ukai Thermal Power Plant - Flyash
Chemical Composition

Content	%
Loss of Ignition	3.75
Silica	52.07
Alumina	28.12
Ferric Oxide	6.11
Calcium Oxide	3.45
Magnesium Oxide	5.20
Sulphuric Anhydride	0.86
$Na_2O + K_2O$	0.31

Physical Properties:-

Property	
Colour	Whitish grey with slight black tinge
Fineness (Blaine's Method)	350 m.sq/kg
Lime Reactivity	4.2 N/mm.sq.
Drying Shrinkage	0.07 %
Sp. Gravity	2.24
Dry Bulk Density	870kg/m/cu/

5.   Building Elements From Flyash

Several binders were experimented from time to time by researchers, like
lime, gypsum, cement and polymers etc.

5.1   Bricks/Blocks

Gardin, M.D.[6] (1978) made bricks with 5-20% lime and 95.80% of flyash
plus cinder ash, and obtained a 6 day strength as 1.6 to 4.3 N/mm.sq., having
a water absorption of 19 to 32%.

Bose M.C.[1] (1984) mentioned that 10% lime plus 10% sand plus 80%
flyash with o.2% accelerator, pressed at 15.75 N/mm.sq., bricks were moulded
and steam cured, meeting the code requirements. The costing worked out to be
Rs 600/1000 bricks and they are approved by West Bengal/Tamil Nadu State
Governments.

Neyveli Lignite corporation, Tamil Nadu prepared bricks (size 225 x 112
x 75mm) and blocks (size 390 x 190 x 190 mm) with flyash-lime-gypsum
proportioning are 92-4½-3½% volumetrically. The compressive strength will be
3.5 MN/m.sq. with 28 days wet and 30 days air curing and the cost price will
be Rs 200/1000 bricks. They had built their complete township with these
blocks as well as flooring and observed satisfactory performance since a
decade.

Authors have summarized[12] a series of experimental work in Table-4

Table 4:-

Size of Block 200 x 100 x 100 mm

Sr No.	%Proportion by volume			Water additon %	Bulk Density (KN/m³)	Apparent Porousity %	Compressive Strength (MN/m²)	Efflorescence	Cost of 1000 Blocks Rs
	Lime	Gypsum	Flyash						
1	22	13	65	29	9.72	19.62	8.2	Moderate	282
2	17	13	70	30.5	9.65	21.81	7.3	Moderate	238
3	15	10	75	31.5	9.62	24.16	6.9	Moderate	221
4	12	8	80	33	9.54	24.77	6.2	Moderate	191
5	10	5	85	34	9.47	26.41	5.4	Moderate	169
6	8	4	88	35.5	9.51	27.71	4.6	Heavy	137
7	6	3	91	36	9.58	27.93	4.2	Heavy	123
8	3	2	95	38	9.42	28.11	2.9	Heavy	81

Note: Except for Sr. No. 7 and 8 the dimensional tolerances were within limits. Curing by sprinkling water for 1 week and then water curing for 3 weeks.

5.2   Flooring. Roofing Tiles

Authors have reported that flooring/roofing tiles are prepared with volumetric proportion as ½ : ½: 1:1 of cement: lime: flyash: fine sand and obtained flexural lod of upto 460 N. The tiles were pressed and water cured for 3 weeks and air curing for 2 weeks. The cost of such unit will be 155 Rs/ brass. An accelerator was used for quick demoulding process.

5.3   Mortars/Finishing/Panelling

Authors have executed mortars/panelling work of proportioning 1:8 to 1:10 as cement, flyash volumetrically which showed significantly well performance.

As an appropriate technology to rural regions, authors have produced blocks with unskilled labour and indigeneous technology. For mass production of such flyash blocks, adopting either CINVA-RAM block press or Elson's wire cut block making machine having a capacity of 20000 blocks per day.

6.   Conclusions

Utilization of flyash in building components is an activity of pollution control and producing cheap/durable elements. A kiln process (used for production of burnt clay bricks) is totally eliminated and the flyash structure are very light weight. Use of flyash and gypsum, both bye-products whose prices are going to remain unchanged for a few years into the future should be accepted as basic materials for urban/rural dwellings on a self help basis. Construction amongst the binders, lime needs longer curing and extension search is for cheap accelerator to gain quicker bond.

Sense of Thanks

Authors are indebted to authorities of college for providing the facilities of the laboratory setup and students who were associated with this projects for a period of a decade.

Bibliography

1)   Bose, M.C., "Utilization of Ash from Thermal Power Station", Jour. of the Inst. of Engineers (Gen. Engg. Div.) June, 1984.
2)   Chitharanjan, N, Ramakrishna, V.S. and Ganeshan S., "A New Production Technology for Manufacture of Cellular Concrete" Indian Concrete Journal Oct. 1987.
3)   Chitharanjan, N., "Compressed Lime-Flyash-Gypsum Blocks", Indian Concrete Journal, June 1983.
4)   Desai, M.D., "Flyash as Building Material", A Technical Report submitted to Rural Tech. Institute, Ahmedabad on Flyashes in Gujarat, April 1974.

5) Desai, M.D., "Geotechnical Aspects of Residual Soils of India", Proc.
    of ISSMFE Committee on sampling and testing of Residual Soils, San
    Fransisco, USA., March 1985.
6) Gardin, M.D., Prasad, K.K., and Chatterjee, Amit., "Manufacture of
    Building Bricks with Flyash", Indian Concrete Journal, May-June 1978.
7) Goodman, L.J., Pama, E.G., et al "Low Cost Housing Technology- An East-
    West Perspective" Pergamon Press, 1979.
8) IS: 6491-1972, "Methods of Sampling Flyash".
9) IS: 3812-1981, "Specification for Flyash for use as Pozzolana and
    Admixture".
10) Pitre, S.G., Batra, V.S., and Kukreja, C.B., "Investigations on Motors of
    Lime and other Waste Materials", Jour. of Inst. of Engineers, (Civil Engg.
    Div), March 1985.
11) Raijiwala, D.B., and Desai, M.D., "Technoeconomic Feasibility of Bricks
    made from Industrial Waste of a Thermal Power Plant" National Seminar on
    Housing for the Rural poor in india, University of Annamalai, Tamil-Nadu,
    June 1985.
12) Raijiwala, D.B. and Desai, M.D., "Criteria for Seismic Design of Adobe
    Low Cost Housing" 8th Symposium on Earthquake Engineering, Roorkee,
    December 1986
13) Rehsi, S.S., "Building Materials from Indian Flyashes", Central Building
    Research Institute, Roorkee, February 1981.
14) Shastri, V.R. "Feasibility of Brick manufacturing from Ukai Flyash", M.E.
    (Struct.) Thesis at South Gujarat University 1978.
15) Verma, C.L., Tehri, S.P., and Mohan Rai. "Technoeconomic feasibility study
    for manufacturing of Lime-Flyashe cellular Concrete Blocks", Indian
    Concrete Journal., March 1983.